Developments in Soil Science 28A

SOIL MINERAL–ORGANIC MATTER–MICROORGANISM INTERACTIONS AND ECOSYSTEM HEALTH

Dynamics, Mobility and Transformation of Pollutants and Nutrients

Volume 28A

Developments in Soil Science
Series Editors: A.E. Hartemink and A.B. McBratney

Titles currently available in this Series

11A PEDOGENESIS AND SOIL TAXONOMY. I. Concepts and Interactions
L.P. Wilding, N.E. Smeck and G.F. Hall (Editors)
ISBN: 0-444-42100-9

13 ELEMENTS OF SOIL PHYSICS
P. Koorevaar, G. Menelik and C. Dirksen
ISBN: 0-444-42242-0

21 VOLCANIC ASH SOILS: Genesis, Properties and Utilization
S. Shoji, M. Nanzyo and R.A. Dahlgren
ISBN: 0-444-89799-2

23 SOIL CONSERVATION AND SILVICULTURE
J. Dvorák and L. Novák (Editors)
ISBN: 0-444-98792-4

24 VERTISOLS AND TECHNOLOGIES FOR THEIR MANAGEMENT
N. Ahmad and A. Mermut (Editors)
ISBN: 0-444-88789-X

25 SOIL QUALITY FOR CROP PRODUCTION AND ECOSYSTEM HEALTH
E.G. Gregorich and M.R. Carter (Editors)
ISBN: 0-444-81661-5

26 ENZYMOLOGY OF DISTURBED SOILS
S. Kiss, D. Paşca and M. Drãgan-Bularda
ISBN: 0-444-50057-X

27 FRACTALS IN SOILS SCIENCE
Y. Pachepsky, J.W. Crawford and W.J. Rawls (Editors)
ISBN: 0-444-50530-X

28A SOIL MINERAL–ORGANIC MATTER–MICROORGANISM INTERACTIONS AND ECOSYSTEM HEALTH: Dynamics, Mobility and Transformation of Pollutants and Nutrients
A. Violante, P.M. Huang, J.-M. Bollag and L. Gianfreda
ISBN: 0-444-51038-9

28B SOIL MINERAL–ORGANIC MATTER–MICROORGANISM INTERACTIONS AND ECOSYSTEM HEALTH: Ecological Significance of the Interactions Among Clay Minerals, Organic Matter and Soil Biota
A. Violante, P.M. Huang, J.-M. Bollag and L. Gianfreda
ISBN: 0-444-51039-7

Developments in Soil Science 28A

SOIL MINERAL-ORGANIC MATTER-MICROORGANISM INTERACTIONS AND ECOSYSTEM HEALTH

Dynamics, Mobility and Transformation of Pollutants and Nutrients

Volume 28A

Edited by

A. Violante
Dipartimento di Scienze Chimico-Agrarie
Università di Napoli Federico II, Portici (Napoli)
Italy

P.M. Huang
Department of Soil Science
University of Saskatchewan, Saskatoon
Canada

J.-M. Bollag
Laboratory of Soil Biochemistry
Center for Bioremediation and Detoxification
The Pennsylvania State University
University Park, PA
USA

L. Gianfreda
Dipartimento di Scienze Chimico-Agrarie
Università di Napoli Federico II, Portici (Napoli)
Italy

2002
ELSEVIER
Amsterdam – Boston – London – New York – Oxford – Paris
San Diego – San Francisco – Singapore – Sydney – Tokyo

ELSEVIER SCIENCE B.V.
Sara Burgerhartstraat 25
P.O. Box 211, 1000 AE Amsterdam, The Netherlands

First edition 2002

Library of Congress Cataloging in Publication Data
A catalog record from the Library of Congress has been applied for.

British Library Cataloguing in Publication Data
A catalogue record from the British Library has been applied for.

ISBN: 0-444-51038-9 (Volume 28A)
ISBN: 0-444-51039-7 (Volume 28B)
ISBN: 0-444-51040-0 (Set comprising volumes 28A and 28B)

ISSN: 0166-2481 (Series)

♾ The paper used in this publication meets the requirements of ANSI/NISO Z39.48-1992 (Permanence of Paper).
Printed in The Netherlands.

PREFACE

The Working Group M.O. (*Interactions of Soil Minerals with Organic Components and Microorganisms*) of the International Union of Soil Sciences (IUSS) was founded in 1990 at the 14th World Congress of Soil Sciences (Kyoto, Japan). It organized two International Symposia in Edmonton (Canada) in 1992 and in Nancy (France) in 1996. Specialized and shorter co-sponsored meetings were also held in Acapulco (Mexico) in 1994, in St Louis (USA) in 1995, in Charlottetown (Canada) in 1999 and in Guelph (Canada) in 2001. As a result of these Symposia and Meetings, seven volumes or special books were published in the last 6 years.

The present book consists of two volumes presenting 59 of a total of 175 invited and volunteered papers presented at the 3rd Symposium on "Soil Mineral-Organic Matter-Microorganism Interactions and Ecosystem Health". Organized by the Working Group MO it was held in Naples-Capri (Italy) from May 22 to 26, 2000. All papers accepted for publication in this book have been subjected to critical peer review. Volume A deals with the dynamics, mobility and transformation of pollutants and nutrients and Volume B covers the ecological significance of the interactions among clay minerals, organic matter and soil biota.

The aim of the 3rd Symposium was to provide a forum for the interaction of soil chemists, soil mineralogists, soil microbiologists, soil biochemists and environmental scientists with the intention of promoting discussions and exchanging information on many topics of mutual interest in this important area of science. It was also meant to stimulate research leading to an integration of knowledge on "soil minerals-natural organics-microorganisms" and their impact on agricultural production and environmental protection.

Soil is a dynamic system in which soil minerals constantly interact with organic matter and microorganisms. The close association and interactions between abiotic and biotic entities in soil environments govern (1) mineral weathering reactions, aggregate formation, and surface reactivity of soil minerals with respect to nutrients and environmental pollutants, (2) the dynamics and transformation of metals, metalloids, and natural and anthropogenic organics, and metabolic processes, growth and ecology of microbes, and (3) has an impact on soil development, agricultural production, environmental protection, and ecosystem integrity. Among soil processes, chemical and biogeochemical reactions have an important role in the speciation, bioavailability, toxicity, transformations and transport of metals and anthropogenic organics.

The 3rd Symposium was sponsored by the Commissions II (Soil Chemistry), III (Soil Biology), VII (Soil Mineralogy) and VIII (Soils and the Environment) of the IUSS, by the University of Naples Federico II and by the Working Group "NAMOX" of the Società Italiana di Scienza del Suolo (SISS). More than 220 scientists of 32 different countries (Australia, Austria, Bangladesh, Canada, Chile, China, Columbia, Czech Republic, Denmark, Egypt, Finland, France, Germany, Hungary, Israel, Italy, Japan, New Zealand, Poland, Spain, The Netherlands, Rumania, Russia, USA, UK, South Africa, Korea, Sri Lanka, Switzerland, Taiwan, Sweden, and Zimbabwe) participated at this scientific event. One hundred and seventy five papers were presented at the meeting. The participants of the Symposium represented several subdisciplines of Soil Sciences as well as Ecology, Environmental Science, Toxicology and Health Science.

We are highly appreciative of the response of the authors to our request for the preparation of updated and original manuscripts and are grateful to the external referees for their expert critiques and inputs to maintain the quality of this book. Sincere appreciation is extended to the Dipartimento di Scienze Chimico-Agrarie, University of Naples Federico II for support during the preparation of these volumes. Mrs. I. Crovella and Mr. M. Clumez deserve a special mention for their help in organizing the Symposium.

The Editors acknowledge with deep gratitude Dr. M. A. Rao for her active collaboration in the preparation of the programme and during the Symposium and mainly for her tireless effort and qualified help in editing this book.

The Editors are also grateful to Mrs Joy Drohan for her excellent technical help in editing the papers in respect to the English style and the typeset format used in these volumes.

Finally, the Editors express their gratitude to the Ministero delle Politiche Agricole e Forestali (Rome), to the University of Naples Federico II, to the Società Italiana di Scienza del Suolo (SISS), to the Gruppo Italiano AIPEA, to the Banco di Napoli, to Shimadzu, Perkin Elmer, Dionex for financial support for organizing the Symposium and for publishing these volumes.

A. Violante
P.M. Huang
J.-M. Bollag
L. Gianfreda

ABOUT THE EDITORS

Antonio Violante is Professor of Agricultural Chemistry at the University of Naples (Italy). He took his Ph.D. in Chemistry at the University of Naples in 1969. He was awarded postdoctoral fellowships from the University of Wisconsin, USA (1976-1977) and the University of Saskatchewan, Canada (1981-1982) and was invited professor at the Department of Soil Science, University of Saskatchewan, Canada in 1985 and 1992.

Dr. Violante was head of the Dipartimento di Scienze Chimico-Agrarie and is Coordinator of the *Doctoral School in Agricultural Chemistry* of the University of Naples Federico II. He has served on many committees of the Italian Society of Soil Science (President of the Session Soil Chemistry), Italian Society of Agricultural Chemistry. He is vice-president and liaisons officer of Gruppo Italiano AIPEA. He was the scientific chairman and chief organizer of International and National Congresses.

Dr. Violante has contributed to promote research on the interface between soil chemistry and mineralogy and soil biology. The areas of research include the formation mechanisms of Al-hydroxides and oxyhydroxides, the surface chemistry and reactivities of short-range ordered precipitation products of Al and Fe, the influence of biomolecules on the sorption/desorption of nutrients and xenobiotics on/from variable charge minerals and soils and on the factors which influence the sorption and residual activity of enzymes on phyllosilicates, variable charge minerals, organo-mineral complexes, and soils.

Dr. Violante is the author or co-author of 135 refereed research articles and book chapters and over 70 additional scientific contributions. He presented papers at many scientific Congresses and Symposia and gave invited lectures at Universities and Research Institutes worldwide. Dr. Violante has international research/teaching experience in Canada, USA, Europe, China and Chile. He has trained students for Master Degree, Ph.D., and postdoctoral fellows and received visiting scientists worldwide. He serves on the editorial board of three international journals.

Pan Ming Huang received his Ph.D. degree in Soil Science at the University of Wisconsin, Madison, in 1966. He is currently Professor of Soil Science at the University of Saskatchewan, Saskatoon, Canada. His research work has significantly advanced the frontiers of knowledge on the nature and surface reactivity of mineral colloids and organomineral complexes of soils and sediments and their role in the dynamics, transformations, and fate of nutrients, toxic metals, and xenobiotics in terrestrial and aquatic environments. His research findings, embodied in over 270 refereed scientific publications, including research papers, book chapters, and 10 books, are fundamental to the development of sound strategies for managing land and water resources.

He has developed and taught courses in soil physical chemistry and mineralogy, soil analytical chemistry, and ecological toxicology. He has successfully trained and inspired M.Sc. and Ph.D. students and postdoctoral fellows, and received visiting scientists world-wide. He has served on numerous national and international scientific and academic committees. He has served as a member of many editorial boards such as the *Soil Science Society of America Journal, Geoderma, Chemosphere, and Advances in Environmental*

Science. He is currently a titular member of the Commission of Fundamental Environmental Chemistry of the International Union of Pure and Applied Chemistry and is the founding and current Chairman of the Working Group MO "*Interactions of Soil Minerals with Organic Components and Microorganisms*" of the International Union of Soil Sciences. He received the distinguished Researcher Award from the University of Saskatchewan and the Soil Science Research Award from the Soil Science Society of America, the American Society of Agronomy, and the American Association for the Advancement of Science.

Jean-Marc Bollag is Professor of Soil Biochemistry and Director of the Center of Bioremediation and Detoxification, Environmental Resources Research Institute, at The Pennsylvania State University. He is the author or coauthor of more than 220 professional papers and serves on the editorial board of five international journals. He is a frequent lecturer at conferences and seminars throughout the world. A recipient of the Julius Baer Fellowship, the Gamma Sigma Delta Research Award, and the Badge of Merit from the Polish Ministry of Agriculture. Dr. Bollag is a Fellow of the American Academy of Microbiology, the Soil Science Society of America, and the American Society of Agronomy. He is also recipient of the "Environmental Quality Research Award" from the American Society of Agronomy.

Dr. Bollag received the Ph.D. degree in Plant Physiology from the University of Basel, Switzerland, and conducted postdoctoral work at the Weitzmann Institute of Science, Rehovoth, Israel, and at Cornell University, Ithaca, New York. He was also a Visiting Scientist in the Biochemistry Section of Agrochemicals at Ciba-Geigy, Basel, Switzerland. Most of his research is related to the fate of pollutants in the environment and to bioremediation problems (incorporation of pollutants into soil organic matter as a detoxification method and application of enzymes for pollution control).

Liliana Gianfreda is Professor of Agricultural Biochemistry and Soil Biochemistry at University of Naples Federico II. She is author or coauthor or more than 150 professional papers and has participated as invited lecturer to several National and International conferences and symposiums.

Dr. Gianfreda received her Ph.D. in Chemistry from the University of Naples, Italy and performed most of her post-doctoral work at the University of Naples. She was Visiting Scientist in the Laboratory of Soil Biochemistry of the Environmental Resources Research Institute of the Pennsylvania State University, USA, and in the Institute of Chemistry, Universidada Estadual de Campinas, Sao Paolo, Brasil.

She is President of the Commission Soil Biology of the Italian Soil Science Society. She is also member of several National and International Scientific Societies (Italian Society of Chemistry, Italian Society of Biochemistry, Italian Society of Soil Sciences, International Union of Soil Sciences, and Italian Society of Agricultural Chemistry

Dr. Gianfreda was Coordinator of several National and International Research projects. She has a large experience in soil enzymology and interactions between enzymatic and non-enzymatic proteins and soil organic and inorganic components. She is also expert of the use of biological agents for the restoration and bioremediation of polluted sites.

REFEREES

The following scientists gave their time and talent as technical referees of manuscripts submitted for publication in this book . Their devotion is greatly appreciated.

Alexander M.	(U.S.A)	**Farini A.**	(Italy)
Andreux F.	(France)	**Fent G.**	(Germany)
Arshad M.A.	(Canada)	**Filip Z.**	(Germany)
Barton C.	(U.S.A.)	**Gadd G.M.**	(U.K.)
Baveye P.	(U.S.A.)	**Gaillard J-F.**	(France)
Berthelin J.	(France)	**Gennari M**	(Italy)
Bespalova A.	(Russia)	**Germida J.**	(Canada)
Bollag J.-M.	(U.S.A.)	**Gerzabek M.H.**	(Austria)
Braun J.	(U.S.A.)	**Gianfreda L.**	(Italy)
Brookes P.C.	(U.K.)	**Gigliotti C.**	(Italy)
Burns R.G.	(U. K.)	**Hsu Pa Ho**	(U.S.A.)
Buurman P.	(The Netherlands)	**Insam H.**	(Austria)
Chenu C.	(France)	**Jackson T.**	(Canada)
Chiu C.Y.	(Taiwan)	**Kandeler E.**	(Germany)
Chorover J.	(U.S.A.)	**Kirchmann H.**	(Sweden)
Colin P.	(France)	**Knackmuss H-J.**	(Germany)
Curtin D.	(Canada)	**Krishnamurti G.**	(U.S.A.)
Davies G.	(U.S.A.)	**Kubicki J.D.**	(U.S.A.)
De Freiteis R. J.	(Canada)	**Kurek E.**	(Poland)
de Kimpe C.	(Canada)	**Leifeld J.**	(Germany)
Dec J.	(U.S.A.)	**Leyval C.**	(France)
Dick R.	(U.S.A.)	**Markkola A.M.**	(Finland)
Djurhuus J.	(Denmark)	**Matschonat G.**	(Germany)
Dubbin W.	(U.K.)	**Mbagwu J.**	(Nigeria)
Ehrlich E.	(Canada)	**McGrath S.P.**	(U. K.)
Eriksson J.	(Sweden)	**Naidja A.**	(Canada)
Etana A.	(Sweden)	**Nannipieri P.**	(Italy)

Olsen J.E.	(Denmark)	**Shindo H.**	(Japan)
Page A.L.	(U.S.A.)	**Speir T.**	(New Zealand)
Pampura T.	(Russia)	**Staunton S.**	(France)
Piccolo A.	(Italy)	**Stotzky G.**	(U.S.A.)
Quiquampoix H.	(France)	**Tabatabai M.A.**	(U.S.A)
Reinhold-Hurek B.	(Germany)	**Tani M.**	(Japan)
Ristori G.	(Italy)	**Torrent J.**	(Spain)
Robert M.	(France)	**Trasar-Cepeda C.**	(Spain)
Ruggiero P.	(Italy)	**Violante A.**	(Italy)
Sakurai K.	(Japan)	**Violante P.**	(Italy)
Schaeffer A.	(Germany)	**Walia S.**	(India)
Schnitzer M.	(Canada)	**Wang M.C.**	(Taiwan)
Schulten H.-R.	(Germany)	**Wilson J.**	(U. K.)
Segat A.	(Argentina)	**Xing G.X.**	(China)
Senesi N.	(Italy)	**Yang J.**	(Korea)
Sequi P.	(Italy)	**Yuan G.**	(New Zealand)

CONTRIBUTORS

Amiotte Suchet P. UMR A 111 Microbiologie des Sols-GeòSol INRA, Université de Bourgogne Centre des Sciences de la Terre 6, Boulevard Gabriel 21000 Dijon, France

Andreux F. UMR A 111 Microbiologie des Sols-GeòSol INRA, Université de Bourgogne Centre des Sciences de la Terre 6, Boulevard Gabriel 21000 Dijon, France

Aylmore L.A.G. Soil Science and Plant Nutrition, The University of Western Australia, Netherlands, 6907 Western Australia

Berg B. Lehrstuhl für Bodenökologie, Universität Bayreuth, Dr Hans Frisch Strasse 1-3, DE-944 4, Bayeruth, Germany

Birkel U. Georg-August-University Göttingen, Institute of Geography, Department of Landscape Ecology, Goldschmidstr. 5, 37077 Göttingen, Germany

Borowska K. Department of Biochemistry, University of Technology and Agriculture, 6 Bernardynska St., 85-029 Bydgoszcz, Poland

Catalano L. Dipartimento di Produzione Vegetale e Tecnologie Agrarie, Università di Udine, Via della Scienza 208, 33100 Udine, Italy

Chenu C. Unitè de Science du Sol, INRA Versailles, France

Churchman G.J. CSIRO Land and Water, Private Mail Bag No. 2, Glen Osmond, South Australia 5064, Australia

Conte P. Dipartimento di Scienze Chimico-Agrarie, Università di Napoli Federico II, Via Università 100, 80055 Portici (NA), Italy

De Marco A. Dipartimento di Biologia Vegetale, Università di Napoli Federico II, Via Foria 223, 80139 Napoli, Italy

De Nobili M. Dipartimento di Produzione Vegetale, Via della Scienza 208, 33100 Udine, Italy

Deiana S. DISAABA, Università di Sassari, V.le Italia 39, 07100 Sassari, Italy

Eriksson J. Department of Forest Ecology, Swedish University of Agricultural Sciences, S-90183 UMEA, Sweden

Fierro A. Dipartimento di Biologia Vegetale, Università di Napoli Federico II, Via Foria 223, 80139 Napoli, Italy

Figliolia A. Experimental Institute for Plant Nutrition, Via della Navicella 4, 00184 Roma, Italy

Fornasier F. Istituto Sperimentale per la Nutrizione delle Piante, S.O.P. di Gorizia, Via Trieste 23, 34170 Gorizia, Italy

Francaviglia R. Istituto Sperimentale per la Nutrizione delle Piante, Via della Navicella 2, 00184 Roma, Italy

Franco I. Dipartimento di Produzione Vegetale e Tecnologie Agrarie, Università di Udine, Via della Scienza 208, 33100 Udine, Italy

Gallardo J.F. Consejo Superior de Investigaciones Cientificas Aptdo. 257. Salamanca 370171, Spain

Gerold G. Georg-August-University Göttingen, Institute of Geography, Department of Landscape Ecology, Goldschmidstr. 5, 37077 Göttingen, Germany

Gerzabek M.H. Austrian Research Centers, Division of Life Sciences, A-2444 Seibersdorf, Austria

Gianfreda L. Dipartimento di Scienze Chimico-Agrarie, Università di Napoli Federico II, Via Università 100, 80055 Portici (NA), Italy

González M.I. Consejo Superior de Investigaciones Cientificas – Centro de Ciencias Medioambientales Serrano, 115 dpdo, 28006 Madrid, Spain

Hanudin E. Laboratory of Environmental Soil Science, Ehime University, 3-5-7 Tarumi, Matsuyama 790-8566, Japan

He J.Z. Department of Resource, Enviroment and Agrochemistry, Huazhong Agricultural University, Wuhan 430070, P.R. China

Henmi T. Laboratory of Environmental Soil Science, Ehime University, 3-5-7 Tarumi, Matsuyama 790-8566, Japan

Hermosín M.C. Istituto de Recursos Naturales y Agrobiologià de Sevilla, CSIC, Avd. De Reina Mercedes 10, E-41080 Sevilla, Spain

Hu H.Q. Department of Resource, Enviroment and Agrochemistry, Huazhong Agricultural University, Wuhan 430070, P.R. China

Huang P.M. Department of Soil Science, University of Saskatchewan, 51 Campus Drive Saskatchewan SK S7N 5A8, Canada

Jackson T.A. National Water Research Institute 867 Lakeshore Road, P.O. Box 5050 Burlington, Ontario l7R 4A6, Canada

Karathanasis A.D. University of Kentucky, Department of Agronomy, N-122K Ag. Science-North Lexington KY, U.S.A.

Kirchmann H. Swedish University of Agricultural Sciences, Department of Soil Sciences, Box 7014, S-750 07 Uppsala, Sweden

Leita L. Istituto Sperimentale per la Nutrizione delle Piante, Via Trieste 23, 34170 Gorizia, Italy

Leppard G.G. National Water Research Institute 867 Lakeshore Road, P.O. Box 5050 Burlington, Ontario l7R 4A6, Canada

Lévêque J. UMR A 111 Microviologie des Sols-GeòSol INRA, Università de Bourgogne Centre des Sciences de la Terre 6, Boulevard Gabriel 21000 Dijon, France

Li X.Y. Department of Resource, Enviroment and Agrochemistry, Huazhong Agricultural University, Wuhan 430070, P.R. China

Linglois N. UMR A 111 Microbiologie des Sols-GeòSol INRA, Université de Bourgogne Centre des Sciences de la Terre 6, Boulevard Gabriel 21000 Dijon, France

Loffredo E. Dipartimento di Biologia e Chimica Agroforestale e Ambientale, Università di Bari, Via Amendola 165a, 70126 Bari, Italy

Manunza B. DISAABA, Università di Sassari, V.le Italia 39, 07100 Sassari, Italy

Matschonat G. Institute of Soil Science and Land Evaluation, University of Hohenheim, D- 70593 Stuttgart, Germany

Matsue N. Laboratory of Environmental soil Science, Ehime University, 3-5-7 Tarumi, Matsuyama 790-8566, Japan

Ming D.W. USDA Forest Service, SRS Center for Forested Wetlands Research c/o Savannah River Ecology Lab, Drawer E. Aiken, SC 29802, U.S.A.

Molinu M.G. DISAABA, Università di Sassari, V.le Italia 39, 07100 Sassari, Italy

Mori A. Istituto Sperimentale per la Nutrizione delle Piante, Via Trieste 23, 34170 Gorizia, Italy

Nguyen Thi-Kim-Ngân UMRA 111 Microbiologie des Sols-GeòSol INRA, Université de Bourgogne Centre des Sciences de la Terre 6, Boulevard Gabriel 21000 Dijon, France

Niemeyer J. University of Trier, Geosciences/Geography, Department of Soil Science, Universitätsring 15, 54286 Trier, Germany

Palma A. DISAABA, Università di Sassari, V.le Italia 39, 07100 Sassari, Italy

Parfitt R.L. Landcare Research, Private Bag 11052, Palmerston North, New Zealand

Pennelli B. Experimental Institute for Plant Nutrition, Via della Navicella 4, 00184 Roma, Italy

Percival H.J. Landcare Research, Private Bag 11052, Palmerston North, New Zealand

Piccolo A. Dipartimento di Scienze Chimico-Agrarie, Università di Napoli Federico II, Via Università 100, 80055 Portici (NA), Italy

Pigna M. Dipartimento di Scienze Chimico-Agrarie, Università di Napoli Federico II, Via Università 100, 80055 Portici (NA), Italy

Premoli A. DISAABA, Università di Sassari, V.le Italia 39, 07100 Sassari, Italy

Ricciardella M. Dipartimento di Scienze Chimico-Agrarie, Università di Napoli Federico II,
Via Università 100, 80055 Portici (NA), Italy

Rossi G. Experimental Institute for Plant Nutrition, Via della Navicella 4, 00184 Roma, Italy

Roux F. UMR A 111 Microbiologie des Sols-GeòSol INRA, Université de Bourgogne Centre des Sciences de la Terre 6, Boulevard Gabriel 21000 Dijon, France

Rutigliano F.A. Dipartimento di Scienze Ambientali, Seconda Università di Napoli, Via Vivaldi 43, 81100 Caserta, Italy

Schulten H.-R. University of Rostock, Agricultural Faculty, Institute of Soil Science, Justus-von-Liebig-Weg 6, 18051 Rostock, Germany

Senesi N. Dipartimento di Biologia e Chimica Agroforestale e Ambientale, Università di Bari, Via Amendola 165a, 70126 Bari, Italy

Sequi P. Istituto Sperimentale per la Nutrizione delle Piante, Via della Navicella 2, 00184 Roma, Italy

Skyllberg U. Department of Forest Ecology, Swedish University of Agricultural Sciences, S-90183 UMEA, Sweden

Socciarelli S. Experimental Institute for Plant Nutrition, Via della Navicella 4, 00184 Roma, Italy

Solinas V. DISAABA, Università di Sassari, V.le Italia 39, 07100 Sassari, Italy

Soulas G. INRA – Microbiologie des Sols, Bd Sully, BV 1540, 21034 Dijon Cedex, France

Sposito G. Division of Ecosystem Sciences, Hilgard Hall 3110, University of California Berkeley, California 94720-3110, U.S.A.

Staunton S. Unitè Sol & Environment, pl Viala, 34060 Montpellier Cedex, France

Theng B.K.G. Lancare Research, Private Bag 11052, Palmerston North, New Zealand

Turrión M.B. University of Valladolid. Area de Edafologia y Quimica Agricola, Palencia 34004, Spain

Vieublé L. INRA - Science du Sol, Rte de St Cvr, 78026 Versailles Cedex, France

Violante A. Dipartimento di Scienze Chimico-Agrarie, Università di Napoli Federico II, Via Università 100, 80055 Portici (NA), Italy

Virzo De Santo A. Dipartimento di Biologia Vegetale, Università di Napoli Federico II, Via Foria 223, 80139 Napoli, Italy

Vrdoljak G. Electron Microscope Lab 26 Giannini Hall, University of California Berkeley, California 94720-3330, U.S.A.

Webb K.M. Soil Science and Plant Nutrition, The university of Western Australia, Netherlands, 6907 Western Australia

Yuan G. Lancare Research, Private Bag 11052, Palmerston North, New Zealand

Zsolnay Á. Institut fur Bodenokologie, GSF, D-85764 Neuherberg bei Munchen, Germany

CONTENTS

Foreseeable Impacts of Soil Mineral-Organic Component-Microorganism Interactions on Society: Ecosystem Health
P.M. Huang 1

Sorption of Copper and Cadmium by Allophane-Humic Complexes
G. Yuan, H.J. Percival, B.K.G. Theng and R.L. Parfitt 37

Colloid-Mediated Transport of Metals Associated with Lime-Stabilized Biosolids
A.D. Karathanasis and D.W. Ming 49

Heavy Metals and Litter Decomposition in Coniferous Forests
A. Virzo De Santo, A. Fierro, B. Berg, F.A. Rutigliano and A. De Marco 63

Direct and Indirect Effects of Organic Matter on Metal Immobilisation in Soil
S. Staunton 79

Effects of Medium-Term Amendment with Sewage Sludges on Heavy Metal Distribution in Soil
G. Rossi, B. Pennelli, S. Socciarelli and A. Figliolia 99

Uptake and Accumulation of Selenium and Sulfur by Plants as Related to Soil Factors in Poland
K. Borowska 109

The Role of Soil Organic Matter and Water Potential in Determining Pesticide Degradation
K.M. Webb and L.A.G. Aylmore 117

Variability of Pesticide Mineralization in Individual Soil Aggregates of Millimeter Size
L. Vieublé, C. Chenu and G. Soulas 127

The Effect of Soil Mineral-Organic Matter Interaction on Simazine Adsorption and Desorption
Á. Zsolnay, M.C. Hermosín, A. Piccolo and L. Gianfreda 137

Sorption and Release of Endocrine Disruptor Compounds onto/from Surface and Deep Horizons of Two Sandy Soils
E. Loffredo and N. Senesi 143

Distribution of Trinitrotoluene between Aqueous and Solid Phase Soil Organic Matter
J. Eriksson and U. Skyllberg 161

Retention and Mobility of Chemicals in Soil
M. De Nobili, R. Francaviglia and P. Sequi 171

Soil Aggregate Hierarchy in a Brazilian Oxisol
G. Vrdoljak and G. Sposito 197

Energy Dispersive X-Ray Microanalysis and its Applications in Biogeochemical Research
T.A. Jackson and G.G. Leppard 219

Influence of pH and of Several Organic Acids on the Interaction between Esculetine and Iron (III)
S. Deiana, B. Manunza, M.G. Molinu, A. Palma, A. Premoli and V. Solinas 261

Adsorption of Phosphate on Variable Charge Minerals and Soils as Affected by Organic and Inorganic Ligands
A. Violante, M. Pigna, M. Ricciardella and L. Gianfreda 279

Relationships between Organic and Inorganic P Fractions with Soil Fe and Al Forms in Forest Soils of Sierra de Gata Mountains (Western Spain)
M.B. Turrión, J.F. Gallardo and M.I. González 297

Effects of Organic Ligands on Adsorption of Phosphate on a Noncrystalline Al Hydroxide
H.Q. Hu, J.Z. He and X.Y. Li 311

Reactions of some Short-Range Ordered Aluminosilicates with Selected Organic Ligands
E. Hanudin, N. Matsue and T. Henmi 319

The Role of Clays in the Restoration of Perturbed Ecosystems
G.J. Churchman 333

New Approaches to the Molecular Structure and Properties of Soil Organic Matter: Humic-, Xenobiotic-, Biological-, and Mineral-Bonds
H.-R. Schulten 351

Impact of Changing Forest Management on Soil Organic Matter in Low Mountain Acid Media
F. Andreux, F. Roux, N. Linglois, Thi-Kim-Ngân Nguyen, P. Amiotte Suchet and J. Lévêque 383

Effect of Concentration on the Self-Assembling of Dissolved Humic Substances
P. Conte and A. Piccolo 409

Pore Size Changes in a Long-Term Field Experiment with Organic Amendments
H. Kirchmann and M.H. Gerzabek 419

Capacity of Organically Complexed Aluminum, Ionic Strength, and pH to Affect the CEC of Organic Samples
G. Matschonat 425

Abiotic Reactions of Organics on Clay Mineral Surfaces
U. Birkel, G. Gerold and J. Niemeyer 437

The Interaction between Ferricyanide Ion and Unfractionated Humic Substances
A. Mori, F. Fornasier, L. Catalano, I. Franco and L. Leita 449

Index 457

Developments in Soil Science, Volume 28A
Editors: A. Violante, P.M. Huang, J.-M. Bollag and L. Gianfreda

FORESEEABLE IMPACTS OF SOIL MINERAL-ORGANIC COMPONENT-MICROORGANISM INTERACTIONS ON SOCIETY: ECOSYSTEM HEALTH

P.M. Huang

Department of Soil Science, University of Saskatchewan, Saskatoon SK Canada

Soil is the central organizer of the terrestrial ecosystem. Minerals, organic components, and microorganisms, which are three major solid components of the soil, should not be considered as separate entities but rather as a united system constantly in close association and interactions with each other in the terrestrial environment. Foreseeable impacts of these interactions include microbial events, global ion cycling, global climatic changes, biodiversity, biological productivity and human nutrition, geomedicine, development of biotechnology, ecotoxicology and human health, ecosystem risk assessment, and ecosystem risk management and restoration. Therefore, interactions of these three major solid components of soils have enormous impacts on reactions and processes critical to environmental quality and ecosystem health. Fundamental understanding of these interactions at the atomic, molecular, and microscopic levels is essential for restoring, sustaining and enhancing ecosystem health, which include human health, on a global scale.

1. INTRODUCTION

Soil is the skin of the planet Earth. It is the pedosphere which overlaps with the lithosphere, hydrosphere, atmosphere, and biosphere, and is, thus, an integral part of the environment. Therefore, what happens in soil should have a profound impact not only on soil quality and agricultural production, but also on ecosystem health which is defined in terms of ecosystem sustainability as a function of activity, organization, and resilience.

Soil is a focal point of the ecosystem [1, 2]. Soil components, be they minerals, organic matter, or microorganisms, profoundly affect the physical, chemical, and biological processes in the soil [3]. During the past decades, scientific accomplishments in individual sub-disciplines of physics, chemistry, and biology of the soil were impressive. But, information on interactions of soil minerals with organic components and microorganisms is fragmentary and scattered in the literature of soil and environmental sciences. Yet these three groups of soil components are not separate "domains" but constantly in close association and interactions with each other in the ecosystem [4], and thus forming a "united domain".

Interactions among these soil components have enormous impacts on physics, chemistry, and biology of soils. Fundamental understanding of mineral-organic matter-microorganism interactions at the atomic, molecular, and microscopic levels is essential to understanding and regulating their impacts on soil behavior. In 1990, the International Society of Soil Science established the Working Group MO "Interactions of Soil Minerals with Organic Components and Microorganisms" with the objective to promote research and education on the interactions of soil minerals with organic components and microorganisms and their impacts on the production of food and fibers, the sustainability of the environment, and human health on the global scale.

Activities of the Working Group MO have substantially contributed to bringing together scientific endeavors and the fragmented literature on the impacts of the interactions of inorganic, organic, and microbial factors of soils on agricultural and environmental sustainability. Such information serves to identify gaps in our knowledge and provides future direction to stimulate scientific research in this area of science. Physical, chemical and biological processes are interacting in soil environments, and are governed by soil mineral-organic component-microorganism interactions. The foreseeable impacts of these interactions on ecosystem health (Figure 1) are discussed below.

2. MICROBIAL ECOLOGY

Mineral colloids can influence microbial activity through direct and indirect effects [5]. Direct effects are those effects that involve a surface interaction between mineral colloids and microorganisms. Indirect effects are defined as those effects of mineral colloids that modify the environment in which the microorganisms are residing. The distinction between direct and indirect effects of mineral colloids on microorganisms in soil is not always clear, because many microbial activities and interactions are likely affected simultaneously, both directly and indirectly.

Mineral colloids can promote the activity of microorganisms in their vicinity by keeping the pH of microhabitats within the optimal physiological range and by sorbing microbial metabolites that would otherwise be detrimental to growth [6]. On the other hand, mineral colloids may accumulate toxic substances and immobilize extracellular enzymes, and, thus, exert a depressive effect on microbial activity. The mineral colloid-microorganism interaction is, on the whole, beneficial to bacterial survival in soil environments by protecting the organisms against desiccation, exposure to hypertonic osmotic pressure, and ultraviolet radiation. Further, of greater importance in terms of microbial survival is that clays can alter the structure of soil aggregates by creating pores with necks of less than 6 μm in diameter, which are freely accessible to bacteria but not to their predators, notably soil protozoa. Effects of surface-active particles (clays, humic acids, clay-humic acid complexes) on microbial events have been discussed in depth by Stotzky and co-workers [5, 7]. These include spore germination, metabolism, competition and amensalism, parasitism, organic and inorganic nutrition, pathogenesis, toxicity (heavy metals, organic pollutants, gases and volatiles), and transfer of genetic information among bacteria by conjugation, transformation, and transduction in soil.

In view of the indirect effects that surface-active particles can exert in altering the milieu of microorganisms in soil, in addition to the surface interactions between these particles and microorganisms, the role of mineral colloids and their interactions with organic matter in mediating a wide range of the above-mentioned microbial events in the ecosystem [5, 7] deserves increasing attention. Microorganisms, substrate, enzyme and product association with soil surface and within biofilm structures present challenges to those who wish to understand the microbial ecology of the soil microenvironment [8].

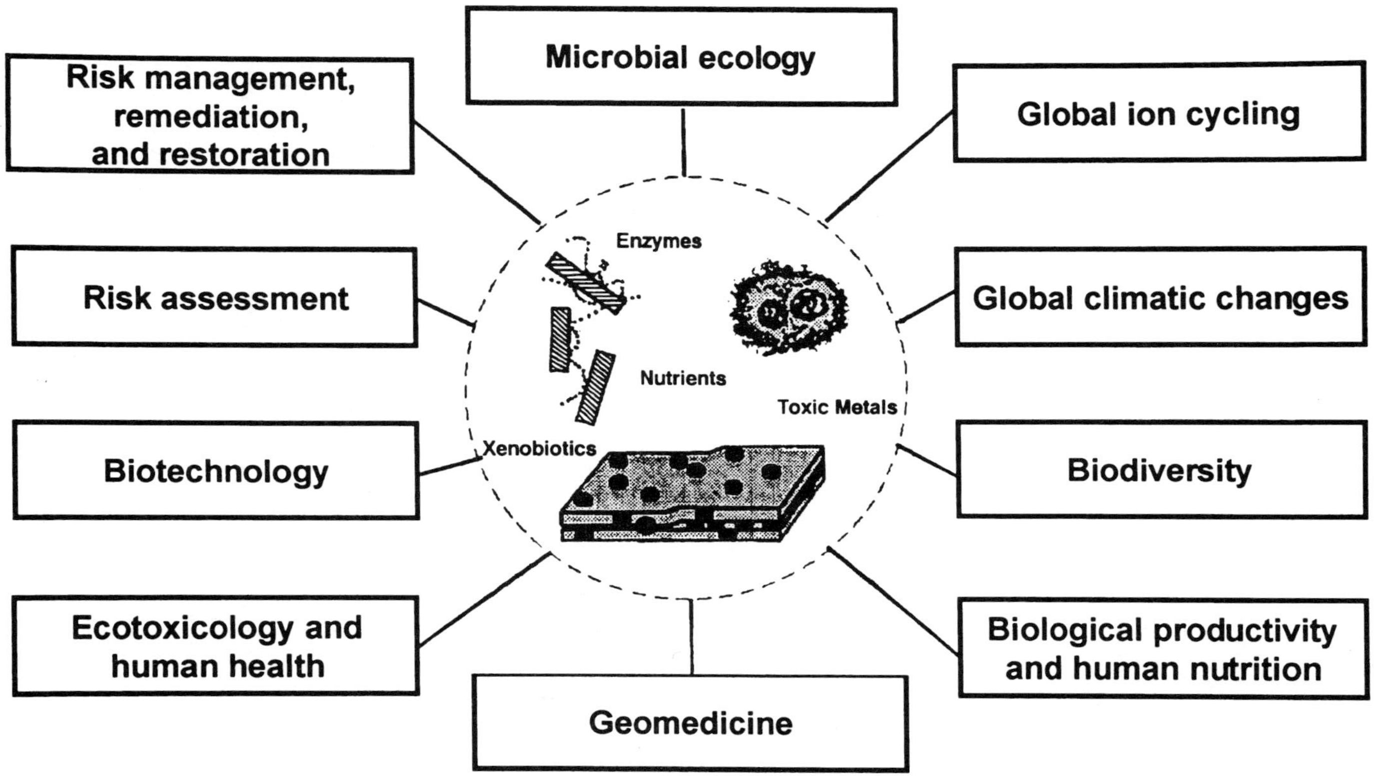

Figure 1. Impact of soil mineral-organic component-microorganism interactions on ecosystem health.

3. GLOBAL ION CYCLING

Major biogeochemical transformations of elements include the cycling of C, N, P, S, and metals [9] which should be very much influenced by interactions of soil minerals with organic matter and microorganisms.

Transformation of soil organic matter are closely related to the mobilization of C, N, P, and S into the soil solution or their immobilization from the solution [10, 11] and to the release or fixation of trace gases such as CH_4, CO_2, OCS (carboxide sulfide), H_2, NO_2, and NO (12, 13). These microbially-mediated processes are the basis for the interrelationship of the C, N, P, and S cycling within the soil-plant system (Figure 2). Interactions of soil minerals with microorganisms and organic components have an important role in influencing the stability and degradation of soil organic matter and its associated nutrients [15], thus, directly affecting the global cycling of C, N, P, and S.

Although scientific research in the past decades has significantly improved our understanding of the types and directions of interrelationships between transformation of soil organic matter and changes in the mineral phase at different scales and the influence on turnover of soil C, N, P, and S, future research should be directed more to a detailed quantification of these processes, the reaction mechanisms involved, and the impact on ecosystem health. One such example is the catalytic role of minerals in the degradation of organic components [16, 17] and in enhancing the cycling of C, N, P, and S in soil environments in situ [15]. Another example that merits further attention is the quantification of the diffusion of organic substrates into different pore size classes (micro- and mesopores) of mineral clusters and aggregates, their accessibility to degrading exoenzymes and the impact on the global cycling of C, N, P, and S.

Metals are part of natural biogeochemical cycles. One of the characteristics of the cycle of metal mobilization and deposition is that the form of the metal is changed. This change in speciation of a metal has a profound effect on its fate and impact on ecosystem health [18, 19]. Metals are found in the environment in solid, solution, and gaseous phases, associated with thousands of different compounds. These associations often reflect the affinity of metal ions for other atoms with free electron pairs, particularly O, N, and S. The critical processes controlling global metal cycling are adsorption-desorption, precipitation-dissolution, complexation, and volatilization [18-21]. Transport in solution or aqueous suspension is the major mechanism for metal movement in the ecosystem. This transport process is greatly influenced by adsorption-desorption on surfaces of minerals and organic matter, precipitation-dissolution especially in reduced environments where sulfide concentration is sufficiently high, and a series of inorganic and organic complexation reactions in dissolved and particulate phases. Through their effect on the chemical environment in soils and sediments, microorganisms can help dissolve, complex, or precipitate metals [20] and can also directly mediate reactions involving metals, such as initiation of fine-grain mineral formation [21].

Transport of particles suspended in the air is an important process for transporting many metals to regions far from their sources. A few metals, most notably Hg, can exist as gases at ambient temperatures. In the case of Hg, reduction of Hg^{2+} to Hg° and alkylation to form methyl- or dimethylmercury can result in the loss of Hg from the aqueous phase (Figure 3). Microorganisms can also convert the methylated forms to Hg°, which is more volatile and less toxic. Several other metals such as As and Se also form organometallic compounds, which can be mediated by microorganisms [23]. These volatile organometallic compounds can dominate transport of the metal in local environments. However, mediation of alkylation

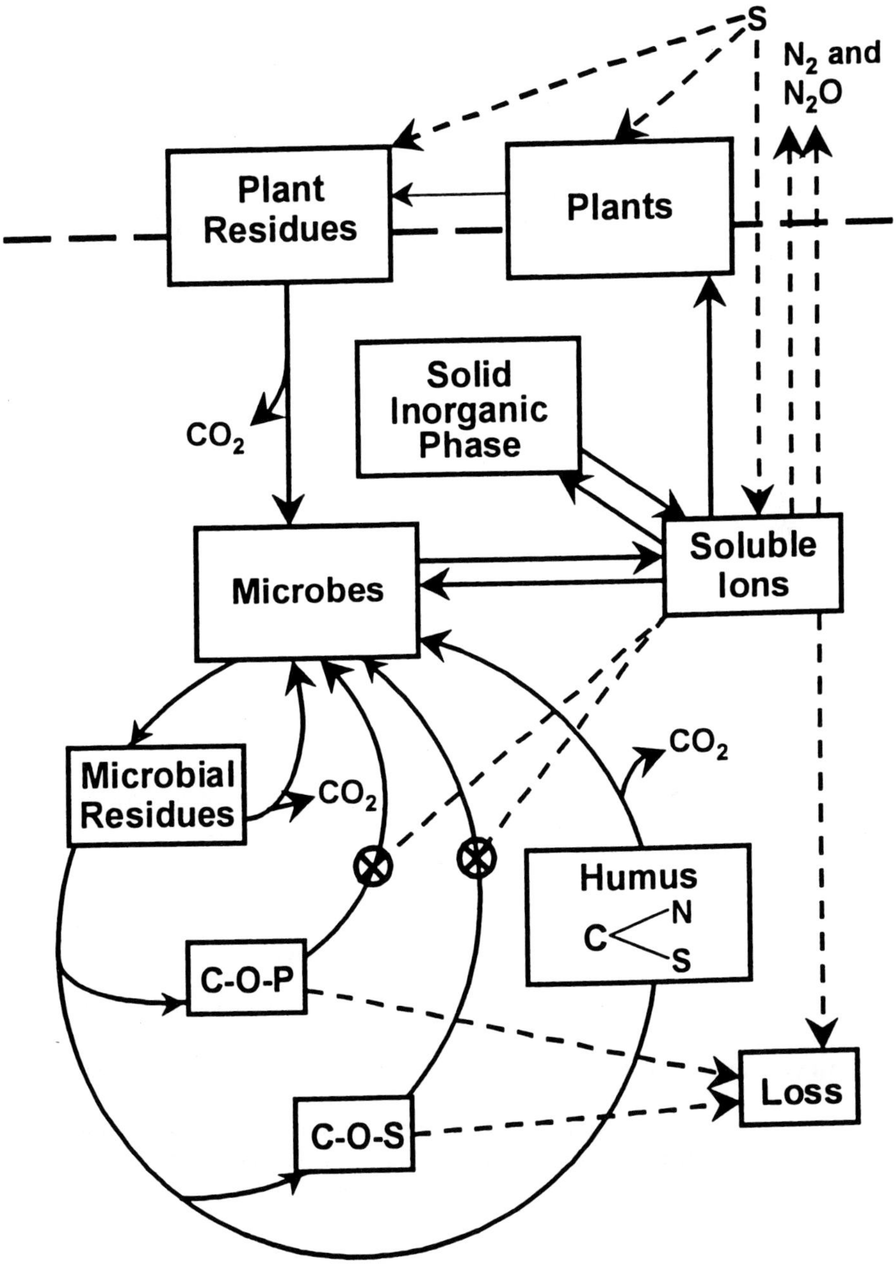

Figure 2. Schematic illustration of interrelations of C, N, S, and P cycling in soil-plant systems. Reprinted from [14].

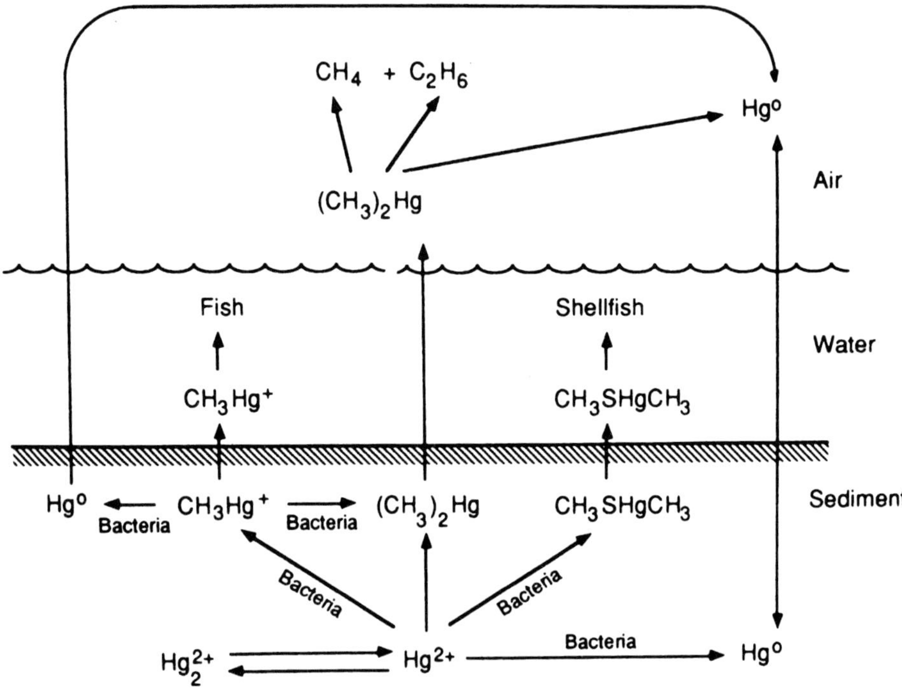

Figure 3. The mercury cycle, demonstrating the bioaccumulation of mercury in fish and shellfish. Reprinted from [22].

of metals such as Hg by bacteria is substantially influenced by Hg speciation on surfaces of mineral colloids (Table 1). Soil minerals, organic matter, and microorganisms have their respective roles in influencing metal speciation and toxicity [19, 25, 26]. Nevertheless, impacts of mineral-organic component-microorganism interactions on metal speciation, toxicity, and cycling in the ecosystem remain to be uncovered.

4. GLOBAL CLIMATIC CHANGE

Many biogeochemical and physical processes are involved in determining the climate of the Earth [27]. Some of these processes are being significantly perturbed by human activity. Of particular importance are reactions and processes in the atmosphere, through which all energy enters and leaves the Earth. The physical and chemical composition of the atmosphere determines the transmission, absorption, and reflection of incoming solar radiation and outgoing terrestrial radiation, and the resulting energy balance determines the surface temperature. The biogeochemical cycles of C, N, and S are central to the radiative properties

Table 1
Biomethylation of Hg(II) adsorbed on mineral colloids common in freshwater sediments, by *P. fluorescens* isolate BPL85-48 during a 25-h incubation period. Reprinted from [24].

Sample ID	Hg(II) Source†	Optical Density Absorbance at 530 nm	RGI‡	CH_3Hg^+(ngl^{-1})††
Blank‡‡	—	0.551	1.28 a	—
Control#	$Hg(NO_3)_2$	0.430	1.00 d	32.86±0.67 a
KGa-1	Kaolinite	0.423	0.98 d	30.53±1.64 ab
STx-1	Montmorillonite	0.451	1.05 c	25.96±4.17 b
MnO_2	Birnessite	0.478	1.11 b	ND¶ c

† Total concentration of Hg(II), added as $Hg(NO_3)_2$ or in adsorbed form, was 6 μmol/100ml.
‡ Relative growth index = optical density of colloid-amended medium/optical density of the control. Values followed by the same letter are not significantly different *(P*≤ 0.05; least significant difference test, LSD = 0.04).
†† Values followed by the same letter are not significantly different (*P*≤ 0.05; least significant difference test, LSD =5.30 ng $CH_3Hg^+ l^{-1}$).
‡‡ Isolate grown in the M-IIY medium in the absence of Hg(II).
\# Isolate grown in M-IIY medium supplemented with $Hg(NO_3)_2$; total Hg(II) concentration = 60 μM.
¶ ND, not detectable.

of the atmosphere. Carbon and N form radiatively important gases. Sulfur is a crucial component of clouds and most aerosols. Figure 4 sketches the two major processes, namely, the greenhouse effect and aerosol/cloud formation, by which chemical cycles affect climate. Water vapor, carbon dioxide, methane, and nitrous oxide are the radiatively important naturalatmospheric trace gases, whereas chlorofluorocarbons (CFC) are the radiatively important anthropogenic trace species. This group of gases, which are produced from a variety of natural and human processes, affect the cycles of water, C, N, and halocarbons, absorb infrared radiation in the atmosphere, and changes the global heat balance. The other important climate-affecting process is aerosol and cloud formation which appears to be dominated by the S cycle. Sulfur gases are produced and then oxidized to sulfuric acid in the atmosphere, forming new aerosol particles. Some of these particles have direct radiative effects to backscatter solar radiation and some may act as cloud condensation nucleic (CCN) to affect cloud albedo which reflect solar radiation. The S cycle, thus, influences the short-wave radiation properties of the atmosphere, whereas the cycles of water, C, N, and trace halocarbons contribute to the long-wave properties. With the exception of water vapor, all of these cycles are severely perturbed by human activity (Table 2).

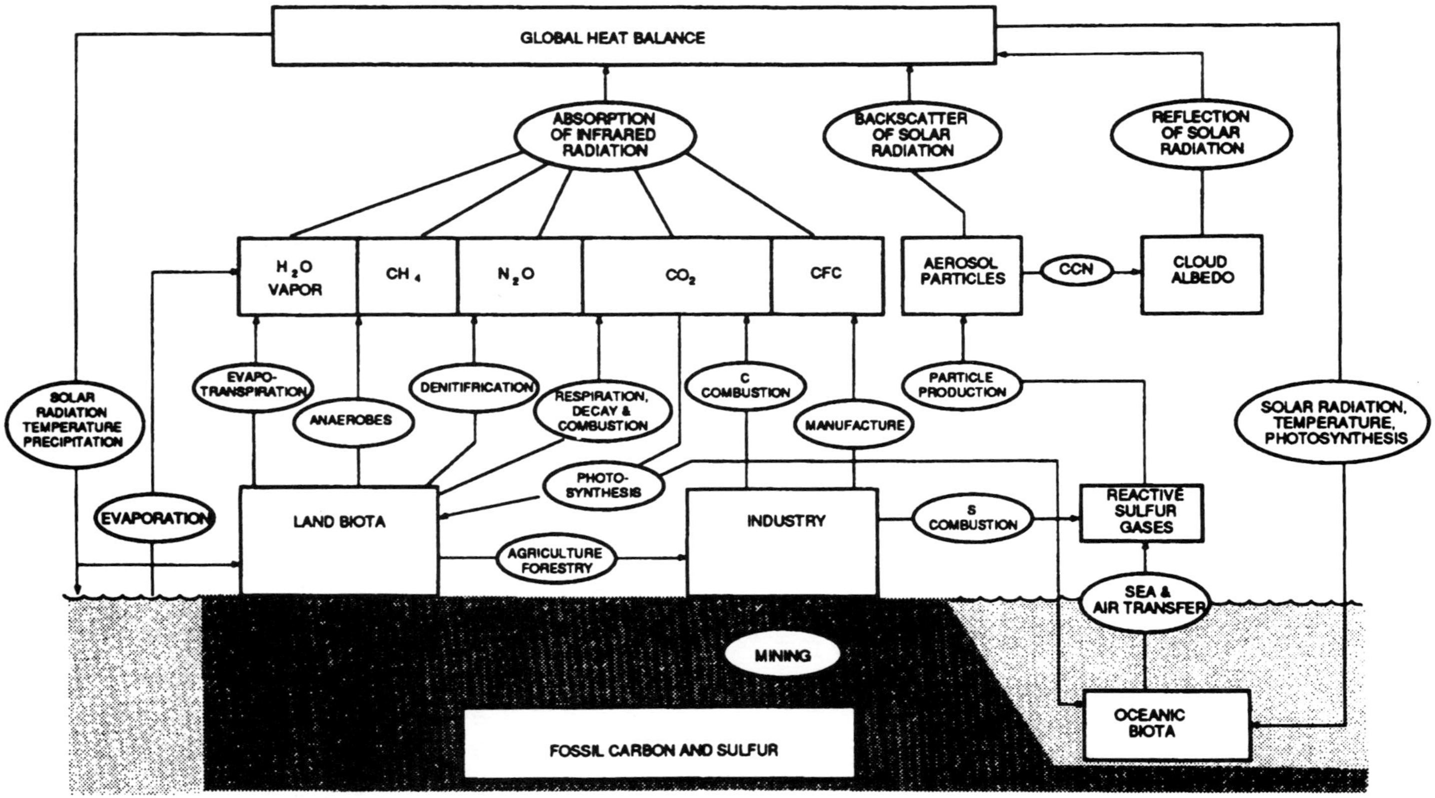

Figure 4. Schematic of the processes that connect global biogeochemical cycles and climate. Boxes denote observables and ovals indicate processes that affect them. Modified from [27].

Table 2
Radiatively important trace species in the atmosphere: Percent change in flux measured relative to the pre-industrial age. Reprinted from [27].

Cycle change	Species	% change
Long-wave absorbers		
Water	H_2O (vapor)	Not known
Carbon	CO_2	+50%
	CH_4	>+65%
Nitrogen	N_2O	+25%
Halogens	Chlorofluorocarbons	$+\infty$
Short-wave reflectors		
Sulfur	SO_4^{-2}	+230%

Transformation of C, N, and S in soils as influenced by land management and the impact on their ion cycling and global climatic change should not be overlooked [15, 28]. Jenkinson et al. [29] estimated the additional degradative effects on soil organic matter if the global annual mean temperature rises during the next 60 years by 3°C. According to their estimate, about 100 Gt C (1 Gt = 10^9t) should be additionally evolved as CO_2 from soil organic matter (1600 Gt C). This will increase the present CO_2 concentration in the atmosphere by 14%, whereas the combustion of fossil fuels (5.4 Gt C yr^{-1}) should add during this 60-year period 330 Gt C to the atmosphere.

Microbial by-products and resistant plant residues adsorbed on soil particles have turnover times in terms of years. Fulvic acids (FAs) have turnover times in terms of hundreds of years, whereas humic acids (HAs) and humins usually approach thousand years in their turnover time [30]. The distribution and annual transfers of C in the various fractions for a grassland Chernozem are shown in Figure 5. Although the HAs and humins constitute by far the majority of the organic C in a system, they contribute only a small proportion to the annual cycling of C within the soil because of their very slow turnover rate. The undecomposed litter (Figure 5) also includes the soil biomass and microbial metabolites. These, together with the plant residues, constitute the active fraction of organic matter that has a prominent role in the cycling of elements such as C, N, and S annually. The influence of crystalline and noncrystalline mineral colloids, which differ in structural configuration and surface properties, on the biodegradation, turnover, and stabilization of organic components, the cycling of C, N, and S, and the impact on global climatic changes merits close attention [3, 4, 15, 17].

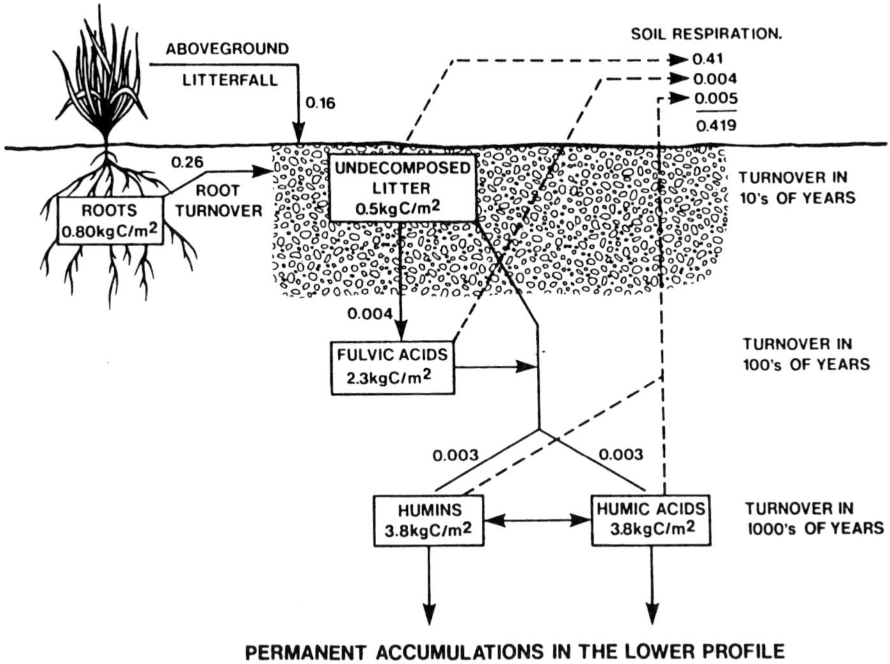

Figure 5. Detrital carbon dynamics for the 0 to 20 cm layer of a chernozem grassland soil. Carbon pools (kg C m^{-2}) and annual transfers (kg C m^{-2} yr^{-1}) are indicated. Total profile content of C to 20 cm is 10.4 kg C m^{-2}. Reprinted from [31].

5. BIODIVERSITY

The functioning and stability of terrestrial ecosystems are determined by plant biodiversity and species composition [32-34]. However, the ecological mechanisms by which they are regulated and maintained are not well understood. These mechanisms need to be identified to ensure successful management for conservation and restoration of diverse natural ecosystems.

Van der Heijden et al. [35] recently reported that below-ground diversity of arbuscular mycorrhizal fungi (AMF) is a major factor contributing to the maintenance of plant biodiversity and to ecosystem functioning. These results emphasize the need to protect AMF and to consider these fungi in future management practices in order to maintain diverse ecosystems. Their research findings highlight the essentially interactive nature of those mechanisms. They also show that conservation of the fungal gene pool is likely to be a prerequisite for maintenance of flouristic diversity in grasslands and other ecosystems such as boreal forests, where the fungal web is known to influence allocation of resources between plant species. Mycorrhizal community is sensitive to perturbations, particularly those associated with cultivation and nutrient enrichment [36]. Their results demonstrated that microbial interactions can drive ecosystem functions such as biodiversity and variability.

Although debate continues over the contribution of diversity to ecosystem functions, empirical studies provide support for the view that flouristically rich systems are more productive [33], show greater stability under stress [37], and are more likely to provide alleviation of global problems posed by atmospheric CO_2 enrichment [38]. A recognition of these properties, coupled with an increasing awareness that the diversity of terrestrial vegetation systems is everywhere under stress, has encouraged biologists to investigate the mechanisms that determine and affect species composition in plant communities.

The above discussions indicate that below-ground microbial diversity substantially influences plant biodiversity, and ecosystem variability. Further, microbial events are significantly affected by surface-reactive particles [5, 7, 8]. However, our knowledge on the effect of mineral-organic component-microorganism interactions on below-ground microbial diversity and the impact on above-ground biodiversity remains to be advanced.

6. BIOLOGICAL PRODUCTIVITY AND HUMAN NUTRITION

Soil is the life-sustaining material which is the structurally porous and biologically active medium that has developed on the continental land surface on our planet (Figure 6). This material is created and continues to evolve through weathering processes driven by biological, climatic, geological, topographic, and chronological influences.

Figure 6. Soil as the life-sustaining material which is the structurally porous and biologically active medium. Reprinted from [39].

Our early ancestors relied on the natural vegetation for their food. The beginning of agriculture and permanent settlements millennia ago was accompanied by an increasing awareness of soil and human's ability to manage soils. With the advent of the industrial revolution, there is an increasing pressure on soil to produce raw materials in demand by commerce and trade. This has resulted in an increased use and abuse of soils that are vital to the life cycles of terrestrial vegetation, and a vast array of soil-inhabiting organisms. Soil is the cradle of agriculture including crop and animal production and a fountain for sustaining human nutrition. However, it is the fragile epidermis of the planet Earth that can sustain human nutrition or cause starvation for humans depending on our management of soil resources.

Interactions of soil minerals with organic components and microorganisms exert enormous influences on the transformation and dynamics of soil organic matter [4, 40, 41], nutrient cycling [11], nutrient bioavailability [42], efficacy and toxicity of pesticides [43, 44], microbial metabolic processes, growth, adhesion, and ecology [5, 6] enzymatic activity [45-47], and soil physical properties [6, 48]. Therefore, interactions of soil minerals with organic components and microorganisms should have great impacts on plant nutrition and biological productivity of soils.

In the rhizosphere, the narrow zone of soil surrounding a living plant that is subject to influence by the root and its exudates, more intense microbial activity and larger microbial populations occur than in the bulk soil [49]. Up to 18% of the C assimilated through photosynthesis can be released from roots. Since the rhizosphere is rich in root exudates, microbial population can be 10 to 100 times larger than the population in bulk soil [39]. The rhizosphere typically extends away from the root for up to 2 mm, but some organisms (e.g., fungi) may be stimulated up to 5 mm away. The rhizosphere is bathed in root exudates and microbial metabolites. Both the amounts and proportion of organic compounds of root exudates vary substantially with plant species and cultivars. Further, the same plant cultivar grown in different soils varies in the kind and amount of low-molecular-weight organic acids (LMWOAs) present in the rhizosphere (Table 3). The chemistry and biology at the soil-root interface, thus, differs significantly from soil to soil. The soil rhizosphere is the bottleneck of the nutrient-feeding zone in soils. Therefore, the dynamics, transformation, and bioavailability of nutrients are bound to be influenced greatly by the chemistry and biology at the soil-root interface. The intense soil mineral-organic component-microorganism interactions in the rhizosphere, thus, deserve close attention in the development of innovative management strategies for land resources to increase biological productivity.

Van der Heijden et al. [35] reported that microbial interactions can influence not only plant biodiversity but also productivity. Both the plant biodiversity, as measured by Simpson's diversity index (Figure 7a), and productivity above and below ground (Figure 7b, c) increased with increasing AMF-species richness. The lowest plant productivity were found in those plots without AMF or with only a few AMF species. In contrast, plant productivity was highest when eight or fourteen AMF species were present. The results showed that plant productivity in a given ecosystem can be dependent on the diversity of fungal symbionts. The results also indicated a mechanistic explanation for the effects of mycorrhizal-species richness on plant productivity. Increased AMF-species richness led to a significant increase in the length of mycorrhizal hyphae in the soil (Figure 7d), to a decreased soil phosphorus concentration (Figure 7e) and to an increased phosphorus content in plant material (Figure 7f). Therefore, increasing AMF biodiversity resulted in more efficient exploitation of soil phosphorus and to a better use of the resources available in the system. The loss of AMF biodiversity, which occurs in agricultural systems [38, 51], could, therefore, decrease ecosystem productivity. The research findings of van der Heijden et al. [35] demonstrated the impact of the loss of biodiversity on the decrease of biological productivity of soils.

Table 3
Amount of low-molecular-weight organic acids (μg/kg dry soil) in rhizosphere soil of durum wheat cv. Kyle grown in three different soils. Reprinted from [50].

	Soil		
Acid	Yorkton	Sutherland	Waitville
Malonic	99a	56a	68a
Succinic	22a	35476c	10826b
Fumaric	12a	150b	71ab
Malic	45a	898c	370b
Tartaric	ND†	665b	214a
trans-Aconitic	ND	13a	3a
Citric	ND	195b	81a
Acetic	865a	29245c	12240b
Propionic	ND	499a	ND
Butyric	ND	7604b	2127a
Total	1043a	74801c	26000b

† ND = not detected.
Means withing the same row having the same letter are not significantly different ($p \leq 0.05$).

This represents an understudied field of research which requires more attention. The present reduction in biodiversity on earth and its potential threat to ecosystem productivity [33] to supply food to sustain human nutrition can be reversed or stopped only if innovative management strategies are developed to protect the ecosystem. Since microbial activity is substantially influenced by surface-reactive soil particles [5, 6], the effect of mineral-organic component-microorganism interactions on microbe species richness and the impact on biological productivity need to be studied.

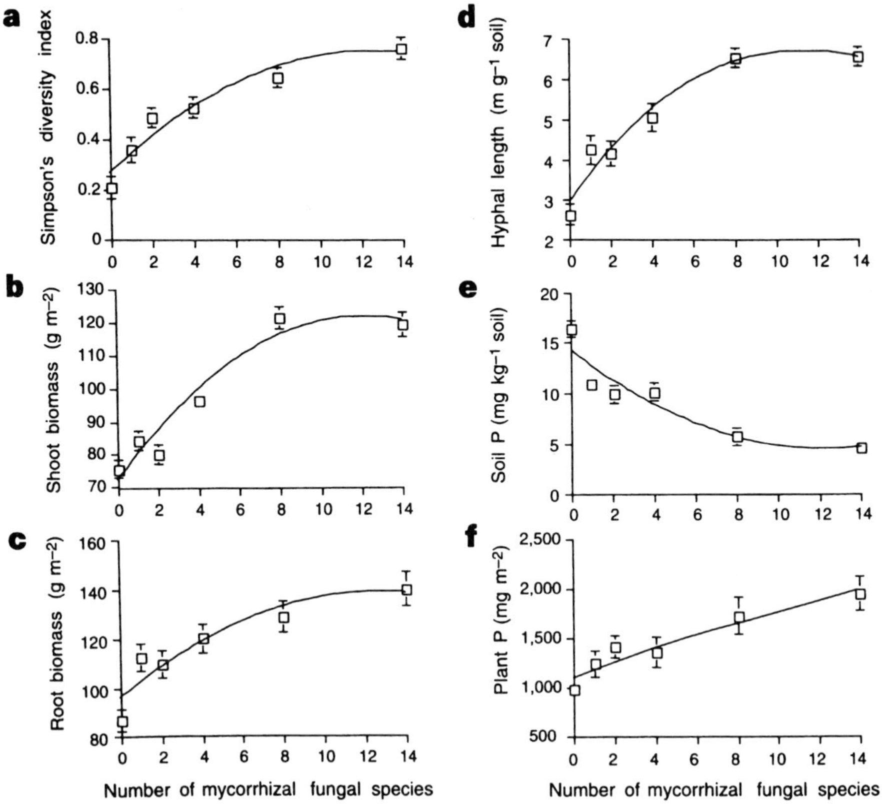

Figure 7. The effect of AMF-species richness on different parameters. Effects on: a) the plant biodiversity, as measured by Simpson's diversity index (fitted curve is $y = 0.271 + 0.077x - 0.003x^2$; $r^2 = 0.63$; $P \leq 0.0001$); b) shoot biomass ($y = -0.334x^2 + 8.129x + 72.754$; $r^2 = 0.69$; $P \leq 0.0001$); c) root biomass ($y = -0.265x^2 + 6.772x + 96.141$; $r^2 = 0.55$; $P \leq 0.0001$); d, length of external mycorrizal hyphae in soil ($y = 0.001x^3 - 0.046x^2 + 0.756x + 2.979$; $r^2 = 0.60$; $P \leq 0.0001$); e) soil phosphorus concentration ($y = 0.065x^2 - 1.593x + 14.252$; $r^2 = 0.67$; $P \leq 0.0001$); and f) total plant phosphorus content (linear relationship; $y = 61.537x + 1156.281$; $r^2 = 0.48$; $P \leq 0.001$), in macrocosms simulating North American old-field ecosystems. Squares represent means (± s.e.m.). Reprinted from [35].

7. GEOMEDICINE

Geomedicine may be defined as the science dealing with the environmental factors which influence the geographical distribution of pathological and nutritional problems relating to human and animal health [52].

Soil-related geomedicine may be termed "edafo-geomedicine" [52]. When the geomedical problems are connected to aquatic environments, the term "hydrogeomedicine" may be used.

The study of geomedical problems related to the atmosphere may be termed "atmo-geomedicine" [53]. These three areas of geomedicine belong to biogeomedicine.

Geomedicine is a young science with very old roots. Knowledge on soil science is needed for solving many geomedical problems [54]. Hunger and malnutrition are serious issues for large groups of populations, especially in developing countries. In addition to prevention of starvation, promotion of better nourishment quality of food and feed is important. Pollution of the environment and the related health problems have increased rapidly in many industrialized countries. Effects of many soil chemical, physical and biological factors and processes on geomedical problems should be studied in depth as they impact on the quality of vegetation and the food and feed produced.

Biogeochemistry of trace elements is greatly influenced by soil-plant-microbe interactions [55]. Therefore, soil mineral-organic component-microorganism interactions deserve close attention in the transformation, dynamics, and bioavailability of trace elements, many of which are of concern to animal nutrition and health and well-being of humankind [56, 57]. They include Se, Fe, I, Zn, Cu, Mn, Mo, Cr, F, Co, Si, V, Ni, As, and Mg. One trace element may serve in one, several or dozens of different metalloenzymes or tissue constituents.

Among trace elements which are of concern in human and animal nutrition, Se is considered one of the most active and versatile [58]. Evidence shows that the catalytic capacity of Se in cell metabolism plays a key role in the maintenance of cell integrity. Selenium is essential for antioxidation of lipids by enzymes. Deficiency of Se is related to cardiomyopathy and cancer. The diverse Se status of different soil areas by city, county, state, or province-wide localities, or in international comparisons, causes variations in age-adjusted annual heart death rate (HDR) and cancer death rate (CDR) per 100,000 (10^{-5} yr^{-1}), despite a commonly held view that food transportation and human migration would preclude such variations [58]. Furthermore, several studies [59] show that the prevalence of multiple sclerosis can be related to geographical latitude and the availability of Se appears to be a decisive factor. The status of soil Se in relation to human health and animal nutrition is of concern in many parts of the world. Soil Se status seems to be, in part, related to leaching intensity and soil parent materials. Further, the dynamics of soil Se is influenced by the nature and concentration of organic acids [60]. Therefore, cropping systems should influence the chemistry and biology of Se of the soil rhizosphere where soil mineral-organic component-microorganism interactions are especially intense.

All the trace elements appear to have vital metabolic roles in critical steps in health risks built into human genetics. To facilitate fundamental understanding of the linkage of trace elements in the soil-plant-environment-animal-human systems, and to provide practical solutions to their deficiency and toxicity problems, we need to advance the knowledge on soil mineral-organic component-microorganism interactions affecting the transformation, dynamics, and bioavailability of trace elements in soils and associated environments.

8. ECOTOXICOLOGY AND HUMAN HEALTH

Soil plays the central role as the organizer of terrestrial ecosystem [2]. Furthermore, it may be perceived as the center of the ecosystem which evolves as a result of interactions of the lithosphere, hydrosphere, atmosphere, and biosphere [61]. A factor of central importance of soil to ecological studies is that soils on a global scale have a range of characteristics which enable an enormous array of microorganisms, plants, animals, and humans to co-exist and thrive. Humans have exploited the ability of soils to provide massive amounts of food. About >40% of the net primary production of the world are exploited by humans [62]. The exploitation is increasing with the addition of 87.6 million people to the global population every year, with the rate addition steadily increasing [63]. The impact of population and the accompanying intensification of agriculture and industrialization on ecotoxicology and human health is, thus, of increasing concern.

Among the environmental compartments, about 90% of environmental pollutants are bound with soil particles and 9% of the pollutants are bound with aquatic sediments (Table 4). These soil- and sediment-bound pollutants are in dynamic equilibrium with the hydrosphere, atmosphere, and biosphere. Soil mineral-organic component-microorganism interactions may enhance the release of environmental pollutants from soils and sediments and, thus, pose a threat to ecosystem health, including human health.

Ecosystems consist of all organisms in a given space or volume, interacting with all of the abiotic factors such as energy, light, soil components, nutrients, pollutants, and others. The system is open and freely exchanges matter and energy across the real or perceived boundaries [1]. An ecosystem is a functional rather than a structural concept. It can be considered at any spatial scale from a soil crum to the biosphere. The ecosystem concept focuses on processes which occur in nature. Ecosystem health is defined in terms of ecosystem sustainability as a function of activity, organization, and resilience. An ecological system is healthy and free from "distress syndrome" if it is stable and sustainable—that is if it is active and maintains its organization and autonomy over time and resilient to stress [65]. Special attention should be paid to the processes which occur in various forci of activity, which have been termed "hot spots" by Beare et al. [66]. "Hot spots" of activity within soils include, in ascending order of size, micro- and macro-aggregates, rhizospheres, and macropores where intense soil mineral-organic component-microorganism interactions occur (Figure 8).

Table 4
Theoretical pollutant distribution in the environment at equilibrium. Reprinted from [64].

Compartment	Concentration (mol/m^3)	% distribution in compartment
Air	4×10^{-10}	0.35
Water	10^{-7}	0.01
Sediment	10^{-2}	9.10
Soil	10^{-2}	90.50
Aquatic biota	10^{-7}	0.01

Assumes approximately 100 kg of pollutant (MW 100) introduced into 10 km^3 of the environment.

Soil scientists should be concerned with the management of a complete range of ecosystem functions, within or encompassing many of these hot spots. This is of central

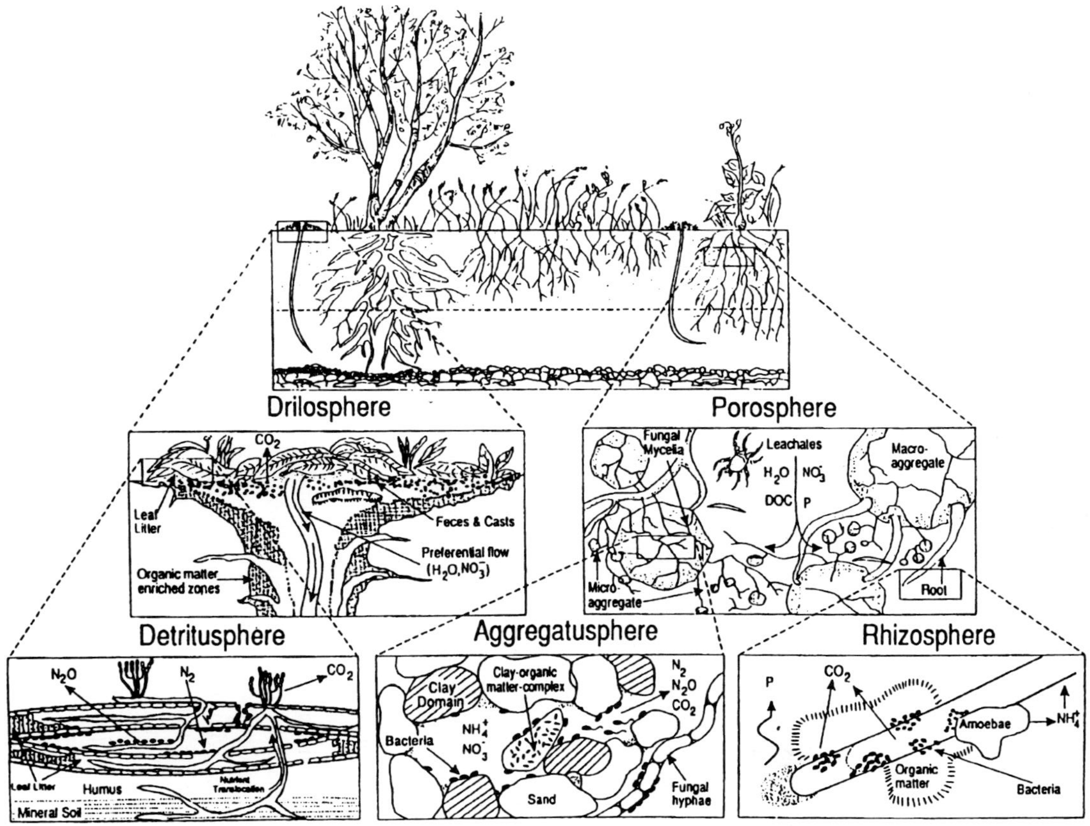

Figure 8. Hot spots of activity within soils, ranging in size from aggregates to rhizosphere, detritus, drilosphere [earthworm burrows and macropores in general (porosphere)] (66). [Reprinted from: M.H. Beare et al., Plant Soil, 170 (1995) 5].

importance to sound ecosystem management. Humans are an integral part of the ecosystem. Land resource management to sustain the capacity of a soil to function within ecosystem boundaries to improve biological productivity, maintain environmental quality, and promote plant and animal growth and human health is the central role of soil scientists in contributing to human welfare. Interactions of soil minerals with organic components and microorganisms have vital roles in governing transformation, speciation, transport, bioavailability, and toxicity of organics, metals, and other inorganics of agricultural and ecological concerns (Figure 9). These interactions are fundamental to understanding and regulating the ecosystem at the molecular level.

Reaction pathways involved in transformations of inorganic and organic pollutants in soil environments have enormous impacts on ecosystem health. Two examples on transformations of metal and organic pollutants are discussed below.

8.1. Cadmium

Cadmium is one of the three heavy metals (Cd, Hg, and Pb) of the highest concern in the environment [68]. It is receiving increased international attention because of its association with health problems, especially bone disease (itai-itai disease) and the renal cortex problem of the kidney [69, 70]. Ninety-five percent of a human's Cd uptake comes from diet, mostly plant products. Therefore, the transfer of Cd from soils to plants is an important pathway of Cd contamination of the food chain to endanger human health.

Transformation mechanisms of Cd in the soil-plant system (Figure 10) influence its chemical reactivity, mobility, bioavailability, and toxicity [72]. Soil processes, which control Cd speciation and its subsequent ecological impact, include complexation, adsorption-desorption, and precipitation-dissolution reactions, which are, in turn, controlled by the activity of Cd species in the soil solution, the nature of the surfaces of soil minerals and organic matter, microbial activities, and related environmental factors.

The LMWOAs secreted by plant roots and produced in microbial metabolites enhance the mobility of Cd through the formation of soluble complexes in the soil rhizosphere, which is reflected in the increase in the Cd release from the soils with the increase in the log $K_{Cd\text{-}LMWOAs}$ (Figure 11).

The kinetics of Cd release by LMWOAs is a diffusion controlled process. The overall diffusion coefficients of Cd release from the soils by LMWOAs [73] follow the same trend as the Cd availability index (1M NH_4Cl-extractable Cd) of the soils [74], the total LMWOAs in the rhizosphere aoils, and grain Cd concentration of durum wheat [75, 76].

Compared with the soils from the control plot, the Cd availability index values of the plots treated with the mono-ammonium phosphate fertilizer are 2-9 times higher in the rhizosphere soil at the 2-week growth stage (Table 5), apparently due to the combined effects of Cd introduced into the soil rhizosphere from the phosphate fertilizer, and Cd complexation with phosphate, and LMWOAs formed through root exudation and microbial activity. At the 7-week stage, such differences were not observed. The fundamentals of the soil-root-microbe interaction and its impact on the nature of Cd labile pool and Cd uptake at various stages of plant growth warrant in-depth research.

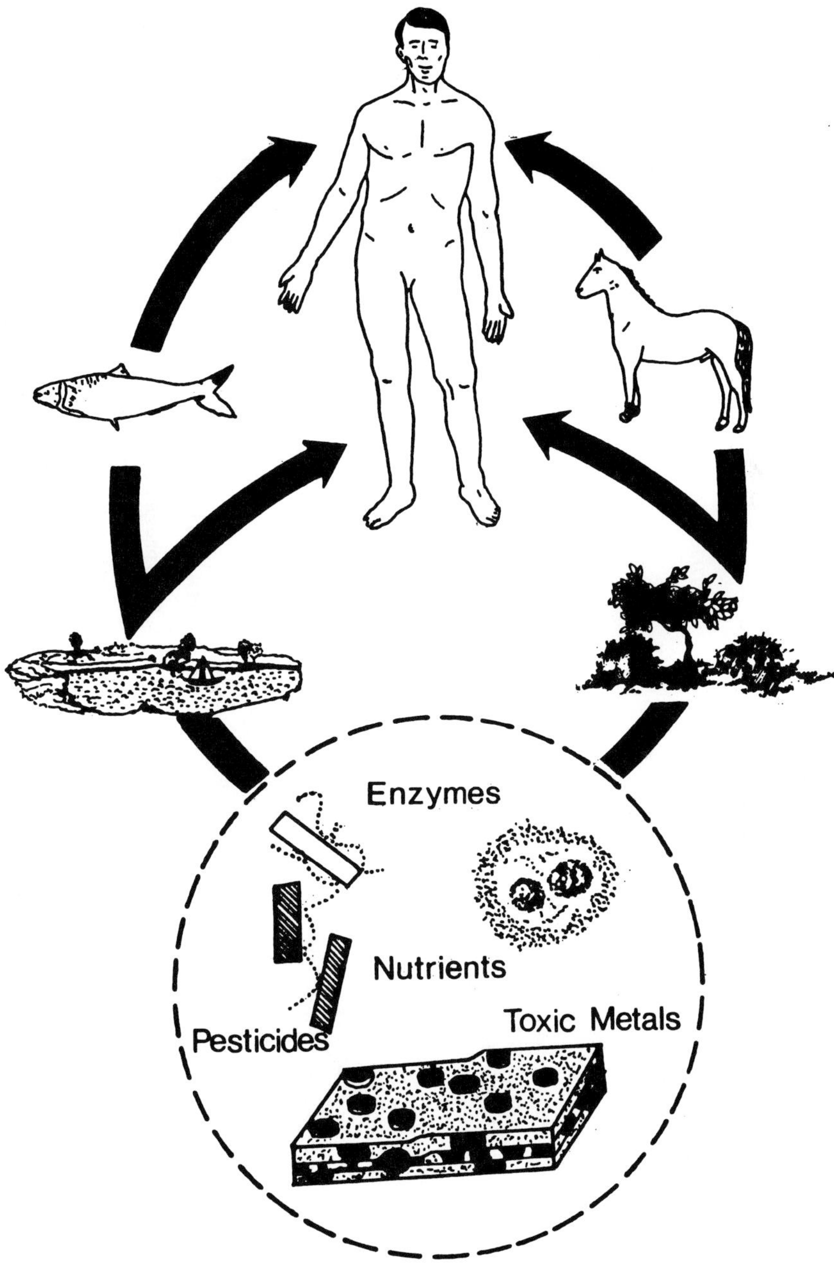

Figure 9. Impact of environmental soil chemistry on agricultural sustainability and the ecosystem health. Environmental soil chemistry is governed by mineral-organic matter-microorganism interactions. Reprinted from [67].

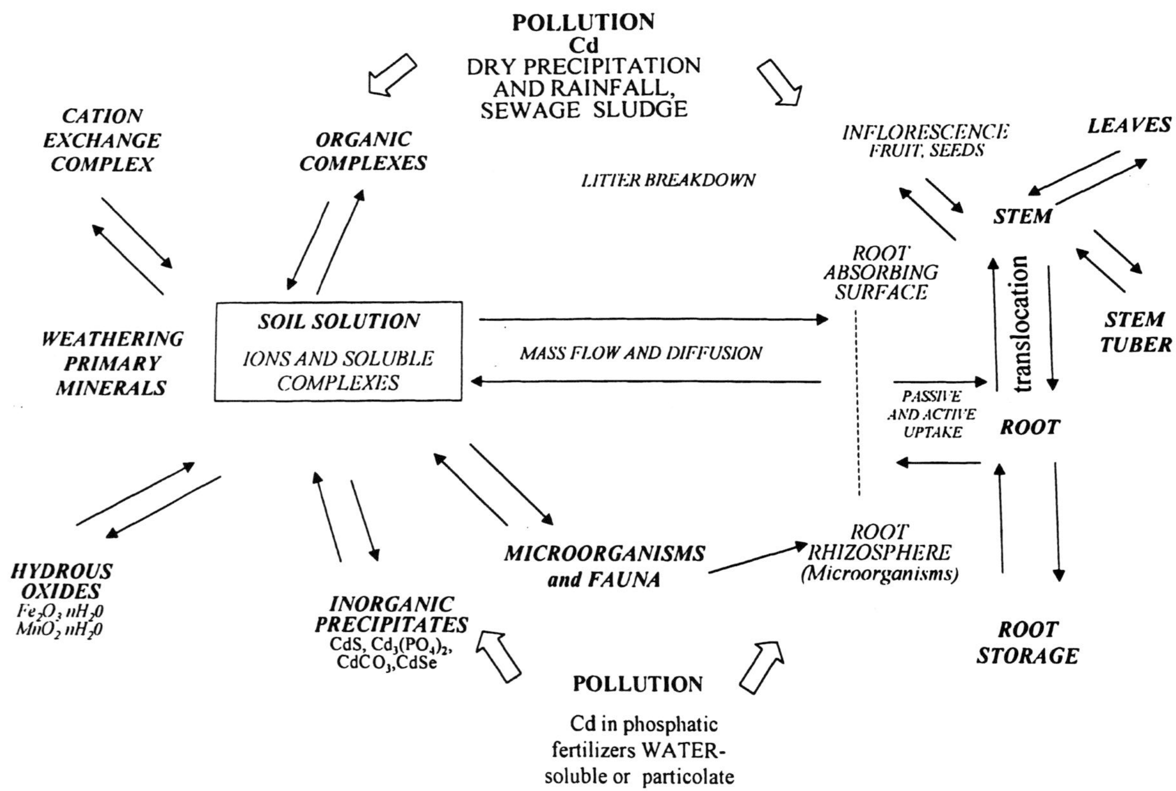

Figure 10. Pathways of cadmium in the soil-plant system. Modified from [71].

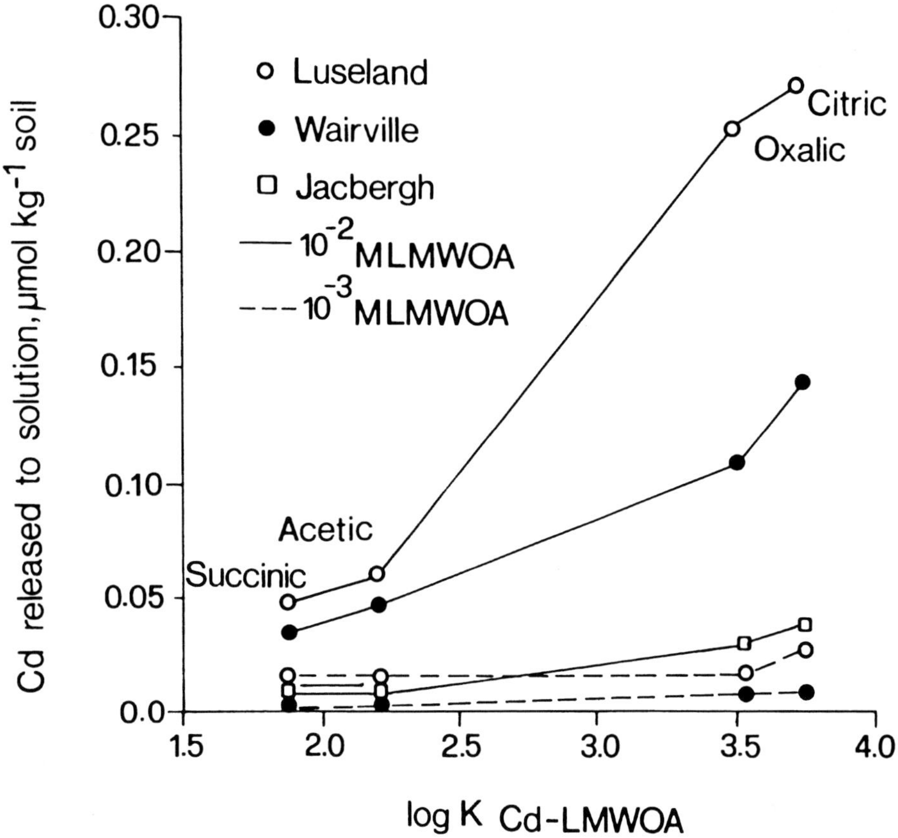

Figure 11. Relationship between Cd released from the soils by selected low molecular weight organic acids (LMWOAs) during the reaction period of 0.25 h and the logarithm of the stability constant of Cd-LMWOA complexes. Reprinted from [73].

The kinetics of Cd release by LMWOAs is a diffusion controlled process. The overall diffusion coefficients of Cd release from the soils by LMWOAs [73] follow the same trend as the Cd availability index (1M NH_4Cl-extractable Cd) of the soils [74], the total LMWOAs in the rhizosphere aoils, and grain Cd concentration of durum wheat [75, 76].

Compared with the soils from the control plot, the Cd availability index values of the plots treated with the mono-ammonium phosphate fertilizer are 2-9 times higher in the rhizosphere soil at the 2-week growth stage (Table 5), apparently due to the combined effects of Cd introduced into the soil rhizosphere from the phosphate fertilizer, and Cd complexation with phosphate, and LMWOAs formed through root exudation and microbial activity. At the 7-week stage, such differences were not observed. The fundamentals of the soil-root-microbe interaction and its impact on the nature of Cd labile pool and Cd uptake at various stages of plant growth warrant in-depth research.

Table 5
The influence of application of Idaho monoammonium phosphate fertilizer on pH and cadmium availability index [CAI]† of the bulk and rhizosphere soils at 2-week and 7-week crop growth stages. Reprinted from [77].

Soil and cultivar		Bulk soil	Rhizosphere soil: Control‡ 2-W	Control‡ 7-W	Idaho‡‡ 2-W	Idaho‡‡ 7-W	LSD# 0.01	LSD# 0.05
Luseland Soil								
Kyle	pH	7.95	7.75	7.90	7.38	7.9	0.11	0.08
	CAI	87	97	88	152	88	85	63
Arcola	pH	7.90	7.80	7.85	7.38	7.85	0.11	0.08
	CAI	87	102	87	208	89	85	63
Jedbergh Soil								
Kyle	pH	8.15	7.95	8.10	7.38	8.10	0.14	0.10
	CAI	9	12	9	80	10	27	19
Arcola	pH	8.10	7.90	8.05	7.70	8.05	0.14	0.10
	CAI	9	16	9	84	8	27	19

† CAI = Cadmium availability index (in μg kg^{-1}) determined by 1 M NH_4Cl extraction method.
‡ Without application of Idaho phosphate fertilizer.
‡‡ With application of Idaho phosphate fertilizer.
# LSD = Least significant difference at $p = 0.01$ and $p = 0.05$.

8.2. Pesticides

The fate of pesticides in the environment is governed by transformation and transport processes and the interaction of these processes (Figure 12) and it affects their efficacy and the impact on environmental quality and human health. Pesticides can be transformed by biotic and abiotic processes in soils [43]. Although it is often assumed that biotic transformations control the degradation of pesticides in soil environments, they are bound to be influenced by abiotic transformation processes, especially their interactions with soil minerals. The major processes that affect the efficacy and fate of pesticides in soils and their impact on ecosystem health are their retention by soil minerals and organic matter, abiotic catalysis, biotic transformation, and transport in soil to the surface water and groundwater, and to the atmosphere [41, 43, 78-80].

Mineralogical properties affect the degradability of pesticides adsorbed on clay surfaces [81]. Additions of montmorillonite to diquat solutions containing a mixed soil microbiota completely inhibit diquat mineralization when a sufficient amount of clay for total removal of diquat from the solution is present (Table 6). The microbial degradation of diquat is not

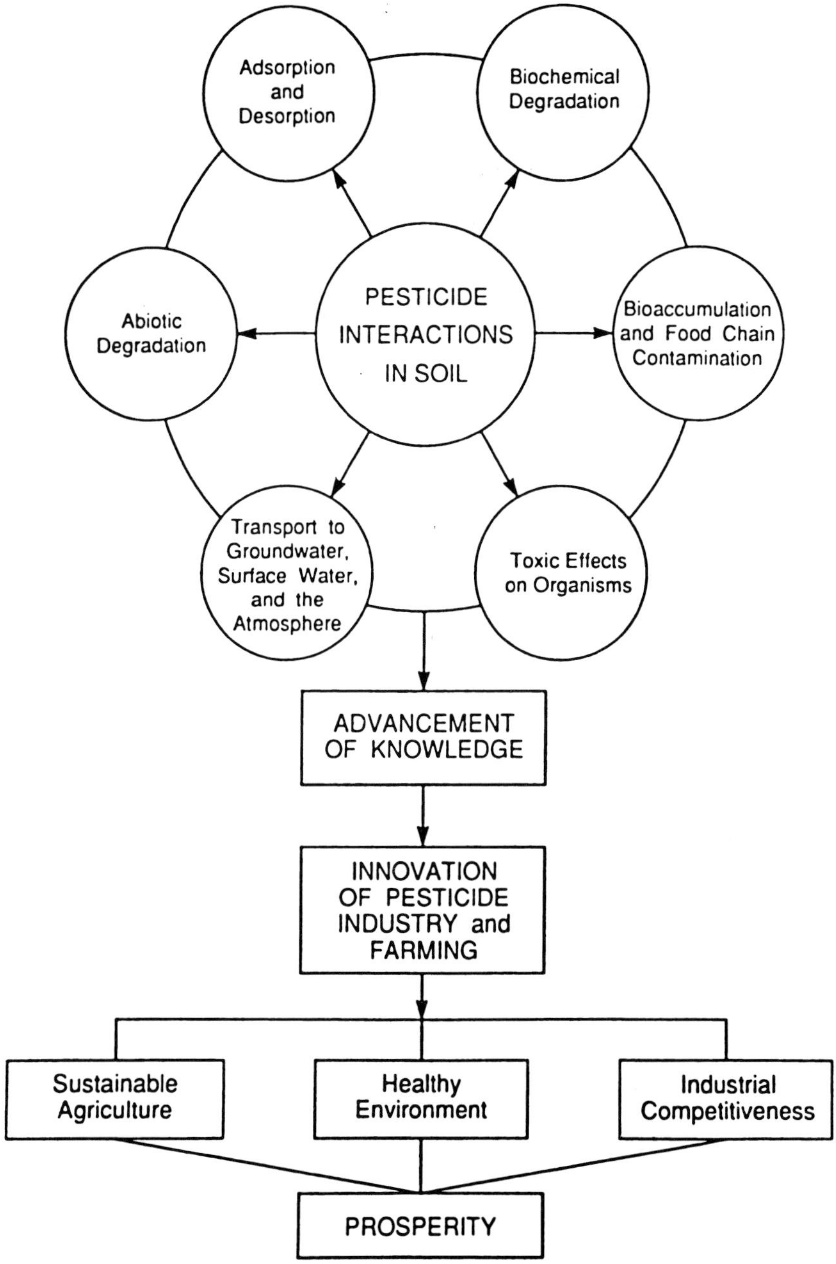

Figure 12. Transformation and transport processes of pesticides in the soil environment and their impacts on environmental quality and human health and prosperity. Reprinted from [44].

influenced by adsorption on kaolinite, as diquat is readily desorbed from the surface of this clay. It is, thus, not possible to distinguish between microbial degradation in the solution and on kaolinite surfaces. Research on the impact of mineral-pesticide interactions on microbial degradation of pesticides deserves increasing attention.

The significance of soil mineral-catalyzed abiotic transformations in the fate of pesticides in the environment has become widely recognized [17, 43]. Even in soil environments with intense biological activity, abiotic transformations of pesticides deserves attention. Abiotic and biotic transformations often take place simultaneously. The transformation of pesticides as catalyzed by soil minerals include reactions that are heterogeneously catalyzed on mineral surfaces and those that occur in the solution under the influence of the electric field charged surface. Some products of the surface-catalyzed transformations of pesticides remain chemically and physically adsorbed on the surfaces of soil particles. Products of abiotic transformations of pesticides may be less toxic or may still be as toxic or even more toxic than the parent compounds.

Table 6
Cumulative Release of $^{14}CO_2$ Resulting from Application of [^{14}C] Diquat to Nutrient Solutions containing Montmorillonite (M) and Kaolinite (K). Reprinted from [81].

Diquat added (µmol)	Clay added† (mg)	Radioactivity in evolved $^{14}CO_2$ (cpm)‡				
		1 week	2 weeks	4 weeks	6 weeks	8 weeks
			Experiment 1			
3.7	0	1,450	15,800	89,700	178,000	
3.7	4 M	1,660	28,000	81,100	115,000	
3.7	9 M	90	1,510	4,020	5,310	
3.7	60 K	13,100	69,600	183,000	262,000	
3.7	160 K	12,300	72,900	139,000	180,000	
			Experiment 2			
7.4	0	2,090	8,710	191,000	424,000	
7.4	10 M	658	8,760	87,600	134,000	
7.4	20 M	0	0	0	0	
7.4	160 K	1,880	58,200	380,000	589,000	
7.4	340 K	775	65,000	426,000	628,000	
			Experiment 3			
3.7	0	37,000	80,800	159,000	217,000	254,000
3.7	5 M	16,300	27,300	44,300	60,200	72,800
3.7	10 M	0	0	0	0	0
3.7	80 K	38,800	58,700	106,000	150,000	187,000
3.7	170 K	24,400	46,000	94,500	152,000	204,000

†In experiment 1, 10 mL of deionized water containing 10 µc (3.7 µmoles) of ^{14}C-diquat were added to varying amounts of each of the clay minerals; experiments 2 and 3 were performed in a manner similar to experiment 1, except that the amounts of clay and/or ^{14}C-diquat were different.

‡Corrected for $^{14}CO_2$ derived from sterile control (< 200 cpm).

Soil organic matter can induce surface-catalyzed reactions of adsorbed pesticides, but it can also hinder the degradation of pesticides by decreasing the availability to microbial attack [82]. Microbial degradation of pesticides adsorbed on organomineral complexes, which are very reactive components of soils and sediments, and the catalytic role of these complexes in the transformation of pesticides in the environment are less understood and, thus, merit in-depth research.

9. BIOTECHNOLOGY

The probability and frequency of the transfer of genetic information, both intra-and interspecifically, between microorganisms, especially bacteria, in soil and other natural habitats have become important concerns [5]. Although genetic transfer has been extensively studied in pure culture systems, few studies have been conducted in natural microhabitats [5, 7, 83]. Even fewer studies have investigated the role of mineral colloids and organic components and their interactions in the transfer of genetic information between microorganisms in soils. It is obvious from these studies and their contradictory results and interpretations that more studies are necessary to clarify the role of mineral colloids, organic matter, and organo-mineral complexes in not only the survival of genetically engineered microorganisms but also the transfer of genetic information by conjugation in soil. Further, because of the potential hazards associated with the transfer of genes—whether introduced accidentally or deliberately—in soil and contiguous habitats (e.g., groundwater, surface water, atmosphere, plants, animals, human beings), the importance of mineral-organic matter-microbe interactions in influencing the fate of genetically engineered microbes in soil and related environments should receive increasing attention.

Surface-active particles, primarily mineral colloids and humic substances, in soil and other natural habitats are important in the persistence of biomolecules that in the absence of such particles are rapidly degraded or inactivated by the indigenous microbiota. Many biomolecules are important in the ecology, biodiversity, evolution of microorganisms, environmental protection, and biological control of pests [7]. Insecticidal proteins produced by *Bacillus thuringiensis* and by transgenic plants containing toxin genes, enzymes, bacterial transforming DNA, and viruses are examples of the persistence to biodegradation and the retention of biological activity of biomolecules when bound on such particles. Therefore, these surface-active particles affect the transfer of genetic information among bacteria by conjugation, transformation, and transduction in soil. All these aspects affect ecosystem health.

The gains in food production provided by the Green Revolution have reached their ceiling while world population continues to rise. The application of advances in plant biotechnology is going to be essential especially in developing countries if farmers' yields and yield ceilings are to be raised, excessive pesticide use reduced, the nutrient value of basic foods increased, and farmers on less favored lands provided with varieties better able to tolerate drought, salinity and lack of soil nutrients [84]. However, when biotechnology is applied at the field scale, the impact of soil mineral-organic matter-microorganism interactions on the effectiveness of growing genetically modified crops in different soil environments so as to achieve the objectives of being more productive, yet less damaging to the ecosystem and consequently more beneficial to human health remains to be uncovered. We need to assure that biotechnology is applied to agriculture only when this can be done safely and effectively in helping to achieve future food security and protect ecosystem health for the world.

10. RISK ASSESSMENT

One of the great ecological challenges today is to determine if soil contamination will lead to adverse effects on beneficial organisms. The mere presence of a chemical in air, water, and soil, biota, or food can raise public fears and a call for regulation of that chemical or its removal from the environment in which it was found. Environmental monitoring may be able to show that the chemical is a contaminant because concentrations are above some established background. Contaminants have background levels in soils and biological tissues as a result of their persistence and global dispersion. What is not known is if the measured concentration of a contaminant can have an adverse effect on an organism. To answer this question, we must use the tool of risk assessment which combines exposure assessment (how is the organism exposed to the chemical and how much exposure is there?) and effect assessment (how does the chemical harm the organism? how much exposure to the chemical is required to cause harm?).

Assessing exposure to contaminants in soil environments includes the determination of the pathways to exposure (inhalation, drinking water, food or direct soil ingestion) and the extent of contaminant transfer from soil to the various pathway components [85]. Possible soil pathways of human contaminant exposure is presented in Table 7. In environmental exposure assessment, a transfer coefficient (K) must be determined for each component of the pathway, namely, the fraction of the component that is transferred from one component to another (e.g., from soil to plant, soil to air, soil to surface water). Such an approach is used in Mackay's general fugacity model [87]. The transfer coefficient can be viewed as an availability index, although other terms (extractability, solubility, accessibility, volatility) can also be used. This will depend on the mechanism of transfer, which should be greatly influenced by interactions of soil minerals with organic components and microorganisms. These interactions should profoundly influence the transformation, dynamics, bioavailability and toxicity of contaminants in soil and related environments. Because of these interactions, some pollutants that would be otherwise at an acceptable level may be released and enter the pathway affecting health.

Chemicals in soil can exist in solid, liquid, and gaseous phases, and inorganic and organic forms (Figure 13). The solid phase includes primary and secondary minerals, undecomposed and stable organic matter, biomass, and sorbed species. The liquid phase includes dissolved ions, uncharged molecules, dissolved organic matter (humic and fulvic acids and discrete organic compounds), dissolved xenobiotics (e.g., herbicides and insecticides), and small inorganic and organic polymers, and microparticulates (e.g., fine colloidal metal oxides). The soil gaseous phase contains normal soil gases (N_2, O_2, CO_2, NO_x), NH_3, volatile organic compounds (natural and anthropogenic organics), and under reducing conditions, H_2S, H_2 and CH_4. These components interact by physical, chemical, and biological processes and influence the transformation of contaminants. Therefore, soil mineral-organic component-microorganism interactions merit attention in soil pathways of human contaminant exposure (Table 7) and risk assessment.

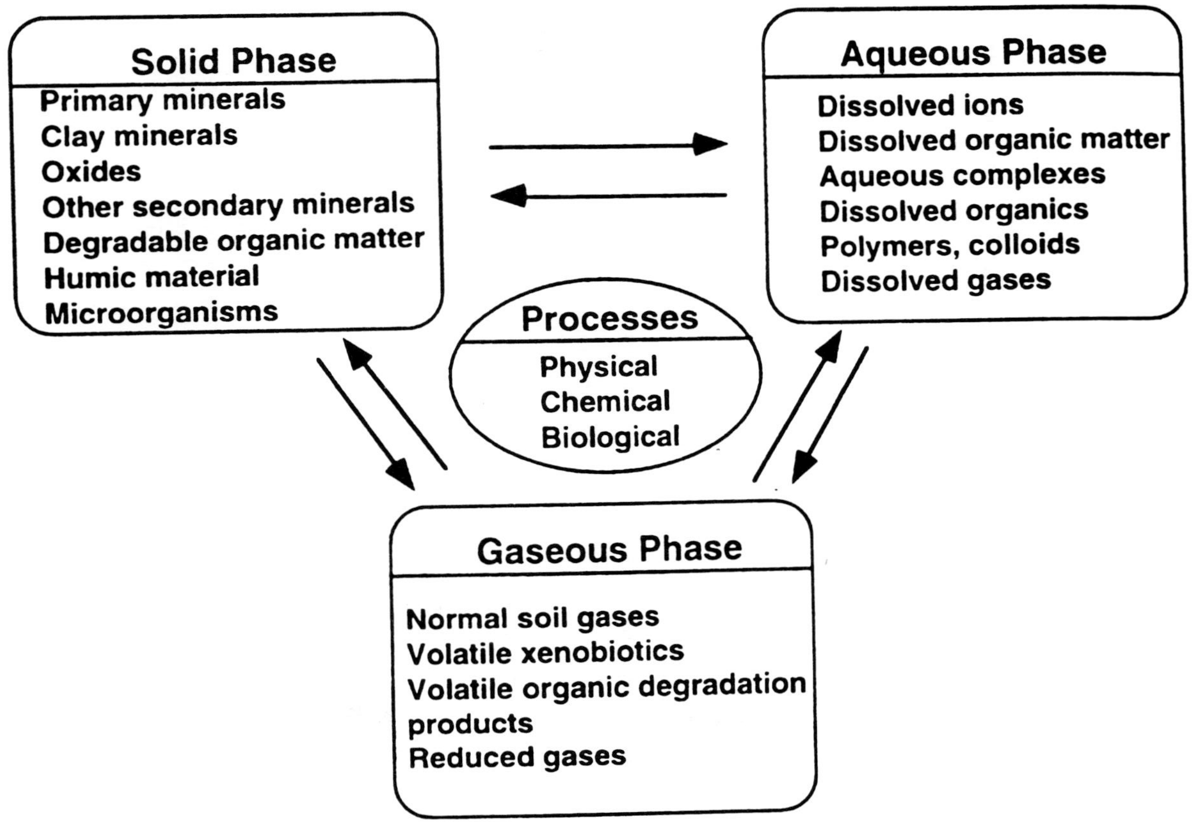

Figure 13. Chemical composition of soil solid, aqueous and gaseous phases. Reprinted from [85].

Table 7
Pathways for human exposure to soil-borned contaminants [85]. The most exposed individual (MEI) of the population is also identified. Based on U.S. Evnrironmntal Protection Agency exposure assessment for land applied contaminants in sewage sludge. Reprinted from [86].

Pathway	MEI
Soil →Plant→Human	Human lifetime plant ingestion; general population
Soil→Plant→Human	Human lifetime plant ingestion; home gardener
Soil→Human	Child
Soil→Plant→Animal→Human	Human lifetime ingestion of animal products; animals raised on forage
Soil→Animal→Human	Human lifetime ingestion of animal products; animals ingest soil
Soil→Dust→Human	Human lifetime dust inhalation
Soil→Surface water→Human	Human lifetime ingestion of surface water and fish
Soil→Ground water→Human	Human lifetime ingestion of groundwater
Soil→Air→Human	Human lifetime inhalation of volatilized contaminants

11. RISK MANAGEMENT, REMEDIATION, AND RESTORATION

Ecosystem health is largely influenced by human activities. Soil has an important role as an environmental buffer and central organizer of the ecosystem. Contamination of soils from anthropogenic chemicals and their subsequent degradation has become a major concern because of the critical role of soil resources in sustaining agricultural production, economic development, and ecosystem health. Both inorganic and organic anthropogenic compounds in soils may not only adversely affect their production potential but also deteriorate the quality of the food chain, the air, the surface water, and the underlying groundwater. This scenario may require risk assessment and the subsequent risk management through use of remediation technology to restore ecosystem health. Land use, past and present, critically influences the extent and intensity of soil contamination (Figure 14). The logical junctures where intervention measures could be employed are also indicated.

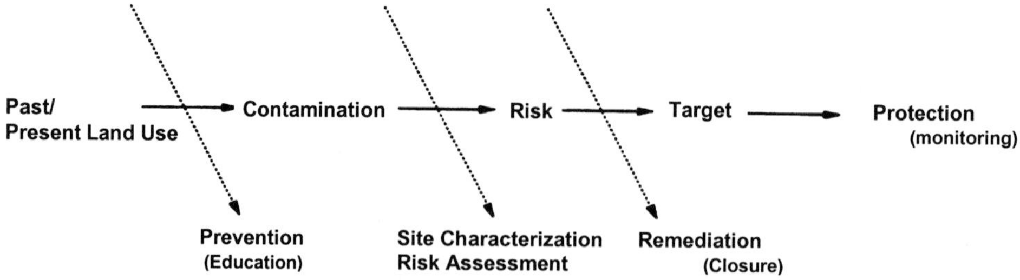

Figure 14. Sequence of soil contamination-remediation indicating critical junctions for action. Reprinted from [88].

Soil scientists, especially soil chemists, have used selected ameliorants to remediate metal contaminated soils (Table 8). These ameliorants are rather inexpensive and readily available on a global basis. They may provide only interim solutions in stabilizing contaminants. Soil processes involved in remediation by these ameliorants include precipitation, sorption, ion exchange, fixation, and complexation. These reactions may be substantially influenced by soil mineral-organic component-microorganism interactions especially in the rhizosphere soils where these interactions are intense. Therefore, the efficacy of these ameliorants, particularly over the long term, may vary with the nature and properties of soils and cropping systems.

Table 8
Selected ameliorants that are adapted to metal contaminated soils. Reprinted from [88].

Technique	Target contaminants	Soil processes involved	Constraints
Limestone	Metals, radionuclides	Precipitation, sorption	Ineffective for oxyanions; certain crops (lettuce, spinach, tuber, and others); short term
Zeolite	Metals, radionuclides	Ion exchange, sorption, fixation	Insufficient data; short term
Apatite	Metals	Sorption, precipitation, complexation	Selective; insufficient data
Clay mineral	Metals, radionuclides	Ion exchange, sorption, fixation	Type of clay; short term

Plants have been described in engineering terms as solar driven pumps and filtering systems that extract and concentrate elements from the environment [89, 90]. Some of the metals that may accumulate in plants are those that have nutritional value. Certain plants also have the ability to accumulate metals (e.g., Cd, Cr. Pb, Ni, Co, Se, and Hg) that have no known biological function [91]. The value of metal-accumulating plants has recently been realized [89, 92]. Further, plant-assisted bioremediation are used to remediate soil contaminated with organic compounds [93, 94]. In phytoremediation, a wide range of processes are involved in the soil rhizosphere where rhizospheric microorganisms interact intensely with soil minerals and root exudates. These interactions should have a profound influence on transformations of metals and anthropogenic organic compounds and the efficacy of phytoremediation.

Xenobiotics can also be degraded by oxidative coupling reactions which are catalyzed either biologically by polyphenol oxidases and peroxides [95] or abiotically by metal oxides and clay minerals [17, 96]. These naturally occurring processes are believed to result in the detoxification of xenobiotics and have been suggested as a means of decontamination. While indigenous enzymes are usually not likely to provide a satisfactory decontamination of polluted sites, amending soil with enzymes derived from specific microbial cultures or plant material may stimulate detoxification processes. Abiotic and biotic catalysts co-exist in soil environments. Abiotic catalysts can influence microbial formation of enzymes and enzymatic activity [97]. Many soil abiotic catalysts can also immobilize enzymes and alter their performance [98, 99]. Further, many soil abiotic catalysts influence the activity of the desorbed enzymes [100]. Therefore, the influence of interactions of abiotic and biotic catalysts on the transformation and toxicity of xenobiotics and the impact on risk management, remediation, and restoration of ecosystem health is an issue that will be of intense interest for years to come.

12. CONCLUSIONS

Soil evolves through interactions of the lithosphere, hydrosphere, atmosphere and biosphere. It is an integral part of the ecosystem where human beings sustain life and flourish. It is a dynamic system in which soil minerals constantly interact with organic matter and microorganisms. Close association and interactions of these components mediated by soil solution and atmosphere govern mineral weathering reactions, mineral-organic complex and structural formation, profile development, and physical, chemical, and biological processes of soils pertaining to the dynamics, transformations, and fate of ions and uncharged molecules either vital or deleterious to life and the sustainability of ecosystem integrity.

Interactions of minerals with organic matter and microorganisms in soil and related environments have at least the following foreseeable impacts: (1) microbial events, (2) global ion cycling, (3) global climatic changes, (4) biodiversity, (5) biological productivity and human nutrition, (6) geomedicine, (7) industrialization, waste disposal, ecotoxicology, and human health with respect to the bioavailability and toxicity of metals, metalloids, other inorganics, anthropogenic organic pollutants, biomolecules, and pathogens, (8) development of biotechnology, (9) ecosystem risk assessment, and (10) ecosystem risk management, remediation, and restoration.

Therefore, we have an enormously challenging mission to fulfill to restore, sustain, and enhance ecosystem health, which includes human health, in the new millennium. This mission can only be fulfilled through (1) advancement of knowledge on fundamentals of mineral-organic component-microorganism interactions, (2) innovation of undergraduate and graduate curriculum programs to incorporate the vital knowledge advanced on this subject, (3) technology transfer for the development of innovative management strategies for ecosystem sustainability, and (4) development of continuing education and information technology programs to communicate the discoveries to society and to ensure the scientific knowledge is 'socially robust' and that its production is seen by society to be both transparent and participative.

REFERENCES

1. Odum, E.P., 1989. Ecology and our endangered life-support systems. Sinauer Association, Sunderland, MA.
2. Coleman, D.C., Hendrix, P.F., Odum, E.P., 1998. Ecosystem health: An overview. *In*: Huang, P.M, Adriano, D.C., Logan, T.J., Checkai, R.T. (Eds.), Soil Chemistry and Ecosystem Health. SSSA Special Publication 52, Soil Science Society of America, Madison, WI, pp. 1-20.

3. Huang, P.M., Schnitzer, M. (Eds.), 1986. Interactions of soil minerals with natural organics and microbes. SSSA Special Publication 17, Soil Science Society of America, Madison, WI.
4. Huang, P.M., Berthelin, J., Bollag, J.-M., McGill, W.M., Page, A.L. (Eds.), 1995. Environmental impact of soil component interactions. Vol. I Natural and anthropogenic organics and Vol. II Metals, other inorganics, and microbial activities. Lewis Publishers/CRC Press, Boca Raton.
5. Stotzky, G., 1986. Influence of soil mineral colloids on metabolic processes, growth, adhesion, and ecology of microbes and viruses, *In*: Huang, P.M., Schnitzer, M. (Eds.), Interactions of soil minerals with natural organics and microbes. SSSA Special Publication 17, Soil Science Society of America, Madison, WI, pp. 305-428.
6. Theng, B.K.G., Orchard, V.A., 1995. Interactions of clays with microorganisms and bacterial survival in soil: A physicochemical perspective. *In*: Huang, P.M., Berthelin, J., Bollag, J.-M., McGill, W.B., Page, A.L. (Eds.), Environmental impact of soil component interactions. Vol. II Metals, Other Inorganics, and Microbial Activities, CRC Press/Lewis Publishers, Boca Raton, pp.123-143.
7. Stotzky, G., 2000. Effect of the binding of biomolecules (DNA, proteins, viruses) on surface-active particles (clays, humic acids, clay-humic acid complexes) on their persistence and biological activity in the environment. *In*: Violante A., Gianfreda, L., (Eds.), Third International Symposium of the Working Group MO Interactions of Soil Minerals with Organic Components and Microorganisms of the International Union of Soil Sciences, Naples-Capri, Italy, May 22-26, 2000, p.31.
8. Burns, R.G., Stach, J.E.M., 2000. Microbial ecology of soil biofilms: substrate bioavailability, bioremediation and complexity. *In*: Violante, A., Huang, P.M., Bollag, J.-M., Gianfreda, L. (Eds.), Soil Mineral-Organic Matter-Microorganism Interactions and Ecosystem Health. Ecological Significance of the Interactions among Clay Minerals, Organic Matter and Soil Biota. Elsevier, The Netherlands. This book,Volume B.
9. Butcher, S.S., Charlson, R.J., Orians, G.H., Wolfe, G.V. (Eds.), 1992. Global Biogeochemical Cycles, Academic Press, London.
10. Schoenau, J.J., Bettany, J.R., 1987. Organic matter leaching as a component of carbon, nitrogen, phosphorus, and sulfur cycles in a forest, grassland, and gleyed soil. Soil Sci. Soc. Am. J. 51, 646-651.
11. Duxbury, J.M., Harper, L.A., Mosier, A.R., 1993. Contribution of agroecosystems to global climate change, In: Harper, L.A. et. al. (Eds.), Agricultural Ecosystem Effects on Trace Gases and Global Climate Change, ASA Special Publication 55, Soil Science Society of America, Madison, WI, 1-18.
12. Cicerone, R.J., Oremland, R.S., 1988. Biogeochemical aspects of atmospheric methane, Global Biogeochem. Cycles, 2, 299-327.
13. Conrad, R., 1995. Soil microbial processes involved in production and consumption of atmospheric trace gases. *In*: Jones, J.G. (Ed.), Advances in Soil Ecology,. Vol. 1, Plenum Press, New York, 207-250.
14. McGill, W.B., Cole, C.V., 1981. Comparative aspects of cycling of organic C, N, S and P through soil organic matter. Geoderma 26, 267-286.
15. Guggenburger, G., Haider, K.M. 2002. Effect of mineral colloids on biogeochemical cycling of C, N, P, and S in soil. *In*: Huang, P.M., Bollag, J.-M., Senesi, N. (Eds.), Interactions between Soil Particles and Microorganisms and their Impact on the Terrestrial Ecosystem. John Wiley & Sons, Chichester, UK, (in press).

16. Huang, P.M. 1995. The role of short-range ordered mineral colloids in abiotic transformations of organic components in the environment. *In*: Huang, P.M., Berthelin, J. Bollag, J.-M., McGill, W.B., Page, A.L. (Eds.), Environmental Impact of Soil Component Interactions, Vol. I Natural and Anthropogenic Organics, CRC Press/Lewis Publishers, Boca Raton, pp. 135-167.
17. Huang, P.M., 2000. Abiotic catalysis. *In*: Sumner, M.E. (Editor-in-Chief) Handbook of Soil Science, CRC Press, Boca Raton, B303-332.
18. Benjamin, M.M., Honeyman, B.D., 1992. Trace metals. *In*: Butcher, S.S., Charlson, R.J., Orians, G.H., Wolfe, G.V. (Eds.), Global Biogeochemical Cycles. Academic Press, London, pp. 317-352.
19. Hayes, K.F., Traina, S.J. 1998..Metal ion speciation and its significance in ecosystem health, *In*: Huang, P.M., Adriano, D.C., Logan, T.J., Checkai, R.T. (Eds.), Soil Chemistry and Ecosystem Health. SSSA Special Publication 52, Soil Science Society of America, Madison, WI., pp. 45-84.
20. Kurek, E., 2002. Microbial mobilization of metals from soil minerals under aerobic conditions. *In*: Huang, P.M., Bollag, J.-M., Senesi, N. (Eds.), Interactions of Soil Particles with Microorganisms and their Impact on the Terrestrial Ecosystem. John Wiley & Sons, Chichester, UK, (in press).
21. McLean, J.S., Lee, J.-U., Beveridge, T.J., 2002. Interactions of bacteria and environmental metals, fine-grained mineral development and bioremediation strategies. *In*: Huang, P.M., Bollag, J.-M., Senesi, N. (Eds.), Interactions of Soil Particles with Microorganisms and their Impact on the Terrestrial Ecosystem. John Wiley & Sons, Chichester, UK, (in press).
22. National Academy of Sciences, 1978. An Assessment of Mercury in the Environment. National Academy of Sciences, Washington, DC.
23. Gadd, G.M.,1993. Microbial formation and transformation of organometallic and organometalloid compounds. FEMS Microbial. Rev. 11, 297-310.
24. Farrell, R.E., Huang, P.M., Germida, J.J., 1998. Biomethylation of mercury (II) adsorbed on mineral colloids common in freshwater sediments, Appl. Organomet. Chem. 12, 613-620.
25. Stotzky, G., Babich, H., 1980. Physicochemical factors that affect the toxicity of heavy metals to microbes in aquatic habitats. *In*: Colwell, R.R., Foster, J., (Eds.), Proc. ASM Conf. Aquatic Microbial Ecology. University of Maryland, College Park, MD, pp.181-203.
26. Stotzky, G., Babich, H., 1986. Physicochemical environmental factors affect the response of microorganisms to heavy metals: Implications for the application of microbiology to mineral exploration. *In*: Carlisle, D., Berry, W.L., Kaplan, I.R., Watterson, J.R. (Eds.), Mineral Exploration: Biological Systems and Organic Matter. Prentice-Hall, Englewood Cliffs, NJ, pp. 238-264.
27. Charlson, R.J., Orians, G.H., Wolfe, G.V., Butcher, S.S., 1992. Human modification of global biogeochemical cycles. *In*: Butcher, S.S., Charlson, R.J., Orians, G.H., Wolfe, G.V. (Eds.), Global Biogeochmical Cycles. Academic Press, pp. 353-361.
28. Lal, R., 1998. Soil Processes and the Carbon Cycle. CRC Press, Boca Raton.
29. Jenkinson, D.S., Adams, D.E., Wild, A., 1991. Model estimates of CO_2 emissions from soil in response to global warming. Nature 351, 304-306.
30. Paul, E.A., Van Veen, J.A., 1978. The use of tracers to determine the dynamics of organic matter. Trans. Int. Soc. Soil Sci. (Edmonton) Vol. 3, pp. 61-102.

31. Schlessinger, W.H., 1977. Carbon balance in terrestrial detritus. Ann. Rev. Ecol. Syst. 8, 51-81.
32. Schulze, E.D., Mooney, H.A., (Eds.), 1993. Biodiversity and Ecosystem Function. Springer, Berlin.
33. Tilman, D., Wedlin, D., Knops, J. 1996. Productivity and sustainability influenced by biodiversity in grassland ecosystems. Nature 379, 718-720.
34. Hooper, D.U., Vitousek, P.M., 1997. The effect of plant composition and diversity on ecosystem processes. Science 277, 1302-1305.
35. van der Heijden, M.G.A.,. Klironomos, J.N., Ursic, M., Moutoglis, P., Streitwolf-Engel, R., Boller, T., Wiemken, A., Sanders, I.R., 1998. Mycorrhizal fungal diversity determines plant biodiversity, ecosystem variability and productivity. Nature 396, 69-72.
36. Helgason, T., Daniell, T.J., Husband, R., Fitter, A.H., Young, J.P.Y., 1998. Ploughing up the wood-wide web? Nature 394, 431.
37. Tilman, D., Downing, J.A., 1994. Biodiversity and stability in grassland. Nature 367, 363-365.
38. Naeem, S., Thompson, L.J., Lawler, S.P., Lawton, J.H., Woodfin, R.M., 1994. Declining biodiversity can alter the performance of ecosystems. Nature 368, 734-737.
39. Sposito, G., Reginato, R.J., 1992. Opportunities in Basic Soil Science Research. Soil Science Society of America, Madison, WI.
40. Jenkinson, D.S., Rayner, J.H., 1977. The turnover of soil organic matter in some of the Rothamsted classical experiments. Soil Sci. 123, 298-305.
41. Huang, P.M., 1990. The role of soil minerals in influencing transformations of natural organics and xenobiotics in the environment. *In*: Bollag, J.-M., Stotzky, G. (Eds.), Soil Biochemistry Vol. 6, Marcel Dekker, New York, 29-115.
42. Barber, S.A., 1995. Soil Nutrient Bioavailability: A Mechanistic Approach. Second Edition, John Wiley & Sons, New York.
43. Cheng, H.H., 1990. Pesticides in the Soil Environment: Processes, Impact, and Modeling. Soil Science Society of America, Madison, WI.
44. Huang, P.M., 1994. Transformation of pesticides as influenced by soil minerals in the environment. *In*: McDuffie, H.H., Dosman, J.A., Semchuk, K.M., Olenchock, S.A., Senthilsevan, A. (Eds.), Agriculture, Health and Safety: Workshop, Environment, Sustainability. Center for Agricultural Medicine, University of Saskatchewan, Saskatoon, pp. 165-175.
45. Burns, R.G., 1986. Interactions of enzymes with soil minerals and organic colloids. In: Huang, P.M., Schnitzer, M. (Eds.), Interactions of Soil Minerals with Natural Organics and Microbes. SSSA Special Publication 17, Soil Science Society of America, Madison, WI, pp. 429-451.
46. Naidja, A., Huang, P.M., Bollag, J.-M., 1997. Activity of tyrosinase immobilized on hydroxyaluminum-montmorillonite complexes, J. Mol. Catal. A: Chemical 115, 305-316.
47. Naidja, A., Huang, P.M., Bollag, J.-M., 1998. Comparison of reaction products from transformation of catechol catalyzed by δ-MnO_2 or tyrosinase, Soil Sci. Soc. Am. J. 62, 188-195.
48. Emerson, W.W., Foster, R.C., Oades, J.M., 1986. Organo-mineral complexes in relation to soil aggregation and structure. *In*: Huang, P.M., Schnitzer, M. (Eds.), Interactions of Soil Minerals with Natural Organics and Microbes. SSSA Special Publication 17, Soil Science Society of America, Madison, WI, 521-548.

49. Waisel, Y., Eshel, A., Kafkafi, U., 1996. Plant Roots. The Hidden Half, Second Edition, Marcel Dekker, New York.
50. Szmigielska, A.M., Cieslinski, G., Van Rees, K.C.J., Huang, P.M., 1997. Low molecular weight dicarboxylic acids in rhizosphere soil of durum wheat. Commun. Soil Sci. Plant Anal. 28, 99-111.
51. Streitwolf-Engel, R., Boller, T., Wiemken, A., Sanders, I.R., 1997. Clonal growth traits of two Pruncella species are determined by co-occurring arbuscular mycorrhizal fungi from a calcareous grassland. J. Ecol. 85, 181-191.
52. Låg, J., 1980. Survey of geomedical problems: Including some examples from investigations carried out in Norway, Geomedical Aspects in Present and Future Research. Universitatsforlaget, Oslo, pp.11-19.
53. Huang, P.M., 1987. Impact of mineralogical research on soil and environmental sciences, *In*: Boersma, L.L. (Ed.), Future Developments in Soil Science Research. Soil Science Society of America, Madison, WI, 485-498.
54. Låg, J., 1994. Geomedicine, an expanded application of soil science. *In*: Transactions 15th World Congress Soil Science, Acapulco, Mexico, Vol. 3a, Commission II: Symposia, pp. 557-567.
55. Chang, A. C., Page, A. L., Bon-Jun Koo, 2002. Biogeochemistry of phosphorus, iron, and trace elements in soils as influenced by soil-plant-microbial interactions. *In*: Violante, A., Huang, P.M., Bollag, J.-M., Gianfreda, L. (Eds.), Soil Mineral-Organic Matter-Microorganism Interactions and Ecosystem Health. Ecological Significance of the Interactions among Clay Minerals, Organic Matter and Soil Biota. Elsevier, The Netherlands. This book, Volume B.
56. National Academy of Sciences (U.S.), 1974. Geochemistry and the Environment. Vol. I, The Relation of Selected Trace Elements to Health and Disease, Washington, D.C..
57. National Academy of Sciences (U.S.,), 1977 Geochemistry and the Environment. Vol. II, The Relation of Other Selected Trace Elements to Health and Disease, Washington, D.C.,.
58. Jackson, M.L, 1986. Geochemical characteristics of land and its effect on human heart and cancer death rates in the United States and China. Applied Geochem. 1, 175-180.
59. Schalin, G., 1980. Multiple sclerosis and selenium. *In*: Låg, J. (Ed.), Geomedical Aspects in Present and Future Research, Universitatsforlaget, Oslo, pp. 81-102.
60. Dynes, J.J., Huang, P.M., 1997. Influence of organic acids on selenite sorption on poorly ordered aluminum hydroxides. Soil Sci. Soc. Am. J. 61, 772-783.
61. Pessarakli, M. (Ed.), 1994. Handbook of Plant and Crop Stress. Marcel Dekker, New York.
62. Vitrousek, P.M.,. Ehrlich, P.R , Ehrlich, A.H., Matson, P.A., 1986. Human appropriation of the products of photosynthesis. BioScience 36, 368-373.
63. Brown, L.R., Flavin, C. 1996. State of the World, W.W. Norton & Co., New York.
64. Crosby, D.G., 1982. Environmental Chemistry: An overview, Environ. Toxicol. Chem. 1, 1-8.
65. Constanza, R.,. Norton, B.G, Haskell, B.D., 1992. Ecosystem Health. Island Press, Washington, D.C..
66. Beare, M.H., Coleman, D.C., Crossley, D.A. Jr., Hendrix, P.F., Odum, E.P., 1995. A hierarchical approach to evaluating the significance of soil biodiversity to biogeochemical cycling, Plant Soil 170 , 5-22.

67. Huang, P.M., 2000. Environmental Soil Chemistry and Its Impact on Agriculture and the Ecosystem. Università degli Studi di Napoli Federico II, Portici, Italy.
68. McBride, M.B., 1994. Environmental Chemistry of Soils. Oxford University Press, Oxford.
69. Frieberg, L., Nordbert, G.F., Vouk, V.B., 1979. Handbook of the Toxicology of Metals, Elsevier, North-Holland, New York.
70. Huang, P.M., 1987. Agricultural soils: Fate of toxic substances, *In*: Singh, M.G. (Editor-in-Chief) Systems and Control Encyclopedia. Pergamon Press, Oxford, pp. 143-148.
71. Peterson, P.J., Alloway, B.J., 1979. Cadmium in soils and vegetation, *In*: M. Webb (Ed.), The Chemistry, Biochemistry, and Biology of Cadmium. Elsevier, North-Holland, Biomedical Press, Amsterdam, pp. 45-92.
72. Huang, P.M., 1998. Distinguished Researcher Award Lecture. The University of Saskatchewan, Saskatoon, SK, Canada.
73. Krishnamurti, G.S.R., Cieslinski, G., Huang, P.M., Van Rees, K.C.J., 1997. Kinetics of cadmium release from soils as influenced by organic acids: Implication in cadmium availability. J. Environ. Qual. 26, 271-277.
74. Krishnamurti, G.S.R., Huang, P.M., Kozak, L.M., Rostad, H.P.W., Van Rees, K.C.J., 1977. Distribution of cadmium in selected soil profiles of Saskatchewan, Canada: Speciation and availability. Can. J. Soil Sci. 77, 613-619.
75. Cieslinski, G., Van Rees, K.C.J., Huang, P.M., Kozak, L.M., Rostad, H.P.W., 1996. Cadmium uptake and bioaccumulation in selected cultivars of durum wheat and flax as affected by soil type. Plant Soil 182, 115-124.
76. Cieslinski, G., Van Rees, K.C.J., Szmigielski, A.M., Krishnamurti, G.S.R., Huang, P.M., 1998. Low-molecular-weight organic acids in rhizosphere soils of durum wheat and their effects on cadmium bioaccumulation. Plant Soil 203, 109-117.
77. Krishnamurti, G.S.R., Huang, P.M., Van Rees, K.C.J., 1996. Studies on soil rhizosphere: Speciation and availability of Cd. Chem. Spec. Bioavailab. 8, 23-28.
78. Saltzman, S., Yaron, B., 1986. Pesticides in Soil. Van Norstrand Reinhold Co., New York.
79. Grover, R., (Ed.), 1988. Environmental Chemistry of Pesticides, CRC Press, Boca Raton, Vol. 1, pp. 1-19.
80. Huang, P.M., Bollag, J.-M., 1998. Minerals-organics-microorganisms interactions in the soil environment, *In*: Huang, P.M., Senesi, N., Buffle, J. (Eds.), Structure and Surface Reactions of Soil Pesticides. Vol. 4, IUPAC Series on Analytical and Physical Chemistry of Environmental Systems, John Wiley & Sons, Chichester, England, pp. 3-39.
81. Weber, J.B,.Coble, H.D., 1968. Microbial decomposition of diquat adsorbed on montmorillonite and kaolinite clays. J. Agric. Food Chem. 16, 475-478.
82. Dec, J., Huang, P.M., Senesi, N., Bollag, J.-M., 2002. Impact of interactions of soil colloids and microorganisms on transformation of organic pollutants. *In*: Huang, P.M., Senesi, N., Buffle, J. (Eds.), Interactions between Soil Particles and Microorganisms and their Impact on the Terrestrial Ecosystem. Vol. 8, IUPAC Series on Analytical and Physical Chemistry of Environmental Systems, John Wiley & Sons, Chichester, England, (in press).
83. Graham, B.J., Istock, C.A., 1979. Gene exchange and natural selection cause *Bacillus subtilis* to evolve in soil culture. Science 204, 637-639.
84. Conway, G., Toenniessen, G., 1999. Feeding the twenty-first century. Nature 402, C55-C58.

85. Logan, T.J., 1998. Impact of soil chemical reactions on food chain contamination and environmental quality, *In*: Huang, P.M., Sparks, D.L., Boyd, S.A. (Eds.), Future Prospects for Soil Chemistry. Soil Science Society of America, Madison, WI, pp. 191-204.
86. U.S. Environmental Protection Agency, 1995. A guide to the biosolid risk assessments for the USEPA Part 503 rule. Office of Wastewater Management, USEPA 832-B-93-005, USEPA, Washington, D.C..
87. Mackay, D., Paterson, S., 1991. Evaluating the multimedia fate of organic chemicals: A level III fugacity model,. Environ. Sci. Technol. 25, 427-436.
88. Adriano, D.C., Chlopecka, A., Kaplan, D.I., 1998. Role of soil chemistry in soil remediation and ecosystem conservation. *In*: Huang, P.M., Adriano, D.C., Logan, T.J., Checkai, R.T (Eds.), Soil Chemistry and Ecosystem Health. Soil Science Society of America, Madison, WI, pp. 361-386.
89. Cunningham, S.D., Berti, W.R., 1993. Remediation of contaminated soils with green plants: An overview, In Vitro. Cell. Dev. Bio. 29P, 207-212.
90. Salts, D.E., Blaylock, M., Kumar, N., Dushenkov, V., Ensley, B.D., Chet, I., Raskin, I, 1995. Phytoremediation: A novel strategy for the removal of toxic metals from the environment using plants. Biotechnology 13, 468-474.
91. Bakers A.J.M., Brooks, R.R., 1989. Terrestrial higher plants which hyperaccumulate metallic elements: A review of their distribution, ecology and phytochemistry. Biorecovery 1, 81-126.
92. Pierzynski, G.M., Sims, J.T., Vance, G.F., 2000. Soil and Environmental Quality, CRC Press, Boca Raton.
93. Walton, B.T., Anderson, T.A., 1992. Plant-microbe treatment systems for toxic waste. Current Opinion Biotech. 3, 267-270.
94. Anderson, T.A., Guthrie, E.A., Walton, B.T., 1993. Bioremediation. Environ. Sci. Technol. 27, 2630-2636.
95. Bollag, J.-M., 1992. Decontaminating soil with enzymes. Environ. Sci. Technol. 26, 1876-1881.
96. Huang, P.M., Wang, M.C., Wang, M.K., 1999. Catalytic transformation of phenolic compounds in the soils. *In*: Inderjit, K, Dakshini, M.M., Foy, C.L. (Eds.), Principles and Practices in Plant Ecology. CRC Press, Boca Raton, pp. 287-306.
97. Claus, H., Filip, Z.K., 1990. Effects of clays and other solids on the activity of phenoloxidase produced by some fungi and antinomycetes. Soil Biol. Biochem. 22, 483-488.
98. Gianfreda, L., Bollag, J.-M., 1994. Effects of soils on the behaviour of immobilized enzymes. Soil Sci. Soc. Am. J. 58, 1672-1681.
99. Naidja, A., Huang, P.M., Bollag, J.-M., 2000. Enzyme-clay interactions and their impact on transformations of natural and anthropogenic organic compounds in soil. J. Environ. Qual. 29, 677-691.
100. Claus, H., Filip, Z.K., 1988. Behaviour of phenoloxidases in the presence of clays and other soil-related adsorbents. Appl. Microbiol. Biotechnol. 28, 506-511.

Developments in Soil Science, Volume 28A
Editors: A. Violante, P.M. Huang, J.-M. Bollag and L. Gianfreda

SORPTION OF COPPER AND CADMIUM BY ALLOPHANE- HUMIC COMPLEXES

G. Yuan, H.J. Percival, B.K.G. Theng and R.L. Parfitt

Landcare Research, Private Bag 11052, Palmerston North, New Zealand

We have investigated the sorption of Cu and Cd by allophane and its complexes with a soil humic acid (HA), varying in carbon content from 14–123 g kg^{-1}. The sorption of Cu and Cd was measured at pH 5.0 and 5.5, using the batch technique and a single concentration (2mM) of Cu or Cd in the presence of $LiClO_4$. The sorbed metals were partly desorbed by treatment with 0.1 M KNO_3. Allophane sorbed 71 and 179 mmol Cu kg^{-1} at pH 5.0 and 5.5, respectively. By comparison, only 3 mmol kg^{-1} of Cd was sorbed at pH 5.0 and 9 mmol kg^{-1} at pH 5.5. Irrespective of pH, complex formation with HA led to a marked increase in Cu and Cd sorption. Indeed, sorption increased linearly with the amount of HA sorbed. Less than 20% of the sorbed Cu, and 49–94% of sorbed Cd, was desorbable by KNO_3. Copper appears to be specifically sorbed by allophane through coordination with exposed hydroxyl groups because sorption can occur even when allophane was positively charged. Moreover, the sorption capacity for Cu increased with pH, and the majority of the sorbed Cu was not desorbable. In contrast, the sorption of Cd by allophane largely occurs by electrostatic interactions. The relative contribution to metal sorption of allophane and HA in allophane–HA complexes was derived from the linear relationship between sorption and HA content of the complexes. The HA content at which there is an apparently equal (50/50) contribution of allophane and HA to Cu sorption is 28 and 87 g C kg^{-1} allophane at pH 5.0 and 5.5, respectively. Above these apparently "equal-contribution indicator" values the HA component in the complex contributes more to sorption than does the allophane counterpart. For Cd the apparently "equal-contribution indicator" values are 1.9 and 5.2 g C kg^{-1} allophane at pH 5.0 at pH 5.5, respectively. This observation has implications for soil management in terms of permissible loads for heavy metals and practical means of increasing the sorption capacity of soil.

1. INTRODUCTION

Copper (Cu) and cadmium (Cd) in soils may be derived from parent materials or introduced through such human activities as timber preservation [1] and the application of phosphate fertilizers [2]. The mobility of heavy metals in soil, and their availability to plants, are greatly influenced by sorption to soil constituents, of which clay and humus are the most active [3-4].

There is circumstantial evidence to indicate that the organic matter (humus) in soil plays an important role in heavy metal sorption [5–7]. However, as McBride et al. [8] have pointed out, metal sorption experiments often fail to reveal a strong correlation between organic matter content and either metal solubility or metal sorption capacity. This is possibly because the effect of organic matter is difficult to separate from that of pH, the soils used have a range of organic matter contents, and soil organic matter is chemically heterogeneous [8].

Often the dominant mineral constituent in the clay fraction of New Zealand soils derived from volcanic ash, allophane is a short-range ordered hydrous aluminosilicate mineral. Its primary or unit particle is a hollow spherule with an outer diameter of 3.5–5.0 nm and a wall thickness of ca 0.7 nm [9–10]. The spherule wall is composed of a curved Al–O,OH octahedral (gibbsitic) sheet to which orthosilicate [(O_3) Si(OH)] groups are attached on the inside. Structural defects within the wall give rise to ≈0.3 nm wide perforations [11] where (OH)Al(H_2O) groups are exposed. These groups can acquire protons on the acid side, and lose protons on the alkaline side, of the point of zero charge, and hence are responsible for the pH-dependent variable-charge characteristics of allophane.

Allophane can sorb metal cations, including Cu and Cd ions, through electrostatic (coulombic) interactions and by inner-sphere coordination involving exposed hydroxyl groups [12–13]. Since soil organic matter can also sorb metal ions, allophanic surface soils have a large propensity for taking up Cu and Cd ions [14]. Clay and organic matter in soil, however, are closely associated to form a clay-organic complex. As such, their individual involvement in heavy metal sorption, and contribution to the sorption capacity of soil, are difficult to assess. Here we investigate the sorption of Cu and Cd by allophane and its complexes with humic acid (HA). By using allophane complexes with varied amounts of sorbed HA, we can evaluate the apparent contributions of allophane and HA in the allophane-HA complex to metal sorption based on HA quantity.

2. MATERIALS AND METHODS

Both the allophane and HA used in this study were obtained from New Zealand soils using conventional methods of clay separation and organic matter extraction. The allophanic clay was obtained from the C horizon of the One Tree Point Sand which classifies as an Aeric Alaquod in Soil Taxonomy [15] and as a Humus-pan Perch-gley Podzol in the New Zealand Soil Classification [16]. The humic acid was obtained from the Koputaroa Sand which is a Typic Udipsamment in Soil Taxonomy or an Acidic Sandy Brown Soil in the New Zealand Soil Classification. Details of the location of the sampling sites, together with the procedures for clay separation and HA extraction/purification have been given by Yuan et al. [17].

The allophane content of the clay, estimated both by infrared spectroscopy and acid-oxalate extraction [18], was 850 g kg^{-1}. The allophanic clay (subsequently referred to as "allophane") contained trace amounts of iron (3 g kg^{-1}) together with quartz and kandite impurities. The carbon content of the allophane was 57 g kg^{-1} and the nitrogen content was 1.1 g kg^{-1}. The surface charge of the allophane was determined by ion adsorption [17] and the results are given later. The purified HA contained 530 g C kg^{-1}, 60.7 g N kg^{-1}, 8.3 g S kg^{-1}, 12 g ash kg^{-1}, and had a COOH acidity of 260 $cmol_c$ kg^{-1} as determined by direct titration [19].

Allophane–HA complexes with HA contents of 14–123 g OC kg^{-1} were obtained by adding a solution of HA to a suspension of allophane in 2 mM $CaCl_2$ [17]. Sorption of Cu and Cd by allophane and allophane–HA complexes was measured at pH 5.0 and 5.5, using the batch technique and a single concentration (2 mM) of Cu or Cd. Briefly, allophane (100 mg) or allophane–HA complexes containing 100 mg of allophane were mixed with 30 mL of 2mM $Cu(ClO_4)_2$ or 2mM $Cd(ClO_4)_2$, in the presence of 2mM $LiClO_4$ as background electrolyte. The pH of the suspensions was initially adjusted to 5.0 and 5.5 by adding $HClO_4$ or LiOH and re-adjusted twice during the 24-h period of shaking at 20 ± 1 °C. Adjustment of pH caused < 2% change of the initial 30 mL volume, and corrections for the quantity of Cu and Cd present in the final solution were made accordingly. At the end of shaking, the suspension was centrifuged for 12 min at 18,000 rpm (38,700 × *g*). Concentrations of Cu and Cd in the supernatant solutions were determined by atomic absorption spectrophotometry (AAS). The amount of Cu and Cd sorbed were calculated from the difference between the amount of Cu and Cd initially added to, and that measured at equilibrium with, the allophane and allophane-HA complexes. Blanks at pH 5.0 and 5.5 were prepared to allow for any precipitation. Sorption experiments were conducted in triplicate.

To assess the possibility of formation of a hydroxide or hydroxy carbonate precipitate during sorption experiment, the logarithms of the Cu^{2+} and Cd^{2+} activities were calculated using the extended Debye-Hückel equation and were compared with the solubilities of several Cu and Cd minerals. Solubility lines of Cu and Cd minerals were determined from the data of Lindsay [20], assuming atmospheric partial pressure of CO_2.

Desorption was carried out (in triplicate) by dispersing the allophane and allophane–HA complexes manually in 30 mL of 0.1 M KNO_3 solution, shaking for 24 h at 20 ± 1 °C, and centrifuging the suspension for 12 min at 18,000 rpm (38,700 × *g*). The concentration of Cu and Cd in the supernatant solutions was determined by AAS, allowing the amount of Cu and Cd desorbed to be calculated after correcting for entrained Cu and Cd.

The composition of the HA and the organic matter associated with the allophane sample was analysed by pyrolysis-field ionization mass spectrometry (Py-FIMS) as described by Schulten and Leinweber [21]. Sample crucible was heated automatically with a heating rate of 10°C/scan in the temperature range from 100 to 700 °C. The thermal degradation of the sample is indicated by the thermogram which reflects the purity, thermal stability and weight loss (volatile matter). Ten classes of organic compounds (EX1 – EX10) were identified according to their marker signals (mass-to-charge ratios). The contents of each class was estimated from the respective ratio of ion intensity to total ion intensity.

3. RESULTS

Figure 1 shows that all metal–allophane/complex systems were undersaturated with respect to the least soluble hydroxide or hydroxy carbonate phase, precluding the possibility of precipitate formation involving Cu or Cd hydroxide or hydroxy carbonate. This conclusion was further supported by the lack of a detectable change in metal concentrations of the blanks.

Figure 2 shows that allophane by itself can sorb much Cu. As pH increased from 5.0 to 5.5, sorption increased from 71 to 179 mmol kg^{-1}. Irrespective of pH, complex formation with HA led to a marked increase in sorption. Indeed, sorption (S) increased linearly as the amount of HA sorbed increased from 14 to 123 g organic carbon (OC) kg^{-1} allophane. The best-fit relations were:

$$S\ (\text{mmol Cu kg}^{-1}\ \text{allophane}) = 65.4 + 2.38\ OC \qquad (1)$$
(at pH 5.0, $R^2 = 0.992$, $P<0.01$)

$$S\ (\text{mmol Cu kg}^{-1}\ \text{allophane}) = 192.6 + 2.22\ OC \qquad (2)$$
(at pH 5.5, $R^2 = 0.992$, $P<0.01$)

Figure 3 shows that much less Cd than Cu was sorbed by allophane and its complexes under the same experimental conditions. For allophane by itself, 3.1 mmol kg^{-1} of Cd was sorbed at pH 5.0

and 9.2 mmol kg $^{-1}$ at pH 5.5. These values are more than one order of magnitude smaller than for Cu. The effect of pH on Cd sorption was also relatively small. The linear relationships between Cd sorption and the amount of HA complexed with allophane may be expressed by Equations (3) and (4):

$$S\ (\text{mmol Cd kg}^{-1}\ \text{allophane}) = -5.1 + 1.61\ OC \quad (3)$$
(at pH 5.0, $R^2 = 0.996$, $P<0.01$)

$$S\ (\text{mmol Cd kg}^{-1}\ \text{allophane}) = 10.1 + 1.78\ OC \quad (4)$$
(at pH 5.5, $R^2 = 0.998$, $P<0.01$)

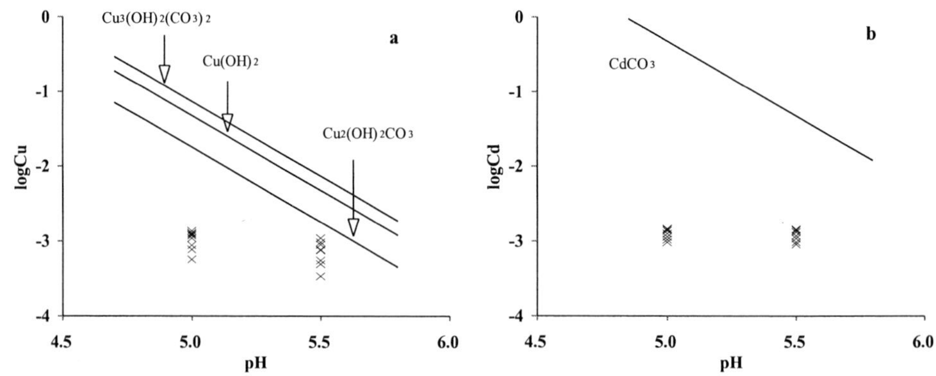

Figure 1. Solubility of Cu and Cd ($\log(M^{2+})$) after 24-h equilibration of allophane and allophane-HA complexes with (a) 2mM $Cu(ClO_4)_2$, and (b) 2mM $Cd(ClO_4)_2$.

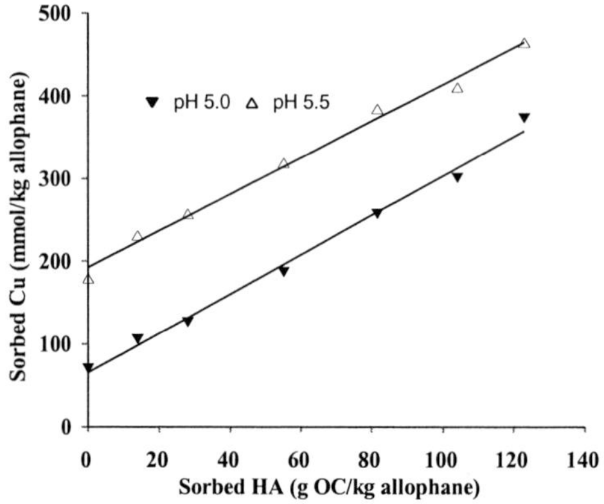

Figure 2. Cu sorption by allophane and allophane-HA complexes with various amounts of sorbed humic acid.

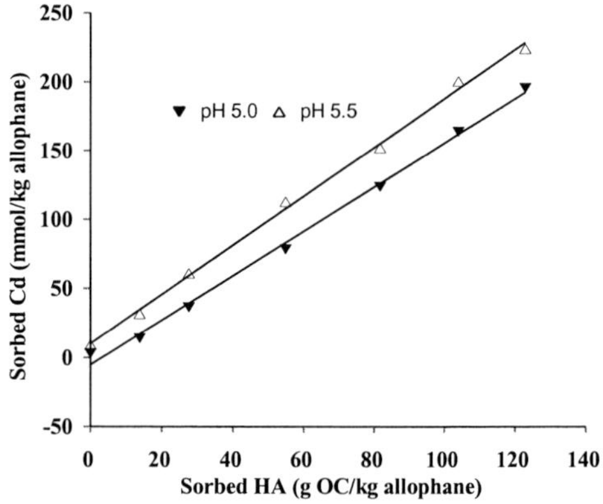

Figure 3. Cd sorption by allophane and allophane-HA complexes with various amounts of sorbed humic acid.

Figure 4 shows that part of the sorbed Cu and Cd can be displaced (desorbed) from allophane and allophane–HA complexes by treatment with KNO_3. However, the ratios of desorbed to sorbed metal ions were substantially different for Cu and Cd. In the case of Cu less than 20% was desorbable by this means, whereas 49–94% of sorbed Cd could be displaced. For Cu there was a clear effect of pH, with a lower desorption ratio at higher pH, whereas with Cd the pH effect was minimal. The desorption ratio of Cd decreased as the HA content increased (from A to F) in allophane–HA complexes.

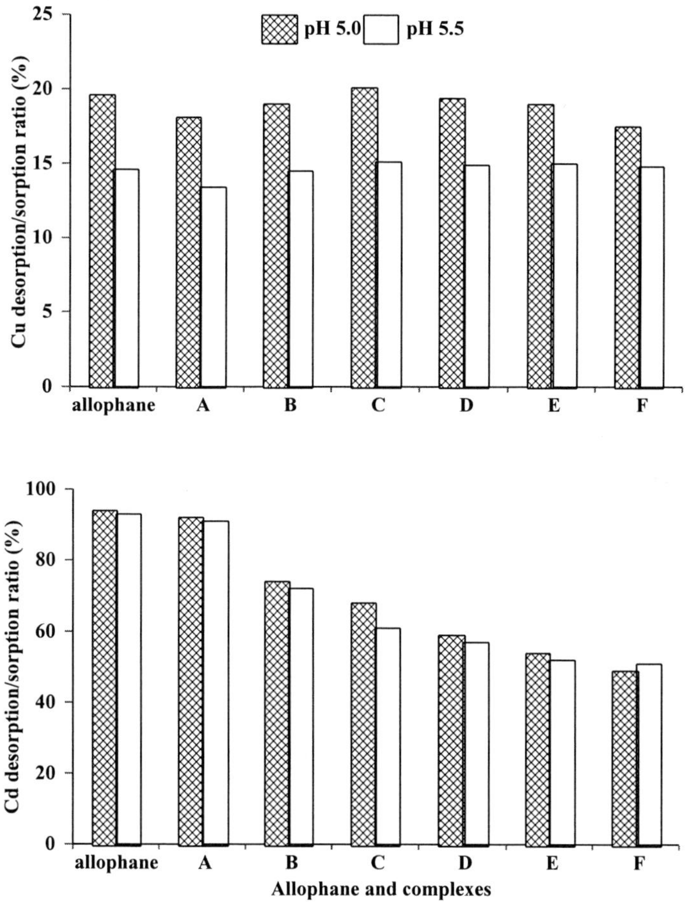

Figure 4. Desorption of Cu and Cd from allophane and its complexes with various amounts of sorbed humic acid (Cu and Cd were previously sorbed at pH 5.0 and 5.5); A, B, C, D, E, F refer to complexes with sorbed humic acid content of 13.9, 27.9, 55.1, 81.7, 104 and 123 g OC/kg allophane, respectively.

The surface charge characteristics of the allophane are shown in Figure 5. At pH 5.0, the allophane had a positive charge of 73 and a negative charge of 9 $mmol_c kg^{-1}$, giving a net (positive) surface charge of 64 $mmol_c kg^{-1}$. At pH 5.5, the positive charge decreased to 57, while the negative charge increased to 12 $mmol_c kg^{-1}$, resulting in a net (positive) charge of 45 $mmol_c kg^{-1}$. The point of zero charge (PZC) of allophane in 0.01 M NaCl was (pH) 6.4.

Table 1 summarizes the composition of the HA and the organic matter associated with the allophane derived from Py-FIMS. The organic matter associated with the allophane contained proportionally more lipids and alkylaromatics but less N-compounds and fatty acids than the HA sample.

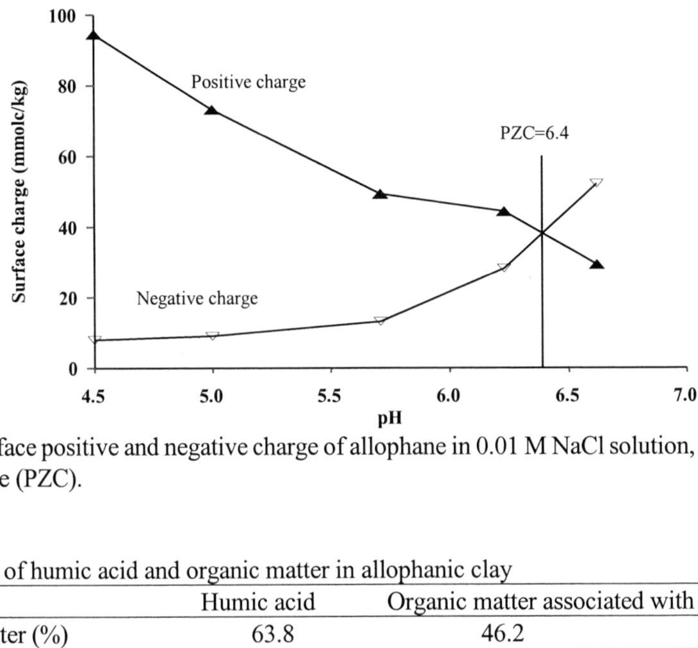

Figure 5. Surface positive and negative charge of allophane in 0.01 M NaCl solution, and the point of zero charge (PZC).

Table 1
Composition of humic acid and organic matter in allophanic clay

	Humic acid	Organic matter associated with allophane
Volatile matter (%)	63.8	46.2
	% of total ion intensity*	
EX1 – carbohydrates	5.3 (11)	2.0 (4)
EX2 – phenols	9.0 (18)	8.1 (16)
EX3 – lignin dimers	3.5 (7)	5.8 (11)
EX4 – lipids	7.8 (16)	13.3 (25)
EX5 – alkylaromatics	9.0 (18)	16.6 (32)
EX6 – N-compounds	7.6 (15)	4.0 (8)
EX7 – sterols	1.3 (3)	0.5 (1)
EX8 – peptides	3.1 (6)	1.2 (2)
EX9 – suberin	0.1 (0)	0.0 (0)
EX10 – fatty acid	2.8 (6)	0.8 (2)
Sum – EX1-10	49.5	52.3

* numbers in parentheses refer to percentages of assigned mass signals.

4. DISCUSSION

4.1. Sorption characteristics of allophane and its complexes for Cu and Cd

Cu^{2+} is under-saturated with respect to the least soluble hydroxide and hydroxy carbonate phases (Figure1). Although it seems unlikely that precipitation of these phases contributes to the sorption of Cu at pH 5.0 and 5.5, the possibility of layered double hydroxide formation as suggested by Sparks [22] for systems of hydrolysed Al^{3+} and Ni^{2+}, Co^{2+}, or Zn^{2+}, cannot be excluded. Furthermore, sorption increases with pH (Figure 2) and far exceeds the amount of negative charge on allophane at either pH (Figure 5). These observations strongly suggest that Cu is specifically sorbed by allophane. At pH $\leq$ 5.5, Cu^{2+} is the dominant species in solution. Charge balance would require two units of negative surface charge for one Cu^{2+} ion. The allophane has negative surface charge of 9 and 12 $mmol_c\ kg^{-1}$ at pH 5.0 and 5.5, respectively. The negative charge at pH 5.0 (9 $mmol_c\ kg^{-1}$) can only account for 6% of the Cu^{2+} sorbed (equivalent to 142 $mmol_c\ kg^{-1}$). The corresponding value at pH 5.5 is 3% of the sorbed Cu^{2+}(equivalent to 358 $mmol_c\ kg^{-1}$). Thus, cation exchange is not a major process responsible for Cu sorption by allophane. Instead, inner-sphere coordination of Cu^{2+} by surface AlOH groups apparently plays a key role in sorption. Clark and McBride [12, 23] have suggested that Cu is preferentially coordinated to adjacent AlOH groups, exposed on the surface of allophane particles, through a binuclear (bidentate) mechanism, although weaker binding to isolated AlOH and SiOH groups is possible. The pH-dependence of Cu sorption by allophane (Figure 2) is consistent with this suggestion, since an increase in pH would facilitate neutralisation of the protons released when Cu^{2+} coordinates with surface hydroxyl groups of allophane. This type of coordination also accounts for the low percentages (< 20 %) of Cu desorbable by KNO_3 (Figure 4) since K^+ does not coordinate with surface AlOH groups.

Similarly, the sorption of Cu^{2+} by humic substances occurs through coordination with oxygen-containing functional groups (e.g., COOH, phenolic OH, C=O). Amino and imino groups may also be involved [24]. Because of the coordinating ability of HA with metals, it is expected that Cu sorption by allophane–HA complexes would increase with the HA content of the complexes. The COOH content of the HA is 2600 mmol kg^{-1}. This is equivalent to 4.91 mmol COOH g^{-1} C, as the HA has a C content of 530 g kg^{-1}. If we assume that Cu coordinates with COOH in a bidentate manner, this would equate to Cu sorption of 2.45 mmol g^{-1} C, a value in close agreement with the coefficients of 2.22 and 2.38 in Equations (1) and (2). This is good evidence for a copper-organic bidentate formation.

Under the same experimental conditions allophane and its complexes sorb smaller amounts of Cd (Figure 3) than Cu. In other words, the affinity of allophane and its complexes for Cd is lower than for Cu. The same sequence of affinity was observed for Al hydroxide [25] and humic acid [26].

Allophane sorbed 3.1 mmol kg^{-1} of Cd at pH 5.0 and 9.2 mmol kg^{-1} at pH 5.5. Being divalent at pH $\leq$ 5.5, these values correspond to 6.2 and 18.4 $mmol_c\ kg^{-1}$ of Cd^{2+} sorbed, close to the negative surface charge of allophane (9 $mmol_c\ kg^{-1}$ at pH 5.0 and 12 $mmol_c\ kg^{-1}$ at pH 5.5). This leads to the conclusion that, in contrast to the situation for Cu, electrostatic interaction (cation exchange) is mainly responsible for Cd sorption by allophane. Using a synthetic allophane Denaix et al. [13] have also proposed that Cd sorption at pH < 7 occurs by cation exchange. The high ratio of desorbed to sorbed Cd (94% at pH 5.0 and 93% at pH 5.5, as shown in Figure 4), and the minimal effect of pH on this ratio, suggest that Cd is not strongly bound by allophane, in keeping with the electrostatic interaction mechanism. Because the negative surface charge increases with pH so does the sorption of Cd. This extra Cd, however, is also desorbable by KNO_3.

The HA in allophane–HA complexes contributes to further sorption of Cd by allophane (Figure 3). However, the contribution of HA to Cd sorption is substantially lower than that to Cu sorption because the slopes in Equations (3) and (4), 1.78 and 1.61, are smaller than the corresponding values for Cu in Equations (1) and (2). This is in agreement with the higher affinity of Cu for HA than that of Cd for HA [26]. The Cd sorbed by the HA in allophane–HA complexes is less desorbable than that sorbed by allophane itself, as indicated by the decreasing desorption/sorption ratio of Cd (Figure 4). This suggests that Cd may also coordinate with HA.

4.2. Relative contribution of HA and allophane to Cu and Cd sorption

Understanding the relative contributions of organic matter and inorganic components of soil to contaminant sorption is important. It provides a means of predicting how a potential change in soil attributes, as a result of environmental change and/or soil management practice, would affect the performance of the soil as an environmental filter for contaminants.

Because of the intrinsic complexity of the clay-HA interaction the question whether the sorption of metal cations to humic acid-modified clay conforms to an additive, competitive, or synergistic model has not been resolved. The linear additivity model in which sorption by clay-organic complex reflects the sum of its component parts has been favoured by some workers while alternative sorption models have been advocated by others [27, and references therein]. Little information is available on the sorption capacity of allophane-humic complex for heavy metals. Because of its relative simplicity, however, the linear additivity model may serve as a reference model for the sorption of metal ions by clay-humic complexes.

On this basis, we derive an empirical indicator to compare the apparent contributions of humic acid and allophane to the sorption of metal ions by the allophane-HA complex. Because both components of the complex contribute to Cu and Cd sorption, there is a certain value of HA for which the allophane and HA makes an apparently equal (50/50) contribution to the sorption of Cu and Cd by the complexes. In other words, the allophane-HA complex has twice capacity as allophane for sorbing metal ions. Below this "equal-contribution indicator" value allophane contributes more to the sorption than does HA, whereas above this value the opposite is true. The equal-contribution indicator value for Cu at pH 5.0 and 5.5 is 28 and 87 g OC kg^{-1} allophane, respectively. These values were obtained from Figure 2 by doubling the value of Cu sorption by allophane itself and then finding out the corresponding X-axis value through the regression line. Similarly, the equal-contribution indicator value for Cd is 1.9 g OC kg^{-1} allophane at pH 5.0 and 5.2 at pH 5.5. The equal-contribution indicator value for Cu is one order of magnitude higher than that for Cd because the amount of Cu sorbed by allophane is one order of magnitude greater than that of Cd under the same conditions (and the slopes in Equations (1) to (4) are comparable).

Although organic matter is capable of sorbing much Cu and Cd there is often no strong correlation between the organic matter content of soils and their metal sorption capacity [8, 28]. This might be because the composition of organic matter varies from soil to soil. Thus, empirical relations between sorption capacity and certain soil attributes obtained for a particular soil should not be extrapolated to other soils without due caution.

The large difference in equal-contribution indicator values for Cu and Cd has implications for soil management. It would be much more difficult to increase the sorption capacity of allophanic soils for Cu by increasing the organic matter content, than it would be for Cd since the equal-contribution indicator value for Cu is much higher than that for Cd. This is particularly true at the higher pH of 5.5. Thus, prevention of acidification may be more effective than organic matter addition in maintaining the sorption capacity of allophanic soils for Cu. On the other hand, the large sorption capacity of allophane with respect to Cu implies that allophanic soils can take up large amounts of Cu. In addition, these soils are capable of retaining sorbed Cu against leaching. In contrast, allophane is not very effective in sorbing Cd. However, the sorption capacity of allophanic soils for Cd can be quickly raised by addition of organic matter, although much of the sorbed Cd is desorbable.

The sorption behaviour of allophane and its HA complexes towards Cu and Cd may have other implications for soil management. Since allophanic soils may retain much more Cu than their negative charge would indicate, the use of a charge-related parameter such as cation exchange capacity (CEC) to indicate permissible Cu loadings may be inappropriate. Our data suggest that the allophane and organic carbon content (of the soil) would provide a better index of its capacity to sorb Cu.

5. CONCLUSIONS

Allophane can sorb much greater amounts of Cu than Cd under similar conditions. Cu that is sorbed is also strongly retained, probably because of coordination with active surface groups. In contrast, Cd is relatively weakly held by allophane because it is largely sorbed by a cation exchange mechanism. Cu and Cd sorption by allophane–HA complexes increases linearly with HA content. At a certain pH value, the relative contribution of the allophane and HA components to metal sorption may be evaluated using an empirical indicator value that denotes an equal (50/50) contribution of allophane and HA to metal sorption. Above this value, the HA in the complexes contributes more to sorption than the allophane component. The equal-contribution indicator value for Cu is 28 g C kg^{-1} allophane at pH 5.0, and 87 g C kg^{-1} at pH 5.5. The corresponding values for Cd are only 1.9 g C kg^{-1} at pH 5.0 and 5.2 g C kg^{-1} at pH 5.5. Thus, it would be much more difficult to increase the sorption capacity of allophanic soils for Cu by addition of organic matter than it would be for Cd. As allophane can sorb much more Cu than its negative surface charge would indicate, the use of a charge-related parameter (*e.g.*, CEC) as a guide to permissible Cu loadings for allophanic soils might not be appropriate.

ACKNOWLEDGEMENTS

This research was supported by the New Zealand Foundation for Research, Science and Technology (CO9811). We are grateful to Professors H.-R. Schulten and P. Leinweber of the University of Rostock, Germany, for the Py-FIMS analyses. We thank Matthew Taylor of Landcare Research, Hamilton, for valuable discussions. The constructive comments by two anonymous reviewers are appreciated.

REFERENCES

1. Carey, P.L., McLaren, R.G., Cameron, K.C., Sedcole, J.R., 1996. Leaching of copper, chromium, and arsenic through some free-draining New Zealand soils. Aust. J. Soil Res. 34, 583-597.
2. Loganathan, P., Hedley, M.J., Gregg, P.E.G., Currie, L.D., 1996. Effect of phosphate fertilizer

type on the accumulation and plant availability of cadmium in grassland soils. Nutrient Cycling in Agroecosystems 46,169-178.
3. McBride, M.B., 1989. Reactions controlling heavy metal solubility in soils. Adv. Soil Sci. 10, 1-56.
4. Alloway, B.J., 1995. Heavy Metals in Soils. 2nd ed. Blackie Academic and Professional, London,
5. Elliot, H.A., Liberati, M.R., Huang, C.P., 1986. Competitive adsorption of heavy metals by soils. J. Environ. Qual. 15, 214-219.
6. Yuan, G., Lavkulich, L.M., 1997. Sorption behaviour of copper, zinc, and cadmium in response to simulated changes in soil properties, Commun. Soil Sci. Plant Anal. 28, 571-587.
7. Zhang, M., Alva, A.K., Li, Y.C., Calvert, D.V., 1997. Chemical association of Cu, Zn, Mn, and Pb in selected sandy citrus soils. Soil Sci. 162, 181-188.
8. McBride, M., Sauve, S., Hendershot, M., 1997. Solubility control of Cu, Zn, Cd, and Pb in contaminated soils. Eur. J. Soil Sci. 48, 337-346.
9. Hall, P.L., Churchman, G.J., Theng, B.K.G., 1985. Size distribution of allophane unit particles in aqueous suspensions. Clays Clay Miner. 33, 345-349.
10. Wada, K., 1989. Allophane and imogolite. *In*: Dixon, J.B., Weed, S.B. (Eds.), Minerals in Soil Environment. 2nd ed. Soil Sci. Soc. Am., Madison, WI. pp.1051-1087.
11. Wada, S.-I., Wada, K., 1977. Density and structure of allophane. Clay Minerals. 12, 289-298.
12. Clark, C.J., McBride, M.B., 1984. Chemisorption of Cu(II) and Co(II) on allophane and imogolite. Clays Clay Miner. 32, 300-310.
13. Denaix, L., Lamy, I., Bottero, J.Y., 1999. Structure and affinity towards Cd^{2+}, Cu^{2+}, Pb^{2+} of synthetic colloidal amorphous aluminosilicates and their precursors. Colloids and Surfaces A: Physicochem. Eng. Aspects. 158, 315-325.
14. Abd-Elfattah, A., Wada, K., 1981. Adsorption of lead, copper, zinc, cobalt, and cadmium by soils that differ in cation-exchange materials. J. Soil Sci. 32, 271-283.
15. Soil Survey Staff, 1998. Keys to Soil Taxonomy. 8th ed. Natural Resources Conservation Service, United States Department of Agriculture, Washington, D.C.
16. Hewitt, A.E., 1998. New Zealand Soil Classification. 2nd ed. Manaaki Whenua Press-Landcare Research, Lincoln, New Zealand.
17. Yuan, G., Theng, B.K.G., Parfitt, R.L., Percival, H.J., 2000. Interactions of allophane with humic acid and cations. Eur. J. Soil Sci., 51, 35-41.
18. Parfitt, R.L., 1990. Allophane in New Zealand - A review. Aust. J. Soil Res. 28, 343-360.
19. Swift, R.S., 1995. Organic matter characterization. *In*: Sparks, D.L. (Ed.), Methods of Soil Analysis: Part 3, Chemical Methods. Soil Science Society of America, Madison, WI. pp.1011–1070.
20. Lindsay, W.L., 1979. Chemical Equilibria in Soils. John Wiley & Sons, New York.
21. Schulten, H.-R., Leinweber, P., 1999. Thermal stability and composition of mineral-bound organic matter in density fractions of soil. Eur. J. Soil Sci. 50, 237-248.
22. Spark, D.L. 2000. Advances in understanding the kinetics and mechanisms of metal sorption at the mineral/water interface. *In*: Violante A., Gianfreda, L. (Eds.) Soil Mineral-Organic Matter-Microorganism Interaction and Ecosystem Health. Abstracts of 3rd ISMOM, May 2000, Naples, Italy. p.34.
23. Clark, C.J., McBride, M.B., 1985. Adsorption of Cu(II) by allophane as affected by phosphate. Soil Sci. 139, 412-421.
24. Stevenson, F.J., 1982. Humus Chemistry: Genesis, Composition, Reactions. John Wiley &

Sons, New York.
25. Kinniburgh, D.G., Jackson, M.L., Syers, J.K., 1976. Adsorption of alkaline earth, transition, and heavy metal cations by hydrous oxide gels of iron and aluminum. Soil Sci. Soc. Am. J. 40,796-799.
26. Stevenson, F.J., 1977. Nature of divalent transition metal complexes of humic acids as revealed by a modified potentiometric titration method. Soil Sci.123, 10-17.
27. Murphy, E.M., Zachara, J.M. 1995. The role of sorbed humic substances on the distribution of organic and inorganic contaminants in groundwater. Geoderma 67, 103-124.
28. Gray, C.W., McLaren, R.G., Roberts, A.H.C., Condron, L.M., 1999. Solubility, sorption and desorption of native and added cadmium in relation to properties of soils in New Zealand. Eur. J. Soil Sci. 50,127-137.

Developments in Soil Science, Volume 28A
Editors: A. Violante, P.M. Huang, J.-M. Bollag and L. Gianfreda

COLLOID-MEDIATED TRANSPORT OF METALS ASSOCIATED WITH LIME-STABILIZED BIOSOLIDS

A.D. Karathanasis[a] and D.W. Ming[b]

[a] University of Kentucky, Department of Agronomy. N-122K Ag. Science-North, Lexington, KY 40546-0091, USA.

[b] NASA-Johnson Space Center, Houston, TX 77058, USA.

This study investigated the role of colloid particles dispersed from lime-stabilized biosolid amendments to mediate the transport of associated metals through intact soil columns in laboratory leaching experiments. Water-dispersible colloids of <1 μm in size and concentrations of 100 and 50 mg/L were fractionated from lime-stabilized biosolid materials by centrifugation and stabilized at a pH range of 7.0 to 11.5 and electrical conductivity (EC) of 100 to 1500 μS/cm. Colloid composition ranged from 5-45% in carbonates, 5-15% in aluminosilicate minerals, 20-38% in organic carbon, and about 100 mg g^{-1} in trace metals. The stabilized biosolid colloids were applied to undisturbed soil columns of a Paleudalf under steady rate (2.2 cm/hr) gravity flow conditions. De-ionized water spiked with metals at levels similar to the total load (soluble plus sorbed) carried by the colloids was used as a control leaching treatment. The eluents were monitored for colloid, Cu, Zn, and Pb breakthrough concentrations. Biosolid colloid elution was irregular throughout the leaching cycle, ranging from 0.1 to > 1.0 C/Co and increasing with input concentration, pH, and ionic strength. Settling kinetics experiments suggested that the latter conditions enhanced the stability and mobility of the biosolid colloids. Metal elution in the presence of colloids was several orders of magnitude greater than the control (soluble metal phase), following the sequence Pb>Zn>Cu. In all cases, eluted metal levels increased with colloid elution, pH, and ionic strength, suggesting a strong colloid-metal association. Under the most favorable conditions, average metal elution in association with biosolid colloids ranged from 65 to 80% of the input metal concentration. Between 35 and 90% of the eluted metals was colloid-bound, while the rest was soluble. The highest metal-colloid association was observed with Pb, suggesting a predominantly chemisorption mechanism for organic colloid surfaces and a co-precipitation mechanism for carbonate colloid surfaces. The lowest metal-colloid sorption affinity was observed with Cu, suggesting organic complexation as the main transport mechanism. These trends were corroborated by energy dispersive microprobe observations and dissolved organic carbon analysis. The findings of this study suggest a greater potential mobility and contamination risk than originally anticipated from metals associated with lime-stabilized biosolids through colloid mediation and co-transport processes.

1. INTRODUCTION

Application of agricultural and industrial waste products (biosolids) to land is being widely promoted as a cost-effective management alternative. Agricultural land, forest land, and land reclamation sites are increasingly being used for land application throughout the world. In addition to beneficial components such as organic matter and nutrients, these waste amendments also contain trace metals (such as Cd, Cu, Hg, Ni, Pb, and Zn), which may present a human toxicity risk through food chain or water contamination [1, 2].

Proponents of land application of agricultural, municipal, and industrial wastes cite a virtual immobilization of the heavy metals contained in the waste by organic complexation and mineral chemisorption in the soil matrix. However, in many laboratory and field studies, the investigators were unable to achieve mass balances between metals applied and those retained within the soil profile following sludge application. Dowdy et al. [3] could not account for nearly 50% of the applied Cd and Zn in the upper meter of soil, while Streck and Richter [4] could not locate 30% of the applied metal load, even though they observed a migration of 5% below the 70 cm soil depth. Substantial shortfalls in metal mass balances following 10-20 years after biosolid application have also been reported recently by Brown et al. [5], Richards et al. [6], and Baveye et al. [7].

While most of these disparities in metal mass balances were attributed to specific environmental conditions or random experimental errors, recent research findings suggest that preferential flow-path leaching (through soil macropores) of ionic or complexed metals in association with organic or mineral colloids may explain the metal shortfall. Richards et al. [6] demonstrated that metal leaching from land-applied sludges can occur at a much greater rate and quantity than predicted below the 1.5 m soil depth without showing any evidence of metal enhancement in lower soil horizons. This is an alarming observation because in most sludge application studies, the accumulation of metals in the upper soil profile and their minimal concentration in lower soil zones has been interpreted as a presumptive evidence of metal immobility. Studies with undisturbed soil columns [8] showed rapid elution of Cd, Zn, Cu, and Pb, especially in the presence of soluble organics, due to preferential flow paths. In several other field and laboratory studies, soluble and colloidal organics have been shown to facilitate metal transport [9, 10, 11].

Undisturbed soil column leaching experiments in our laboratory have shown that, in addition to metal-organic complexes, metal-mineral colloid associations may dramatically increase Cu, Zn, and Pb transport through soil macropores [12, 13]. The presence of metal-colloid associations (mineral and organic) formed during the leaching process enhanced metal transport by 5 to 50-fold for Cu and Zn and up to 3,000-fold for Pb over control treatments lacking colloids. The greatest metal transport potential was shown by colloids with high surface charge, organic carbon content, and electrophoretic mobility. In contrast, colloids with large particle size, low surface charge, and high Fe-and Al-hydroxy-oxide contents showed considerably lower metal co-transportability.

In recent years, the use of untreated sewage sludges is gradually decreasing in favor of alkaline stabilized biosolids produced by the addition of lime or mixtures of lime with cement or sodium silicate. This process induces metal immobilization through pH increases and $CaCO_3$ precipitation. In addition, the heat generated during the lime addition essentially sterilizes the product rendering it pathogen free [14]. While the alkaline stabilization process may reduce metal solubility, the resulting high pH may enhance organic carbon solubilization and colloid particle dispersion and mobilization. Therefore, even though the soluble metal levels are reduced, there is potential for mobilization of larger pools of metals associated with organic complexes and organic/mineral colloid particles [15, 16, 17, 5].

The objectives of this study were: a) to assess the potential of lime-stabilized biosolid colloids to mediate the transport of Cu, Zn, and Pb through undisturbed soil columns in laboratory leaching experiments, and b) to establish chemical gradients and conditions enhancing or inhibiting mediated transport.

2. MATERIALS AND METHODS

2.1. Biosolid colloid generation and characterization

A composted municipal lime-stabilized biosolid material was dispersed in de-ionized H_2O (1:20 ratio), shaken for one hour, then centrifuged at 130 *g* (750 rpm) for 3.5 min to collect the suspended fraction in a stock suspension [18]. The concentration of the colloid fraction was determined gravimetrically and turbidimetrically. A sub-sample of the stock colloid was collected for physicochemical and mineralogical characterization.

Two biosolid colloid fractions of 50 and 100 mg/L concentration were prepared by proper dilution of the stock suspension with de-ionized water. Another set of biosolid colloid fractions with the same concentration (50 and 100 mg/L, respectively) was generated from the stock suspension after three cycles of washing with de-ionized water. The purpose of the second set of biosolid colloids was to simulate approximate untreated and post application biosolid conditions; and to test their behavior under different pH, DOC, and ionic strength conditions.

Physicochemical properties of the biosolid colloids were determined following methods of the Natural Resources Conservation Service [19]. Analyses were performed on moist colloid samples. Total elemental composition was determined by HNO_3-HCl digestion [20]. Cation exchange capacity was determined with the 1 *M* barium acetate method (19). Elemental concentrations were determined by an Instrumentation Laboratory (Franklin, MA) model S11 AA/AE spectrophotometer. Organic carbon was determined by using a Leco Carbon Analyzer, Model CR-12 (Leco Corp., St. Joseph, MI). Colloid suspension, pH, and EC were measured with a Denver Instruments Model 250-pH-ISE-EC meter.

The mineralogical composition of the biosolid colloids was determined by x-ray and TG analysis using ~ 150 mg samples. A Phillips PW 1840 diffractometer and PW 1729 x-ray generator (Mahwah, NJ) were utilized for x-ray analysis according to procedures described by Karathanasis and Hajek [21]. The diffractometer was equipped with a cobalt x-ray tube, operated at 40kV and 30 mA, and a Bragg-Bretano design goniometer. A scanning rate of 0.05° 2θ per min from 2° to 40° and a scattering slit of 0.1° was used for the analysis. A TA thermo-gravimetric analyzer, with a heating rate of 20°C/min under N_2 atmosphere, was used for TG analysis. Micromorphological and elemental distribution characterization of the biosolid colloids were also performed with a SX100 backscattered electron imaging system interfaced with a microprobe analyzer operated at 15 kV and 30 nA.

Adsorption isotherms were also generated to evaluate the affinity of the biosolid colloids for Cu, Zn, and Pb. One hundred mg/L colloid suspension was added to 10 mL Teflon test tubes with 5 mL adsorbate solution containing 0-5 mg/L metal concentrations. Metal solutions were prepared from $CuCl_2$, $ZnCl_2$, and $PbCl_2$, reagents (>99% purity, Aldrich Chemicals, Milwaukee, WI). Samples were shaken on a reciprocating shaker for 24 hours and centrifuged for one hour at 2750 *g* (3500 rpm). Supernatants were collected and analyzed for Cu, Zn, and Pb by atomic absorption spectroscopy. The Freundlich isotherm equation was used to describe the experimental adsorption data.

2.2. Undisturbed soil columns

Duplicate undisturbed soil columns were taken from the upper Bt horizon of a Maury soil (fine, mixed, mesic Typic Paleudalf). This soil has demonstrated considerable preferential flow due to extensive macroporosity in previous studies [12]. Each column was prepared by carving the soil into a cylindrically shaped pedestal of 13-cm diameter and 20-cm length and encasing with an equal length of polyvinyl-chloride (PVC) pipe of 16-cm diameter. The size of the columns was selected to compensate for spatial variability, especially in soil hydraulic conductivity. The annulus between the intact soil column and the PVC pipe was sealed with expansible polyurethane foam. The columns were left in the field overnight to allow the foam to dry before they were separated from their base and transported to the laboratory. Physicochemical properties of the Maury soil used in the column experiment are reported in Table 1 [12, 13].

2.3. Leaching experiments

Prior to setting up the leaching experiment, the undisturbed columns were trimmed flat from both ends and saturated from the bottom upward with de-ionized H_2O (immersed by about 9/10 in open buckets) to remove air pockets. Then the columns were set up in stands and about five pore-volumes of de-ionized H_2O containing 0.002% (w/w) of NaN_3 (to cease biological activity) were introduced into the top of each column (downward vertical gravity flow) with a peristaltic pump at a constant flux (2.21 cm/hr) to remove loose material from the pores of the soil columns. No additional colloid generation from the soil matrix was detected in the eluent after five pore volumes, suggesting attainment of a steady-state colloid elution concentration of near zero prior to the initiation of the colloid tracer experiments. At that point, a set of duplicate columns was designated for each of the following solution and/or suspension leaching treatments.

1. Control with metal solution containing 10 mg/L Cu, Zn, and Pb as chloride salts, pH ~ 6.0.
2. Unwashed biosolid colloid with a concentration of 50 mg/L, pH ~ 11, EC ~ 600 μS/cm, DOC = 46 mg/L, and total metal load adjusted to about 10 mg/L with chloride salts.
3. D-H_2O washed biosolid colloid with concentration of 50 mg/L, pH ~ 7, EC ~ 100 μS/cm, DOC ~ 45 mg/L, and total metal load adjusted to about 10 mg/L with chloride salts.
4. Unwashed biosolid colloid with concentration of 100 mg/L, pH ~ 11, EC ~ 1500 μS/cm, DOC ~ 180 mg/L, and total metal load adjusted to about 10 mg/L with chloride salts.
5. D-H_2O washed biosolid with concentration of 100 mg/L, pH ~ 8.5, EC ~ 150 μS/cm, DOC ~ 46 mg/L, and total metal load adjusted to about 10 mg/L with chloride salts.

A total of 10 undisturbed columns were used in the experiment.

The leaching solutions/suspensions were applied to the top of each column through a continuous step input of 2.21 cm/hr controlled with a peristaltic pump. This rate was tested in earlier experiments and found to provide consistent free flow conditions without ponding on the top of the columns. All input mixtures were allowed to equilibrate for 24 h before application. For ≈ 10 days, eluents were monitored with respect to volume, colloid, and metal concentration. Breakthrough curves (BTCs) were constructed based on reduced metal and colloid concentrations (ratio of effluent concentration to influent concentration = C/C_o) and pore-volume (flux averaged volume of solution pumped per column pore volume).

Colloid concentrations in the eluent were determined with a Bio-Tek multichannel (optical densitometer with fiber-optics technology; Bio-Tek Instruments, Winooski, VT) microplate reader, pre-calibrated with known concentrations of each colloid at 540 nm. Total metal concentration in the eluents was allocated to solution phase and colloidal phase (colloid-bound contaminant). The eluent samples were centrifuged for 30 min at 2750 *g* (3500 rpm) to separate the soluble contaminant fraction from the colloid-bound contaminant fraction. The absence of colloidal material in the supernatant solution was verified by filtration through a 0.2-μm membrane filter. The soluble metal (Cu, Zn, Pb) fractions were analyzed by atomic absorption- (concentrations > 0.5 mg/L) or inductively coupled plasma (ICP) spectrometry (concentrations < 0.5 mg/L). The colloid fraction was extracted with 1 *M* HNO_3-HCl (20) solution and analyzed with the same methodology used for the soluble fraction. The reproducibility between replicate columns was within ±15%.

3. RESULTS AND DISCUSSION

3.1. Soil and biosolid colloid properties

Physiochemical and compositional properties of the Maury soil and the biosolid colloids used in the study are listed in Table 1. Notable are the striking differences in total organic carbon, pH, CEC, EC, DOC, aluminosilicate minerals, carbonate and total metal content between the soil and the biosolid colloids. Total organic carbon content and CEC was almost twice as high in the unwashed colloid compared to the washed colloid as a result of the water-wash and dissolution/removal of carbonates. This points out the organic origin of the charge in the biosolid colloids. The total organic carbon of the soil was negligible (0.5%) but its CEC was moderate (1/3 of the washed and 2/3 of the unwashed colloid) originating mainly from aluminosilicate minerals. The EC of the unwashed biosolid colloids was 6-10 times higher than that of the washed colloids and more than 20 times greater than the soil EC. Similarly, DOC in the unwashed colloids was about 4 times greater than the washed biosolid colloids and 15 times higher than the soil. These differences are expected to have significant implications on colloid stability and transport phenomena through the soil columns.

Significant differences are also evident in metal sorption affinities between soil and biosolid colloids. For all 3 metals, the K_f values for the Maury soil were lower than those for the biosolid colloids. This indicates that colloids migrating through the pores of the Maury soil will not only maintain their sorbed metal load but would also out-compete matrix soil sites for soluble metals present in the soil macropore space and even metals sorbed in the soil matrix. Even though the increased colloid sorption affinity for Zn and Pb over that for Cu was expected, the extremely high K_f values of the unwashed biosolid colloids for these two metals is surprising, considering their moderate surface charge. Apparently, the high affinity is not only the result of chemisorption, but possibly metal co-precipitation with the abundant carbonates.

An assessment of the above physicochemical and compositional effects on colloid stability was made with settling kinetics experiments (Figure 1). The 50 mg/L washed colloid appeared to be the least stable throughout the experiment. In spite of considerable pH and EC differences, the other 3 colloids appeared to have similar stability. Apparently, the high pH of the unwashed colloids maintains a dispersive environment that overcomes the high ionic

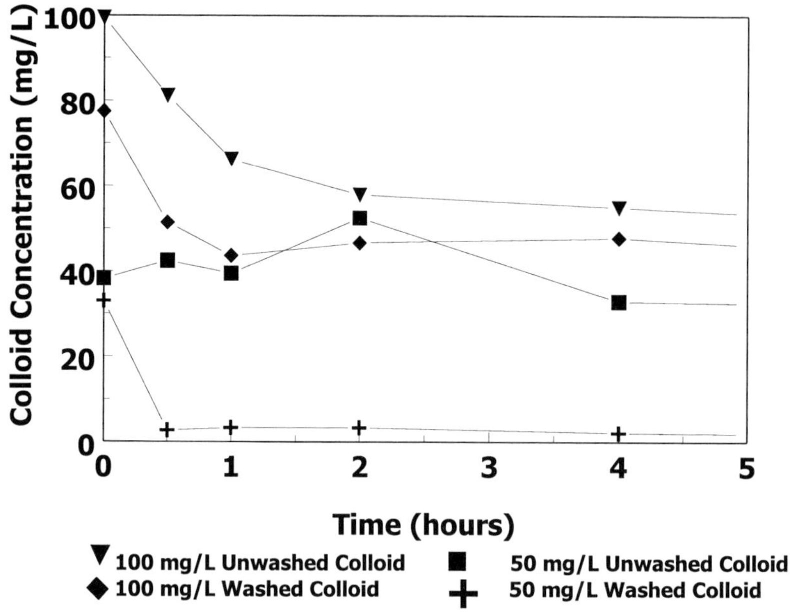

Figure 1. Settling kinetics of the four biosolid colloids used in the study.

strength effects, which under other conditions may have caused coagulation and flocculation. Scanning electron microscope observations confirmed a finely dispersed colloid matrix in both unwashed and washed colloids, comprised primarily of < 300 nm diameter organo-carbonate or organo-mineral particles. These results reinforce the concerns raised earlier about the potential mobilization of other metal pools (DOC-complexed, colloid-sorbed) and transport in subsurface soil environments.

3.2. Colloid elution

Colloid breakthrough curves for the four colloid suspensions used in the leaching experiment are shown in Figure 2. In all cases, colloid breakthrough was highly irregular with several maxima and minima of different intensity throughout the experiment. This pattern is typical of alternating convective cycles, during which colloids are transported by preferential mass flow through soil macropores, and diffusion cycles, during which colloid elution is limited or restricted by physical filtration and/or chemical interaction with the soil matrix. Colloid elution maxima were highest and of longer duration with the unwashed colloids and increased with increasing colloid concentration (higher with 100 mg/L than the 50 mg/L suspensions). Elution maxima of C/C_o >1 observed with the 100 mg/L and 50 mg/L unwashed colloids are probably the result of indigenous soil colloid mobilization or remobilization of already deposited biosolid colloids within the soil matrix as a result of dispersion phenomena caused by the high pH of the biosolid suspensions. This is evident from the pH and EC BTC's (Figure 3) showing good correlation with colloid elution maxima. For the 50 mg/L unwashed colloids and the 100 mg/L washed colloids, convective preferential flow maxima were more prominent during the first 8 pore volumes of elution with

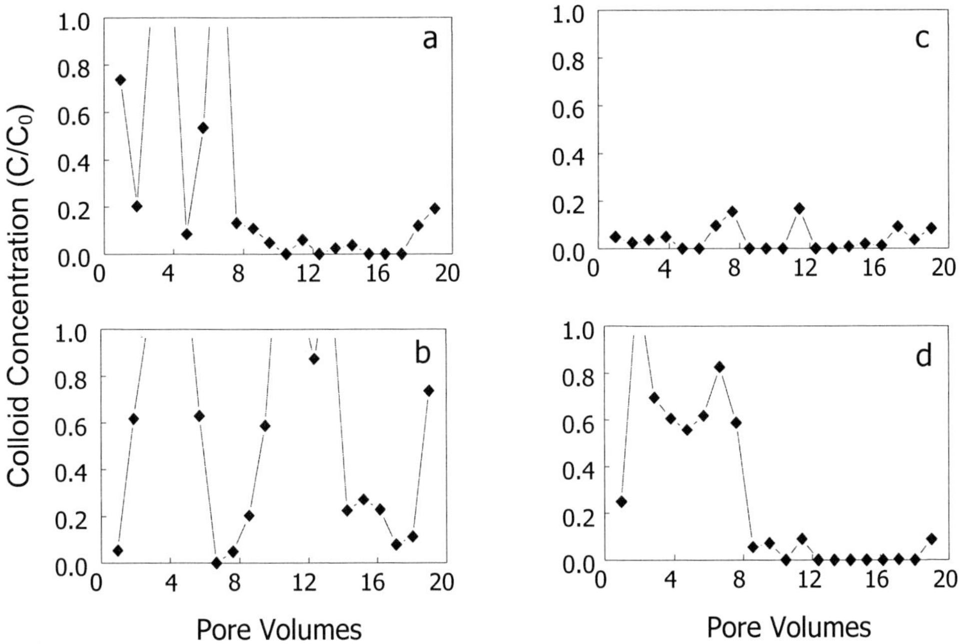

Figure 2. Breakthrough curves (C/C_0= ratio of effluent to influent concentration) for the four biosolid colloids eluted through Maury soil columns: (a) 50 mg/L unwashed colloid, (b) 100 mg/L unwashed colloid, (c) 50 mg/L washed colloid, and (d) 100 mg/L washed colloid.

only minor resurgences afterwards. In contrast, the 100 mg/L unwashed colloid continued to show elution maxima throughout the experiment, while the 50 mg/L washed colloid peak never exceeded 0.2 C/C_o.

The pH of the eluted suspensions appears to be the dominant factor controlling these elution patterns, since it was maintained around 11 throughout the leaching experiment for the unwashed colloids, while it was stabilized at around 7 for the washed colloids (Figure 3). These relationships are also consistent with the colloid stability patterns shown by the settling kinetics experiments (Figure 1). It is interesting that in spite of the high ionic strength of the unwashed colloids, their high pH was able to maintain their stability and promote greater mobility, while the lower buffered pH of the washed colloids apparently induced coagulation and thus, easier filtration by the soil matrix

3.3. Metal Elution

Figures 4, 5, and 6 show BTC's for Cu, Zn, and Pb eluted in the absence (metal solution control) and presence of biosolid colloids. Elution in the presence of colloids is plotted separately for the soluble metal fraction and for the sum of the soluble and colloid-bound (total) fraction. Metal elution was essentially zero for all control treatments, suggesting total attenuation by the soil matrix. The presence of colloids enhanced drastically the elution of both soluble and total metal levels, showing an excellent correlation with colloid elution patterns. This confirms the strong association between metals and colloids and their role as carriers or facilitators in the transport process.

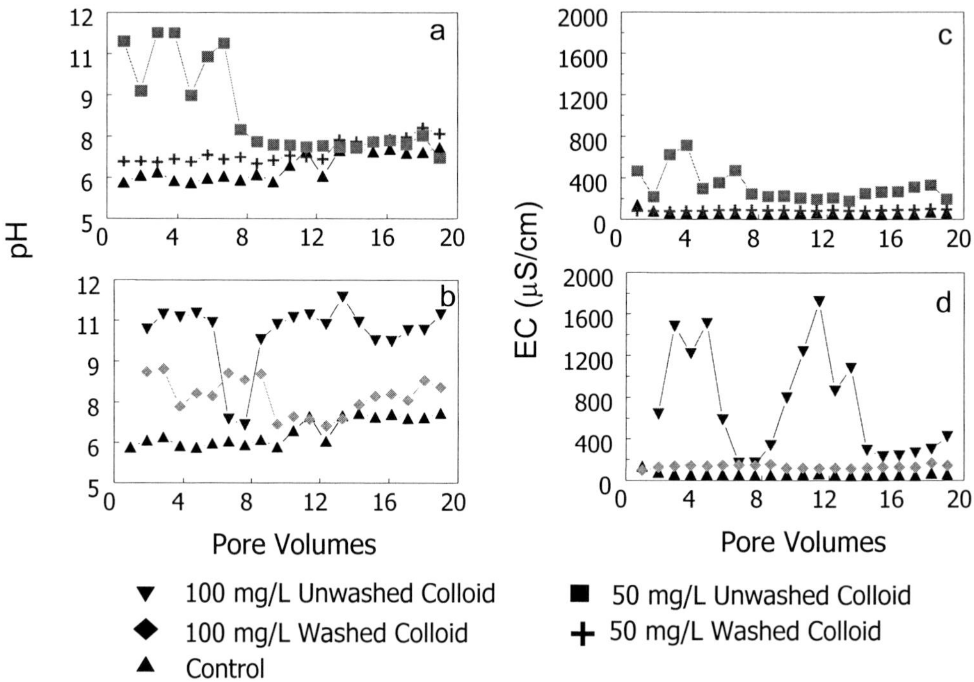

Figure 3. Breakthrough curves (C/C_0= ratio of effluent to influent levels) for pH and EC of the eluted control solutions and biosolid colloid suspensions: (a) and (c) control and 50 mg/L biosolid colloids, respectively; (b) and (d) control and 100 mg/L biosolid colloids, respectively.

Total Cu elution was highest with the 100 mg/L unwashed biosolid colloids, exceeding by several orders of magnitude the control and the 50 mg/L washed colloid treatments, and being more than twice greater than the 50 mg/L unwashed and the 100 mg/L washed colloids. The high pH and high concentration of the unwashed 100 mg/L colloids apparently were responsible for the enhanced mobility of Cu. The high pH not only increased colloid stability but may have also enhanced the solubility of low molecular weight organic complexes associated with the biosolid colloids [22, 5]. The mobilization of the organic complexes and their high affinity for Cu could account for the drastic increases in Cu elution [23]. Nearly 50% of the total eluted Cu was in the soluble form, thus providing strong evidence of the role played by the DOC in the Cu transport process. The similar (1:1) ratio of the eluted soluble to colloid-bound Cu in the different colloid treatments is consistent with the K_f values, which are not that different between the washed and unwashed colloids. Due to its high binding potential with DOC complexes (24) and relatively low K_f for the soil matrix, Cu has been documented to be one of the most mobile metals from land-applied biosolids [25]. This mobility has been found to increase by nearly 50% at elevated pH levels such as those generated during the lime stabilization process [26].

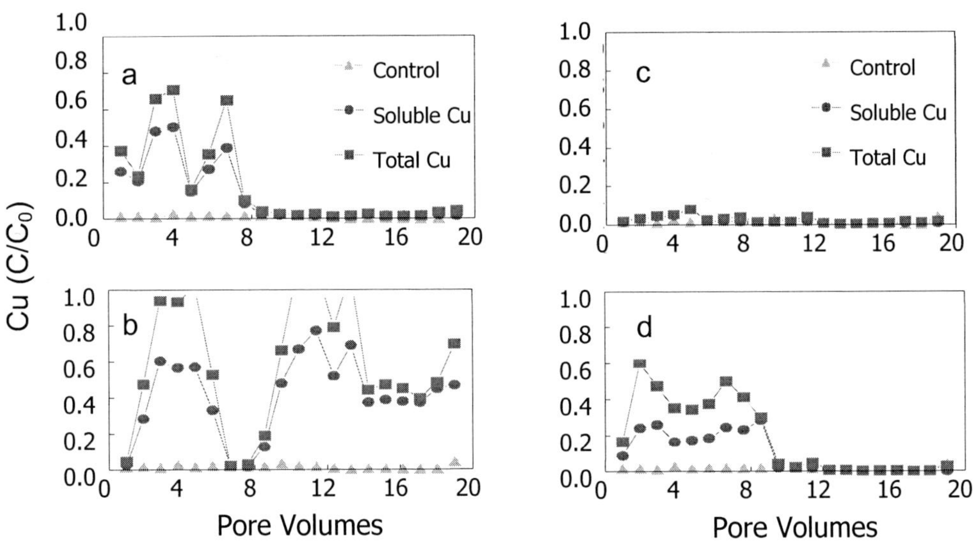

Figure 4. Breakthrough curves (C/C_0= ratio of effluent to influent concentration) for Cu eluted in the absence (control) and presence of biosolid colloids: (a) 50 mg/L unwashed colloid, (b) 100 mg/L unwashed colloid, (c) 50 mg/L washed colloid, and (d) 100 mg/L washed colloid.

Total Zn elution patterns were very similar to those observed for total Cu (Figure 5). However the eluted soluble Zn load was much lower than that for Cu, amounting to <10% of total Zn. Considering the excellent agreement between colloid and Zn BTC's, this implies that > 90% of the eluted Zn was colloid-bound. Therefore, chemisorption appears to be the dominant Zn-transport mechanism. This mechanism is corrobated by the strong affinity of Zn for the biosolid colloids over that of the soil matrix, especially for the unwashed colloids (Table 1). The considerably higher affinity for Zn shown by the unwashed colloids, in spite of their lower surface charge, is attributed to the presence of $CaCO_3$ precipitates. Carbonates have high sorption capacity for Zn, particularly at elevated pH's, through formation of surface hydrated complexes and eventual coprecipitation and incorporation of the metal into the carbonate structure [27]. Therefore, chemisorption processes, involving ion-exchange at organic colloid surfaces and coprecipitation on carbonate colloid surfaces are responsible for the majority of the total eluted Zn. The small contribution by the DOC complexes is consistent with the generally low complexation affinity exhibited by Zn compared to that of Cu and other metals [24].

Total Pb elution was < 0.2 C/C_o in the presence of the 50 mg/L colloid suspensions and exceeded twice 1.0 C/C_o in the presence of both, unwashed and washed, 100 mg/L colloids. The total Pb BTC's were nearly identical within the same colloid concentration treatment. However, the soluble Pb fraction eluted in the presence of unwashed colloids was 5-6 times greater than that of the washed colloids; but considerably lower than the soluble Cu fraction (Figure 6). This suggests that although the transport of small Pb fractions my have been facilitated through DOC-Pb complexes, the largest fraction was transported through chemisorption to colloid particles. The high, but similar, K_f values of unwashed and washed

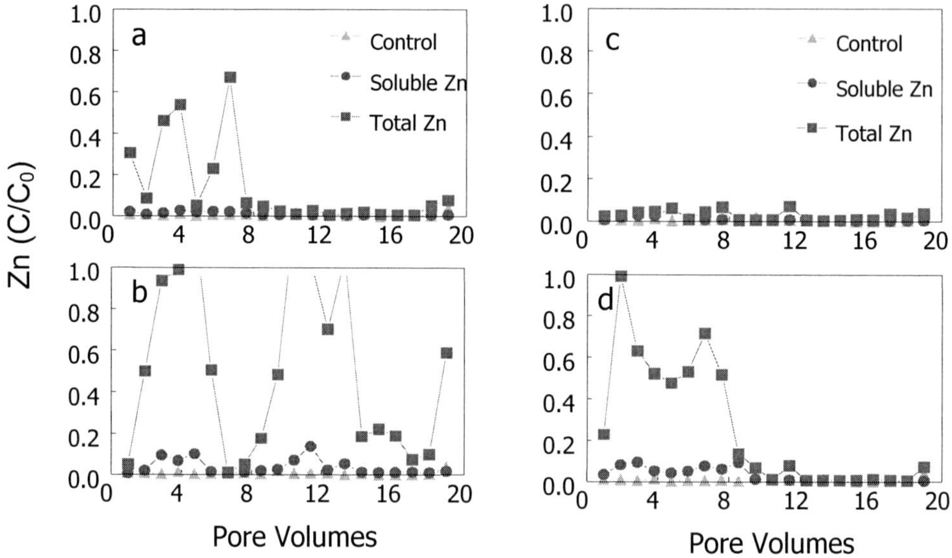

Figure 5. Breakthrough curves (C/C$_0$= ratio of effluent to influent concentration) for Zn eluted in the absence (control) and presence of biosolid colloids: (a) 50 mg/L unwashed colloid, (b) 100 mg/L unwashed colloid, (c) 50 mg/L washed colloid, and (d) 100 mg/L washed colloid.

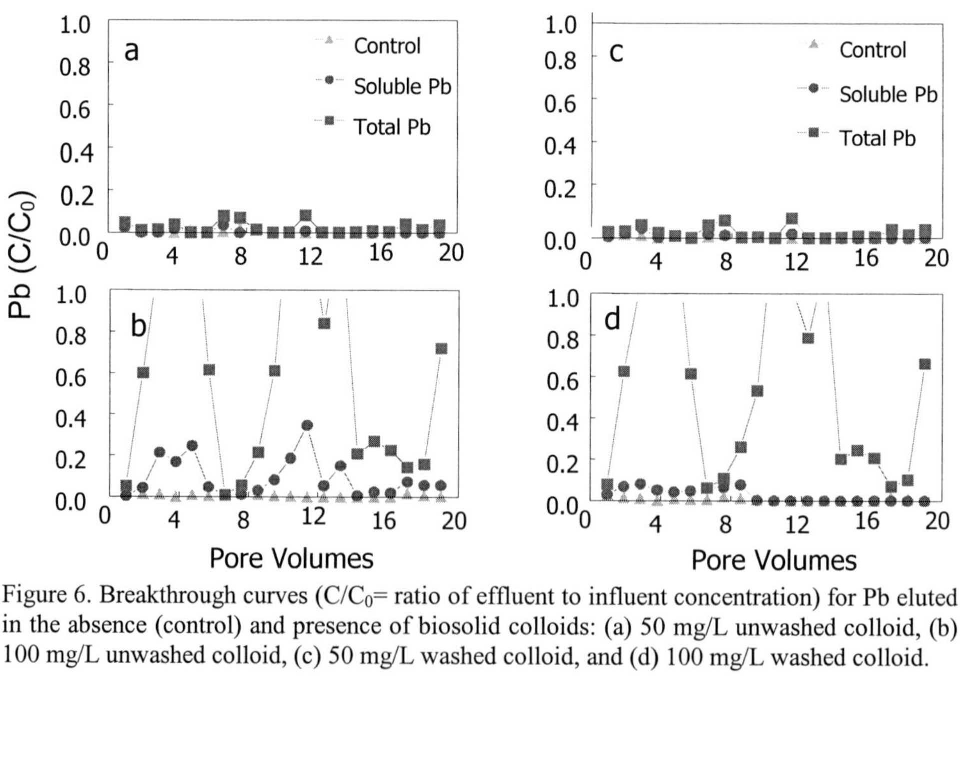

Figure 6. Breakthrough curves (C/C$_0$= ratio of effluent to influent concentration) for Pb eluted in the absence (control) and presence of biosolid colloids: (a) 50 mg/L unwashed colloid, (b) 100 mg/L unwashed colloid, (c) 50 mg/L washed colloid, and (d) 100 mg/L washed colloid.

Table 1. Physicochemical and mineralogical properties of the soil and biosolid colloids used in the study.

Properties*	Maury Soil	Unwashed Biosolid Colloid	Washed Biosolid Colloid
Clay (%)	35	--	--
Hydraulic Conductivity (cm/min)	2.6 ± 0.7	--	--
Bulk density	1.6	--	--
Total organic carbon (%)	0.5	20	38
pH	5.8	11.0-11.5	7.0-8.5
CEC (cmol/kg)	21.9	32	60
EC (μS cm)	65	600-1500	100-150
DOC (mg/L)	12	180	45
HISM (%)	46	5** (HISM + Mica + Kaolinite)	15** (HISM + Mica + Kaolinite)
Mica (%)	10		
Kaolinite (%)	36		
Quartz (%)	8	10	20
Carbonates (%)	--	45	5
Cu, Zn, Pb (mg/g)	<0.1	36	21
K_f(Cu)	1.14	1.29	1.56
K_f(Zn)	0.78	7.44	2.90
K_f(Pb)	1.75	6.61	5.53

* CEC= cation exchange capacity; EC= electrical conductivity; DOC= dissolved organic carbon; HISM = hydroxy-interlayered smectite; K_f= Freundlich coefficient.

** Total amount of aluminosilicate minerals.

colloids for Pb (Table 1) are consistent with the Pb elution patterns. The slightly higher affinity of the unwashed colloids for Pb may be due to carbonate co-precipitation phenomena, while the increased soluble Pb fraction eluted in the presence of the unwashed colloids is probably the result of the elevated pH that mobilized additional DOC complexes. Lead is known to have high sorption affinity for particle surfaces and most of its mobility is realized through association with particulate material [13], while its organic complexing affinity has been documented to be between that of Cu and Zn [24].

4. CONCLUSIONS

The potential migration of biosolid colloids and associated transport of the metals Cu, Zn, and Pb in subsurface soil environments was clearly demonstrated in this experiment. The undisturbed soil column experiments have shown that colloid-mediated transport of the metals may be several orders of magnitude greater than that of control metal solutions considering only a soluble metal phase. In spite of the supposed beneficial reduction of the soluble metal load through lime stabilization processes, the study showed that the resulting high pH may mobilize additional metal pools associated with DOC complexes and organic or carbonate colloid particles. The dispersive environment created under the prevailing alkaline conditions may induce considerable colloid generation and facilitated transport through preferential flow paths within the soil matrix. The greater metal sorptive affinity of the colloids over that of the soil matrix guarantees that the colloid-sorbed metals will migrate to deeper soil depths without experiencing the presumed competitive retention by the soil matrix. Our findings suggest a greater potential mobility and contamination risk than originally anticipated from metals associated with lime-stabilized biosolids through colloid mediation and co-transport processes.

REFERENCES

1. USEPA., 1992. Technical Support Document for Land Application of Sewage Sludge. Volume I, USEPA Publication 822/R-93-00a.
2. Harrison, E.Z., McBride, M.B., Bouldin, D.R., Bouldin., 1999. Land application of sewage sludges: an appraisal of the US regulations. Int. J. Environ. Pollut. 11, 1-36.
3. Dowdy, R.H., Latterell, J.J., Hinesly, T.D., Grussman, R.B., Sullivan D.L, 1991. Trace metal movement in an Aeric Ochraqualf following 14 years of annual sludge applications. J. Environ. Qual. 20, 119-123.
4. Streck, T., Richter J., 1997. Heavy metal displacement in a sandy soil at the field scale I: measurements and parameterization of sorption. J. Environ. Qual. 26, 49-56.
5. Brown, S., Chaney, R., Angle, J.S, 1997. Subsurface liming and metal movement in soils amended with lime-stabilized biosolids. J. Environ. Qual. 26, 724-732.
6. Richards, B.K., Steenhuis, T.S., Peverly, J.H., McBride, M.B., 1998. Metal mobility at an old, heavily-loaded sludge application site. Environ. Pollut. 99, 365-372.

7. Baveye, P., McBride, M.B., Bouldin, D., Hinesly, T.D., Dohdoh, M.S.A., Abdel-Sabour, M.F., 1999. Mass balance and distribution of sludge-borne trace elements in a silt loam soil following long-term applications of sewage sludge. Sci. Tot. Environ. 227, 13-28.
8. Camobreco, V.J., Richards, B.K., Steenhuis, T.S., Peverly, J.H., McBride, M.B., 1996. Movement of heavy metals through undisturbed and homogenized soil columns. Soil Sci. 161, 740-750.
9. van Erp, P.J., van Lune, P., 1991. Long-term heavy-metal leaching from soils, sewage sludge and soil/sewage sludge mixtures. *In:* Hermite, P.L. (Ed.), Treatment and use of sewage sludge and liquid agricultural wastes. Elsevier Science Pub. Co., New York, pp. 122-127.
10. del Castillo, P., Chardon, W.J., Salomons, W., 1993. Influence of cattle-manure slurry application on the solubility of cadmium, copper, and zinc in a manured acidic, loamy-sand soil. J. Environ. Qual. 2, 689-697.
11. Persicani, D., 1995. Analysis of leaching behavior of sludge-amended metals in two field soils. Water Air Soil Pollut. 83, 1.
12. Karathanasis, A.D., 1999. Subsurface migration of Cu and Zn mediated by soil colloids. Soil Sci. Soc. Am. J. 63, 830-838.
13. Karathanasis, A.D., 2000. Colloid-mediated transport of Pb through soil porous media. Intern. J. Environ. Stud. 57, 579-596.
14. Smith, K.A., Gains, L.E., Logan, T.J., 1998. Effect of calcium oxide dose on thermal reactions, lime speciation, and physical properties of alkaline stabilized biosolids. Water Environ. Res. 70, 224-230.
15. Roy, S.B., Dzombak, D.A., 1997. Chemical factors influencing colloid-facilitated transport of contaminants in porous media. Environ. Sci. Technol. 37, 656-664.
16. Elliott, H.A., Brown, G.A., 1989. Comparative evaluation of NTA and EDTA for extractive decontamination of Pb-polluted soils. Water Air Soil Pollut. 45, 361-367.
17. Hsiau, P.C., Lo, S.L., 1997. Characteristics of four alkaline biosolids produced from sewage sludge. Resources Conserv Recycl. 21, 185-197.
18. Seta, A.K., Karathanasis, A.D., 1997. Stability and transportability of water dispersible soil colloids. Soil Sci. Soc. Am. J. 61, 604-611.
19. USDA-NRCS, 1996. Soil Survey Laboratory Methods Manual. Soil Survey Investigations Report No. 42, Version 3.0, National Soil Survey Center, Lincoln, NE.
20. American Public Health Association (APHA), 1989. Standard methods for the examination of water and wastewater. 17th ed. APHA, Washington, DC.
21. Karathanasis, A.D., Hajek, B.F., 1982. Revised methods for quantitative determination of minerals in soil clays. Soil Sci. Soc. Am. J. 46, 419-425.
22. Han, N., Thompson, M.L., 1999. Copper-binding ability of dissolved organic matter derived from anaerobically digested biosolids. J. Environ. Qual. 28, 939-944.
23. Temminghoff, E.J.M., van Der Zee, S.E.A.T.M., DeHaan, F.A.M., 1997. Copper mobility in a Cu-contaminated sandy soil as affected by pH and solid and dissolved organic matter. Environ. Science Technol. 31, 1109-1115.
24. Harter, R.D., 1983. Effect of soil pH on adsorption of Pb, Cu, Zn, and Ni. Soil Sci. Soc. Am. J. 47, 47-51.
25. McBride, M.B., Richards, B.K., Steenhuis, T., Spiers, G., 1999. Long-term leaching of trace elements in a heavily sludge-amended silty clay loam soil. Soil Sci. 164, 613-623.

26. Richards, B.K., Peverly, J.H., Steenhuis, T.S., Liebowitz, B.N., 1997. Effect of processing mode on trace elements in dewatered sludge products. J. Environ. Qual. 26, 782-788.
27. Zachara, J.M., Cowan, C.E., Resch, C.T., 1991. Sorption of divalent metals on calcite. Geochim. Cosmochim. Acta 55, 1549-1562.

Developments in Soil Science, Volume 28A
Editors: A. Violante, P.M. Huang, J.-M. Bollag and L. Gianfreda

HEAVY METALS AND LITTER DECOMPOSITION IN CONIFEROUS FORESTS

A. Virzo De Santo[a], A. Fierro[a], B. Berg[b], F. A. Rutigliano[c] and A. De Marco[a]

[a]Dipartimento di Biologia Vegetale, Università Federico II, Via Foria 223, 80139 Napoli, Italy

[b]Lehrstuhl für Bodenökologie, Universität Bayreuth, Dr Hans Frisch Strasse 1-3, DE-944 48, Bayreuth-Germany

[c]Dipartimento di Scienze Ambientali, Seconda Università di Napoli, Via Vivaldi 43, 81100 Caserta, Italy

The dynamics of Mn, Zn, Fe, Cu, Pb, and Cd during litter decomposition was studied to understand how litter and soil metal concentrations influence the accumulation and/or the release of metals. Three types of leaf litters (green and brown leaves of *Populus tremula* L., green leaves of *Betula pubescens* Ehrh.) and four types of needle litters (green and brown needles of *Pinus sylvestris* L., brown needles of *Pinus contorta* L. and *Pinus pinea* L.), differing in heavy metal concentrations were incubated at two unpolluted coniferous forest sites: a temperate silver fir (*Abies alba* Mill.) forest at site Monte Taburno, southern Italy, and a boreal Scots pine (*Pinus sylvestris* L.) forest at site Jädraås, Sweden. Compared to the Scots pine forest, the humic surface horizon of the silver fir forest had a higher pH, higher concentrations of available Cu and Cd, and lower Fe concentration. After a period of about 900 days, all litter types were more decomposed at M. Taburno than at Jädraås. Zn, Fe, Pb and Cd concentrations in leaf and needle litters increased at both sites; Cu concentrations increased only in the silver fir forest; Mn concentration decreased in the litter richest in Mn and increased in the litters poorest in Mn. Litter decomposition in an early phase (0 up to 528/565 days) was significantly and positively correlated to initial Cu and Cd concentrations in litter. In a late phase (528/565 up to 918/929 days), decomposition was correlated significantly and negatively with litter Zn and Cu, and significantly and positively with litter Mn concentration at the start (528/565 days) of the period. At both sites, increases in absolute amounts of Fe, Cd, and Pb were observed in all litters; however, *P. pinea*, which showed the highest initial concentration of Pb, released Pb during decomposition. All litters released Cu at Jädraås and accumulated Cu at M. Taburno. Zn was released at both sites from all leaf litters and from the needle litter of *P. contorta*. Mn was released at both sites from all litters, except the litter of *P. tremula,* which accumulated Mn at Jädraås. The results indicate that heavy metal accumulation or release may depend on the gradient of metal concentration between litter and soil, on the pH of the soil, and on the capacity of litter to bind metal; atmospheric deposition could account at least partly for the increase of absolute amounts of Pb, Fe, and Cu at M. Taburno.

1. INTRODUCTION

The ever increasing amounts of heavy metals wasted into the biosphere and the consequent accumulation in soils may have far reaching implications for the maintenance of ecosystems due to the toxicity of these elements to essential ecological processes, such as decomposition of organic matter and nutrient cycling. Several literature sources report that high concentrations of heavy metals markedly inhibit soil microbial activity. Microbial biomass in a sewage sludge-amended soil was found to be 50% lower than in adjacent low metal soils [1]. Studies of organic matter decomposition in a gley soil and an andisol showed that soil respiration was inhibited by Cd, Cu, and Pb [2]. The emission over many years of large amounts of Cu, Zn, and other metals as aerosols from a brass foundry in Sweden inhibited conifer litter decomposition in the surrounding forest and caused a reduction of tree growth due to the consequent deficiencies of plant macronutrients [3].

Deposition of heavy metals is not limited to the surroundings of the emitters but is widespread also in remote areas, due to long distance transport of aerosols; thus Wittig and Neite [4] reported an accumulation of Pb in the surface horizon of the soil area affected by the "stem flow" (i.e., the runoff of rain water from the crown down the bark to the ground) in beech woods all over Europe, with the highest level of pollution in the Ruhr area and the lowest level of pollution in the Mediterranean countries. A microcosm experiment demonstrated that oak litter containing high concentrations of Fe, Zn, Cu, Cr, Ni, and Pb, collected from a polluted site inside the city of Naples, showed a significantly lower decomposition and less abundant fungal mycelia as compared to non contaminated litter collected at a control site; moreover, soil respiration and the amount of mycelia correlated negatively with soil Pb, Zn, and Cr concentration [5]. In a study by Berg et al. [6], unpolluted and polluted needle litters of *Pinus sylvestris* were incubated in a heavy metal pollution gradient; decomposition rates were found to be strongly influenced by metal pollution with both litter and soil contributing to the reduced litter decomposition rate.

Although it has been well documented that heavy metals can disturb decomposition processes, knowledge of the dynamics of heavy metals in decomposing litter is relatively scarce. According to Rühling and Tyler [6] and Berg et al. [7], the concentrations of heavy metals increase during decomposition in polluted areas. Laskowski and Berg [8], studying decomposition of Scots pine needles and oak-hornbeam leaves in two unpolluted forests, found that the concentrations of heavy metals in late stages reached levels at or above those reported to retard decomposition. The existing data do not allow the determination of any generalization of the pattern of the dynamics of heavy metals during litter decomposition. In particular, it is not clear how the process is influenced by litter composition and site-related characteristics, such as soil pH and heavy metal content.

The aim of this study was to determine how the concentrations and the absolute amounts of heavy metals of seven litter types, differing in the initial levels of heavy metals, changed during decomposition at two forest soils differing in levels of nutrients, heavy metals, pH, organic matter content, and climate.

2. MATERIALS AND METHODS

Three types of leaf litters (green and brown leaves of *Populus tremula* L., green leaves of *Betula pubescens* Ehrh.) and four types of needle litters (green and brown needles of *Pinus sylvestris* L., brown needles of *Pinus contorta* L. and *Pinus pinea* L.), differing in heavy metal concentrations, were used in this study. Litters were incubated at two sites: 1) a temperate silver fir (*Abies alba* Mill.) forest at an altitude of 1100 m on Monte Taburno (41° 06'N; 14° 36'E, 42 km NE of Naples, Campania), and 2) a boreal Scots pine (*Pinus sylvestris* L.) forest at the village Jädraås in central Sweden (60° 49'N; 16° 30'E, ca 200 km NNE of Stockholm). Both sites are far from heavy metal pollution sources. At high elevation, the average annual precipitation for M. Taburno is 2166 mm, and the long-term average temperature is 7.9°C, based on the records from the Montevergine weather station 20 km south of the site and at the same altitude. The silver fir stand, which is about 100-120 years old and was planted on a soil formerly covered by common beech, has been described in previous papers [9, 10]. The soil, a Typic Hapludand, medial, mesic, formed a deep profile on the pyroclastic parent material. The humic surface horizon (top 5 cm) was relatively rich in essential nutrients, the pH was 5.97 and the C/N ratio was 13 (Table 1).

The Scots pine stand (about 145 years old) was that of the former Swedish Coniferous Forest Project research site Jädraås, located at an altitude of 185 m a.s.l. on a flat area of deep glacifluvial sand sediments. The mean annual precipitation at a nearby village is 609 mm, and the mean annual temperature is 3.8°C. The forest has been described in earlier papers [11, 12]. The soil is a Typic Haplocryod, with a weakly developed A_e horizon (bleached horizon; 2-7 cm) and a typical mor humus. A very loose L horizon (A_{00}), interwoven with living mosses and lichens, covers an F/H horizon (A_{01}-A_{02}) of 5-10 cm. The pH range is 3.9-4.2 in the F/H horizon and 4.6-4.8 in the upper mineral soil. The parent mineral material as well as the whole soil is considered to be very poor in essential nutrients. The C/N ratio of the humic surface horizon (top 5 cm) is 42.3, and the nutrient content is very low (Table 1).

2.1. Litter collection, incubation and mass loss determination

The needle litter of *P. sylvestris* was sampled at Jädraås (Sweden) in the autumn of 1993 from the branches of trees in a stand that was about 25 years old. Brown needles from the needle generation to be shed were taken at abscission from trees growing in an area of about 100 x 100 m. Green needles were sampled from the same area and taken from the second and third needle generations.

Brown leaves of *P. tremula* were sampled at the same site and time as the needles of *P. sylvestris*. Green leaves of *B. pubescens* and *P. tremula* were sampled at the same site in early August. Brown needles of *P. contorta* were sampled in the autumn of 1993 from trees about 25 years old at a research site close to the town of Malung, central Sweden [13]. Brown needles of *P. pinea* from the needle generation to be shed were taken at abscission from the branches of trees in a stand about 50 years old at a research site on Mount Vesuvio close to the village of Terzigno (site description by Virzo De Santo et al. [10]). *P. pinea* needles were sampled in June during the dry period when there was a maximum needle litter fall.

Before weighing, the needles and leaves were air-dried at room temperature to about 5–8% moisture. Dry mass was determined on 25 samples at 85°C and the largest difference in moisture content was less than +/- 0.5% of the average.

Each type of litter was put in separate litter bags and 0.6-1.0 g of litter was enclosed in each bag. Litter bags (8 x 8 cm for needle litter and about 10 x 15 cm for leaf litter) were made of terylene net with a mesh size of about 1 mm. At both sites, the bags were placed on the litter (L) layer in a measurement plot (1 x 1 m) in each of 25 plots in a randomized design. They were fastened to the ground by 10-15 cm–long stainless steel pegs through a 1 cm–wide edge on the bags. The incubations at site M. Taburno were made on May 11, 1994 and those at site Jädraås on November 09, 1994. At Jädraås, samplings took place mainly three times annually and at M. Taburno four times annually. On each sampling occasion, we collected one sample of each litter type from each of the 25 plots. When collected, the 25 samples of each type were transported directly to the laboratory. Plant remains, such as mosses, lichens and lingonberry, were removed after which the loss of dry mass was determined by drying the samples to a constant mass at 85°C. Mean values of mass loss were calculated for each sampling. The 25 samples of each type of litter were combined for chemical analyses.

2.2. Sampling of soil

At both sites, soil was sampled from the humic surface horizon to a depth of ca 5 cm at 8 places within the same area as the bags were incubated.

2.3. Chemical analyses of litter and soil

Soil (2mm mesh) and litter samples were ground in a laboratory mill equipped with a filter allowing particles of less than 1 mm to pass. The samples were digested in a Milestone (mls 1200) Microwave Laboratory System with a mixture of hydrofluoric and nitric acid (HF 50% v/v : HNO_3 65% v/v = 1:2). Element concentrations of digested samples were measured by atomic absorption spectrometry (AAS) (SpectrAA 20 Varian) using standard solutions (STD Analyticals, Carlo Erba) diluted in the same acid matrix for extraction. Fe, Mn, and Zn concentrations in the digestion extracts were measured using flame AAS; Cu, Cd, and Pb concentrations in the digestion extracts were measured by graphite furnace AAS. The extractable fraction of elements in soil was determined as follows: K, Mg, and Ca were measured in 0.4 M $BaCl_2$ extracts at pH 8.1 ± 0.1; Mn, Zn, Fe, Cu, Pb, and Cd were determined in 0.02 M EDTA and 0.5 M $CH_3CO_2NH_4$ extracts at pH 4.65 ± 0.05, according to the method of Lakanen and Erviö [14]. Phosphorus was determined using the method described by Bray and Kurtz [15]. All analyses were carried out on three subsamples.

2.4. Statistics

Correlation coefficients were determined to clarify: 1) in each type of litter the relationship between absolute amounts of metals (as percent of the metal level in fresh litter) and accumulated litter mass loss; 2) the relationship between metal concentration at the beginning of a decomposition period and the rate of decomposition (in $\mu g\ g^{-1} d^{-1}$) during the period; and 3) the relationships among different heavy metals in fresh litter and in decomposing litter after 528/565 and 918/929 days for Taburno/ Jädraås.

3. RESULTS AND DISCUSSION

3.1. Chemical composition of the litter types

As shown in Table 1, the range in nutrient composition among the litter types was large. Leaf litter was generally richer in nutrients as compared to needle litter. Within the same type of litter, brown litter, as compared to green litter, was poorer in N, P, S and K due to retranslocation occurring during leaf senescence; Ca, which accumulates as pectate in cell walls with leaf age, showed instead higher concentrations in brown litters as compared to green ones.

The concentrations of heavy metals in fresh litter are shown in Table 2. Green needles of *P. sylvestris* showed a higher concentration of Cu and a lower concentration of Mn, Fe, and

Pb as compared to brown needles; green leaves of *P. tremula* had lower concentrations of Mn, Zn, and Cd as compared to brown leaves. The differences might be ascribed to the translocation of carbon compounds from the leaf/needle before abscission [16] and/or to the translocation of metals, depending upon their mobility.

Table 1
Litter and soil nutrient concentrations. Values for litter are initial nutrient concentrations (at the start of the decomposition period). Values for soil refer to the total (tc) and to the available (af) nutrient concentration.

Nutrient mg g^{-1} dw	Ca	K	Mg	P	N	S	C
Litter							
P. tremula (g)	8.4	14.2	2.29	2.12	24.2	1.87	503.7
P. tremula (b)	17.1	6.3	2.13	0.63	6.8	1.37	505.9
B. pubescens (g)	9.5	9.0	3.37	1.96	24.3	1.54	483.0
P. contorta (b)	8.7	0.5	1.06	0.29	3.1	0.44	529.7
P. pinea (b)	7.1	5.9	2.40	0.57	3.0	1.36	507.7
P. sylvestris (g)	3.9	5.9	0.79	1.36	12.1	0.81	515.9
P. sylvestris (b)	5.6	0.5	0.34	0.20	3.6	0.44	532.2
Soil (humic surface horizon)							
Taburno tc	19.99	17.7	4.76	2.84	8.5	—	110.2
Taburno af	7.73	0.23	0.23	0.01	—	—	—
Jädraås tc	3.23	10.9	0.98	0.47	10.6	—	448.3
Jädraås af	0.79	0.13	0.06	0.06	—	—	—

g = green leaves and/or needles; b = brown leaves and/or needles.

Manganese concentrations varied by a factor of 10, with *P. contorta* needles having 2030 µg g^{-1} and *P. tremula* leaves and *P. pinea* needles being below 200 µg g^{-1}. The range (100-2030 µg g^{-1}) of Mn concentrations in studied litters (Table 2) was wider than the typical values (40-70 µg g^{-1}) for leaves and needles from relatively unpolluted areas scarcely affected by soil contamination [17].

B. pubescens and *P. tremula* leaves had high Zn concentrations (223-107 µg g^{-1}) whereas pine needles were lower in Zn, with 85-48 µg g^{-1} (Table 2). Zn concentrations in the leaves of *P. tremula* and *B. pubescens* and in the needles of *P. contorta* were somewhat higher than the typical concentrations (30-40 µg g^{-1}) in leaves and needles from relatively unpolluted areas scarcely affected by soil contamination [17]; in the needles of *P. pinea* and *P. sylvestris* (green and brown), Zn concentrations were instead very close to the typical values given by Bargagli [17].

Fe concentration was high in *P. pinea* needles (299 µg g^{-1}), whereas the other litter types ranged between 44 and 79 µg g^{-1} (Table 2). Fe concentrations in all litters except for *P. pinea* were lower than the concentrations (90-100 µg g^{-1}) in leaves or needles from relatively unpolluted areas scarcely affected by soil contamination [17].

Table 2
Litter metal concentrations at the start (Initial), at 528/565 days (m) for Taburno/Jädraås, and at the end (f) of the decomposition period (918/929 days for Taburno/Jädraås).

Metal	*Populus tremula*		*Betula pubescens*	*Pinus contorta*	*Pinus pinea*	*Pinus sylvestris*		Soil (humic surface horizon)	
$\mu g\ g^{-1}$ dw	g	b	g	b	b	g	b	tc	af
Mn									
Initial	100	150	760	2030	190	530	1190		
$Taburno_m$	198	273	597	1400	160	416	660		
$Jädraås_m$	382	357	786	1412	219	673	549		
$Taburno_f$	301	355	713	1242	160	370	599	762	123
$Jädraås_f$	688	802	935	1166	362	663	527	367	167
Zn									
Initial	107	126	223	85	49	49	48		
$Taburno_m$	190	231	363	90	86	144	64		
$Jädraås_m$	171	194	326	103	83	100	92		
$Taburno_f$	161	197	267	141	98	144	166	109	29
$Jädraås_f$	206	237	311	110	99	100	176	59	22
Fe									
Initial	44	46	66	53	299	64	79		
$Taburno_m$	2400	2780	2950	260	420	400	760		
$Jädraås_m$	238	196	330	52.4	310	100	81.3		
$Taburno_f$	4400	4110	4988	1050	770	760	1020	26500	330
$Jädraås_f$	520	490	450	100	427	148	167	9400	500
Cu									
Initial	8.8	8.6	7.1	2.8	5.0	4.7	2.6		
$Taburno_m$	23.2	26.4	30.8	3.9	13.2	12.2	8.2		
$Jädraås_m$	6.6	7.9	6.1	2.6	3.5	3.0	2.8		
$Taburno_f$	34.6	36.6	40.4	13.2	13.6	19.7	16.3	62.6	12.9
$Jädraås_f$	10.6	12.5	7.6	3.8	3.3	4.0	4.3	9.2	1.0
Pb									
Initial	n d	n d	1.0	1.0	3.0	1.0	2.0		
$Taburno_m$	19.9	20.6	34.4	2.6	4.0	5.7	4.8		
$Jädraås_m$	1.5	2.9	1.9	2.4	1.8	1.6	3.0		
$Taburno_f$	43	54.1	63.7	14.9	11.8	14.6	15.2	9.7	4.9
$Jädraås_f$	3.1	4.5	4.5	3.1	2.9	2.6	3.6	8.9	4.3
Cd									
Initial	0.3	0.5	0.4	0.6	0.1	0.1	0.1		
$Taburno_m$	2.4	3.5	1.9	0.5	0.6	1.2	0.3		
$Jädraås_m$	0.9	1.3	1.2	1.2	0.2	0.5	0.4		
$Taburno_f$	1.4	1.5	1.7	1.5	0.7	1.0	1.0	0.9	0.2
$Jädraås_f$	1.0	1.8	1.3	1.1	0.4	0.5	0.5	0.7	0.1

g = green leaves and/or needles; b = brown leaves and/or needles; n d = not detectable. Soil total (tc) and available (af) metal concentrations at the two sites.

Cu concentrations were highest (8.8 and 8.6 μg g^{-1}) in the leaves of *P. tremula*, and lowest (2.8 and 2.6 μg g^{-1}) in the brown needles of *P. contorta* and *P. sylvestris* (Table 2). Both leaf and needle litters showed initial Cu concentrations consistent with the range (4-9 μg g^{-1}) considered typical in leaves and needles from relatively unpolluted areas scarcely affected by soil contamination [17].

Lead was not detectable in the leaves of *P. tremula*, while the needles of *P. pinea* had 3 μg g^{-1} (Table 2). In all litters, Pb concentration was in the range of values (0.8-2.0 μg g^{-1}) considered typical in leaves and needles from relatively unpolluted areas scarcely affected by soil contamination [17].

The needles of *P. contorta* had a relatively high Cd level (0.6 μg g^{-1}); the leaf litters ranged between 0.3 and 0.5, and the needle litters of *P. pinea* and *P. sylvestris* had 0.1 μg g^{-1} (Table 2). Cd concentrations (0.1-0.6 μg g^{-1}) in all litter types (Table 2) were higher than the values (0.05-0.07 μg g^{-1}) given for leaves and needles from relatively unpolluted areas scarcely affected by soil contamination [17].

3.2. Chemical composition of the soil

The chemical composition of the soil from the humic surface horizon (Tables 1 and 2) was very different at the two sites. At M. Taburno, the soil was richer in essential nutrients; however, available P exhibited a lower concentration in the soil of M. Taburno (Table 1), consistent with the high P retention capacity of andic soils.

Total concentrations of Mn and Zn were twice as high at M. Taburno as at Jädraås (Table 2), but their available fractions were similar at the two sites; at both sites, the concentrations of Mn were within the range (200-2000 μg g^{-1}) for mineral soil reported by Allen [18]; as for Zn, even at M. Taburno, the concentration was within the range reported by Allen [18] (20-300 μg g^{-1}) for mineral soils and was consistent with the levels (10-105 μg g^{-1}) considered as background values [19].

Compared with Jädraås, the total concentration of Fe (Table 2) was threefold higher in the soil of M. Taburno, while the available fraction was lower (66%). At both sites, Fe concentrations were in the range of the values for mineral soils (5-100 mg g^{-1}) [18].

Total concentration of Cu in the soil of M. Taburno was six times as high as that of the soil at Jädraås, and the available fraction was one order of magnitude higher (Table 2). The concentration of Cu in the soil of Jädraås was lower than the values of the average Cu range (20-30 μg g^{-1}) for unpolluted soils [20]; at M. Taburno, Cu concentration in the soil was higher than the typical values, as found for other remote sites in Campania, likely due to atmospheric deposition (Virzo De Santo, unpublished results).

Total concentrations of Pb and Cd (Table 2), as well as the available Pb fraction, were very similar at M. Taburno and Jädraås; the available Cd fraction was instead twice as high at M. Taburno. Total Pb concentration was at both sites less than 20 μg g^{-1}, a value considered typical of soils in remote areas [21]. Cd concentrations at both sites were consistent with the expected Cd concentration (< 1μg g^{-1}) in unpolluted soil [22].

3.3. Litter decomposition

At both sites, all leaf litter types had higher initial rates of decomposition than the needles (Figure 1); within the same type of litter, green leaves and/or needles had higher initial rates of decomposition than the nutrient-poorer brown leaves and/or needles (Table 1). After about one year of incubation (352/341 days for Taburno/Jädraås), in spite of the warmer and wetter climate at M. Taburno, the green litter of *P. sylvestris,* as well as the green and the brown litter of *P. tremula,* showed slightly higher accumulated mass losses at Jädraås; the brown litter of *P. sylvestris* and *P. pinea,* as well as the green litter of *B. pubescens,* were slightly more decomposed at M. Taburno. Thus it seems reasonable to conclude that climate is not the most important determinant of decomposition in the early phase in this study.

During the 918 days of incubation at M. Taburno (Figure 1), the litter mass loss reached up to 74% and 73% of the initial mass, respectively, for green and brown leaves of *P. tremula*, 70% for the green leaves of *B. pubescens*, 57% for the needles of *P. contorta*, 54% for *P. pinea* needles, and 65% for green and brown needles of *P. sylvestris*. At Jädraås the decomposition of the seven litter types showed the same pattern, however, the mass loss after 929 days was somewhat lower, i.e., 55% for green and brown leaves of *P. tremula*, 54% for the green leaves of *B. pubescens*, 37% for the needles of *P. contorta*, 43% for *P. pinea* needles, and 58% and 48% for green and brown needles of *P. sylvestris,* respectively. Thus in the late phase, decomposition proceeds further at M. Taburno, likely due to the nutrient richness of the soil; Prescott [23] observed a more complete decomposition of the leaf litter of *Betula papyrifera* Marrsh. incubated on a nutrient-rich soil as compared to the same litter incubated on a nutrient-poor soil.

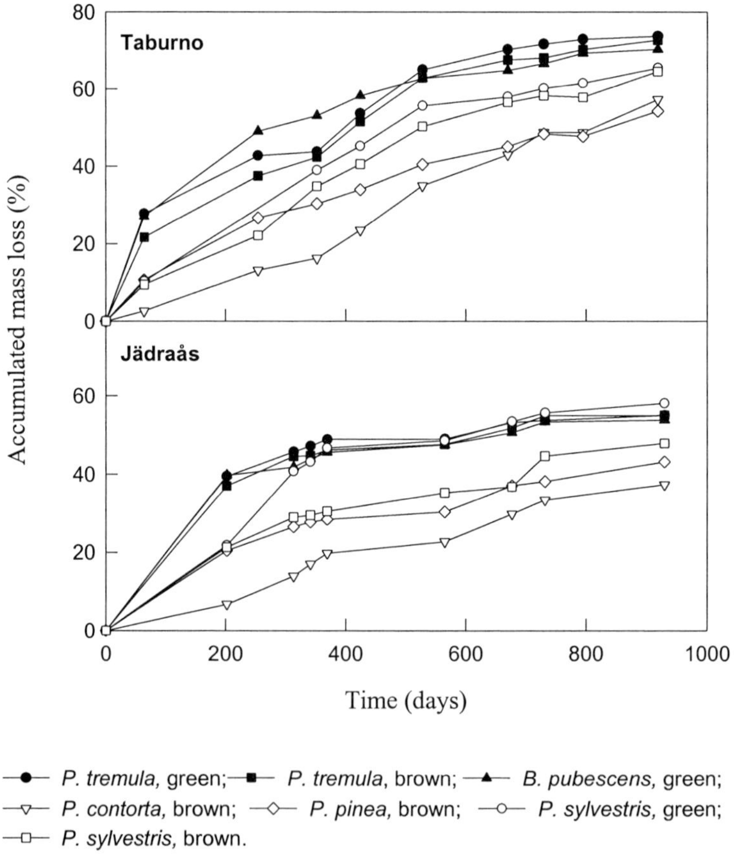

Figure 1. Decomposition of seven litter types incubated in a silver-fir forest (Taburno) and in a Scots pine forest (Jädraås).

3.4. Changes of litter metal concentrations and absolute amounts during decomposition

3.4.1. Manganese

During decomposition, Mn concentrations (Table 2) at both sites decreased in the needles of *P. contorta* and *P. sylvestris* (brown), the two needle litter types richest in manganese; leaf litter of *P. tremula*, the litters poorest in Mn, showed increased Mn concentration at both sites; leaf litter of *B. pubescens* and needle litter of *P. pinea* and *P. sylvestris* (green) increased Mn concentration only at Jädraås.

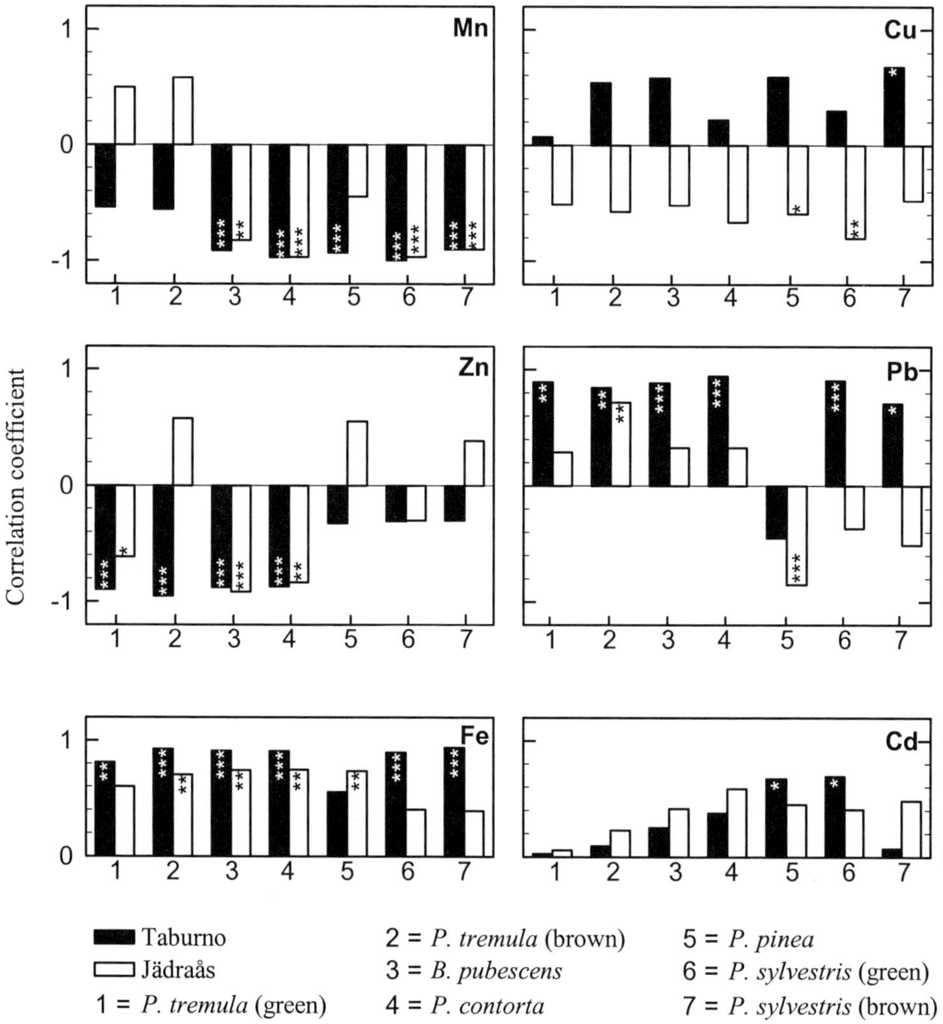

Figure 2. Correlation coefficient (r) for linear regression relating absolute amounts of heavy metals in litter to accumulated mass loss of seven litter types incubated in a silver-fir forest (Taburno) and in Scots pine forest (Jädraås).
Significance level: * $P < 0.05$; ** $P < 0.01$; *** $P < 0.001$.

The absolute amount of Mn decreased during decomposition of all litters at both sites with the exception, at Jädraås, of green and brown litter of *P. tremula,* which tended to increase (not significantly) the absolute amount of Mn (Figure 2). At both sites, the correlations between absolute amounts of Mn and accumulated mass loss of *B. pubescens*, *P. contorta,* and *P. sylvestris* (green and brown) were highly significant and negative, whereas for *P. pinea* the correlation was negative and significant only at M. Taburno.

3.4.2. Zinc

Zn concentration increased during decomposition in all litters and reached values in a range from 99 μg g^{-1} in *P. pinea* to 311 μg g^{-1} in *B. pubescens* (Table 2).

Notwithstanding edaphic and climatic differences between sites, Zn dynamics were similar at the two sites. Changes in absolute amounts of Zn during litter decomposition (Figure 2) showed that Zn was released from all the types of leaf litter (with the exception of *P. tremula*, brown at Jädraås) and from the needle litter of *P. contorta*; for these litter types, the correlation between absolute amounts of Zn and mass loss was significant and negative. No significant change occurred in needle litter of *P. pinea* and *P. sylvestris* (green and brown), which exhibited initial Zn concentrations lower than 50 μg g^{-1}. We might speculate that a critical level of Zn concentration lies around 50 μg g^{-1}; Zn is an essential trace element and under a critical level, it may be suboptimal for decomposers' activity; at higher concentrations, Zn may be toxic.

3.4.3. Iron

At the end of the incubation period, litters decomposed at M. Taburno showed Fe concentrations in the range 760-4988 μg g^{-1}, while litter decomposed at Jädraås showed Fe concentrations in the range 100-520 μg g^{-1}; at M. Taburno, leaf litter exhibited iron concentrations five times higher than those of needle litter.

During decomposition, all litter types increased their Fe content and a significant and positive correlation (Figure 2) was found between absolute amount of Fe and accumulated mass loss with the exception of *P. tremula* (green) and *P. sylvestris* (green and brown) litter at Jädraås and *P. pinea* at M. Taburno.

3.4.4. Copper

During decomposition, Cu concentrations of the different litter types (Table 2) increased at M. Taburno and did not change markedly at Jädraås.

The absolute amounts of Cu in litter (Figure 2) tended to increase at M. Taburno and to decrease at Jädraås; absolute amounts of Cu correlated significantly and positively with decomposition of brown needles of *P. sylvestris* at M. Taburno and negatively and significantly with decomposition of brown needles of *P. pinea* and green needles of *P. sylvestris* at Jädraås.

3.4.5. Lead

At the end of the incubation period, litter decomposed at Jädraås showed Pb concentrations in the range 2.6-4.5 μg g^{-1}, while litter decomposed at M. Taburno showed Pb concentrations in the range 11.8-63.7 μg g^{-1} (Table 2); at M. Taburno, leaf litter exhibited Pb concentrations three times higher than that of needle litter; in contrast, at Jädraås, Pb concentrations were similar in leaf and needle litter. During decomposition (Figure 2) at M. Taburno, the absolute amount of Pb increased significantly in all litter types with the exception of *P. pinea* (the Pb richest litter). At Jädraås, Pb tended to be released from decaying needle litter (with the exception of *P. contorta*) and accumulated in leaf litter; still, a significant correlation between absolute amount of Pb and litter decomposition (Figure 2) was found only for *P. pinea* and for *P. tremula* (brown).

3.4.6. Cadmium

During decomposition, concentrations of Cd (Table 2) increased in all litter types. Absolute amounts of Cd (Figure 2) tended to increase with decomposition, although a positive and significant correlation with decomposition was found only at M. Taburno for *P. pinea* and *P. sylvestris* (green), which showed the lowest initial Cd concentration.

3.5. Relationship between heavy metals concentration and litter decomposition

To understand how the concentrations of heavy metals influence litter decomposition, the correlation coefficients were determined between litter metal concentrations at the start of the period and decomposition rates ($\mu g\ g^{-1}d^{-1}$) in the early phase (0 up to 528/565 days, respectively, at the M. Taburno/Jädraås sites) and in late phase (528/565 up to 918/929 days at M. Taburno/Jädraås).

In the early phase, the rate of decomposition was positively and significantly correlated to the initial Cu concentration (Table 3). Cu is one of the most important essential micronutrients. Unexpectedly, the initial Cd concentrations also showed a positive and significant correlation with the rates of decomposition in the early phase (Table 3). According to Bowen [24], Cd may be essential for animals at very low concentrations, but its possible role in microorganisms is unknown. The inhibitory effect of Cd at relatively high concentration is instead well known [22].

Table 3

Correlation coefficients (r) for linear regressions relating heavy metal concentrations in seven litter types, at the start of two decomposition periods (early phase and late phase), and rates of decomposition in the periods. Level of significance: * $P < 0.05$; ** $P < 0.01$.

	Early phase (0 to 528/565 days)	Late phase (528/565 to 918/929 days)
Mn	-0.52	+0.68**
Zn	+0.39	-0.61*
Fe	-0.36	-0.46
Cu	+0.62*	-0.55*
Pb	-0.50	-0.42
Cd	+0.68**	-0.45

In the late phase, decomposition rates were significantly and negatively correlated to litter Cu concentrations at the start of the period (Table 3). The increase of Cu concentrations in decaying litter could thus reach levels at which Cu is toxic to decomposers (the toxicity of Cu to fungi is well known and its use as fungicide is widespread). A significant and negative correlation was also found between litter Zn concentrations at the start of the late phase and decomposition rates in the period (Table 3). On the contrary, litter Mn concentrations at the start of the late phase were significantly and positively correlated to the rates of decomposition in the period (Table 3); this finding was consistent with the role of Mn in lignin degradation and the consequent control of decomposition in late phase. The rates of decomposition, both in the early and the late phase, decreased with the increase of Fe and Pb concentrations, but the correlation was not statistically significant (Table 3).

Berg and Ekbohm [25] found that decomposition of *P. sylvestris* needle litter decreased at sites with concentrations of heavy metals in fresh litter in the following ranges in $\mu g\ g^{-1}$: Zn, 110-250; Fe, 140-380; Cu, 20-100; Pb, 44-311; Cd, 0.6-1.2. After about 900 days of decomposition at M. Taburno, we have found in all litter types Fe concentrations exceeding these levels; moreover, the concentrations of Cu in leaf litter exceeded and in needle litter approached the lower limit of the range given above. At Jädraås, Fe concentrations reached the values of the given range in all litter types, with the exception of *P. contorta.* Zn concentrations in all litter types, with the exception of *P. pinea* and *P. sylvestris* (green), and Cd concentrations in leaf litter and in *P. contorta*, exceeded the limits given by Berg and Ekbohm [25]. Actually, decomposition rates in the late phase were negatively correlated with Zn, Fe, Cu, Pb, and Cd concentrations in litter after 528/565 days of incubation (Table 3); the correlation was significant for Zn and Cu and could explain 38% and 30%, respectively, of the decrease in decomposition. Thus, together with factors such as accumulation of recalcitrant substances in litter, the high heavy metal concentrations in litter at late stages might be responsible for some suppression of litter decomposition in areas not exposed to severe environmental pollution.

3.6. Dynamics of heavy metals during decomposition and their relationship with soil heavy metal content and site characteristics

During decomposition, a net accumulation of some metals and a net release of other metals may occur (Figure 2). Fe and Cd showed a net accumulation in decomposing litter at both sites (Figure 2). Fresh litter was poorer in Fe and Cd than the soil at the incubation sites (Table 2); the transport by fungal mycelium from soil to litter is likely to contribute to the accumulation of Fe as decomposition proceeds, as suggested by the highly significant correlation between amounts of Fe and accumulated mass loss in all litter types when incubated at M. Taburno and in most of them when incubated at Jädraås. As for Cd, the lack of a correlation between net amount and accumulated mass loss, with the exception of *P. pinea* and *P. sylvestris* (green) at M. Taburno, suggests that the accumulation of Cd in litter is most likely supported by physico-chemical processes, such as a higher capacity of Cd fixation by partially decomposed litter than by more highly decomposed soil organic matter. As decomposition of litter proceeded, Mn and Zn clearly decreased in amount at both sites, with a few exceptions at Jädraås (Figure 2). It is noteworthy that available Mn and Zn concentrations in soil were similar or lower than Mn and Zn concentrations in litters (Table 2) and that the release was more significant when the gradient litter-soil was larger. We may speculate that more decomposed soil organic matter is able to bind Mn and Zn more strongly than the components of decaying litters.

Changes in amounts of Cu showed a contrasting pattern at the two sites, i.e., all litter types accumulated Cu at M. Taburno and released Cu at Jädraås (Figure 2). The two sites showed marked differences in Cu content of the soil; at M. Taburno, the level of total Cu was in fact sixfold higher and that of the available fraction one order of magnitude higher than that at Jädraås (Table 2). The results suggest that litter Cu dynamics are influenced by Cu concentration in the soil with accumulation in or release from litter, depending on high or low Cu level in the soil, respectively. The low pH at Jädraås could also promote Cu release from litter. The lack of a significant correlation with accumulated mass loss in most litters suggests that Cu exchange between litter and soil is controlled by physico-chemical rather than biological processes.

During decomposition at M. Taburno, Pb showed a significant accumulation in all litters except *P. pinea*, while at Jädraås, all types of needle litter showed a clear release, which was highly significant for *P. pinea* (Figure 2). *P. pinea* was the litter with the highest initial content of Pb (Table 2) and the only one showing a net release at both sites; this litter also had the lowest level of Pb concentration at the end of the decomposition period. We may hypothesize that decaying *P. pinea* litter has a lower capacity to bind Pb as compared to the other litters. The level of Pb in the soil was about the same at the two sites (Table 2), thus the different behavior may not be ascribed to soil Pb content. We might speculate about the occurrence of different organic substrates, generated by a different pattern of decomposition at the two climatically contrasting sites, having different abilities to bind Pb. More likely, the differences in Pb concentrations may be ascribed to differences in atmospheric deposition between M. Taburno and Jädraås. The higher increases at M. Taburno of Pb (as well as of Fe and Cu) in leaf litters (with larger interface substrate/air) than in needle litters strengthen this hypothesis. Moreover, we have observed at M. Taburno trunk base phenomena suggesting that atmospheric deposition occurs at this site (Virzo De Santo, unpublished results).

When analyzing the correlation among different heavy metals, it was found that in fresh litter, before incubation, the concentrations of Cu and Mn were significantly and negatively correlated, whereas the concentrations of Pb and Fe were significantly and positively correlated (Figure 3). After 528 and 918 days of incubation at M. Taburno, a significant correlation occurred among most of the heavy metals (Figure 3); in contrast, at Jädraås, only at the end of the incubation (929 days) a significant correlation was found between Pb and Zn and between Pb and Cd, as well as between Cu and Cd. The significant correlation among heavy metals suggests that they have common sources and/or similar dynamics. Thus the highly significant correlation among Pb, Fe, and Cu occurring only in the litter incubated at M. Taburno suggests that their increases in concentration may arise at least partly from atmospheric deposition. On the other hand, the correlation between Pb and Zn, and Pb and Cd found at both sites may reflect mainly the capacity of more decomposed litters to bind the metals. Actually, both Zn and Cd concentrations at the end of the incubation were very similar at M. Taburno and Jädraås.

4. CONCLUSIONS

Decomposition of leaf litter of *P. tremula* and *B. pubescens* and needle litter of *P. contorta, P. pinea* and *P. sylvestris* during a period of about 900 days showed similar patterns at the two incubation sites but proceeded further at the M. Taburno site, which was characterized by a higher annual mean temperature, a soil richer in nutrients and a higher pH as compared to the Jädraås site. In the early phase, decomposition significantly increased with the level of Cu, an important micronutrient, and unexpectedly, also with the level of Cd, for which no role in decomposer metabolism is known. In the late phase, decomposition was significantly and positively correlated to litter Mn concentrations, consistent with the role of

Fresh litter

Mn						
Zn	-0.082					
Fe	-0.276	-0.362				
Cu	-0.760*	0.576	-0.200			
Pb	0.136	-0.457	0.840*	-0.636		
Cd	0.380	0.555	-0.495	0.253	0.599	
	Mn	Zn	Fe	Cu	Pb	Cd

Taburno (528 days)

Mn						
Zn	-0.217					
Fe	-0.375	0.868*				
Cu	-0.548	0.912**	0.947**			
Pb	-0.282	0.968***	0.947**	0.947**		
Cd	-0.478	0.682	0.854*	0.825*	0.711	
	Mn	Zn	Fe	Cu	Pb	Cd

Jädraås (565 days)

Mn						
Zn	0.017					
Fe	-0.539	0.600				
Cu	-0.425	0.706	0.557			
Pb	0.112	-0.109	-0.439	0.005		
Cd	0.453	0.659	0.025	0.607	0.568	
	Mn	Zn	Fe	Cu	Pb	Cd

Taburno (918 days)

Mn						
Zn	0.217					
Fe	-0.112	0.783*				
Cu	-0.208	0.812*	0.974***			
Pb	-0.080	0.862*	0.975***	0.977***		
Cd	0.521	0.777*	0.763*	0.709	0.778*	
	Mn	Zn	Fe	Cu	Pb	Cd

Jädraås (929 days)

Mn						
Zn	0.277					
Fe	-0.217	0.606				
Cu	0.186	0.699	0.741			
Pb	0.315	0.882**	0.463	0.619		
Cd	0.639	0.675	0.456	0.803*	0.771*	
	Mn	Zn	Fe	Cu	Pb	Cd

Figure 3. Correlation matrix among heavy metals in fresh litter and in decomposing litter after 528/565 and 918/929 days of incubation at the Taburno/Jädraås sites.
$^*P < 0.05$; $^{**}P < 0.01$; $^{***}P < 0.001$.

Mn in lignin degradation; on the contrary, decomposition decreased significantly with the increase of Zn and Cu concentrations, indicating that the high concentrations of heavy metals in litter at late stages might be responsible for some suppression of litter decomposition also in areas not exposed to severe environmental pollution.

Concentrations of Zn, Fe, Pb and Cd increased in all litter types at both sites, whereas Cu increased in all litter types only at M. Taburno. The absolute amounts of Fe and Cd in decomposing litters increased at both sites. All litters accumulated Cu at M. Taburno and released Cu at Jädraås. All litter types except *P. pinea* accumulated Pb at M. Taburno; at Jädraås, Pb was accumulated by leaf litter and released from needle litter. The results indicate that heavy metal accumulation or release may depend on the gradient of metal concentration between litter and soil, as well as on the capacity of litter to bind metal. Atmospheric deposition could account at least partly for the increase of absolute amounts of Pb, Fe, and Cu in litter at M. Taburno; the significant correlation among these heavy metals, indicating that they may have a common source, supports the hypothesis.

ACKNOWLEDGMENTS

Financial support by the Consiglio Nazionale delle Ricerche to A. Virzo De Santo (9704310.CT04 and 98.00530.CT04) and to B. Berg for a short-term mobility grant (1999). Financial support by European Union (EU project QLRT-2000-00596) and by German Ministry for Education, Science, Research and Technology (BMBF, Grant No PT BEO-51-0339476) to B. Berg, while working as a guest scientist at BITÖK, University of Bayreuth. Mrs. Sirkka Koivuoja and Mrs. Sylva Ramström are gratefully acknowledged for preparing and cleaning all the litter bags.

REFERENCES

1. Brookes, P.C., McGrath, S.P., 1984. Effects of metal toxicity on the size of the soil microbial biomass. J. Soil Sci. 35, 269-279.
2. Hattori, H., 1992. Influence of heavy metals on soil microbial activities. Soil Sci. Plant Nutr. 38, 93-100.
3. Tyler, G., Balsberg-Pålsson, A.M., Bengtsson, G., Bååth, E., Tranvik, L., 1989. Heavy-metal ecology of terrestrial plants, microorganisms and invertebrates. Water Air Soil Pollut. 47, 189-215.
4. Wittig, R., Neite, H., 1989. Distribution of lead in the soils of *Fagus sylvatica* forests in Europe. *In*: Öztürk, M.A. (Ed.), Plants and Pollutants in Developed and Developing Countries. Ege University Izmir, pp. 199-206.
5. Cotrufo, M.F., Virzo De Santo, A., Alfani, A., Bartoli, G., De Cristofaro, A., 1995. Effects of urban heavy metal pollution on organic matter decomposition in *Quercus ilex* L. woods. Environ. Pollut. 89, 81-87.
6. Rühling, A., Tyler, G., 1973. Heavy metal pollution and decomposition of spruce needle litter. Oikos 24, 402-416.
7. Berg, B., Ekbohm, G., Söderström, B., Staaf, H., 1991. Reduction of decomposition rates of Scots pine needle litter due to heavy metal pollution. Water Air Soil Pollut. 69, 165-177.
8. Laskowski, R., Berg, B., 1993. Dynamics of some mineral nutrients and heavy metals in decomposing forest litter. Scand. J. For. Res. 8, 446-456.
9. Caputo, G., 1966-1967. Ricerche sulla vegetazione forestale del gruppo del Taburno Camposauro (Appenino Campano). Delpinoa, 8-9, 91-128.
10. Virzo De Santo, A., Berg, B., Rutigliano, F.A., Alfani, A., Fioretto, A., 1993. Factors regulating early-stage decomposition of needle litters in five different coniferous forests. Soil Biol. Biochem. 25, 1423-1433.
11. Berg, B., Wessen, B., Ekbohm, G., 1982. Nitrogen level and decomposition in Scotspine needle litter. Oikos 38, 291-296.
12. Berg, B., Ekbohm, G., McClaugherty, C., 1984. Lignin and holocellulose relations during long-term decomposition of some forest litters. Long-term decomposition in a Scots pine forest IV. Can. J. Bot. 62, 2540-2550.
13. Berg, B., Lundmark, J.-E., 1987. Decomposition of needle litter in lodgepole pine and Scots pine monocultural systems - a comparison. Scand. J. For. Res. 2, 3-12.
14. Lakanen, E., Erviö, R., 1971. A comparison of eight extractants for the determination of plant available micronutrients in soils. Suom. Maataloustiet. Seuran Julk. 123, 232-233.

15. Bray, R.H., Kurtz, L.T., 1945. Determination of total, organic and available form of phosphorus in soils. Soil Sci. 59, 39-45.
16. Staaf, H. 1982. Plant nutrient changes in beech leaves during senescence as influenced by site characteristics. Acta Oecol 3, 161-170.
17. Bargagli, R., 1998. Trace elements in Terrestrial Plants. An Ecophysiological Approach to Biomonitoring and Biorecovery. Springer-Verlag, Berlin, Heidelberg, New York.
18. Allen, S.E., 1989. Chemical analysis of ecological materials. Blackwell Scientific Publications.
19. Angelone, M., Bini, C., 1992. Trace elements concentrations in soils and plants of Western Europe. *In*: Adriano, D.C. (Ed.), Biogeochemistry of trace metals. Lewis Publishers, Boca Raton, FL, pp. 19-60.
20. Baker, D.E., Senft, J.P., 1995. Copper. *In*: Alloway, B.J. (Ed.), Heavy metals in soils. Blackie Academic and Professional, Chapman & Hall, pp. 179-205.
21. Davies, B.E., 1995. Lead. *In*: Alloway, B.J. (Ed.), Heavy metals in soils. Blackie Academic and Professional, Chapman & Hall, pp. 206-223.
22. Alloway, B.J., 1995. Heavy Metals in Soils. Blackie Academic and Professional, Chapman & Hall.
23. Prescott, C.E., 1996. Influence of forest floor type on rates of litter decomposition in microcosms. Soil Biol. Biochem. 28, 1319-1325.
24. Bowen, H.J.M., 1979. Environmental chemistry of the elements. Academic Press, London.
25. Berg, B., Ekbohm, G., 1991. Litter mass-loss rates and decomposition patterns in some needle and leaf litter types. Long-term decomposition in a Scots pine forest VII. Can. J. Bot. 69, 1449-1456.

Developments in Soil Science, Volume 28A
Editors: A. Violante, P.M. Huang, J.-M. Bollag and L. Gianfreda

DIRECT AND INDIRECT EFFECTS OF ORGANIC MATTER ON METAL IMMOBILISATION IN SOIL

S. Staunton

Unité Sol & Environnement, pl Viala, 34060 Montpellier Cedex, France

The mechanisms by which soil organic matter influences the immobilisation of trace metals in soils are reviewed. The techniques employed to elucidate these interactions and their limitations and shortcomings are discussed. The experimental approaches are classified as follows: interpretation of pedological data; direct solid phase analysis; solution complexation; selective chemical extraction; soil adsorption properties; addition of organic matter to soil; and adsorption properties of synthetic soils. References are given to both review articles and some representative experimental studies.

1. INTRODUCTION

Both natural sources and human activity result in the presence of numerous metals in soils. Comprehensive review of the sources, levels and fate of many heavy metals in soils are given by Ross [1] and Alloway [2]. Whether they are essential nutrients or toxic elements depends not only on their biological role, but often also on their concentration. Their mobility in soil systems determines to a large extent their assimilation by biological systems and their transfer to surface and ground waters. This depends on their chemical form and their association with soil components, namely their speciation. Metal ions are taken up by biological systems from the solution phase, and in this phase may be transported by mass flow both short distances, e.g. towards roots, or greater distances towards aquifers. The lability of ions associated with the solid phase determines extent of diffusive flux and the ease and rapidity with which a depleted solution phase is replenished. A good knowledge of their speciation is therefore necessary to predict, and if possible to control, their long-term behaviour. This is true whether the aim is to increase or to limit the transfer to food chains, or to minimise the contamination of surface and ground waters.

Organic matter, along with other soil components, particularly clay minerals and the oxides of Fe and Mn [3], plays an important role in the speciation of many metals in soil. Harter & Naidu [4] have reviewed some of the interactions of metals with soil organic matter. The effect of organic matter may be by direct interaction due to their adsorption properties and complexation capacity, or indirectly because of the modification of mineral surface properties with organo-mineral complexes are formed. Given the complexity and diversity of the chemistry of the metals and the multi-component, tri-phasic and heterogeneous nature of soil, a complete review of the effects of organic matter on metal speciation in soil is impossible. The aim of this review is to give an overview of metal-organic matter interactions in soil and some of the methods available to investigate them.

I shall consider metal speciation in solution, the distribution of metal ions between the solid and liquid phases of soil and the associations between metal ions and the soil solid phase. I shall concentrate on the fairly direct effect of organic matter on metal speciation, and give only brief mention of some of the indirect effects of organic matter which may modify soil pH and redox conditions [5] and which, being a substrate for microbes, may in turn modify the oxidation state and hence the adsorption properties of mineral phases [6, 7]. Volatile metal complexes will not be discussed. The various methods currently available to either infer or to measure directly metal-organic matter interactions will be presented and the major advantages and disadvantages of each discussed. Where possible references to more detailed reviews are given, and numerous representative examples of works using these techniques and approaches cited.

2. ROLES OF SOIL ORGANIC MATTER IN METAL SPECIATION

There is a tremendous variety of organic matter in soils. It is therefore impossible to catalogue all the ways in which it can influence the fate of metals in soil. However, simple distinctions can be made at two levels. Firstly, the effect may be direct or indirect and secondly the organic matter may be in the dissolved, solution phase or in the solid phase, either associated with a mineral fraction or not. Much of the study of the effect of soil organic matter on metals has concentrated on direct effects such as adsorption and complex formation. Such effects are often easier to investigate experimentally than indirect effects, which include the modification of pH and redox conditions and the dissolution or weathering of mineral phases. Figure 1 is a schematic representation of the possible effects of organic matter on the immobilisation of metals in soils.

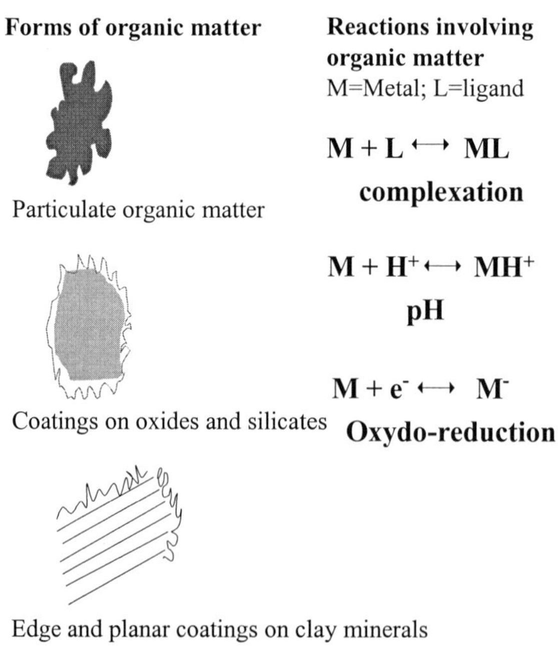

Figure 1. Schematic representation of the interaction of organic matter with other soil constituents, and the physico-chemical factors that it influences, thereby modifying the speciation of trace metals.

Sometimes the distinction between direct and indirect effects is difficult to make and may not be useful. An example of this is when organic coatings modify the surface properties of clay minerals. The effect can be considered to be indirect only if it can be clearly demonstrated that there is no chemical interaction between the metal and the organic component.

The affinity between organic molecules and metals can be predicted or understood from the Hard-Soft Acid-Base Theory (HSAB). This concept, formulated by Pearson [8, 9], is based on the premise that Lewis acids and bases can be classified according to their hardness or softness, and that like attracts like. Sullivan, Sposito, Xu & Harsh and Harter & Naidu [10-13, 4] give fuller descriptions of the theory. In simple terms, softness is related to the polarisability of an atom or molecule. Thus a soft, polarisable acid, such as lead or mercury should be associated with organic ligands, especially sulphur groups, whereas a hard acid such as caesium or manganese (II) will be attracted to hard mineral surfaces and oxygen bearing ligands. Although many transition metals of interest are classified as intermediate (Fe^{2+}, Co^{2+}, Ni^{2+}, Cu^{2+} and Zn^{2+}), HSAB theory may help explain some of the trends in affinity of soil components for various metals).

2.1. Soluble organic matter

A fraction of soil organic matter is present in the solution phase. The proportion in solution depends not only on the nature of the organic matter, but also upon other soil properties, and can easily be modified by changing pH and redox conditions. It is therefore sensitive to inorganic amendments including liming [e.g. 14, 15]. This dissolved organic matter (DOM) consists of both simple monomers and complex macromolecules and polymers. The formation and stability of metal-organic complexes in solution has been extensively studied, both experimentally and by mathematical modelling. The Le Chatelier principle predicts that the formation of metal complexes in solution should lead to the desorption of adsorbed metals, to counterbalance the decrease in concentration of free metal in solution. Most emphasis has been placed on humic and fulvic acids as they are thought to be better characterised than water-soluble organic matter in general. Smaller molecules, including monomers, are easy to study and some have considerable chelating power. However because of their biological degradation and their adsorption, they often have a short lifetime in soil [16], particularly in soil solution, and so their role has been studied to a much lesser extent. Stability constants of complexes involving simple organic ligands and many metals are available in standard textbooks, and have been included in geochemical models such as GEOCHEM and SOILCHEM [17,18].

Indirect effects of DOM on metal speciation are difficult to deduce from traditional investigations. While it is well known that organic matter (both in the solution and solid phases) influence soil pH, which in turn plays an important role in metal speciation, many studies do not emphasise this role. The weathering and dissolution of solid phases that adsorb metals is another possible indirect effect that receives inadequate attention. Although SOM has been implicated in the weathering of soil minerals, some studies do not adequately separate pH effects from the action of the conjugate bases of acidic organic matter. Similarly there is evidence that even small amounts of organic anions such as citrate may solubilise iron oxides [19, 20]. This could lead to a reduction in the adsorption of some metals.

2.2. Solid phase organic matter

As briefly outlined above, the direct association between SOM and metals in the solid phase can be understood from HSAB theory. The complexity of soil solids makes the study of metal speciation and the interpretation of data more difficult than for the solution phase. Organic matter may be present in particulate form, or as coatings on mineral surfaces. Coatings are certainly chemically heterogeneous. There has been some suggestion that they are spatially incomplete [21], but since microscopic techniques do not detect small molecules such observations are liable to erroneous interpretation. Metals may form bridging complexes with organic coatings. In addition, metal-organic ligand complexes, formed in solution are also liable to be adsorbed by soil surfaces.

The definition of exchange reactions involving organic substances and metals is problematic. If DOM displaces adsorbed metal from the soil exchange complex, there is no interaction between the metal and organic matter. Furthermore no analysis of the soil can show that such an exchange took place. The organic matter is no longer dissolved, and so correlations between DOM and metal content will not provide the necessary indications. Such simple exchange reactions may take place between anionic organic matter and oxyanions of metals and metalloids (e.g. MnO_4^-, TcO_4^-, AsO_3^-). While this may also occur for cationic species or involve neutral organic matter, it would be difficult to distinguish the result from the consequences of the formation of organic coatings on mineral surfaces that modify their surface properties.

Since the soil solid phase is more heterogeneous that the solution phase, it is to be expected that the indirect effects of SOM on metal speciation are also more complicated. Complex formation and adsorption phenomena are to varying degrees pH-dependent, thus any change in soil pH induced by SOM will affect them. Another effect is the role of SOM on aggregation. Strictly speaking the aggregation of soil particles does not imply any chemical change, so the nature and strength of interaction should not be affected. However, aggregation may modify the rate of access to adsorption sites thereby modifying the kinetics of apparent reactions, some data may be interpreted as a direct effect on SOM on metal speciation.

3. METHODS OF INVESTIGATION OF THE ROLE OF SOM IN METAL SPECIATION

3.1. Interpretation of pedological data and soil composition

A limited amount of information on metal speciation is available from observations of their distribution in relation to that of other soil constituents. One example is the estimation of the solubilising effect of organic matter by correlating organic matter content (total or dissolved) with either the proportion of a metal in soil solution or a soil-plant transfer factor. This approach requires a large number of soils, preferably with similar metal loadings. Another possibility is to study the vertical distribution of metals in one, or a number of, soil profiles and to compare this with the chemical composition in each layer. Pollutant metals have often been deposited at the soil surface. Movement down the soil profile indicates mobility and therefore weak association with the immobile solid phase. Accumulation in a horizon reflects association with the solid phase.

Wilcke et al. [22] found that while Cr, Ni and Zn migrated to B horizons of an Alpine Podzol, Cu, Cd and Pb remained in the surface horizon that had a greater organic carbon content. Rahman et al. [23] found evidence of a strong association between Cu and humus in

Japanese Andisols, but no such relationship for Zn, Co and Ni. Egli et al [24] have followed the distribution of Pb, Cd and Zn over time in a forest soil. They concluded that the behaviour of major and minor chemical components were largely determined by the decomposition of organic matter. Changes in Pb content correlated strongly with organic matter, a somewhat lesser effect was found for Zn, whereas Cd showed almost no effect of organic matter. Sterckeman et al. [25] report that the mobility of Zn in soil profiles increases with organic carbon in the surface horizon, but for Cd and Pb mobility seemed to depend only on pH and sand content. Balabane et al. [26] have emphasised the importance of the effect of heavy metals on organic matter dynamics. This in turn influences the decomposition of metallophyte-derived organic debris and the subsequent time-dependent vertical distribution of heavy metals within the soil profile.

Interpretation of such observations can be rather subjective. For example, metal depletion and high organic matter content in a layer suggests the solubilising effect of SOM, whereas metal accumulation in an organic layer could be interpreted as a strong association between the metal and organic matter in the solid phase. Both may be correct. After a detailed study of Pb in surface soils, and particularly in soil solution, Wang & Benoit [27] demonstrate that non causal correlations may exist between DOC and dissolved Pb. Usually somewhat more information is gleaned by comparing the distribution patterns of the various metals present in the same system, given their contrasting chemical properties. Another refinement of this approach is to determine the physical distribution of each metal within a layer by size and/or density fractionation [28-29]. Light fractions are usually associated with high organic matter contents [30], and so a relatively large metal concentration in these fractions suggests interaction with organic matter. Likewise a large metal concentration in the clay-sized fraction should suggest the possibility that some vertical transport has occurred by particle movement, and so refine the conclusions of the observations of vertical distribution.

Ultimately, *in situ* studies can only give information that integrates the entire range of phenomena that determine metal mobility and so it is difficult to establish which processes are determinant. However, in favour of in situ studies, it must be recalled that information on indirect effects of organic matter on metal mobility, such as aggregation, are lost when soil is sampled and homogenised prior to laboratory based experiments. On the other hand, it is impossible to distinguish unambiguously direct and indirect effects without complementary controlled studies.

An important drawback to the approaches described in this sub-section is that they tend to be labour intensive and therefore the number of sites that can be investigated and the analyses that are performed are necessarily limited. More importantly, as has been indicated above the interpretation of such observations is very difficult, and strong evidence of organic matter-metal association is never obtained. Misinterpretation is possible unless data are assessed in the light of a good knowledge of the chemistry of each metal. For example, despite the fact that caesium has little or no tendency to associate with organic matter, the mobility of radiocaesium in organic soils has been taken as an indication of direct interaction between this alkali metal and SOM [31, 32]. In fact the reasons are complex, and lie both in the absence of strongly adsorbing minerals in these soils and the modification of mineral surface properties in the presence of organic matter.

3.2. Direct solid phase analysis

Few methods exist that can directly probe the speciation of metals associated with soil surfaces and they are often of even more limited application to the study of the role of organic matter. Among the methods that have been used are electron microbe analysis and X-ray techniques, such as X-ray Absorption Fine Structure Spectroscopy (XAFS). The former used relatively easily available technology, whereas the latter, and related techniques, require highly specialised equipment and considerable expertise. Microprobe analysis involves the mapping of the surface for various elements and comparison of elemental distributions to infer chemical association from physical proximity. A quarter of a century ago it was used with success to investigate the relationships between heavy metals and both iron and manganese [33]. More recently Hiller & Brummer [34, 35] have used microprobe analysis to show that both humic substances and charcoals bind heavy metals (Cu, Cd, Ni, Pb, Zn and Co). Unfortunately organic components are difficult to identify, particularly if they have low molecular weight. The artefacts associated with sample preparation necessary for the detection of organic compounds and the non visibility of low molecular weight compounds are severe drawbacks to this approach.

Fendorf et al. [36] and Schultz & Bertsch [37] have given excellent reviews of X-ray techniques. These techniques allow the local chemical environment of a species to be determined, giving information on nearest neighbours, oxidation state and bond distances. In principal, these are potentially powerful techniques, capable of providing useful information in a non invasive manner. Speciation is assessed by comparing spectra to typical fingerprint spectra obtained with known compounds, and is therefore limited by the availability of such standards.

XAFS and XANES (X-ray Absorption Near-Edge Spectroscopy) have been combined with classical selective extraction technique to elucidate metal speciation [for example, 38-40]. Strawn & Sparks [41] have recently shown that the nearest neighbours of Pb in a contaminated silt-loam soil were O and C, but after chemical removal of SOM with hypochlorite, Pb was associated with O, Si and other Pb atoms. In a few, rather extreme cases, the association of organic matter with metals in heavily contaminated soils (or soil-like materials) has been identified. For example, Manceau and co-workers [42, 43] have demonstrated that lead is associated with organic matter in heavily contaminated soils near smelters. However Morin et al. [44, 45] found rather weak evidence of association between Pb and organic matter in natural.ly Pb-rich and industrial.ly contaminated soils. Xia et al. [46] have used XFAS to demonstrate the complexation of Hg with reduced S functional groups along with carboxyl and phenol groups of humic acid. Hesterberg et al. [47] have used XAFS to demonstrate the association of mercury with humic acid adsorbed on goethite. The XAFS study of Welter et al. [48] did not show evidence of association of Pb with organic matter, but they reported the presence of C, along with O and Si in discrete Pb-containing particles using X-ray energy spectroscopy.

The major drawbacks are that the detection limit is such that large concentrations of the metals must be present and only its average environment is probed, thus only major chemical forms can be detected. Despite these reservations, the increase in number and accessibility of synchrotron facilities and inevitable technical progress will undoubtedly make these techniques more applicable to problems in soil science.

Some information on the nature of the bonding between organic matter and trace metals may also be obtained from spectroscopic analysis, mainly spectrophotometry, infrared spectroscopy (IR), electron spin resonance (ESR) and nuclear magnetic resonance (NMR).

Interference from soil mineral components containing iron make these techniques unsuitable for whole soil samples, and have generally been restricted to the study of humic substances. Senesi and Ross [49, 1] have reviewed such techniques. Using these techniques it may be shown whether complexes are inner or outer sphere, the former involving some degree of covalent bonding [50]. It has also been confirmed that it is mainly the –COOH and phenolic –OH groups of humic substances that are responsible for metal complex formation, as expected from modelling studies [51]. In sewage sludges, however, the large proportion of protein materials lead to the significant involvement of amide-N and amide-O groups [52]. As Cheshire et al. [53] point out, the extraction of soil organic matter may break some of the metal-organic bonds, and so in many cases metals are added to extracted organic materials. Although these techniques do not require very sophisticated equipment, they are relatively little used because of the limited information available, the difficulties in interpreting spectra and the fact that extracted, purified organic matter, no-longer associated with the mineral matrix must be used.

3.3. Solution complexation

In many cases complexation between metals and dissolved organic matter (DOM) is inferred from the increased solubility of metals on addition of simple organic compounds, DOM or purified soluble organic matter such as fulvic acid or soluble humic acid. In natural waters there is considerable competition between organic and inorganic ligands, particularly carbonate for hard water [54]. Spectroscopic data [e.g. 55] and size exclusion separation [e.g. 56] may allow the solution phase speciation to be deduced. A comparison of total solution concentration using Atomic Absorption Spectroscopy or ICP and the free ion solution concentration measured using selective electrodes may allow the degree of complexation to be deduced [e.g. 57]. De Oliveria et al. [58] claim that direct measurement of free metal ion concentration is not always necessary. However many studies of binary metal-organic complexation, usually involving potentiometric titration, have also been published, often applying mathematical models to obtain complex stability constants. There is a wealth of chemical data available for simple organic ligands such as EDTA and oxyanions (e.g. citrate) that may be present in root and microbial exudats. There is little information concerning more complex exudates except for iron and phytosiderophores [e.g. 59]. I shall not provide a comprehensive review of the literature in this field, nor give details of the underlying assumptions behind the models used to describe the pH and ionic strength dependence of binding. Among the most widely reported models, that appear to give satisfactory descriptions of the metal binding properties of DOM (and particulate humic acid) are the NICA, bimodal NICCA-Donnan, WHAM and MINTEQAL2 models [60-66]. Geochemical models, such as GEOCHEM and SOILCHEM [17, 18], although not designed to elucidate the association of metals with organic compounds, also include stability constants of such complexes.

A much-voiced criticism of such work is that metal concentrations are excessively large, by comparison with usually encountered concentrations in soil solution. Measurements made with ion selective electrodes are often in the milli- to micro-molar range. However analytical progress has done much to allay these criticisms, and concentrations as small as the nanomolar range can now be achieved using polarographical techniques [67, 68]. The use of ion selective electrodes in speciation studies of trace metals has been reviewed by Florence [69]. It has recently been reported that the binding properties for lead of humic and fulvic acids from various origins are very similar [70]. This may reflect an artefact in the extraction procedure for humic substances. However, if the reason is that the same binding groups (—COOH and —OH) present in natural soluble organic matter, are responsible for metal complexation, then this finding is of crucial importance. Obviously it is vital to obtain confirmation of this observation for other metals and using DOM that has not undergone the harsh extraction and purification treatment employed for humic substances.

3.4. Selective chemical extraction

One of the most widely used, and most often criticised, methods for assessing metal speciation in soil is so-called selective extraction. For the reasons outlined below, at best, this approach can only give an operationally defined assessment of the forms of metal present. It cannot not allow true speciation. This method, or rather class of methods, is based on the premise that when a soil constituent is solubilised using a chemical extractant, all the metal associated with that component is thereby released into solution where it can be sampled and analysed. A large number of chemical extractants has been reported in the literature and often the precise experimental conditions vary between studies. Often a series of sequential extractions is employed, progressing from the most gentle to the most severe extraction. The best known of these is that developed by Tessier et al. [71] for sediments, but various sequences are commonly used, including those proposed by McLaren & Crawford, Sposito et al., Shuman and Miller et al. [72-75]. Another significant procedure is that developed and tested by the European Soils Bureau [76-79]. These techniques have been reviewed by Beckett, Ross and Sheppard & Stephenson [80, 1, 81].

The organic matter extractants commonly used in either simple or sequential extractions are sodium hydroxide, pyrophosphate, hydrogen peroxide, sodium hypochlorite, and less commonly sulphuric acid. The last, which removes waxes, also dissolves carbonates and is never used to study organic matter-associated metals. It is well known that sodium hypochlorite removes a greater proportion of organic carbon than does hydrogen peroxide, and is less destructive to soil mineral phases [82], but it is not commonly used in speciation studies, and rarely in sequential extractions [83-84]. The investigations that use this approach are too numerous and the conclusions on the speciation of metals in soils too strongly contrasting for me to attempt any review here.

There are several criticisms of this approach. Firstly, none of the chemical extraction procedures is truly selective. As Beckett [80] has reviewed in detail, each extractant attacks phases other than its target to varying extents. Secondly, the target phase may not be completely removed by a single attack by the chosen extractant, and few studies investigate the effect of repeated treatments [1]. Differences in experimental procedure including time, temperature and pH may affect efficiency [for example, 85]. Successive attacks by the same reagents may continue to dissolve the target phase and release metals. The order of attack is also important [86, 87] because of the complex composition of coatings and precipitates, which is responsible for some of the differences between simple and sequential extractions. Thirdly, metals released may be readsorbed by the residual solid phase or precipitated and so not be accounted for [88, 89]. Readsorption may be limited by addition of various compounds to the extraction cocktail [90, 91]. Fourthly, various experimental artefacts, such as the loss of solid material at each stage and the incomplete dispersion during the early stages may give misleading results. Finally, perhaps the most unfortunate consequence of the wide spread use of this approach is that the simplicity of the terminology may beguile some into believing the reality of the fractions identified. Although there is an increasing effort to refer to the "oxidisable-fraction" rather than "organic matter-associated" phase, this is still a risk. At best the size of this operationally defined pool is only a crude approximation.

3.5. Adsorption properties of soils or soil fractions

The measurement of the adsorption properties of a soil is a classical method for assessing the affinity of the soil for a given metal. Adsorption parameters such as the distribution parameter, Kd, or the fitting parameters obtained using one of the various mathematical formulations employed in soil science, the most common being the linear, Freundlich and Langmuir isotherms. Usually measurements are carried out in suspension, with differing amounts of metal added to a series of soil suspensions. The background electrolyte is chosen either to mask variations in solution composition between soils (e.g. 0.01 M $CaCl_2$) or to mimic the true soil solution composition. Adsorption is calculated from the depletion in solution after phase separation. Occasionally adsorption is also assessed from column experiments, but the separation of chemical (adsorption) and physical (solute movement) phenomena is complex.

Once adsorption has been characterised by a small number of parameters, they may be related to their organic matter content, and hence the role of organic matter in immobilising metals inferred. For a large number of samples, a multi-parametric analysis allows the relative statistical weight of organic matter content to be deduced. In order to estimate the role of organic matter for a smaller number of samples, it must be assumed that all other soil properties vary little or are of little consequence for metal speciation. Adsorption properties indicate the capacity of the soil to immobilise the metal, and insofar as adsorption is reversible, also give an estimation of availability.

However this approach, commonly used for the adsorption of organic pollutants, is less applicable for metals, since organic matter is not the single most important soil component in metal adsorption. For example, Sadiq & Zaidi found no significant correlation between the adsorption and organic matter content for Cd or Cu on 27 soils [92] nor did Buchter et al. for 15 elements on 11 soils [93]. However Bolton & Evans [94] did find a significant correlation between the Cd adsorption capacity of Ontario soils and the organic carbon content (but the effect was not as strong as for the iron oxide content). As outlined above, in the consideration of statistical. correlations between metal. content and soil properties, a very large number of soils would have to be investigated, or a more limited series where only the amount and nature of organic matter varied. Sauvé et al.. [95] have compiled data from the literature on solid-solution partitioning of metals in soil. They report a strong correlation with soil pH for most metals, and much less significant effects for organic matter content. Tiller et al. [96] inferred the relative importance of soil organic matter and other adsorbing surfaces from the pH-dependence of the metal (Cd, Ni and Zn) adsorption on clay-sized fractions of soils. They caution against "undue emphasis on any particular sorption process in developing theoretical sorption models". Sauvé et al. [67] have added or chemically removed organic matter to a Pb-contaminated soil and analysed the dissolved and labile Pb by Graphite furnace Atomic Absorption Spectroscopy and Differential Pulse Aniodic Stripping Voltammetry. They showed that Pb solubility is independent of organic matter in the pH range 3-6.5, but at higher pH, solubility is enhanced by the formation of Pb-OM complexes.

Another possibility is to fractionate a given soil and then compare the adsorption properties of the fractions that have different organic matter contents. The fractionation could be physical, but the nature of the mineral components of different soil size fractions vary too much for statistical analysis to be valid.

Another variant on this fractionation approach is to remove differing amounts and types of organic matter using chemical treatments, as for speciation studies by selective extraction. The adsorption properties of the resulting residues can then be compared. Some of the criticisms listed for selective extractions apply here also, namely that the chemical extractant is not completely selective. Care must also be taken to ensure that the composition of the exchange complex has not been modified by the chemical extraction procedure, since metal adsorption is often an ion exchange phenomenon. Ideally both the initial soil and the treated residue should be washed to give the same ion composition. Petruzzelli et al. [97] have compared the Cu and Cd binding properties of four soils at constant pH before and after organic matter removal by $H_2O_2/Na_4P_2O_7$. The data were fitted to Langmuir adsorption isotherms and the binding capacities and affinities compared. The binding capacity for Cu decreased by about a factor of four when organic matter was removed, but the trend was less marked for Cd. When binding capacity was calculated as a fraction of CEC, then the decrease for Cu was only about 20%, whereas the relative binding capacity for Cd was enhanced by a factor of about two. For both cations the binding affinities were markedly reduced by organic matter removal. Shuman [98] also found large decreases in Zn adsorption capacity after removal of organic matter (by NaOCl/DTPA) in five soils, but a observed a decrease for a sixth soil. Lion et al. [99] also report significant decreases in the adsorption of Cd and Pb on sediments after the chemical removal of organic matter ($NaOH/H_2O_2$ or H_2O_2). More recently we showed that the removal or organic matter enhanced radiocaesium adsorption on soil clays [100], although the mechanism involved is of course different for this alkali metal than for transition metals (see below).

I consider that this approach has much potential to help elucidate the role of organic matter, and other soil constituents in the immobilisation of metals in soils. It should of course be applied with caution, since the extraction techniques may change the CEC, the pH and the cationic composition of the soil, all of which may influence subsequent metal adsorption. Other soil components may be removed or weathered by the treatments, and so all changes in adsorption properties cannot be attributed to organic matter removal. However, despite these caveat, it is a useful, if time-consuming procedure that is not sufficiently exploited.

3.6. Addition of OM to soil (adsorption properties or metal solubilisation)

Although it is difficult to remove organic matter from soil it is relatively easy to investigate the influence of additions of organic substances on the metal adsorption properties of a soil. The organic matter may be added in the laboratory in the form of well-characterised low or high molecular weight organic molecules. Alternatively, samples may be taken in the field where the same soil has received various levels of organic amendments. The draw-back to the latter is that the organic amendments are often sewage sludges that contain metal pollutants. In such cases, it is more likely that the extractability of the metals already in the sample will be investigated, rather than the adsorption of further additions of metals. It must always be remembered that the addition of organic material to soil and the subsequent incubation conditions may influence the pH and the redox potential and therefore modify metal speciation both directly, by supplying new adsorption sites and complexing species in solution, and indirectly [see for example, 101]. O'Connor et al. [102] attempted to elucidate the mechanisms determining the change in mobility of Ni and Zn in sludge-amended soil, but found that none of their hypotheses (soil solution ionic strength, inorganic-metal complexes, competition for sorption sites and organic-metal competition) explained the treatment effects. Few studies attempt to resolve mechanisms. The literature on the mobility and availability of metals in sludge amended soils is too abundant to be reviewed here, excellent reviews are given by Alloway [2, 5].

This approach is particularly interesting for the investigation of the likely effect of root and microbial exudates on metal speciation in the rhizosphere. Hamon et al. [103] have published one of the very few studies that attempt to assess the effect of changes in soil composition in the rhizosphere on trace metals. They found that Cd and Zn were largely uncomplexed (the form in which metals are most likely to be taken up by plants) in non rhizosphere soil, but after growth of radishes both cations were mostly in a complexed form, and their solubility was highly influenced by DOC.

Some understanding of the role of organic substances on metal speciation in soils can be obtained from controlled laboratory experiments where high or low weight organic molecules are added to soils. The usual candidates for such studies are humic and fulvic acids as models of soil macromolecules and various simple organic molecules, such as EDTA and the conjugate bases of organic acids including acetate, citrate, oxalate. These studies have been briefly reviewed by Harter & Naidu [4]. Although strongly complexing ligands, such as EDTA always decrease adsorption of metals, and simple organic anions release metals from soils [104] no other general trend emerges form these investigations, which is probably not surprising, given the differences in the soils studied and the chemical characteristics of the various metals. Unfortunately complementary analyses are rarely carried out to determine the indirect effect of the organic substances on metal sorption via the change in charge characteristics of the adsorbing surfaces or the dissolution of mineral phases with the resulting loss of adsorption sites, or the exposure of new adsorption sites on previously coated clays. Chairidchai & Richtie [105,106] emphasise the importance of surface charge characteristics on addition of simple organic molecules to a variable charge soil. Koopal and co-workers [107-109] demonstrate that the association of organic polyelectrolytes with variable charge surfaces modify both pH and surface charge and may thus influence metal-ion adsorption.

3.7. Adsorption properties of synthetic soils (addition of OM to reference minerals)

Another approach is to investigate the adsorption properties of synthetic soils, with known and carefully controlled composition. This allows some of the artefacts associated with the approaches outlined above to be avoided. However it must be recognised that synthetic soils, often simple binary or ternary mixtures of reference soil components, can never reproduce the complexity of real soil. Typically a single mineral component is considered, either a clay or a metal oxide, and this mineral is mixed with varying amounts of soluble or insoluble organic matter. The adsorption properties of the resulting complex can then be studied, and differences between the reference mineral and the coated mineral or mixture attributed to the organic component. Some of these studies are reviewed by Harter & Naidu [4]. The findings of various authors are often contradictory, and depend on the experimental conditions, the organic matter:mineral ratio and the existence of oxide-coatings on the mineral surfaces. The addition of fulvic acid [110] and humic acid [111] to kaolinite lead to increased adsorption of copper. However, Gupta & Harrison [112] report decreased Cu sorption on kaolinite on addition of humic acid. Similarly the addition of humic acid to smectite has been found to decrease Cd adsorption if Al or Fe coatings are present, but have no effect on clean smectite [113], whereas Campbell et al. [114] found an increase in adsorption. The explanation for these differences may lie in the relative roles of metal sorption on a bare or coated clay surface and adsorption of metal-humic acid complex, which in turn depends on the proportions of clay and humic substance. In addition, it may be difficult to distinguish between metal associated with particulate humic acid and humic-clay complexes. Bar-Tal et al. [115] found that fulvic acid decreased the adsorption of zinc on montmorillonite, because the affinity of the Zn-FA complex to clay is lower than that of the aquated cation or the $ZnOH^+$ complex. Copper has a greater affinity for organic matter than for iron oxide, therefore coatings of natural organic matter enhance Cu adsorption [116]. The effect of the coating was found to be both pH dependent and to a lesser extent time dependent. Lamy et al. [117] report a pH-dependent increase in Cd adsorption on goethite in the presence of oxalate. Given the small tendency of Cd to form complexes with oxalate, they attributed their observation to the formation of bridging complexes at the surface.

The effect of organic coatings on oxide surfaces is even more complex because of variable charge characteristics, redox reactions and dissolution at the surface. Harter & Naidu [4] discuss the importance of the creation of cation exchange and metal complexation sites on manganese oxide surfaces by the reduction of Mn(IV) by various low-molecular-weight-organics, but were unable to find examples of this mechanism in the immobilisation of metals.

Recent work has shown that the adsorption on clays of radiocaesium is decreased by organic coatings. No direct effect of organic matter on radiocaesium adsorption was hitherto predicted because of the very weak affinity for organic molecules for Cs in comparison to clay minerals [118-120]. This appears to be a general effect, observed for soil-extracted dissolved organic matter, fulvic and humic acids, soil-extracted polysaccharide and a protein. The decreased affinity is more marked for trace amounts of caesium than at larger loadings, and for illite in comparison to montmorillonite. Furthermore although the effect depends on the amount of a given substance adsorbed, it is not a simple function of surface coverage or molecular size of the organic. Enhanced adsorption on soil fractions when organic matter is chemically removed confirms this trend [100]. We have postulated that organic matter acts by preventing the partial layer collapse, particularly of frayed edge sites in illite, this being the mechanism responsible for the extremely high affinity of caesium for some clays.

4. CONCLUSIONS

The effect of organic matter on the immobilisation of metals in soils is highly complex. It depends on the nature of the organic matter and the other soil components and the chemistry of the metals. All simple generalisations are therefore flawed. No simple, universal method exists to assess both direct and indirect effects. Ideally a combination of different chemical or physical analytical techniques should be used. At all times, the chemical nature of the metal must be considered, to avoid false interpretations of observations and the numerous possible experimental artefacts borne in mind.

REFERENCES

1. Ross, S.M., 1994. Retention, transformation and mobility of toxic metals in soils. *In*: Ross, S.M. (Ed.), Toxic metals in Soil-Plant Systems. John Wiley & Sons, London, pp. 63-152.
2. Alloway, B.J., 1995. Heavy Metals in Soils. Blackie Academic & Professional, London.
3. Jenne, E.A., 1968. Controls on Mn, Fe, Co, Ni, Cu and Zn concentration in soils and water; the significant role of hydrous Mn and Fe oxides. Adv. Chem. 73, 337-387.
4. Harter, R.D., Naidu, R., 1995. Role of metal- organic complexation in metal sorption by soils. Adv. Agron. 54, 219-263.
5. Alloway, B.J., 1997. The mobilisation of trace elements in soils *In*: Prost, R. (Ed.), Contaminated soils. Institut National de la Recherche Agronomique (INRA), Paris, pp. 133-158.

6. Lovley, D.R., 1995. Microbial Reduction of iron, manganese, and other metals. Adv. Agron. 54, 175-231.
7. Kostka, J.E., Stucki, J.W., Nealson, K.H., Wu Jun, 1996. Reduction of structural Fe(III) in smectite by a pure culture of *Shewanella putrefaciens* strain MR-1. Clays Clay Miner. 44, 522-529.
8. Pearson, R., 1968a. Hard and soft acids and bases. HSAB, Part I -Fundamental principles. J. Chem. Educ. 45, 581-587.
9. Pearson, R., 1968b. Hard and soft acids and bases. HSAB, Part II-Underlying theories. J. Chem. Educ. 45, 643-648.
10. Sullivan, P.J., 1977. The principle of hard and soft acids and bases as applied to exchangeable cation selectivity in soils. Soil Sci. 124, 117-121.
11. Sposito, G., 1991. The Thermodynamics of Soil Solutions. Oxford University Press, Oxford.
12. Sposito, G., 1994. The Surface Chemistry of Soils. Oxford University Press, Oxford.
13. Xu, S., Harsh, J.B., 1990. Hard and soft acid-base model verified for monovalent cation selectivity. Soil Sci. Soc. Am. J. 54, 1596-1601.
14. Zhu, B., Alva, A.K., 1993. Trace metal and cation transport in a sandy soil with various amendments. Soil Sci. Soc. Am. J. 57, 723-727.
15. Romkens, P., Hoenderboom, G., Dolfing, J., 1999. Copper solution geochemistry in arable soils: field observations and model application. J. Environ. Qual. 28. 776-783.
16. Bowen, G.D., Rovira, A.D., 1991. The rhizosphere: the hidden half of the hidden half. *In*: Waisel, Y., Eshel, A., Kafkafi, U. (Eds.), Plant Roots: the Hidden half. Marcel Dekker, New York, pp. 641-669.
17. Sposito, G., Mattigod, S.V., 1980. A Computer Program for the Calculation of Chemical Speciation in Soil Solutions and other Natural Water Systems. The Kearney Foundation of Soil Science, University of California, Riverside and Berkeley, USA.
18. Sposito, G., Coves, J., 1988. A Computer Program for the Calculation of Chemical Speciation in Soils. The Kearney Foundation of Soil Science, University of California, Riverside and Berkeley, USA.
19. Pohlman, A.A., McColl, J.G., 1986. Kinetics of metal dissolution from forest soils by soluble organic acids. J. Environ. Qual. 15, 86-92.
20. Jones, D.L., Darrah, P.R., Kochian, L.V., 1996. Critical evaluation of organic acid mediated iron dissolution in the rhizosphere and its potential role in root iron uptake. Plant Soil 180, 57-66.
21. Chenu, C., 1993. Clay- or sand-polysaccharide associations as models for the interface between micro-organisms and soil: water related properties and microstructure. Geoderma 56, 143-156.
22. Wilcke, W., Baumler, R., Deschauer, H., Kaupenjohann, M., Zech, W., 1996. Small scale distribution of Al, heavy metals, and PAHs in an aggregated Alpine Podzol. Geoderma 71, 19-30.
23. Rahman, S., Nagatomo, Y., Inoue, Y., Takaki, H., 1998. Distribution of micro elements in an Andisol profile in Takatoge, Kagoshima, Japan. Soil Sci. Plant Nutr. 44, 9-18.
24. Egli, M., Fitze, P., Oswald, M., 1999. Changes in heavy metal contents in an acidic forest soil affected by depletion of soil organic matter within the time span 1969-93. Environ. Poll. 105, 367-379.
25. Sterckeman, T., Douay, F., Proix, N., Fourrier, H., 2000. Vertical distribution of Cd, Pb and Zn in soils near smelters in the North of France. Environ. Poll. 107, 377-389.

26. Balabane, M., Faivre, D., van Oort, F., Dahmani-Muller, H., 1999. Mutual effects of soil organic matter dynamics and heavy metals fate in a metallophyte grassland. Environ. Poll. 105, 45-54.
27. Wang, E.X., Benoit, G., 1996. Mechanisms controlling the mobility of lead in the Spodosols of a northern hardwood forest ecosystem. Environ. Sci. Technol. 30, 2211-2219.
28. Ducaroir, J., Lamy, I., 1995. Evidence of trace metal association with soil organic matter using particle size fractionation after physical dispersion treatment. Analyst 120, 741-745.
29. Ducaroir, J., Cambier, P., Leydecker, J.P., Prost, R., 1990. Application of soil fractionation methods to the study of the distribution of pollutant metals. Pflanzernähr Bodenk. 153, 349-358.
30. Oades, J.M., Vassallo, A.M., Waters, A.G., Wilson, M.A., 1987. Characterization of organic matter in particle size and density fractions from a red-brown earth by solid-state C N.M.R. Aust. J. Soil Res. 25, 71-82.
31. Tegen, I., Dorr, H., 1996. Mobilization of cesium in organic rich soils: correlation with production of dissolved organic carbon. Water Air, Soil Poll. 88, 133-144.
32. Takenaka, C., Onda, Y., Hamajima, Y., 1998. Distribution of caesium-137 in Japanese forest soils: correlation with the contents of organic carbon. Sci. Total Environ. 222, 193-199.
33. McKenzie, R.M., 1975. An electron microprobe study of the relationships between heavy metals and manganese and iron in soils and ocean floor nodules. Aust. J. Soil Sci. 13, 177-188.
34. Hiller, D.A., Brummer, G.W., 1991. Elektronenmikrostrahlanalysen zur Erfassung der Schwermetallbindungsformen in Boden unterschiedlicher Schwermetallbelastung. Mitteil. Deut. Bodenk. Gesell. 66, 1085-1088.
35. Hiller, D.A., Brummer, G.W., 1997. Mikrosondenuntersuchungen an unterschiedlich stark mit Schwermetallen belasteten Boden. 2. Gehalte an Schwermetallen und anderen Elementen in Huminstoffaggregationen, Streustoffen und Holzkohlepartikeln. Zeit. Pflanzen. Bodenk. 160, 47-55.
36. Fendorf, S.E., Sparks, D.L., Lamble, G.M., Kelley, M.J., 1994. Applications of X-ray absorption fine structure spectroscopy to soils. Soil Sci. Soc. Am. J. 58, 1583-1595.
37. Schulze, D.G., Bertsch, P.M., 1995. Synchrotron X-ray techniques in soil, plant, and environmental research. Adv. Agron. 55, 1-66.
38. Guest, C., Schulze, D., Thompson, I., Scheinost, A., Huber, D., 1999. *In*: Wenzel, W.W., Adriano, D.C., Alloway, B., Doner, H.E., Keller, C., Lepp, N.W., Mench, M., Naidu, R., Pierzynski, G.M. (Eds.), Proceedings of the 5th International Conference on the Biogeochemistry of Trace Elements. Volume II, pp. 1086-1087.
39. Juillot, F., Morin, G., Ildefonse, P., Dumat, C., Benedetti, M., Chevallier, P., Brown, G.E. Jr, Calas, G., 1999. Zinc speciation in a smelter-impacted soil by a combination of EXAFS data and chemical extractions. *In*: Wenzel, W.W., Adriano, D.C., Alloway, B., Doner, H.E., Keller, C., Lepp, N.W., Mench, M., Naidu, R., Pierzynski, G.M. (Eds.), Proceedings of the 5th International Conference on the Biogeochemistry of Trace Elements. Volume II, pp. 1090-1091.
40. Thoming, J., Stichnothe, H., Mangold, S., Calmano, W., 2000. Hydrometallurgical approaches to soil remediation - process optimization applying heavy metal speciation. Land Contam. Reclam. 8, 19-31.

41. Strawn, D.G., Sparks, D.L., 2000. Effects of soil organic matter on the kinetics and mechanisms of Pb(II) sorption and desorption in soil. Soil Sci. Soc. Am. J. 64, 144-156.
42. Manceau, A., Boisset, M.C., Sarret, G., Hazemann, J.L., Mench, M., Cambier, Ph., Prost, R., 1996. Direct determination of lead speciation in contaminated soils by EXAFS spectroscopy. Environ. Sci. Technol. 30, 1540-1552.
43. Manceau, A., Hergé, J.C., Sarret, G., Hazemann, J.L., Boisset, M.C., Mench, M., Cambier, Ph., Prost, R., 1997. Direct determination of heavy metal speciation in soils by EXAFS spectroscopy. *In*: Prost, R. (Ed.), Contaminated soils. Institut National de la Recherche Agronomique (INRA), Paris, pp 99-120.
44. Morin, G., Ostergren, J.D., Juillot, F., Ildefonse, P. Calas, G., Brown, Jr G.E., 1999. XAFS determination of the chemical form of lead in smelter-contaminated soils and mine tailings: Importance of adsorption processes. Am. Miner. 84, 420-434.
45. Morin, G., Juillot, F., Ildefonse, P., Calas ,Ý.G., Samama, J.C., Chevallier, P., Brown, G.E. Jr, 2001. Mineralogy of lead in a soil developed on a Pb-mineralized sandstone (Largentière, France). Am. Miner. 86, 92-104.
46. Xia, K,. Skyllberg,U.L., Bleam, W.F., Bloom, P.R., Nater, E.A., Helmke, P.A., 1999. X-ray absorption spectroscopic evidence for the complexation of Hg(II) by reduced sulfur in soil humic substances. Environ. Sci. Technol. 33, 257-261.
47. Hesterberg, D., Zhou, W., Hutchison, K.J., Sayers, D.E., 1999. EXAFS study of mercury bonding to goethite with adsorbed humic acid. *In*: Wenzel, W.W., Adriano, D.C., Alloway, B., Doner, H.E., Keller, C., Lepp, N.W., Mench, M., Naidu, R., Pierzynski, G.M. (Eds.), Proceedings of the 5th International Conference on the Biogeochemistry of Trace Elements. Volume II, pp. 1089-1090.
48. Welter, E., Calmano, W., Mangold, S., Troger, L., 1999. Chemical speciation of heavy metals in soils by use of XAFS spectroscopy and electron microscopical techniques. Fresenius J. Anal. Chem. 364, 238-244.
49. Senesi, N., 1992. Metal-humic substance complexes in the environment. Molecular and mechanistic aspects by multiple spectroscopic approach. *In*: Adriano, DC (Ed.), Biogeochemistry of Trace Metals. Lewis Publishers, Boca Raton, Florida, pp. 429-496.
50. Lakatos, B., Tibai, T., Meisel, J., 1977. EPR spectra of humic acids and their metal complexes. Geoderma 19, 319-338.
51. Murray, K., Linder, P.W., 1983. Fulvic acids: structure and metal binding. I-A random molecular model. J. Soil Sci. 34, 511-523.
52. Boyd, S.A., Sommers, L.E., Nelson, D.W., 1979. Infrared spectra of sewage sludge fractions: evidence for an amide metal binding site. Soil Sci. Soc. Am. J. 43, 893-899.
53. Cheshire, M.V., Berrow, M.L., Goodman, B.A., Mundie, C.M., 1977. Metal distribution and nature of some Cu, Mn and V complexes in humic and fulvic acid fractions of soil organic matter Geochim.Cosmochim. Acta 41, 1131-1138.
54. Breault, R.F., Colman, J.A., Aiken, G.R., McKnight, D., 1996. Copper speciation and binding by organic matter in copper-contaminated streamwater. Environ. Sci. Technol. 30, 3477-3486.
55. Moulin, C., Wei Jiang, van Iseghem, P., Laszak, I., Plancque, G., Moulin, V., 1999. Europium complexes investigations in natural waters by time-resolved laser-induced fluorescence. Anal. Chim. Acta 396, 253-261.
56. Morrisson, A.R., Park, J.S., Sharp, B.L., 1990. Application of high-performance size-exclusion liquid chromatography to the study of copper speciation in waters extracted from sewage sludge treated soils. Analyst 115, 1429-1433.

57. McBride, M.B., Richards, B.K., Steenhuis, T., Russo, J.J., Sauve, S., 1997. Mobility and solubility of toxic metals and nutrients in soil fifteen years after sludge application. Soil Sci. 162, 487-500.
58. de Oliveira, C.R., Lombardi, A.T., Jardim, W.F., 1995. Copper complexation by naturally occurring organic matter: a multiligand model. Chem. Speci. Bioavail. 7, 125-131.
59. Neubauer, U., Nowack, B., Furrer, G., Schulin, R., 2000. Heavy metal sorption on clay minerals affected by the siderophore desferrioxamine B. Environ. Sci. Technol. 34, 2749-2755.
60. Kinniburg, D.G., Milne, C.J., Benedetti, M.F., Pinheiro, J.P., Filius, J., Koopal, L.K., van Riemsdijk, W.H., 1996. Metal ion binding by humic acid: application of the NICA-Donnan model. Environ. Sci. Technol. 30, 1687-1698.
61. Benedetti, M.F., van Riemsdijk, W.H., Koopal, L.K., 1996. Humic substances considered as a heterogeneous Donnan gel phase. Environ. Sci. Technol. 30, 1805-1813.
62. Temminghoff, E.J.M., van der Zee, S.E.A.T.M., de Haan, F.A.M., 1997. Copper mobility in a copper-contaminated sandy soil as affected by pH and solid and dissolved organic matter. Environ. Sci. Technol. 31, 1109-1115.
63. Kinniburgh, D.G., van Riemsdijk, W.H., Koopal, L.K., Borkovec, M., Benedetti, M.F., Avena, M.J., 1999. Ion binding to natural organic matter: competition, heterogeneity, stoichimetry and thermodynamic consistency. Colloids Surf. A 151, 147-166.
64. Christensen, J.B., Christensen, T.H., 1999. Complexation of Cd, Ni, and Zn by DOC in polluted groundwater: a comparison of approaches using resin exchange, aquifer material sorption, and computer speciation models (WHAM and MINTEQA2). Environ. Sci. Technol. 33, 3857-3863.
65. Christensen, J.B., Botma, J.J., Christensen, T.H., 1999. Complexation of Cu and Pb by DOC in polluted groundwater: a comparison of experimental data and predictions by computer speciation models (WHAM and MINTEQA2). Water Res. 33, 3231-3238.
66. Christensen, J.B., Christensen, T.H., 2000. The effect of pH on the complexation of Cd, Ni and Zn by dissolved organic carbon from leachate-polluted groundwater. Water Res. 34. 3743-3754.
67. Sauve, S., McBride, M., Hendershot, W., 1998. Soil solution speciation of lead(II): effects of organic matter and pH. Soil Sci. Soc. Am. J. 62, 618-621.
68. Sauve, S., Norvell, W.A., McBride, M., Hendershot, W., 2000. Speciation and complexation of cadmium in extracted soil solutions. Environ. Sci. Technol. 34, 291-296.
69. Florence, T.M. 1986. Electrochemical approaches to trace element speciation in waters. A review. Analyst 111, 489-505.
70. Pinheiro, J.P., Mota, A.M., Benedetti, M.F., 1999. Lead and calcium binding to fulvic acids: salt effect and competition. Environ. Sci. Technol. 33, 3398-3404.
71. Tessier, A., Campbell, P.G.C., Bisson, M., 1979. Sequential extraction procedure for the speciation of particulate trace metals. Anal. Chem. 51, 844-851.
72. McLaren, R.G., Crawford, D.V., 1973. Studies on soil copper. I-The fractionation of copper in soils. J. Soil Sci. 24, 172-181.
73. Sposito, G., Lund, L.J., Chang, A.C., 1982. Trace metal chemistry in arid-one field soils amended with sewage sludge. I-Fractionation of Ni, Cu, Zn, Cd and Pb in solid phases. Soil Sci. Soc. Am. J. 46, 260-264.
74. Shuman, L.M., 1985. Fractionation method for soil microelements. Soil Sci. 140, 11-22.
75. Miller, W.P., Martens, D.C., Zelazny, L.W., 1986. Effect of sequence in extraction of trace metals from soils. Soil Sci. Soc. Am. J. 50, 598-601.

76. Ure, A.M., Quevauvillier, Ph., Muntau, H., Griepink, B., 1993. Speciation of heavy metals in soils and sediments. An account of the improvement and harmonization of extraction techniques undertaken under the auspices of the BCR of the Commission of the European Communities. Int. J. Environ. Anal. Chem. 51, 135-151.
77. Davidson, C.M., Thomas, R.P., McVey, S.E., Perala, R., Littlejohn, D., Ure, A.M., 1994. Evaluation of a sequential extraction procedure for the speciation of heavy metals in sediments. Anal. Chim. Acta 291, 277-286.
78. Quevauvillier, P., Rauret, G., Muntau, H., Ure, A.M., Rubio, R., López-Sánchez, J.F., Fiedler, H.D., Griepink, B., 1994. Evaluation of a single extraction procedure for the determination of extractable trace metal contents in sediments. Fres. J. Anal. Chem. 349, 808-814.
79. Ure, A.M., 1996. Single extraction schemes for soil analysis and related applications Sci.Total Environ. 178, 3-10.
80. Beckett, P.H.T., 1989. The use of extractants in studies on trace metals in soils, sewage sludges, and sludge-treated soils. Adv. Soil Sci. 9, 143-176.
81. Sheppard, M.I., Stephenson, M., 1997. Critical evaluation of selective extraction methods for soils and sediments. *In*: Prost, R. (Ed.), Contaminated soils. Institut National de la Recherche Agronomique (INRA), Paris, pp. 69-97.
82. Lavkulich, L.M., Wiens, J.H., 1970. Comparison of organic matter destruction by hydrogen peroxide and sodium hypochlorite and its effects on selected mineral components. Soil Sci. Soc. Am. Proc. 34, 755-758.
83. Kuo, S., Heilman, P.E., Baker, A.S., 1983. Distribution and forms of copper, zinc, cadmium, iron and manganese in soils near a copper smelter. Soil Sci. 135, 101-109.
84. Shuman, L.M., Hargrove, W.L., 1985. Effect of tillage on the distribution of manganese, copper, iron and zinc in soil fractions. Soil Sci. Soc. Am. J. 49, 1117-1121.
85. Coutras, S., Bourgeois, S., Bermond, A., 2000. A critical study of the use of hydrogen peroxide to determine trace elements bound to soil organic matter Environ. Technol. 21, 77-86.
86. Calvet, R., Bourgeois, S., Msaky, J.J., 1990. Some experiments on extraction of heavy metals present in soil. Int. J. Environ. Anal. Chem. 39, 31-45.
87. Berti, W.R., Cunningham, S.D., Jacobs, L.W., 1997. Sequential chemical extraction of trace elements: development and use in remediating contaminated soils. *In*: Prost, R. (Ed.), Contaminated soils. Institut National de la Recherche Agronomique (INRA), Paris, pp. 121-131.
88. Rendell, P.S., Batley, G.E., Cameron, A.J., 1980. Adsorption as a control of metal concentration in sediment extracts. Environ. Sci. Technol. 14, 314-318.
89. Bermond, A., Sommer, G., 1989. Simulation of heavy metals extraction in soil samples compared with experimental results. Environ. Technol. Lett. 10, 989-994.
90. Howard, J.L., Vandenbrink, W.J., 1999. Sequential extraction analysis of heavy metals in sediments of variable composition using nitrilotriacetic acid to counteract resorption. Environ. Poll. 106, 285-292.
91. Tu, Q., Shan, X.Q., Qian, J., Ni, Z.M., 1994. Trace metal redistribution during extraction of model soils by acetic acid/sodium acetate. Anal. Chem. 66, 3562-3568.
92. Sadiq, M., Zaidi, T.H., 1991. The adsorption characteristics of soils and removal of cadmium and nickel from wastewaters. Water Air Soil Poll. 16, 293-299.

93. Buchter, B., Davidoff, B., Amacher, M.C., Hinz, C., Iskandar, I.K., Selim, H.M., 1989. Correlation of Freundlich Kd and n retention parameters with soils and elements. Soil Sci. 148, 370-379.
94. Bolton, K.A., Evans, L.J., 1996. Cadmium absorption capacity of selected Ontario soils. Can. J. Soil Sci. 76, 183-189.
95. Sauvé, S., Hendershot, W., Allen, H.E., 2000. Solid-solution partitioning of metals in contaminated soils: dependence on pH, total metal burden, and organic matter. Environ. Sci. Technol. 34, 1125-1131.
96. Tiller, K.G., Gerth, J., Brümmer, G., 1984. The relative affinities of Cd, Ni and Zn for different soil clay fractions and goethite. Geoderma 34, 17-35.
97. Petruzzelli, G., Guidi, G., Lubrano, L., 1978. Organic matter as an influencing factor on copper and cadmium adsorption by soils. Water Air Soil Poll. 9, 263-269.
98. Shuman, L.M., 1988. Effect of removal of organic matter and iron- or manganese-oxides on zinc adsorption by soil. Soil Sci. 146, 248-254.
99. Lion, L.W., Altmann, R.S., Leckie, J.O., 1982. Trace-metal adsorption characteristics of estuarine particulate matter: evaluation of contributions of Fe/Mn oxide and organic surface coatings. Environ. Sci. Technol. 16, 660-666.
100. Dumat, C., Cheshire, M.V., Fraser, A., Shand, C., Staunton, S., 1997. The effect of removal of soil organic matter and iron on the adsorption of radiocaesium. Eur. J. Soil Sci. 48, 675-683.
101. Wang, G., Chen, J.B., Gu, X.G., Gao, S., Cai, H.Y., 1999. Effect of organic materials on speciation of copper in soil solution. Pedosphere 9, 139-146.
102. O'Connor, G.A., Essington, M.E., Elrashdi, M., Bowman, R.S., 1983. Nickel and zinc sorption in sludge-amended soils. Soil Sci. 135, 228-235.
103. Hamon, R.E., Lorenz, S.E., Holm, P.E., Christensen, T.H., McGrath, S.P., 1995. Changes in trace metal species and other components of the rhizosphere during growth of radish. Plant Cell Environ. 18, 749-756.
104. Naidu, R., Harter, R.D., 1998. Effect of different organic ligands on cadmium sorption by and extractability from soils. Soil Sci. Soc. Am. J. 62, 644-650.
105. Chairidchai, P., Ritchie, G.S.P., 1990. Zinc adsorption by a lateritic soil in the presence of organic ligands. Soil Sci. Soc. Am. J. 54, 1242-1248.
106. Chairidchai, P., Ritchie, G.S.P., 1992. The effect of pH on zinc adsorption by a lateritic soil in the presence of citrate and oxalate. J. Soil Sci. 43, 723-728.
107. Vermeer, A.W.P., Leermakers, F.A.M., Koopal, L.K., 1997. Adsorption of weak polyelectrolytes on surfaces with a variable charge. Self-consistent-field calculations. Langmuir 13, 4413-4421.
108. Vermeer, A.W.P., McCulloch, J.K., van Riemsdijk, W.H., Koopal, L.K., 1999. Metal ion adsorption to complexes of humic acid and metal oxides: deviations from the additivity rule. Environ. Sci. Technol. 33, 3892-3897.
109. Vermeer, A.W.P., Koopal, L.K., 1999. Charge adjustments upon adsorption of a weak polyelectrolyte to a mineral oxide: the hematite-humic acid system. J. Coll Interf. Sci. 212, 176-185.
110. Dalang, F., Buffle, J., Haerdl, W., 1984. Study of the influence of fulvic substances on the adsorption of copper(II) ions at the kaolinite surface. Environ. Sci. Technol. 18, 135-141.
111. Haas, C.N., Horowitz, N.D., 1986. Adsorption of cadmium to kaolinite in the presence of organic matter. Water Air Soil Poll. 27, 131-140.

112. Gupta, G.C., Harrison, F.L., 1982. Effect of humic acid on copper adsorption by kaolin. Water Air Soil Poll. 17, 357-360.
113. Levy, R., Francis, C.W., 1976. Adsorption and desorption of cadmium by synthetic and natural organo-clay complexes. Geoderma 15, 361-370.
114. Campbell, G.D., Galicia, H.F., Schindler, P.W., 1987. Binding of cadmium by montmorillonite-humic acid mixtures: miscible displacement experiments. Aust. J. Soil Res. 25, 391-403.
115. Bar-Tal, A., Bar-Yosef, B., Chen, Y., 1988. Effects of fulvic acid and pH on zinc sorption on montmorillonite. Soil Sci. 146, 367-373.
116. McBride, M.B., Martinez, C.E., Sauve, S., 1998. Copper(II) activity in aged suspensions of goethite and organic matter. Soil Sci. Soc. Am. J. 62, 1542-1548.
117. Lamy, I., Djafer, M., Terce, M., 1991. Influence of oxalic acid on the adsorption of cadmium at the goethite surface. Water Air Soil Poll. 57-58, 457-465.
118. Dumat, C., Staunton, S., 1999. Reduced adsorption of caesium on clay minerals caused by various humic substances. J. Environ. Radioactivity 46, 187-200.
119. Dumat, C., Quiquampoix, H., Staunton, S., 2000. Adsorption of caesium by synthetic clay-organic matter complexes: effect of the nature of organic polymers. Environ. Sci. Technol. 34, 2985-2989.
120. Staunton, S., Dumat, C., Zsolnay, A., 2001. Possible role of soil organic matter in the availability of radiocaesium. J. Environ. Radioactivity, 58, 163-173.

Developments in Soil Science, Volume 28A
Editors: A. Violante, P.M. Huang, J.-M. Bollag and L. Gianfreda

EFFECTS OF MEDIUM-TERM AMENDMENT WITH SEWAGE SLUDGES ON HEAVY METAL DISTRIBUTION IN SOIL

G. Rossi, B. Pennelli, S. Socciarelli and A. Figliolia

Experimental Institute for Plant Nutrition, via della Navicella 4, 00184 Rome, Italy
E-mail: nutrminerale@isnp.it

The effects of medium-term applications of liquid, dehydrated and composted sewage sludges on a silty loam soil were studied in a field trial in northern Italy. Once per year, randomized plots were added of liquid, dehydrated and composted sewage sludges. A sequential extraction of Zn, Cu, Ni and Pb from sludge and soil samples was performed after six years of sludge application to evaluate the heavy metal distribution in the different sewage sludges and their influence on the pool of potentially bioavailable metals in amended soils. Comparison of metal fractions in the three sludges (liquid, dehydrated and composted sludge) showed that the process of composting altered the chemical forms between the metals and the organic matrix, resulting in a shift of metals from the residual to the more labile and potentially bioavailable forms. In particular, an increase of the EDTA-extractable Zn, Ni, Pb and Cu was observed, as well as a significant increase of the NaOH-extractable Cu. With regard to soil, heavy metals were found mostly in the residual form (HNO_3-extractable) and at the sixth year of sludge amendment, except for zinc, the soil fractionation of heavy metals was not affected by the amendment practices. A significant increase of the potentially bioavailable fractions (EDTA and NaOH-extractable) of Zn was observed.

1. INTRODUCTION

The agricultural use of sewage sludge is known to be a valid alternative to landfill or incineration because of the possibility of recycling the nutrients and organic matter contained in these materials [1]. In particular, sludge organic matter added to soil can maintain, and in some cases improve, the level of soil humus [2-3]. This aspect is important particularly for Mediterranean soils, which generally contain low quantities of humified organic matter. Previous studies on Italian farmlands reported losses of organic matter up to 1.5% per year [4], thus a great effort has been made for the conservation and enrichment of organic matter in arable soils [5].

One of the main constraints in the use of sewage sludge in agriculture is its content of heavy metals. The accumulation of heavy metals in soil and their removal by crops and/or their migration to the water table are the major risks when using sewage sludge as a soil amendment [6]. Nowadays, the environmental directives in many countries regulate the use of sewage sludge in agriculture by limiting the total heavy metal concentration allowed for sludge and soil.

This criterion, despite being safe and reliable, does not take into account other important factors, such as mobility and bioavailability of heavy metals. Heavy metals in sludge and soil are in fact distributed in several chemical forms, depending on the type of matrix and the kind and strength of bonds they may form with the soil or sludge components. For instance, Zn^{2+} and Ni^{2+} show a tendency to form organic complexes with amide and amino ligands, thus increasing their mobility in soils with respect to Cu and Pb, which tend to be sorbed by cation exchange, especially by humic and fulvic acids [7].

Briefly, the main forms of potentially toxic metals in soil [8] are: in the soil solution (ionic, chelated and colloidal forms), as readily exchangeable ions on inorganic or organic surfaces, as adsorption complexes, in the form of precipitated sesquioxides and insoluble salts and as ions in crystal lattices of secondary clay minerals.

Previous studies [9-10] indicated that residual forms of Cd, Cu, Ni and Zn were the most important metal fractions in untreated soil. In sludge-amended soils, amounts of exchangeable metals (KNO_3-extractable) were very low, Zn and Pb occurred mainly as carbonate, Ni was mainly bound in sulfide form in the residual fraction and Cu was mainly present in organic forms. These results suggest that long-term soil amendment with sewage sludge might shift the soil heavy metals from residual forms to forms that are potentially more mobile, labile and available to soil organisms and plants [11].

The aim of this research was to investigate the partition of Zn, Cu, Ni and Pb in soil and sludge, by means of chemical speciation, to evaluate any change of compartmentalization between organic and inorganic soil components after six years of sludge application.

2. MATERIALS AND METHODS

The experiment was carried out on a silty loam calcic Cambisol [12] located in northern Italy (region of Emilia Romagna). Maize (*Zea mais* L), wheat (*Triticum aestivum* L.) and sugarbeet (*Beta vulgaris* L.) were cultivated in a three-course rotation. Since 1988, anaerobically digested dehydrated (D), liquid (L) and composted (C) sludges were applied to soil every year at rates of 7.5 Mg ha^{-1} $year^{-1}$ of dry matter (single dose, 1) and 15 Mg ha^{-1} $year^{-1}$ of dry matter (double dose, 2) in randomized plots. Thus the experimental layout consisted up of: Control (untreated soil), D1, D2, L1, L2, C1 and C2.

In 1994, sludge application was decreased to 5 and 10 Mg ha^{-1} y^{-1} of dry matter.

Composted sludge was obtained by composting dehydrated sludge with wheat straw (sludge/straw ratio = 9:1, w/w).

Surface (0-30 cm) soil samples were collected in 1990 and 1996 after the maize harvesting. Soil and sewage sludge samples were air-dried and passed through a 2 mm sieve before chemical analysis [13]. The heavy metal (Zn, Cu, Ni and Pb) fractionation was performed according to the Sposito procedure [9], which consists of a sequential extraction with the following reagents: KNO_3, 0.5 M for 16 hours (exchangeable form); three successive extractions with deionized water of two hours each (adsorbed form); NaOH, 0.5 M for 16 hours (organically bound form); Na_2-EDTA, 0.05 M for 6 hours (carbonate form); and HNO_3, 4 M for 16 hours at 80°C (residual form).

Metal concentrations in the extracts were determined by inductively coupled plasma (ICP) spectroscopy.

The data were analyzed by ANOVA and LSD tests at a confidence level of 95% ($P \leq 0.05$).

3. RESULTS AND DISCUSSION

Chemical and physical properties of the control soil and sludge are listed in Tables 1 and 2.

Table 1
Physical-chemical characteristics of soil. Concentrations are referred to dry weight

Parameter		Value	Parameter	Value
sand	%	23	CEC (cmol kg^{-1})	13.8
silt	%	55	tot. Cu (mg kg^{-1})	68.2
clay	%	22	DTPA-Cu (mg kg^{-1})	15.2
pH		7.8	tot. Ni (mg kg^{-1})	43.8
total $CaCO_3$ (g kg^{-1})		210	DTPA-Ni (mg kg^{-1})	0.6
active $CaCO_3$ (g kg^{-1})		82	tot. Pb (mg kg^{-1})	14.9
organic matter (g kg^{-1})		16	DTPA-Pb (mg kg^{-1})	1.8
available P (mg kg^{-1})		16	tot. Zn (mg kg^{-1})	78.5
tot. N (mg kg^{-1})		1180	DTPA-Zn (mg kg^{-1})	2.3

Table 2
Main parameters of the sludge (average values of the sludge used) and limits set by the Italian legislation. Concentrations are referred to dry weight

Parameter	Liquid Sludge (L)	Dehydrated Sludge (D)	Composted Sludge (C)	Limits*
dry matter (g kg^{-1})	29.91	243.25	622.31	
pH	7.73	8.01	6.96	
organic C (g kg^{-1})	319	286	222	> 200
total Cu (mg kg^{-1})	906	934	666	< 1000
total Ni (mg kg^{-1})	202	221	161	< 300
total Pb (mg kg^{-1})	125	118	109	< 750
total Zn (mg kg^{-1})	1514	1574	1140	< 2500

* Maximum allowable concentration of heavy metals and organic carbon according to the Law 99/92.

3.1. Metals in untreated soil

The trend of heavy metals in the control soil is shown in Figure 1. For zinc, a clear reduction of the residual fraction from 1990 to 1996 occurred (19%). As for Cu, the total amount did not change severely during that time; nevertheless, we observed an increase of the residual fraction (23%) and a similar decrease of the NaOH-extractable fraction. No time effect was evident for nickel. As for lead, besides the decrease observed in 1996, there was a shift towards carbonate-linked fractions (EDTA-extractable). The EDTA-extractable fraction in fact increased by 17%, against a reduction of 30% of the residual.

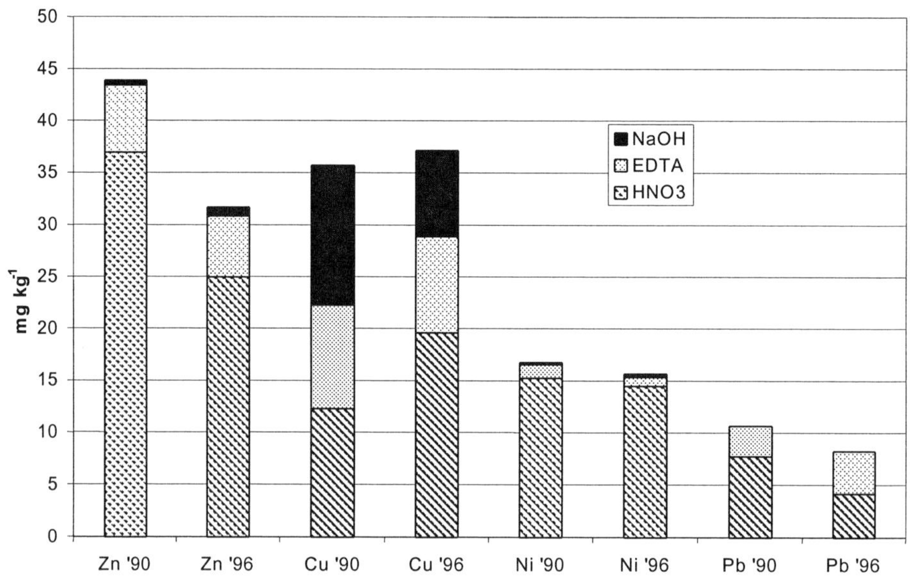

Figure 1. Concentrations of the heavy metal fractions in control soil of 1990 and 1996.

3.2. Metals in the sewage sludge

Table 3 illustrates the mean concentrations (1990-1996) of the heavy metals sequentially extracted from sludges. The values of Least Significant Difference (LSD) refer to the ANOVA performed by "metal" and by "extractant".

For zinc, the main differences were observed in the HNO_3-extractable fraction where the metal extracted from the composted sludge was significantly lower than from liquid (80%) and dehydrated sludges (79%). Conversely, the EDTA-extractable fraction was significantly higher in composted sludge (40% vs. 7.57% in liquid and 6.78% in dehydrated sludge).

The same trend was observed for copper; moreover, the NaOH-extractable fraction also was significantly higher in composted sludge (32%) than in liquid (5%) and dehydrated (5%) forms.

Except for the EDTA-extractable fraction, where the composted sludge gave the highest value (15% vs. 4% in the others) for any other fraction, the nickel concentrations followed the pattern: composted < liquid < dehydrated. By comparing the metal distributions in sludge, Ni was likely to have a stronger affinity to the NaOH-extracted fraction.

Even for lead, in the composted sludge, the residual fraction gave the lowest concentration, while the EDTA-extracted fraction gave the highest.

Table 3
Mean concentrations of Zn, Cu, Ni and Pb in dehydrated (D), composted (C) and liquid (L) sludges obtained through sequential extraction (mg kg^{-1} d.w.)

Sludges	Zn	Cu	Ni	Pb	Extractant
D	2.94	13.86	18.57	0.00	
C	2.72	7.05	5.30	0.00	KNO_3
L	1.56	26.88	15.72	0.00	
*LSD	0.58	4.65	0.83		
D	0.94	2.59	3.90	0.00	
C	0.84	2.76	1.77	0.00	H_2O
L	1.19	3.66	3.19	0.00	
*LSD	0.19	0.30	0.24		
D	203.93	46.93	122.91	3.98	
C	157.50	161.88	83.99	4.82	NaOH
L	182.84	44.38	104.01	2.50	
*LSD	13.08	19.40	5.48	0.88	
D	100.75	27.62	12.53	15.19	
C	384.04	92.81	31.44	44.76	EDTA
L	110.22	36.79	13.05	9.76	
*LSD	46.40	9.30	5.18	6.17	
D	1178.15	749.67	174.00	115.73	
C	405.04	247.17	82.76	41.88	HNO_3
L	1159.45	742.86	160.47	107.80	
*LSD	85.00	40.20	5.94	6.20	

*LSD = least significant differences at $P \leq 0.05$.

The percentage of metals in the exchangeable fraction (KNO_3) plus the adsorbed fraction (H_2O), regardless of the kind of sludge, was zero for Pb. For Zn, ranges were from 0.2 to 0.4%; for Cu, 2-3.4%; for Ni, 3.4-7.3%.

3.3. Metals in amended soil

The results indicated that the soil HNO_3-extractable Zn was unaffected by the amendment with sewage sludges, whilst there was a general increase of the EDTA-extractable Zn in comparison to the control soil (1996) (Figure 2).

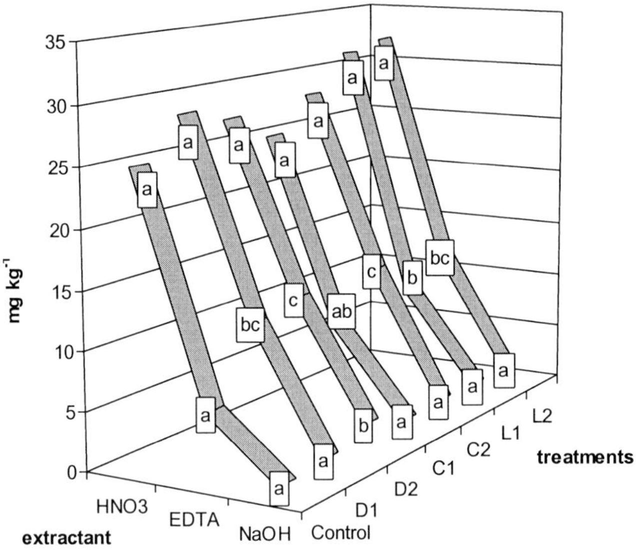

Figure 2. Zinc fractions in soil. Comparisons regard concentrations obtained by the same extractants. Different letters indicate significant differences ($P \leq 0.05$).

For copper, as shown in Figure 3, there was no effect due to the kind of sludge, no dose effect and no sludge effect with respect to control, for any fraction. It is noticeable that the NaOH-extractable fraction (organic bound form) of copper is the highest of all the NaOH-extractable fractions--more than 10 ppm on average in all treatments.

By comparing the nickel fractions in soil (Figure 4) and in sludge (Table 3), the distribution of Ni in soil looks unrelated with its distribution in the sludge. The HNO_3-extractable fractions in fact did not differ among the treatments; EDTA-extracted Ni increased significantly in plots amended with dehydrated sludge (17 and 21% in D1 and D2 vs. 13% of control). NaOH-extracted Ni was significantly higher than control in D2 and C2 only. In addition, there was no more than a 50 ppm difference between EDTA- and NaOH-extracted Ni in sludge, on average, in soil.

The predominant fractions of lead in the control soil (Figure 5) were the HNO_3-and EDTA-extractable fractions, accounting for 50% and 48%, respectively, of the extracted lead; as for the residual Pb, no significant differences among treatments occurred. The soil EDTA-extractable fractions instead gave significantly higher concentrations in C2 (21.1%) and L2 (18.4%).

Taking into account the natural trend of Pb in soil (Figure 1), where Pb fractionates in residual and carbonate forms equally, it is presumable that composted sludge firstly, and liquid sludge secondly, at double dose, would boost this natural shift.

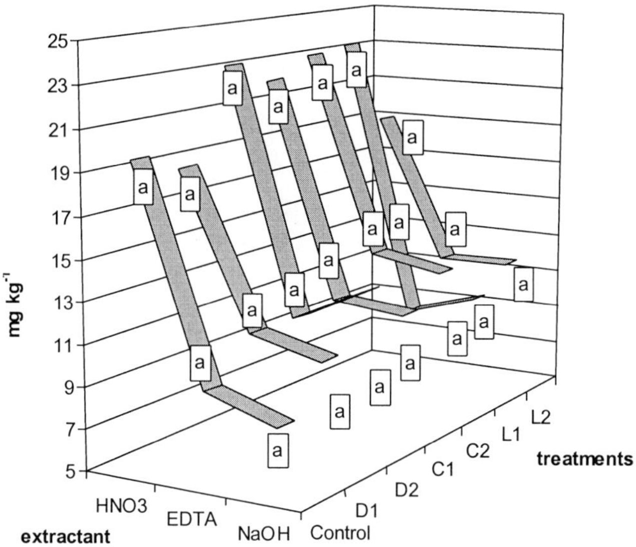

Figure 3. Copper fractions in soil. Comparisons regard concentrations obtained by the same extractants. Different letters indicate significant differences ($P \leq 0.05$).

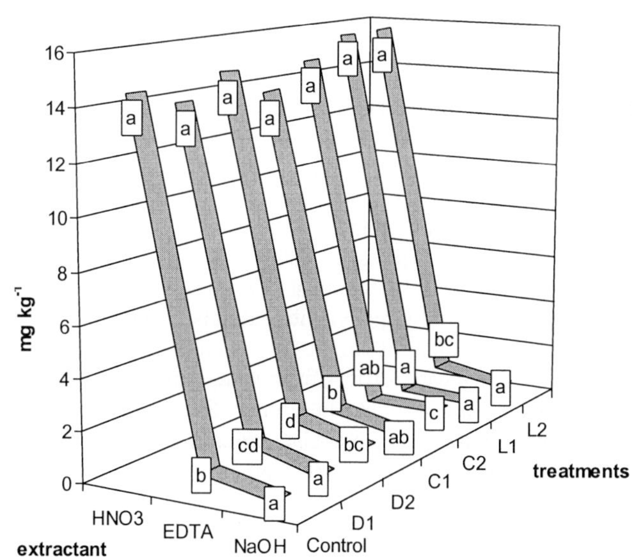

Figure 4. Nickel fractions in soil. Comparisons regard concentrations obtained by the same extractants. Different letters indicate significant differences ($P \leq 0.05$).

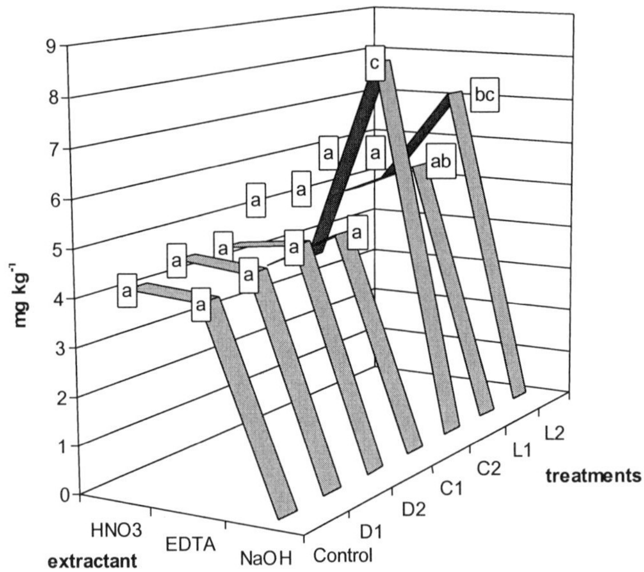

Figure 5. Lead fractions in soil. Comparisons regard concentrations obtained by the same extractants. Different letters indicate significant differences ($P \leq 0.05$).

4. CONCLUSIONS

The results show that different sludge processing techniques are likely to result in different heavy metal partitions in the sludge.

The composted sludge differs from the liquid and dehydrated sludges for in having higher amounts of EDTA-extractable metals; in contrast, the liquid and dehydrated sludges do not show strong differences with regard to the distribution of heavy metals. Besides this, in the composted sludge, the sum of EDTA and NaOH fractions exceeds the HNO_3 extractable fraction for Cu, Zn and Pb. Adversely, the most of Nickel in sludge is extracted by NaOH. In spite of the abundance of potentially bioavailable metals in the composted sludge, there was not a direct increase of metal availability in the soil amended with this sludge.

The partition of heavy metals clearly differs from sludge to soil. In the Control soil, after six years, the main changes in the heavy metal forms distribution involved copper and lead. Copper showed a shift towards the residual fraction; lead shifted towards potentially bioavailable forms (EDTA and NaOH fractions).

In all amended soils, except for lead, metals showed a decreasing trend with extractant: HNO_3 > EDTA > NaOH extracted. In the case of lead, residual and potentially available forms are of the same order of magnitude in both natural and treated soils.

Among the treatments, no differences were recorded among residual fractions. With regard to the potentially bioavailable fraction, the copper concentration in soil was not influenced by the treatments, but the other metals (zinc, nickel and lead) always showed some significant increases.

Finally, the application to soil of differently processed sludge over a medium-term period (six years) does not affect severely the natural heavy metal distribution of the soil. The few significant changes of potentially bioavailable fractions observed in amended soil are not necessarily linked to the type of sewage sludges applied (composted, dehydrated and liquid).

REFERENCES

1. Sequi, P., Figliolia, A., Benedetti, A., 1992. Utilizzo di Acque Reflue e Fanghi in Agricoltura: la chiusura del ciclo degli elementi nutritivi. *In*: A. Frigerio (Ed.), Acque Reflue e Fanghi. Innovazioni nel Trattamento e nello Smaltimento. 12 B-26 B. Centro Scientifico Internazionale, Milano.
2. Levi-Minzi, R., Riffaldi, R., Giudi, G., Poggio, G., 1985. Chemical characterization of soil organic matter in a field study with sewage sludge and compost. *In*: Williams, J.H., Guidi, G., L'Hermite, P.H. (Eds.), Long-term Effect of Sewage Sludge and Farm Slurries Applications. Applied Science Publication , London, pp. 151-160.
3. Rossi, G., Soldati, P., Penneli, B., Figliolia, A. 1999. Evaluation of the sewage sludge effects on organic matter and heavy metal accumulation in an Italian agricultural soil. *In:* Bech, J. (Ed.). Proceedings of 6th International Meeting on Soils with Mediterranean Type of Climate. Universitat de Barcelona, Barcelona., pp. 929-930.
4. Benedetti, A., 1985. About nitrogen fertilization in the model for potential yield. Annali dell'Istituto Sperimentale per la Nutrizione delle Piante - Roma. Vol. XII. Pubblicazione n. 5, 1-51.
5. Sequi, P., Benedetti, A., 1995. Management techniques of organic materials in sustainable agriculture. Integrated plant nutrition systems. FAO Fertil. Plant Nutr. Bull. 12, 139-154.
6. Figliolia, A., Izza, C., Mangione, D., Leita, L., Bragato, G., de Nobili, M., 1995. Evaluation of the effect on soil-plant system of heavy metals in sludge amended soil. *In:* CD-Proceedings of the Third International Conference on the Biogeochemistry of Trace Elements. Theme A. Biogeochemistry of Trace Elements. Paris., Symposium A3/09.
7. Senesi, N., Sposito, G., Holtzclaw, K. M., Bradford, G.R., 1989. Chemical properties of metal-humic acid fractions of a sewage sludge-amended aridisol. J. Environ. Qual. 18, 186-194.
8. Berrow, M.L., Burridge, J.C., 1980. Trace elements levels in soil: effects of sewage sludge. *In*: MAFF Reference Book N. 326 (Ed.). Inorganic Pollution and Agriculture. HMSO. London. pp. 159-190.
9. Sposito, G., Lund, L.J. Chang, A.C., 1982. Trace metal chemistry in arid-zone field soils amended with sewage sludge. I-Fractionation of Ni, Cu, Zn, Cd and Pb in solid phases. Soil Sci. Soc Am. J. 46, 260-264.
10. Dudka, S., Chlopecka, A., 1990. Effect of solid-phase speciation on metal mobility and phytoavailability in sludge-amended soil. Water Air Soil Poll. 51, 153-160.
11. Ross, S.M., 1994. Retention, transformation and mobility of toxic metals in soils. *In*: Ross, S.M. (Ed.), Toxic Metals in Soil Plant Systems. Wiley, Chichester. pp. 63-152.
12. FAO-UNESCO. 1998. Soil Map of the World. Revised Legend. World Soil Resources Report 60, Rome.
13. Genevini, P.L., Manstretta, M., Mecella, G., 1994. Preparazione del campione. *In*: Ministero delle Risorse Agricole, Alimentari e Forestali. Metodi ufficiali di analisi chimica del suolo. Osservatorio Nazionale Pedologico e per la Qualità del Suolo. Roma., pp. 33-36.

Developments in Soil Science, Volume 28A
Editors: A. Violante, P.M. Huang, J.-M. Bollag and L. Gianfreda

UPTAKE AND ACCUMULATION OF SELENIUM AND SULFUR BY PLANTS AS RELATED TO SOIL FACTORS IN POLAND

K. Borowska

Department of Biochemistry, University of Technology and Agriculture,
6 Bernardynska St., 85-029 Bydgoszcz, Poland

Plant and soil samples were collected from lucerne (*Medicago sativa* L.) plantations in the fourth year of cultivation, from the Kujawy and Pomorze regions of Poland. Total selenium content of plants and soils was determined spectrofluorometrically with the method of Watkinson and total sulfur content was assayed nephelometrically using the Bardsley and Lancaster method. Total selenium content in soils ranged from 0.204-0.238 $mg \cdot kg^{-1}$. Total sulfur content in black earths fluctuated from 0.40-0.67 $g \cdot kg^{-1}$ (mean 0.55 $g \cdot kg^{-1}$). Mean values (0.27 $g \cdot kg^{-1}$) for total sulfur content in brown and lessivé soils are lower compared to black earths. Selenium absorption by plants was not dependent on its total concentration in soils. Plant total sulfur concentrations ranged from 2.18-2.26 $g \cdot kg^{-1}$. Total selenium content in black earths and brown soils was negatively correlated with total sulfur content in lucerne. Statistical analysis showed a significant negative correlation between total selenium content in lessivé soils and total sulfur content in leaves, stems and roots of lucerne. In the present study the sulfur concentration in aerial parts and roots of dandelion was related to the total sulfur content of soils.

1. INTRODUCTION

In a global context, selenium (Se) is a complex but interesting element. The boundaries between animal toxicity and deficiency of Se are relatively narrow, and both phenomena are common around the globe [1]. Owing to the similarities between sulfur and selenium, the transport and the metabolism in the plant of the two elements could be expected to follow similar routes. In soil and plant samples selenium can exist in inorganic forms as elemental selenium, metal selenides, selenite and selenate and as volatile or non-volatile organic species with direct Se-carbon bonds such as methylated compounds, selenoamino acids, and selenoproteins [2]. In plants, selenium partly resembles sulfur in its biological properties and is able to replace sulfur in amino acids as well as in several biological processes [3]. The main source of selenium for animal fodder and human food is the plant-soil system. The levels of selenium in foodstuffs are dependent on soil Se forms and amounts, soil characteristics, agroclimatic conditions, cultivation practises, and the type of crop cultivated [4].

The purpose of this study was to gather more information about possible relationship between selenium and sulfur in soils and plants growing on them, using lucerne and dandelion from selected lucerne plantations.

2. MATERIALS AND METHODS

Plant and soil samples were collected from lucerne plantations in the fourth year of cultivation from Kujawy and Pomorze regions of Poland. The soils represented major soil types and, according to United Nations' Food and Agriculture Organization (FAO) classification, were classified as Gleyic Phaeosems (black earths), Mollic Cambisols (typical brown soils) and Luvisols (lessivé soils). They developed on the glacial till of the ground moraine of the Baltic glaciation (Würm) [5].

Soil surface samples were taken from the depth of 0-20 cm, air-dried and sieved through a 1mm screen. Plant material was sampled before bloom stage, rinsed in deionized water to remove soil particles and dried. Lucerne was separated into leaves, stems and roots and dandelion was separated into aerial parts and roots.

The total selenium content of plants and soils was determined with the method of Watkinson [6] using a Hitachi F-2000 spectrofluorometer. Plant and soil samples were wet-digested with a mixture of concentrated nitric and perchloric acids. The different forms of selenium in the samples were reduced to Se(IV) by boiling with 20% HCl. The selenium was complexed with 2,3-diaminonaphtalene (DAN) to give the fluorescent compound, which was extracted with cyclohexane and read on a spectrofluorometer at excitation and emission wave lengths of 376 and 519 nm, respectively.

Total sulfur content was assayed using the Bardsley and Lancaster method [7]. The samples were analyzed for granulometric composition according to the Bouyoucoss-Cassagrande method [8], organic carbon by wet oxidation with potassium dichromate [8], and pH in distilled water and 0.1 M KCl potentiometrically. All analyses were performed in duplicate.

3. RESULTS AND DISCUSSION

General properties of the soils under study are given in Table 1. The investigated soils had textures of fine sandy loam, sandy clay loam and sandy loam. The pH values of soils were in the neutral range. The soils differed in organic matter content: the organic matter content in surface horizons of black earths, brown soils and lessivé soils ranged from 1.48-2.64%, 1.05-1.59 and 0.07-1.72%, respectively.

Total selenium content in black earths ranged from 0.210-0.282 $mg{\cdot}kg^{-1}$ and was significantly positively correlated with organic matter content (r=0.38) and soil particle size fraction <0.02mm (r = 0.54). Total selenium content in the soils (Table1) was similar to average selenium content in surface horizons in soils of Poland (0.27 $mg{\cdot}kg^{-1}$) [3].

Total sulfur content in black earths ranged from 0.40-0.67 $g{\cdot}kg^{-1}$ (mean 0.55 $g{\cdot}kg^{-1}$). Mean values (0.27 $g{\cdot}kg^{-1}$) for total sulfur content in brown and lessivé soils are lower compared to black earths. Mocek et al. [9] reported concentrations of total sulfur in soils of Poland from 1.0-2.0 $g{\cdot}kg^{-1}$. Harward and Reisenauer [10] and Terelak et al. [11] obtained significantly higher amounts of total sulfur: 0.82-7.84 $g{\cdot}kg^{-1}$ and 1.5-3.7 $g{\cdot}kg^{-1}$, respectively. There was a significant correlation between total sulfur content in soils and soil particle size fraction <0.02mm (r = 0.48) and organic matter content (r = 0.43).

Table 1
Selected properties of soils studied

	Black earths	Typical brown soils	Lessivé soils
	range mean	range mean	range mean
Total selenium [mg·kg^{-1}]	0.210-0.282 0.232	0.095-0.284 0.204	0.183-0.281 0.238
Total sulfur [g·kg^{-1}]	0.40-0.67 0.55	0.24-0.33 0.27	0.10-0.75 0.28
Organic matter [%]	1.48-2.64 2.23	1.05-1.59 1.68	0.07-1.72 0.81
pH in H_2O	6.7-7.8	7.4-7.8	6.7-7.5
pH in KCl	6.1-7.2	6.8-7.0	6.0-6.8
Soil particle size fraction <0.002 mm [%]	9-16 13	10-15 13	5-18 9
Soil particle size fraction <0.02 mm [%]	16-31 24	17-26 21	12-27 18

The content of selenium in lucerne (Figure 1) gathered from black earths and lessivé soils approached the mean level of 0.047 mg·kg^{-1} and 0.095 mg·kg^{-1} for plants from brown soils. The selenium distribution in lucerne was of the following order: leaves>roots>stems. In earlier studies [12], the total concentration in lucerne stems collected in the first year of the lucerne stand was slightly higher than in roots. According to Kishchak [13], the quantity of Se in plants depends on the ability of plants to accumulate Se in their biomass and on the degree of fixation in soils. Selenium absorption by plants does not depend on its total concentration in soils. Plants absorb Se easily from alkaline soils, where it often exists in water-soluble forms. Although acid soils may contain high selenium concentrations, plants assimilate only small amounts because the Se is bound by insoluble iron compounds. Mayland [14] reported that on moderately low selenium soil, alfalfa accumulated more selenium than many other forage plants. This characteristic may be related to differences in rooting depth and to genetic traits that affect the absorption and translocation of selenium to shoots.

The distribution of total sulfur in lucerne is similar to the distribution of selenium (Figure 2). The highest amounts were obtained in the leaves, followed by the stems with the lowest concentration in the roots. According to Howarth [15], total sulfur in plant material generally ranged from 1.0-15.0 g·kg^{-1}. For herbage, extreme values of total sufhur content varied from 0.2-21.1 g·kg^{-1}, but under agricultural conditions, it was usually within the range of 2.0-4.5 g·kg^{-1}.

Total selenium content in black earths and brown soils was negatively correlated with total sulfur content in lucerne ($r = -0.54$ and $r = -0.73$, respectively). Statistical analysis confirmed the significant correlation between total selenium content in investigated lessivé soils and total sulfur content in leaves ($r = -0.72$), stems ($r = -0.58$) and roots ($r = -0.50$) of lucerne.

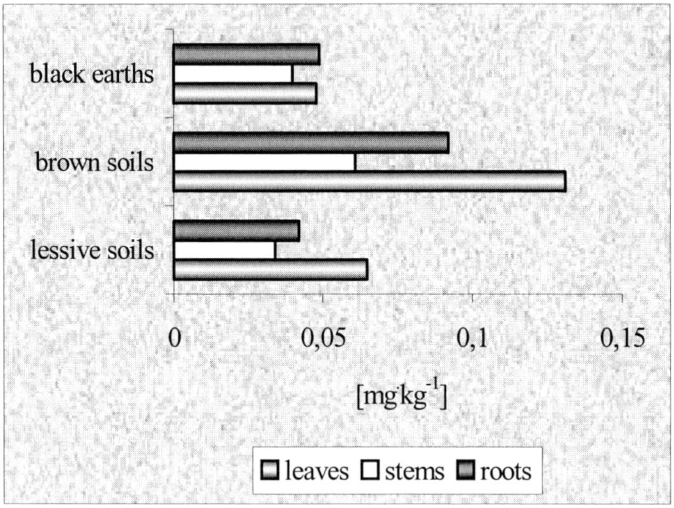

Figure 1. The selenium content [mg· kg^{-1}] in investigated parts of lucerne.

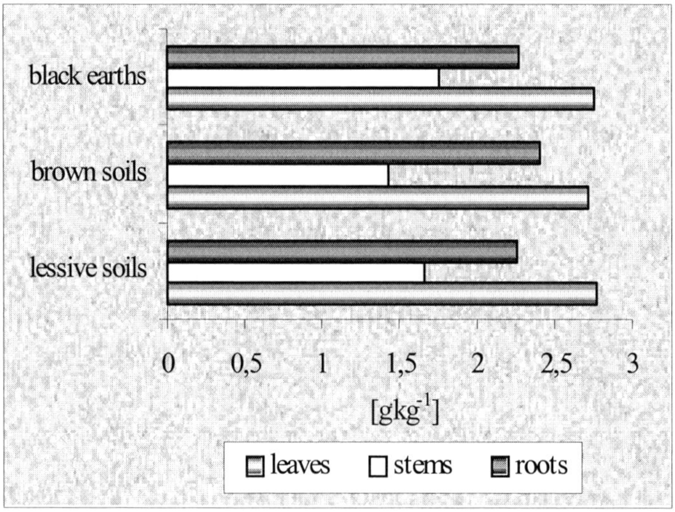

Figure 2. The sulfur content [g· kg^{-1}] in investigated parts of lucerne.

Because of many chemical and physical similarities between the two elements and the fact that selenium is present in the soil in minute quantities (< 10 mg·kg^{-1}), it is reasonable to assume that application of sulfur to the soil could reduce the uptake of selenium by plants [16].

Severson and Gough [17] reported that sulfur interferes with selenium uptake by plants and also with selenium metabolism in ruminants. Therefore, sulfur fertilization of pastures and rangelands could potentially predispose animals to selenium deficiency. According to Terry and Zayed [18], the absorption and translocation of selenate in plants is believed to resemble closely the uptake and movement of sulfate. Selenate is absorbed by the same carrier in root cell membranes as is sulfate. Sulfate ions have been shown to be antagonistic to the uptake of selenate because they compete for the same carrier on the root membrane. Competition between sulfate and selenate may also depend on the concentration of the two ions. Spencer [16] found no antagonism between the two elements when sulfur fertilizers were used at appropriate rates to correct sulfur deficiency on low selenium pastures. He concluded that a lowering of selenium concentration in the plant as a result of sulfur fertilization was due to a dry matter response to sulfur and a subsequent dilution of the selenium taken up by the plant. Sulfate levels may also affect the transport of selenium from roots to shoots. Singh et al. [18] showed that in the absence of sulfur, selenium tends to accumulate in the roots of *Brassica juncea* and in the presence of sulfur more selenium is translocated to the shoot.

The selenium and sulfur content and their distribution in the investigated parts of dandelion are presented in Figure 3 and Figure 4. In general, the content of both elements in aerial parts of dandelion was higher than that in roots. In the present study, the sulfur concentration in aerial parts ($r = -0.38$) and roots ($r = -0.44$) of dandelion was influenced by total sulfur content of black earths. As suggested earlier [12], a characteristic of dandelion as a Se-accumulating plant was confirmed. The importance of this is that dandelion is used more often for screening the environmental status of the soil, especially with respect to pollution with heavy metals.

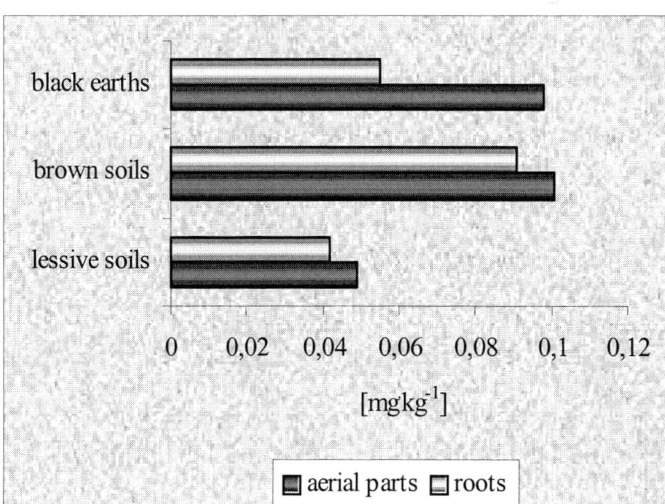

Figure 3. The selenium content [mg· kg^{-1}] in investigated parts of dandelion.

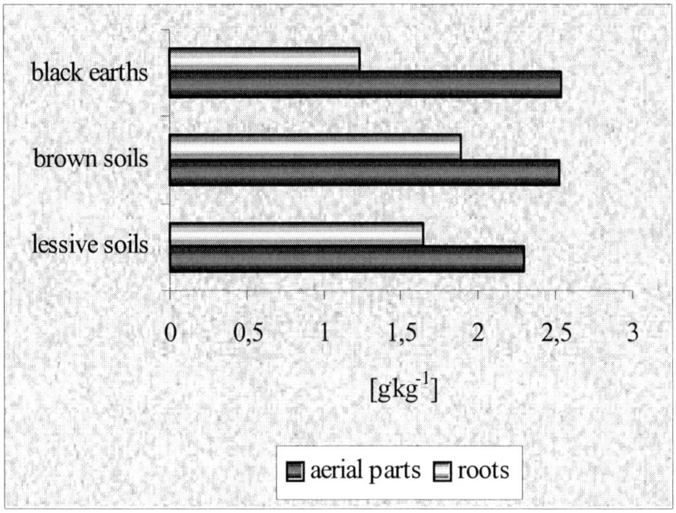

Figure 4. The sulfur content [g· kg^{-1}] in investigated parts of dandelion.

4. CONCLUSIONS

1. Total selenium content in soils under a fourth year of lucerne is similar to average selenium content in surface horizons in soils of Poland (0.27 mg·kg^{-1}) and in black earths was significantly positively correlated with organic matter and soil particle size fraction <0.02mm contents.
2. The mean total sulfur content in black earths was 0.55 g·kg^{-1}, whereas for brown and lessivé soils it was lower (0.27 g·kg^{-1}). Total sulfur content in soils was significantly positively correlated with organic matter and soil particle size fraction <0.02mm contents.
3. Total selenium and sulfur distribution in lucerne was in the following order: leaves> roots> stems. In general, the content of both elements in aerial parts of dandelion was higher than that in roots. Statistical analysis confirmed the significant negative correlation between total selenium content in investigated soils and total sulfur content in leaves, stems and roots of lucerne.

REFERENCES

1. Haygarth, P., 1994. Global importance and global cycling of selenium. *In:* Frankenberger Jr., J.W., Benson, S. (Eds.), Selenium in the Environment. Marcel Dekker, New York, pp. 1-27.
2. Martens, D.A., Suarez, D.L., 1998. Sequential extraction of selenium oxidation states. *In:* Frankenberger, Jr., J.W., Engberg, R.A. (Eds.), Environmental Chemistry of Selenium. Marcel Dekker, New York, pp. 61-79.
3. Kabata-Pendias, A., Pendias, H., 1999. Biogeochemia Pierwiastkow Sladowych. 3rd Ed. PWN, Warszawa.

4. Kumar, A., Krishnaswamy, K., 1997. Selenium content of common Indian cereals, pulses, and spices. J. Agric. Food Chem. 45, 2565-2573.
5. Ciesla, W., 1968. Geneza i Wlasciwosci Gleb Uprawnych Wytworzonych z Gliny Zwalowej na Wysoczyznie Kujawskiej. Rocz. WSR, Poznan.
6. Watkinson, J.H., 1966. Fluorometric determination of selenium in biological material with 2,3-diaminonaphtalene. Anal. Chem. 38, 92-97.
7. Bardsley, C.E., Lancaster, J.D., 1960. Determination of reserve sulfur and sulfates in soils, Soil Sci. Soc. Am. Proc. 24, 265-268.
8. Litynski, T., Jurkowska, H., Gorlach, E., 1976. Analiza Chemiczno-Rolnicza. PWN, Warszawa.
9. Mocek, A., Drzymala, A., Mazur, P., 1997. Geneza, Analiza i Klasyfikacja Gleb. AR Poznan.
10. Harward, M.E., Reisenauer, H.M., 1966. Reactions and movement of inorganic soil sulfur. Soil Sci. 101, 326-338.
11. Terelak, H., Motowicka-Terelak, T., Pasternacki, J., Wilkos, S., 1988. Zawartosc form siarki w glebach mineralnych Polski. Pam.Pul. suppl, 91, 5-14.
12. Borowska, K., 1998. The selenium content in lucerne from stands located on the Kujawy black earths. Proc. 17th EGF Meeting, Debrecen, Hungary, pp. 643-646.
13. Kishchak, I.T., 1998. Supplementation of selenium in the diets of domestic animals. 1998. *In:* Frankenberger Jr., J.W., Engberg, R.A. (Eds.), Environmental Chemistry of Selenium. Marcel Dekker, New York, pp.143-152.
14. Mayland, H.F., 1994. Selenium in plant and animal nutrition. In: Frankenberger Jr., J.W., Benson, S. (Eds.). Selenium in the Environment, Marcel Dekker, New York., pp. 29-46.
15. Howarth, R.W., 1992. The interactions of sulfur with other element cycles in ecosystems. *In:* Howarth, R.W., Stewart, J.W.B., Ivanov, M.V. (Eds.), Sulfur Cycling on the Continents. Wiley & Sons, Chichester, pp.67-84.
16. Murphy, M.D., Quirke, W.A., 1997. The effect of sulfur / nitrogen / selenium interactions on herbage yield and quality. I. J. Agric. Food Res. 36, 31-38.
17. Severson, R.C., Gough, L.P., 1992. Selenium and sulfur relationships in alfalfa and soil under field conditions, San Joaquin Valley, California. J.Environ.Qual. 21, 353-358.
18. Terry, N., Zayed, A.M., 1994. Selenium volatilization by plants. In: Frankenberger Jr., J.W., Benson, S. (Eds.), Selenium in the Environment. Marcel Dekker, New York, pp.343-367.

Developments in Soil Science, Volume 28A
Editors: A. Violante, P.M. Huang, J.-M. Bollag and L. Gianfreda

THE ROLE OF SOIL ORGANIC MATTER AND WATER POTENTIAL IN DETERMINING PESTICIDE DEGRADATION

K.M. Webb and L.A.G. Aylmore

Soil Science and Plant Nutrition, The University of Western Australia, Nedlands, Western Australia 6907

Four pesticides (Simazine, Metribuzin, Fenamiphos and Metalaxyl) chosen for their known or suspected potential to contaminate soil and groundwater, were investigated in terms of their degradation properties in a sandy soil from Medina, Western Australia. Samples were incubated for up to 190 days in a batchwise arrangement with periodic monitoring and sampling. A range of soil physical, chemical and biological properties were simultaneously measured on the samples. Measurements obtained principally allow comparison of residual soil pesticide concentrations with soil water potential and soil respiration rate over time. Results show variation across pesticide type and soil conditions (principally moisture), however trends do not always relate well to a first-order degradation approach and treatment of the data. The surface soil containing the highest organic matter and highest moisture content produced the greatest rates of degradation and generally showed very good agreement with first order degradation behaviour (corroborated with other measures of soil activity). In sub-surface soils conditions apparently conducive to pesticide breakdown were not always realized and the data significantly challenge the view that first order degradation can be used to describe pesticide breakdown in this soil profile.

1. INTRODUCTION

The potential for pesticides to contaminate soil profiles and groundwater has become a major concern throughout the world. Leaching of pesticides is primarily determined by the recharge rate, pesticide sorption and degradation properties in the soil. Knowledge of these factors is essential to the successful development of practical management models [1-3].

The key soil component responsible for sorption of non-ionic organic compounds is soil organic matter. In most agricultural and horticultural soils, sorption is regarded as a hydrophobic partition of the non-ionic organic compounds from the soil aqueous phase to soil organic matter [4]. Sorption retards the downward leaching velocity of pesticides by water in the soil profile.

The rate of degradation essentially determines the rate at which the pesticide compound is eliminated from the soil environment. However, in the case of pesticides, some intermediate metabolites may be just as toxic as, or more toxic than, the parent compound (e.g. fenamiphos nematicide which transforms to its sulfoxide and sulfone forms). In agricultural and horticultural soils, degradation is primarily due to microbial processes [5]. The rate of microbial degradation for a given pesticide depends on two fundamental variables: (1) availability of the pesticide for degradation; and (2) density and activity of the degrading microbes or extra-cellular enzymes. Other soil and environmental variables, such as soil organic matter content, moisture, temperature, pH and aeration status, affect the degradation rate by modifying these two fundamental variables. In addition, individual pesticides differ significantly in their vulnerability to degradation due to their intrinsic structural differences.

Degradation of pesticides in soil is commonly assumed to obey first order kinetics , without regard to soil biomass, organic matter or moisture content, such that the fraction of pesticide remaining un-degraded as it leaches in the soil is expressed as

$$F = \exp(-kt) = \exp(-0.693t/t_{1/2}) \quad (1)$$

where k is the degradation rate constant, t is time and $t_{1/2}$ is the degradation half-life.

Recent studies have raised questions as to the validity of this assumption [6].

Most of the half-life values that have been studied and reported, whether obtained from field or laboratory investigations, are for surface soils [7]. However, whether or not a pesticide will reach ground water in significant amounts is not only affected by the degradation rate in the surface layer, but by the degradation rate in every soil layer above the ground water (the unsaturated zone). As it is impractical to measure the degradation rate of every pesticide in every soil layer of the profile, it is usually assumed when modelling the pesticide ground water pollution potential, that the half-life increases with depth in response to decreases in microbial density and activity, caused, in particular, by decreases in the concentration of organic substrate [2,8]. This assumption, however, has not been experimentally verified. The product of soil microbial biomass content and soil microbial activity is essentially an index of the metabolic state of the soil biomass. By measuring these quantities in the same soil sub-samples from which residual pesticide is determined, valuable corroborating evidence as to the significance and meaning of measured degradation half-lives (in a specific soil sample) can be obtained. The present investigation sought to clarify the interaction of these factors in determining pesticide degradation with depth in a sandy soil of Western Australia.

2. MATERIALS AND METHODS

2.1. Soil properties

Soil samples were taken from the Agriculture Western Australia Medina Vegetable Research Station, Western Australia. The site contained remnant native vegetation, had no previous history of pesticide application and had been partially disturbed by the itinerant movement of and provision for penned emus. The soil, which is a Karrakatta sand of the Spearwood Association was collected as 2 fractions - above approximately 20cm and below 25cm to approximately 50cm. A distinct change in soil composition is noted at approximately 25cm. A buffer zone from approximately 20-25cm was not collected. The upper and lower fractions were ascribed the labels Karrakatta Topsoil (KT) and Karrakatta Subsoil (KS). After collection, two separate batches of KT and KS were air dried (at 30°C), sieved (4mm) and re-bagged for storage. Organic carbon was determined by analysis by LECO® 1000 CHN analyser, pH by measurement on 0.01M $CaCl_2$ solution extract and soil moisture-potential characteristics by gravimetric analysis of soil samples equilibrated with tension plate and pressure plate apparatus. Relevant properties of the soil are summarized in Table 1.

Table 1
Measurements made on the soils selected for incubation studies

Soil	Texture	pH (0.01M $CaCl_2$)	Organic C (w/w%)	Field Capacity (v/v %)	Bulk Density (g/cm^3)
Topsoil (0-25 cm)	Sand	5.5 ± 0.1	1.1± 0.1	4.8 ± 0.1	1.31 ± 0.03
Subsoil (25-50 cm)	Sand	5.4 ± 0.1	0.14 ± 0.02	2.1 ± 0.1	1.47 ± 0.01

2.2. Laboratory incubation

Four pesticides widely used in horticulture and agriculture in the region were studied (Table 2). These pesticides represent a significant range in water solubility and sorption coefficient [4].

Table 2
Pesticides studied and their properties

Common name	Chemical name	Uses	Water solubility† (mg/L)	k_{oc}†
Simazine	2-chloro-4,6-bis(ethylamino)-1,3,5-triazine	Herbicide	3.5	130
Fenamiphos	Ethyl 4-methylthio-*m*-tolyl isopropylphosphoramidate	Nematicide	400	100
Metalaxyl	Methyl *N*-(2-methoxyacetyl)-*N*-(2,6-xylyl)-DL-alaninate	Fungicide	8400	61
Metribuzin	4-amino-6-*tert*-butyl-3-methylthio-1,2,4-triazine-5(4*H*)-one	Herbicide	1220	268

† From Worthing [9] and Wauchope et al. [10].

For each pesticide, amounts of moist soil equivalent to 1 kg oven-dry soil were used. The moisture contents of soil samples were initially adjusted to those equivalent to water potentials (Ψ) of -5 and -100 kPa respectively. Moisture content was then maintained by periodic (weekly to fortnightly) monitoring and amendment with the required amount of water. Some difficulty in maintaining the water potential at –100kPa was experienced, owing to the tight range in soil moisture-potential characteristic for this soil below ψ of –10kPa. 10 mg active pesticide ingredient per kg, simulating usual application rates, was thoroughly mixed with the soil and incubated at constant 18°C under aerobic conditions. Replicate bags were set up for each pesticide and whole bags removed for sampling at different intervals up to 190 days after application. Sub-samples from each removed bag were bulked together for each sampling time, and thoroughly mixed for subsequent extraction and analysis.

Detailed analytical procedures can be found in Kookana et al. [7]. The pesticides were extracted by shaking 5 g moist soil overnight with 10 mL of methanol. The solution was then centrifuged and filtered through a 0.22 μm membrane filter. Pesticide concentration in the solution was then determined on a high performance liquid chromatograph (HPLC, Waters®) equipped with a multiple wavelength UV/visible detector (model 490), an autosampler (model 717plus) and an automated pump controller (model 600E). Reversed-phase chromatography was employed (isocratically) using a Waters® μBondapak-C18 column with acetonitrile (50-70% v/v) in water (50-30% v/v) as mobile phase. Analyses were carried out at 20 °C with 10-20 μL injections and a flow rate of 1mL min^{-1}. In addition to the five pesticides, the two toxic metabolites of fenamiphos - fenamiphos sulfoxide and fenamiphos sulfone, were also analysed. Separation of F. sulfoxide and F. sulfone peaks was obtained by monitoring multiple wavelengths with the detector [11]. The recoveries varied with pesticides. Reproducible recoveries better than 80% were obtained for all of the pesticides studied. Pesticide concentrations in the soil were calculated on the basis of the recoveries at day zero. Soil microbial biomass was obtained by measurement of Ninhydrin positive compounds following chloroform fumigation and extraction with 0.5M K_2SO_4 [12] and soil respiration rates by back titration (with HCl) of KOH which had been reacted with CO_2 evolved from 7 day incubation in a sealed container.

3. RESULTS AND DISCUSSION

For all the chosen pesticides, the surface soil containing the highest organic matter and highest moisture content produced the greatest rates of degradation. Those shaded in Table 3 were able to be fitted to a first order degradation model (R^2 of 0.67 to 0.96).While residual pesticide breakdown often followed first order degradation behaviour, in some cases conditions conducive to pesticide breakdown were not fully realized, for example for simazine, fenamiphos and metalaxyl in the sub-soils (for these, reliable half-lives could not be estimated, Table 3).

Figures 1 to 5 illustrate the residual pesticide concentrations, soil respiration rates and water potentials measured over the incubation period together with first order degradation fits to the data where relevant. Soil Microbial Biomass (SMB) was unaltered in all soil samples as a consequence of the presence of the pesticides and did not vary significantly between treatments in comparison to controls. Initial SMB carbon was approximately 200 and 50 g kg^{-1} soil for topsoil and subsoil respectively and decreased to approximately half these values over time.

Biomass respiration rates have an effective detection limit of 50 mg CO_2-C/kg soil and so respiration data presented should be considered in light of this limit.

Table 3
First-order degradation data fits for all experiments

Pesticide	Sample*	(Ψ,kPa)	R^2	$t_{1/2}$ (d)
Fenamiphos	Topsoil KT	-5	0.96	35
	Topsoil	-100	0.61	27
	Subsoil KS	-5	0.13	
	Subsoil	-100	0.02	
Metribuzin	Topsoil	-5	0.94	145
	Topsoil	-100	0.89	212
	Subsoil	-5	0.93	222
	Subsoil	-100	0.96	193
Metalaxyl	Topsoil	-5	0.96	181
	Topsoil	-100	0.59	428
	Subsoil	-5	0.19	
	Subsoil	-100	0.54	
Simazine	Topsoil	-5	0.67	84
	Topsoil	-100	0.94	107
	Subsoil	-5	0.02	
	Subsoil	-100	0.26	

For fenamiphos, the soil conditions which most clearly demonstrated degradation were those in the higher organic matter topsoil (KT)/higher moisture treatment (-5kPa, Figure 1) with first order fit R^2 0.96 giving $t_{1/2}$ of 35 days). Here soil respiration rate and residual pesticide concentrations over the incubation period related well to one another indicating a clear dependence of pesticide degradation on microbial activity. In the sub-soil (KS) respiration rates and corresponding degradation rates were substantially lower and did not conform to first order degradation kinetics (Table 3).

Metribuzin demonstrated the most consistent degradation behaviour with depth, (Table 3 and Figure 2). For this pesticide, good first order degradation fits (R^2 0.88-0.96) applied to all treatments giving half-lives between 145 and 222 days (Table 3). Note that the degradation half-life for the soil with the lowest organic matter and moisture contents is lower than for all but the highest organic matter and moisture soil treatment. This suggests the possibility of an abiotic degradation mechanism, since there is little difference in the rates of degradation with different treatments. While residual metribuzin concentration was closely related to respiration rate in the surface soil, this was less evident in the sub-soil.

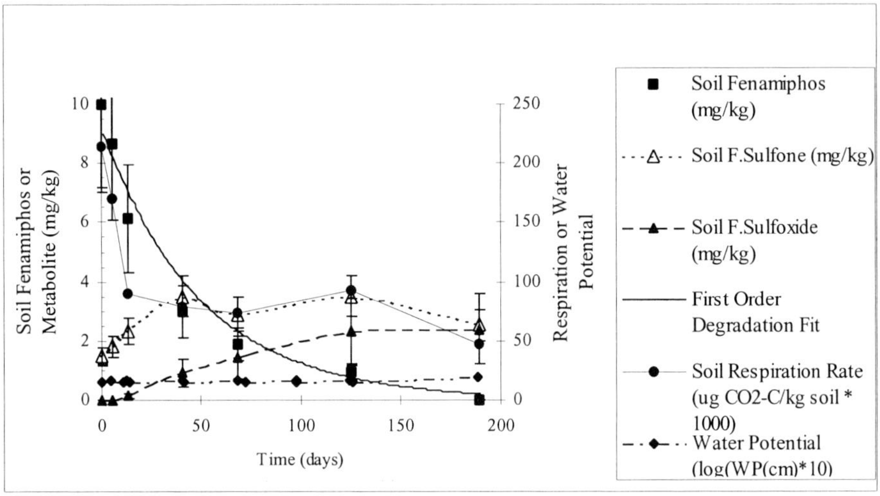

Figure 1. Fenamiphos degradation – Karrakatta surface soil (ψ-5kPa).

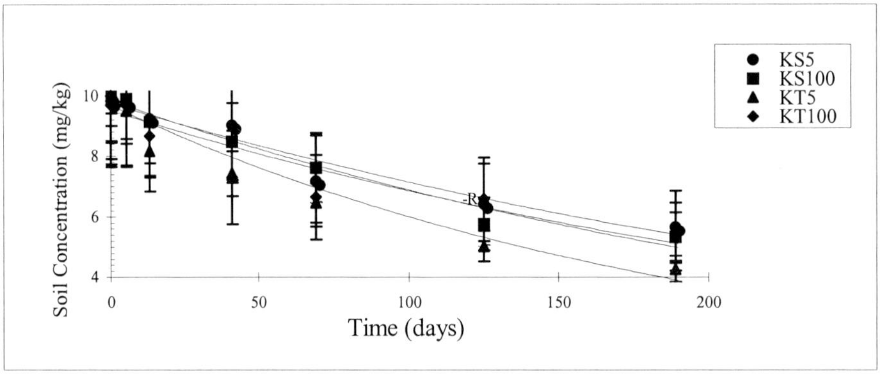

Figure 2. Metribuzin Degradation - Karrakatta Sand.

Metalaxyl demonstrated decreasing degradation with depth (Figure 3) and soil moisture potential but only conformed to first order degradation kinetics at the high moisture potential in the topsoil (Table 3).

Simazine demonstrated the least consistent degradation behaviour compared to metribuzin, fenamiphos and metalaxyl, conforming to first order degradation in the topsoil (Figure 4) but exhibiting essentially no degradation in the subsoil (Figure 5).

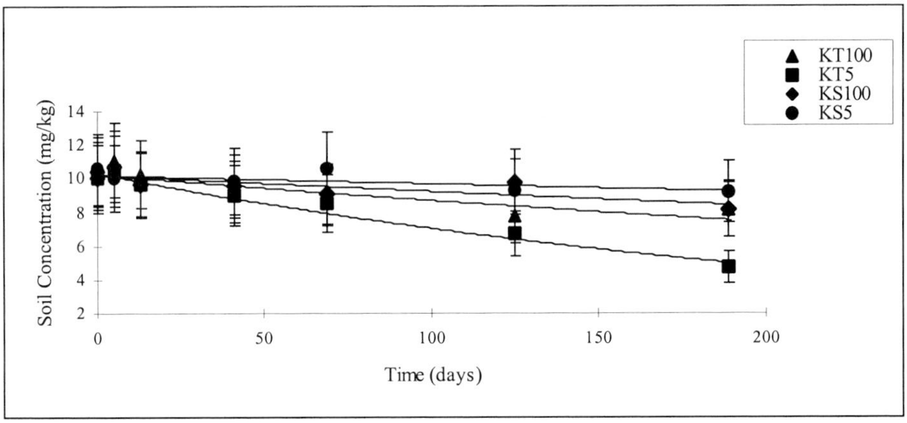

Figure 3. Metalaxyl degradation – Karrakatta sand.

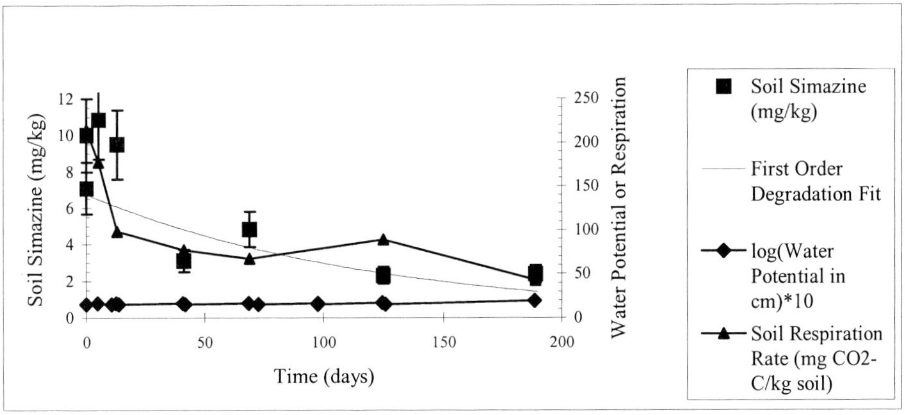

Figure 4. Simazine Degradation - Karrakatta Soil (KT Ψ -5kPa).

Previous studies in these laboratories have show that degradation half-lives of pesticides in subsoil layers are not always longer than in the surface layer despite significantly lower organic matter content in the subsoil layers [8]. There may be several causes for these unexpected results. Firstly, soil organic matter may affect both fundamental variables that determine the degradation rate. On the one hand, it may increase the microbial density and activity by providing organic substrate; on the other hand, it may reduce the availability of pesticide compound for degradation by sorbing the pesticides [13]. The magnitude of the two opposing effects may vary depending on a number of factors: the sorption capacities of the organic materials; the capacity of the organic material to stimulate microbial activities; and the water solubility and sorption coefficient of the pesticide. Secondly, microbial activities also depend on

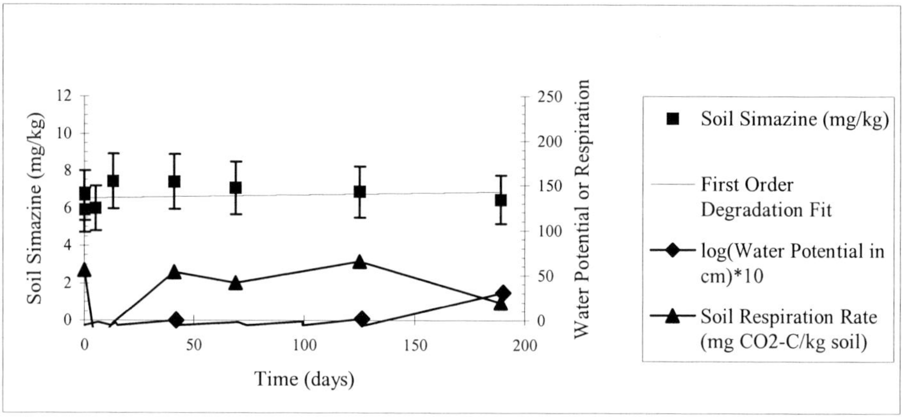

Figure 5. Simazine Degradation - Karrakatta Soil (KS Ψ -5kPa)

other environmental conditions, e.g. moisture, temperature, pH and aeration status which may vary between soil layers. In the present study decreasing soil moisture generally resulted in a significant decrease in degradation rate, although this is not always the case (e.g. fenamiphos in the topsoil and metribuzin across all treatments).

Thirdly, contribution by abiotic processes (e.g. chemical degradation) may also differ between soil layers. The concentration and composition of organic substrate at different layers of the soil profile and moisture content are thus key factors which affect both the availability of pesticide for degradation by sorption and the microbial activity by providing substrate and conducive conditions. These experiments illustrate the interaction between moisture potential, organic matter, microbial biomass and respiration of soil in the degradation of pesticides.

4. CONCLUSIONS

The surface soil containing the highest organic matter and highest moisture content produced the greatest rates of degradation and were able to be fitted to a first order degradation model (R^2 of 0.67 to 0.96). Pesticide degradation rates were in most cases generally much lower, sometimes negligible, in the sub-surface soil compared to the surface soil, although in many cases respiration rates appeared comparable at both depths. Degradation rates generally decreased with decreasing soil moisture content over the potential range -5kPa to -100kPa. For all pesticides other than metribuzin, there appeared to be good correlation between soil respiration rate and pesticide degradation rate in the topsoil. Thus soil respiration rate appears a valid gauge of the presence of a predominantly biotic pesticide degradation process. The results of these experiments also show that fluctuating soil moisture potentials have a major effect on the degradation rate of pesticides, particularly those thought to undergo biotic degradation. This has implications for the realistic representation of pesticide degradation rates (*viz.* half-lifes) in predicting pesticide fate and transport in landscapes which experience major fluctuations in soil water potential over time. The results challenge the view that first order degradation kinetics can invariably be applied to pesticide breakdown in soil. Recognition of these complexities is essential for the development and successful application of practical management models.

ACKNOWLEDGMENTS

Funding of this research by the Australian Research Council is gratefully acknowledged.

REFERENCES

1. Rao, P.S.C., Davidson, J.M., 1980. Estimation of pesticide retention and transformation parameters required in nonpoint source pollution models. *In*: Overcash, M.R., Davidson, L.M (Eds.), Environmental Impact of Nonpoint Source Pollution. 67 Ann Arbor Science Publishers: Ann Arbor, MI. pp. 23.
2. Jury, W.A., Focht, D.D., Farmer, W.J., 1987. Evaluation of pesticide groundwater pollution potential from standard indicies of soil-chemical adsorption and biodegradation. J. Environ. Qual. 16, 422-428.
3. Aylmore, L.A.G., Di,. H.J., 2000. Predicting the groundwater pollution potential of pesticides under variable recharge, Aust. J. Soil Res. 38, 591-602.
4. Green, R.E., Karickhoff, S.W., 1990. Sorption estimates for modeling, *In*: Cheng, H.H. (Ed.), Pesticides in the Soil Environment: Processes, Impacts and Modeling. Soil Sci. Soc. Amer., Madison, WI, pp.79-101.
5. Tortensson, . N.T.L., 1987. Microbial decomposition of herbicides in soil. *In*: Hutson, D.H., Roberts, T.R. (Eds.) Herbicides. John Wiley and Sons, Ltd. New York. pp. 249-270.
6. Di, H.J., Aylmore, L.A.G., Kookana, R.S., 1998. Degradation rates of eight pesticides in a sandy soil from laboratory incubation, field study and simulation. Soil Sci. 163, 404-411.
7. Kookana, R.S., Di, H.J., Aylmore, L.A.G., 1995. A field study of leaching and degradation of nine pesticides in a sandy soil. Aust. J. Soil Res. 33, 1019-1030.
8. Di, H.J., Kookana, R.S., Aylmore, L.A.G., 1995. Application of a simple model to assess the groundwater contamination potential of pesticides. Aust. J. Soil Res. 33, 1031-1040.
9. Worthing C.R., 1983. The Pesticide Manual, 7th Ed. The British Crop Protection Council, Suffolk, UK.
10. Wauchope, R.D., Buttler, T.M., Hornsby, A.G., Augustjin-Beckers, P.M.W. Burt, J.P., 1992. The SCS/ARS/CES pesticide properties database for environmental decision-making. Rev. Environ. Contamin. Toxicol. 123, 1-164.
11. Singh, R., 1989. Simultaneous determination of Fenamiphos, its sulfoxide and sulphone in water by high performance liquid chromatography. Analyst 114, 425-427.
12. Amato, P.L., Ladd, J.N., 1988. Assay for microbial biomass based on ninhydrin-reactive nitrogen in extracts of fumigated soils. Soil Biol.Biochem. 20, 107-114.
13. Hamaker, J.W., 1972. Decomposition: quantitative aspects. *In*: Goring, C.A.I., Hamaker. J.W. (Eds), Organic Chemicals in the Soil Environment. Vol. 1. Marcel Dekker, New York, pp. 253-340.

Developments in Soil Science, Volume 28A
Editors: A. Violante, P.M. Huang, J.-M. Bollag and L. Gianfreda

VARIABILITY OF PESTICIDE MINERALIZATION IN INDIVIDUAL SOIL AGGREGATES OF MILLIMETER SIZE

L. Vieublé[a, c], C. Chenu[a] and G. Soulas[b]

[a]INRA- Science du Sol, Rte de St Cyr, 78026 VERSAILLES CEDEX

[b]INRA-Microbiologie des Sols, Bd Sully, BV 1540, 21034 DIJON CEDEX

[c]Corresponding author: Fax: +33 1.30.83.32.59; E-mail: lvieuble@versailles.inra.fr

The aim of this study was to compare the 2,4-D mineralization in individual aggregates of millimeter size (3 size classes and 96 aggregates per size class). $^{14}CO_2$ coming from the mineralization of ^{14}C ring-labeled 2,4-D and evolved by each aggregate incubated in microtiter plates was trapped with barium hydroxide on filters and measured using a Phosphorimager. We observed an important variability of mineralization in aggregates of each size class and in the different size classes of aggregates. The main factors responsible for these fluctuations could be a heterogeneous distribution of degrading microorganisms or of available carbon necessary for cometabolism.

1. INTRODUCTION

Microbial degradation of pesticides in soil is a key process controlling the fate of these toxic compounds into soil, water and air. Pesticides that should be readily biodegradable have been found to persist for years in soils, even though a degrading population is present and environmental conditions such as temperature, water and oxygen availability are not limiting [1]. Among the factors controlling the biodegradation of xenobiotics in soils, such persistence is generally attributed either to the formation of bound residues, *i.e.*, the incorporation of pesticide molecules into soil organic matter, or to limitations in the availability of pesticides to soil microorganisms. The latter can be due to adsorption of pesticides to soil particles or to mass transfer limitations. The biodegradation of pesticides by soil microorganisms requires contact between the molecule and extracellular enzymes and/or the cell of the decomposer. The relative locations of pesticide, degrading microorganisms and the possibilities of diffusion are likely to control the mineralization rate [1].

Soil is extremely heterogeneous as a habitat for microorganisms, and its chemical composition, as well as the physical and chemical conditions vary at a micro-scale. Several studies were able to demonstrate a heterogeneous distribution of soil microorganisms using either microscopic methods [2-4] or fractionation methods [5, 6]. However, most of these studies deal with the total microbial population; only a few address specific microbial populations [7, 8]. In particular, little is known about the distribution of pesticide-degrading microbial communities in soil structure. A patchy spatial distribution of such populations is expected to influence the dissipation rate of pesticides in soils.

This study was undertaken to assess the spatial distribution of microbial pesticide degraders at millimeter scale in soil. We selected as a model 2,4-dichlorophenoxyacetic acid (2,4-D), which is a soil herbicide that has been used extensively for decades. We measured the mineralization rate of 2,4-D in soil aggregates of various sizes. Specific questions were:

- Do aggregates of different sizes exhibit different 2,4-D mineralization rates?
- Are all aggregates within a size class able to mineralize 2,4-D?
- If heterogeneity in mineralization is observed, is it related to soil characteristics?

2. MATERIALS AND METHODS

Soil samples were taken from the plowed layer (0-30 cm) of a cultivated orthic cambisol at the experimental fields of INRA situated in the park of Versailles palace (France). The soil is a silt loam (sand 33%, silt 50%, clay 17%) with 13.5 g kg^{-1} carbon content, 1.27 g kg^{-1} nitrogen, and a pH of 6.8. The plot is under continuous wheat and had never received 2,4-D previously. The soil was sieved moist into 2-3.15 mm, 3.15-5 mm and 5-7 mm size classes. Free roots and plant debris were discarded during the sieving. Prior to use, the aggregates were preincubated for 7 days at 20°C.

For each aggregate size class, 96 aggregates were placed individually in the wells of microtiter plates. Each aggregate was amended with an aqueous solution of ^{14}C ring-labeled 2,4-D (Isotopchim, specific activity of 15 mCi/mmol) at the concentration of 7.8 μg 2,4-D g^{-1} oven dry aggregate and the water content of 0.41 g water g^{-1} oven dry soil (2.24 times the field capacity). The microtiter plates were incubated at 20°C, ensuring no loss of water from the system. The evolved $^{14}C\text{-}CO_2$ was locally trapped in a paper filter impregnated with barium hydroxide (56 g L^{-1}), which was placed on top of the microtiter plates. After 5, 8 and 12 days, the filters were changed and the radioactive spots were analyzed using a Phosphorimager with reference to a standard curve, as proposed by Tabor et al. [9] and as adapted to soils by Fulthorpe et al. [10].

The microbial biomass was determined by fumigation extraction according to Vance et al. [11]. 5 g oven dry soil were used for each aggregate size class. Soluble carbon and chloroform-solubilized carbon were measured with a DC 190 elemental analyzer. A conversion factor of 2.64 was used to calculate biomass C [11]; nine replicates were performed for each aggregate size class.

The C and N contents of the aggregate classes and of 10 individual aggregates within each size class were measured by combustion in a C&N Carlo Erba Elemental analyzer. Particulate organic matter was separated after dispersion of 15 g of aggregates for each size class by sieving at 50, 100, 200 and 500 μm and flotation-panning [12]. The separation was done in triplicate for each aggregate size class.

3. RESULTS

3.1. Mineralization rates of 2,4-D in aggregates of different size classes

The mean mineralization rates were calculated for ^{14}C-CO_2 evolved from 96 individual aggregates of each size class and the kinetics were plotted (Figure 1). At the beginning of the incubation mineralization was faster in 5-7 mm and 3-5 mm aggregates than in 2-3 mm aggregates and for the 3 aggregate classes, a maximum was attained after 8 days of incubation.

After 12 days of incubation, the extent of 2,4-D mineralization was still in the order 5-7 mm = 3.15-5 mm > 2-3.15 mm and the differences were significant.

3.2. Mineralization of 2,4-D in individual soil aggregates

The extent of mineralization was very heterogeneous in individual aggregates of each size class (Figure 2). We considered that mineralization was nil when it was less than 0.05 μg 2,4-D g^{-1} soil. All aggregates >3.15 mm exhibited positive 2,4-D mineralization, whereas 11.5% of 2-3.15 mm aggregates were below this threshold after 5 days of incubation. After 5 days, the distribution of ^{14}C-CO_2 evolved from positive aggregates ranged from 0.05-to 7.8-μg 2,4-D g^{-1} soil, *i.e.*, a factor of 160. For all aggregates the distributions were normal with maximums at 3-4 μg 2,4-D mineralized g^{-1} soil (3.15-5 mm) and 2-3 μg 2,4-D mineralized g^{-1} soil (5-7 mm). Most 2-3.15 mm aggregates mineralized less than 1 μg 2,4-D g^{-1} soil. After 12 days of incubation, the heterogeneity decreased in aggregates with coefficients of variation (standard deviation/average value of 2,4-D mineralization) of 40-70% (after 5 days) to 20-60% (after 12 days) according to size classes of aggregates and distributions remained normal. After 12 days, all aggregates of the 2-3.15 mm size class were able to mineralize 2,4-D.

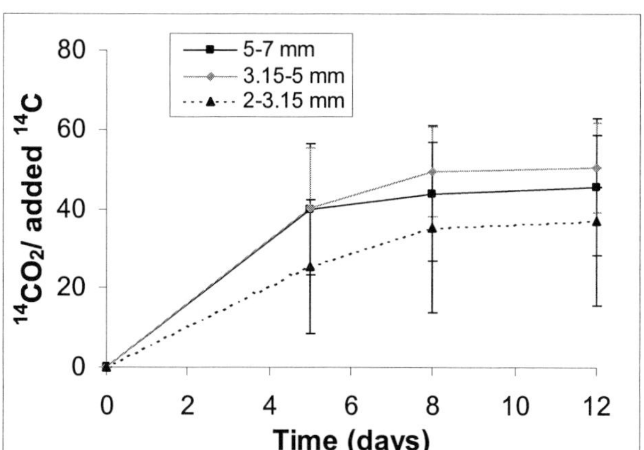

Figure 1. Mean 2,4-D mineralization with time in the 3 size classes of aggregates.

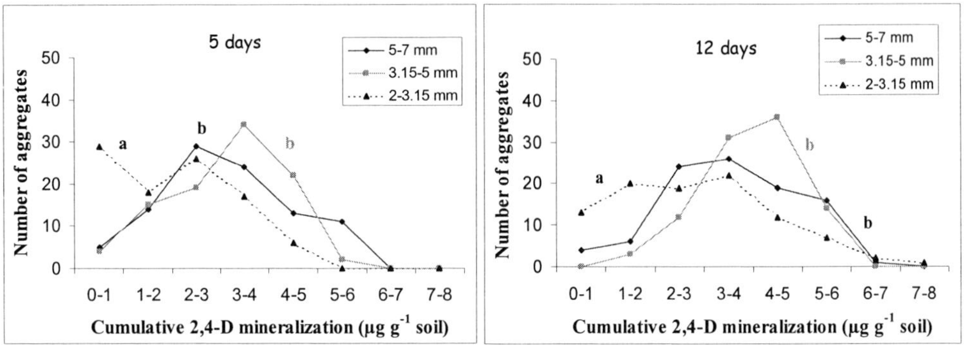

Figure 2. Heterogeneity after 5 and 12 days of 2,4-D mineralization in individual aggregates.

3.3. Characteristics of aggregate classes

Microbial biomass increased with aggregate size (r^2=0.84) (Table 1). The C and N contents and the abundance of particulate organic matter of different sizes did not vary significantly among aggregate size classes (ANOVA P=0.05) (Table 2, Figure 3). It was not possible to quantify either the microbial biomass or particulate organic matter on individual aggregates, but we measured the C and N content of 10 individual aggregates per size class. C and N contents were heterogeneous in individual aggregates (Figure 4), with coefficients of variation (standard deviation/average value of C or N content) of 11 to 20% for C and of 5 to 10% for N that increased as the aggregate size decreased. Aggregates with high C content and low N content probably contained plant residues.

Table 1
Total microbial biomass of the 3 size classes of aggregates

Size classes of aggregates	*Microbial biomass (μg C g^{-1} soil)*
Total soil	207 +/- 37
5-7 mm	166 +/- 24
3.15-5 mm	163 +/- 13
2-3.15 mm	133 +/- 13

Table 2
Carbon (C), nitrogen (N) and particulate organic matter (POM) contents of the 3 size classes of aggregates

Size classes of aggregates	*C (mg g^{-1})*	*N (mg g^{-1})*	*C/N*	*POM (mg g^{-1})*
5-7 mm	**12.4** ± 0.2	**1.19** ± 0.00	**10.4** ± 0.2	**6.35** ± 1.01
3.15-5 mm	**13.2** ± 0.9	**1.23** ± 0.02	**10.7** ± 0.8	**6.54** ± 1.37
2-3.15 mm	**14.1** ± 1.7	**1.23** ± 0.03	**11.3** ± 1.2	**4.84** ± 0.52

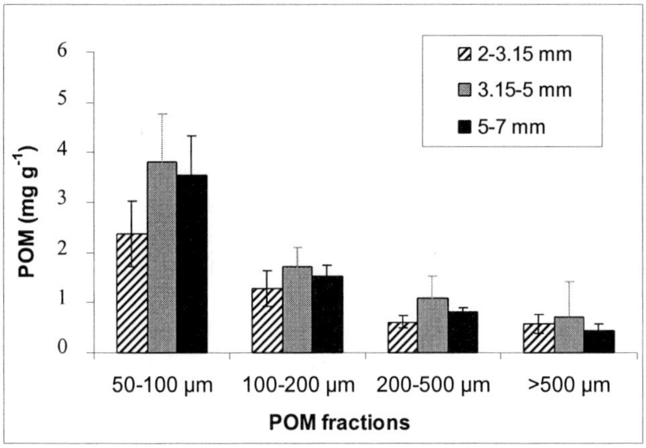

Figure 3. Particulate organic matter (POM) (mg g^{-1}) in different granulometric fractions.

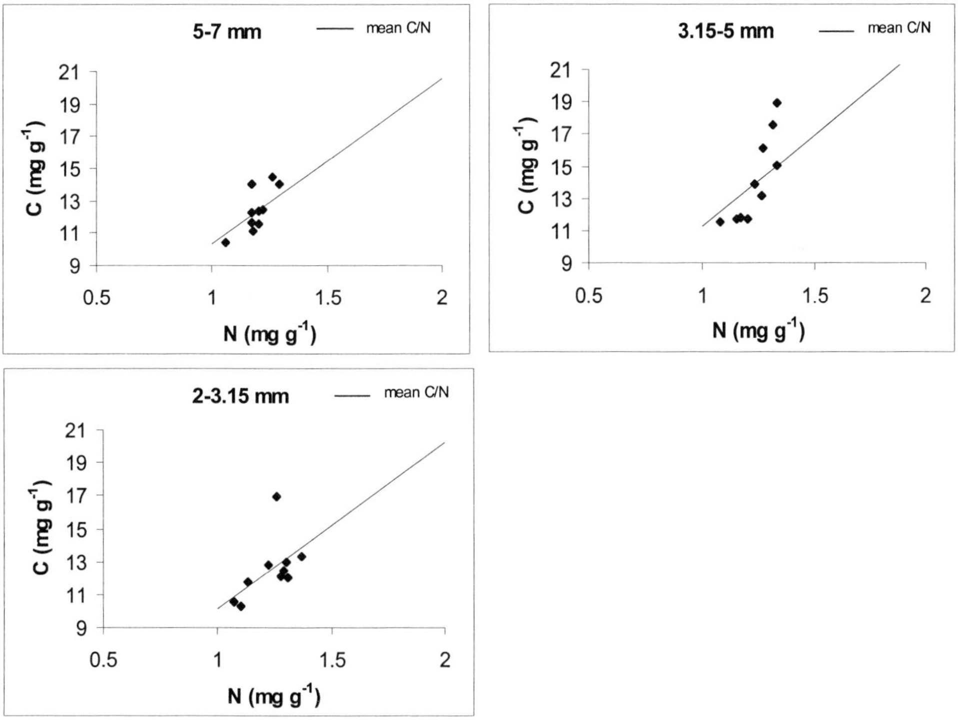

Figure 4. Heterogeneities of C and N contents in 10 aggregates of each size class of aggregates.

4. DISCUSSION

4.1. Mineralization of 2,4-D

The investigated soil mineralized 2,4-D rapidly even though it had never received any 2,4-D previously. In a side experiment, (data not shown), we showed that whatever the class of size of aggregates considered, mineralization was preceded by a lag phase of at least two days corresponding to the growth of degrading microorganisms. If one takes into account this, aggregates would rather exhibit sigmoid mineralization curves. The presence of a lag phase confirms that with a concentration of 7.8 μg/g of soil, the degradation of the 2,4-D is partly under the control of a relatively specific microflora which is naturally present in the aggregates but initially not very abundant. The microflora would be more limiting in the smallest aggregates. The extent of mineralization, *i.e.*, a plateau of 2,4-D mineralization at 40-50% of added ^{14}C, was in agreement with what is usually described for 2,4-D [13].

4.2. Spatial heterogeneity of 2,4-D mineralization

We demonstrated a large distribution of 2,4-D mineralization in aggregates of a given size class and in aggregates of different sizes. Several studies have shown that the mineralization of soil organic matter was different for aggregates of different sizes. Soil respiration was generally found to decrease with aggregate size [14-16]. This could be explained by higher contents of labile soil organic matter and particulate organic matter [17-19] and by a greater soil microbial biomass in large aggregates [6, 14, 16, 20, 21]. However, to our knowledge, no studies have been performed on pesticide mineralization in aggregates of different sizes. Vallaeys et al. [22], using a technique similar to that of ours, found that only 8% of soil aggregates from a cultivated soil never exposed to 2,4-D were able to degrade chain-labeled 2,4-D. However, these authors did not compare the kinetics of 2,4-D mineralization in the aggregates. Lors [23], performing kinetics of 2,4-D mineralization, studied effect of the dinitro-o-cresol on the 2,4-D degraders, but on only one size class of aggregates.

4.3. Factors of the heterogeneity of 2,4-D mineralization

The main factors controlling the biodegradation rates of pesticides in soil are, besides their chemical nature, the environmental conditions (temperature, water availability, oxygen availability), the accessibility and availability of the pesticide (entrapment and diffusion effects, adsorption) and the presence and size of the microbial degrader population. 2,4-D is partly degraded in soil by co-metabolism, which can account for 30% of the mineralization [24]. Co-metabolizing microorganisms rely on the availability of easily degradable carbon sources to degrade the pesticide.

Our incubation set up provided the same physical environmental conditions to all aggregates, and these can be considered to be non-limiting (especially water and oxygen availability). Since 2,4-D was added to each aggregate, the accessibility of 2,4-D could be considered to be more or less the same in a given size class.

Heterogeneity of 2,4-D mineralization among individual aggregates and among aggregate size classes is thus expected to result from: (i) the size of the degrading population, *i.e.*, the distribution of microorganisms in aggregates; (ii) the availability of carbon for co-metabolizing microorganisms and (iii) the extent of 2,4-D or 2,4-D metabolites adsorption to soil aggregates.

We observed that the larger the aggregates were, the more important microbial biomass C and mineralization were. Voos and Groffman [25] found that 2,4-D degradation was correlated with microbial biomass and Veeh et al. [26] found that it was correlated with the total bacterial population as estimated by plate counts. It is expected that the size of the 2,4-D-degrading population, especially that of the zymogene cometabolitic population [27], increases with the total microbial population. The degradation of 2,4-D involves different microbial species able to accomplish the mineralization of 2,4-D's ring or lateral chain [22]. Furthermore, microorganisms do not always have all the enzymes required for complete degradation of the 2,4-D. Ka et al. [28] and Fulthorpe et al. [10], who studied the degradation of 2,4-D in different soils, observed an important diversity of decomposer microorganisms. Particularly in soils with no history of 2,4-D applications, consortia of bacteria may play a major part in the degradation of the herbicide. Vallaeys et al. [22] analyzed the size and the genetic diversity of microbial populations able to degrade 2,4-D in 3 soil aggregates of mm size. Using PCR-RFLP analysis of 16S rRNA, these authors identified 17 different strains able to accomplish a step in 2,4-D chain degradation, but only 4 among the 17 were present in the each of the 3 aggregates. Their results also demonstrated the involvement of diverse genes in 2,4-D degradation.

We observed no significant differences in C, N or particulate organic matter contents in different aggregate classes, but found that individual aggregates differed in their C and N contents. Hence, we could not verify the idea that aggregate classes differ in the availability of carbon sources for co-metabolism, but the hypothesis still holds at the scale of individual aggregates.

According to Ogram et al. [29] and Greer and Shelton [30], the fraction of 2,4-D that is present in solution is preferentially mineralized by soil microorganisms, but the adsorbed fraction can also be decomposed. 2,4-D is known to adsorb to soil constituents and particularly to organic matter by hydrophobic links [31]. In a high organic matter soil, degradation rates were lower than in a low organic matter soil [30]. Mineralization depends on sorption-desorption kinetics and on the nature of the sorbent [32]. However, when the organic carbon content was more than 12%, both the adsorption and the rate of degradation of 2,4-D increased, probably because of increased activity by the co-metabolic microorganisms [33]. The availability of 2,4-D as a consequence of its adsorption to soil constituents may contribute to the observed heterogeneity if differences in organic matter content or quality among individual aggregates or aggregate classes result in significantly different adsorptions. Furthermore, products of 2,4-D metabolism like 2,4-dichlorophenol or chlorophenol are known to adsorb to soil constituents too [13]. Thus, the metabolites of the 2,4-D would be less available for the microorganisms: in this case, the 2,4-D degradation would not be total and would not go to the ultimate stage of mineralization.

5. CONCLUSIONS

The use of microtiter plates allowed us to measure 2,4-D mineralization rates in soil structure units as small as 2-3.15 mm, or 20 mg in mass. We demonstrated large fluctuations in 2,4-D mineralization potential among aggregates. Such fluctuations could be ascribed to the heterogeneous distribution, at this scale, of degrading microbial populations, of labile organic substrates necessary for co-metabolism, or the adsorption potential of 2,4-D and its metabolites. Our results emphasize the need for careful soil sampling, taking large enough samples, during pesticide degradation studies. Furthermore, the impact of heterogeneous distribution of 2,4-D-degrading microorganisms among individual aggregates on the biodegradation rate of this pesticide remains to be assessed.

ACKNOWLEDGMENTS

We are grateful to G. Catroux for fruitful discussions and to N. Rouard for technical assistance.

REFERENCES

1. Scow, K. M., 1993. Effect of sorption-desorption and diffusion processes on the kinetics of biodegradation of organic chemicals in soil. *In:* Linn, D. M. (Ed.), Sorption and Degradation of Pesticides and Organic Chemicals in Soil. Soil Sci. Soc. Am. (SSSA), Madison, USA, pp. 73-114.
2. Kilbertus, G., 1980. Etude des microhabitats contenus dans les agrégats du sol. Leur relation avec la biomasse bactérienne et la taille des procaryotes présents. Rev. Ecol. Biol. Soil 17, 543-557.
3. .Foster, R. C., 1988. Microenvironments of soil microorganisms. Biol. Fertil. Soils 6, 189-203.
4. Gaillard, V., Chenu, C., Recous, S., Richard, G., 1999. Carbon, nitrogen and microbial gradients induced by plant residues decomposing in soil. Europ. J. Soil Sci. 50, 567-578.
5. Hattori, T., 1988. Soil Aggregates as Microhabitats of Microorganisms. Institute for Agricultural Research Tohoku University, Japan, pp. 23-36.
6. Jocteur-Monrozier, L., Ladd, J. N., Fitzpatrick, R., Foster, R. C., Raupach, M., 1991. Physical properties, mineral and organic components and microbial biomass content of size fraction in soils of contrasting aggregation. Geoderma 49, 37-62.
7. Nishiyama, M., Senoo, K., Wada, H., Matsumoto, S., 1992. Identification of soil micro-habitats for growth, death and survival of a bacterium, 1,2,3,4,5,6-hexachlorocyclohexane-assimilating *Sphingomonas paucimobilis*, by fractionation of soil. FEMS Microbiol. Ecol. 101, 145-150.
8. Kabir, M., Chotte, J. L., Rahman, M., Bally, R., Monrozier, L. J., 1994. Distribution of soil fractions and location of soil bacteria in a vertisol under cultivation and perennial grass Plant Soil 163, 243-255.
9. Tabor, H., Tabor, C. W., Haffner, E. W., 1976. Convenient method for detecting $^{14}CO_2$ in multiple samples application to rapid screening for mutants. J. Bacteriol. 128, 485-486.
10. Fulthorpe, R. R., Rhodes, A. N., Tiedje, J. M., 1996. Pristine soils mineralize 3-chlorobenzoate and 2,4-dichlorophenoxyacetate via different microbial populations. Appl. Environ. Microb. 62, 1159-1166.
11. Vance, E. D., Brookes, P. C., Jenkinson, D. S., 1987. An extraction method for measuring soil microbial biomass C. Soil Biol. Biochem. 19, 703-707.
12. Balesdent, J., Pétraud, J. P. P., Feller, C., 1991. Effets des ultrasons sur la distribution granulométrique des matières organiques des sols. Sci. Sol 29, 95-106.
13. Soulas, G., Fournier, J. C., 1987. Cinétiques comparées des dégradations dans le sol du 2,4-D et du 2,4-dichlorophénol seuls ou en mélange. Conséquences sur le comportement des biomasses microbiennes dégradantes correspondantes. Agronomie 7, 193-199.

14. Gupta, V. V. S. R., Germida, J. J., 1988. Distribution of microbial biomass and its activity in different soil aggregate size classes as affected by cultivation. Soil Biol. Biochem. 20, 777-786.
15. Scheu, S., 1992. Automated measurement of the respiratory response of soil microcompartments: Active microbial biomass in earthworm faeces. Soil Biol. Biochem. 24, 1113-1118.
16. Franzluebbers, A. J., Arshad, M. A., 1997. Soil microbial biomass and mineralizable carbon of water-stable aggregates. Soil Sci. Soc. Am. J. 61, 1090-1097.
17. Beare, M. H., Hendrix, P. F., Coleman, D. C., 1994. Water-stable aggregates and organic matter fractions in conventional- and no-tillage soils. Soil Sci. Soc. Am. J. 58, 777-786.
18. Puget, P., Chenu, C., Balesdent, J., 1995. Total and young organic matter distributions in aggregates of silty cultivated soils. Europ. J. Soil Sci. 46, 449-459.
19. Puget, P., Besnard, E., Chenu, C., 1996. Une méthode de fractionnement des matières organiques particulaires des sols en fonction des agrégats. Comptes Rendus à l'Académie des Sciences, Paris, série II, 322, 965-972.
20. Chotte, J. L., Monrozier, L. J., Villemin, G., Albrecht, A., 1993. Soil microhabitats and the importance of the fractionation method. *In:* Mulongoy, K., Merckx, R. (Eds.). Soil Organic Matter Dynamics and the Sustainability of Tropical Agriculture. John Wiley & Sons, Leuven, Belgium, pp. 39-45.
21. Singh, S., Singh, J. S., 1995. Microbial biomass associated with water-stable aggregates in forest, savanna and cropland soils of a seasonally dry tropical region, India. Soil Biol. Biochem. 27, 1027-1033.
22. Vallaeys, T., Persello-Cartieaux, F., Rouard, N., Lors, C., Laguerre, G., Soulas, G., 1997. PCR-RFLP analysis of 16S rRNA, tfdA and tfdA and tfdB genes reveals a diversity of 2,4-D degraders in soil . FEMS Microbiol. Ecol. 24, 269-278.
23. Lors, C., 1997. Impact des produits phytosanitaires sur la diversité spécifique et fonctionnelle de la microflore du sol: Cas du dinitro-*o*-crésol, Thèse de doctorat, Université de Metz, UFR Sciences fondamentales et appliquées.
24. Robertson, B. K., Alexander, M., 1994. Growth-linked and cometabolic biodegradation: Possible reason for occurrence or absence of accelerated pesticide biodegradation. Pestic. Sci. 41, 311-318.
25. Voos, G., Groffman, P. M., 1997. Relationships between microbial biomass and dissipation of 2,4-D and dicamba in soil. Biol. Fertil. Soils 24, 106-110.
26. Veeh, R. H., Inskeep, W. P., Camper, A. K., 1996. Soil depth and temperature effects on microbial degradation of 2,4-D. J. Environ. Qual. 25, 5-12.
27. Soulas, G., 1993. Evidence for the existence of different physiological groups in the microbial community responsible for 2,4-D mineralization in soil. Soil Biol. Biochem. 25, 443-449.
28. Ka, J. O., Holben, W. E., Tiedje, J. M., 1994. Genetic and phenotypic diversity of 2,4-dichlorophenoxyacetic acid (2,4-D)-degrading bacteria isolated from 2,4-D-treated field soils. Appl. Environ. Microb. 60, 1106-1115.
29. Ogram, A. V., Jessup, R. E., Ou, L. T., Rao, P. S. C., 1985. Effects of sorption on biological degradation rates of 2,4-dichlorophenoxyacetic acid in soils. Appl. Environ. Microb. 49, 582-587.
30. Greer, L. E., Shelton, D. R., 1992. Effect of inoculant strain and organic matter content on kinetics of 2,4-dichlorophenoxyacetic acid degradation in soil. Appl. Environ. Microb. 58, 1459-1465.

31. Benoit, P., Barriuso, E., Houot, S., Calvet, R., 1996. Influence of the nature of soil organic matter on the sorption-desorption of 4-chlorophenol, 2,4-dichlorophenol and the herbicide 2,4-dichlorophenoxyacetic acid (2,4-D). Europ. J. Soil Sci. 47, 567-578.
32. Benoit, P., Barriuso, E., Soulas, G., 1999. Degradation of 2,4-D, 2,4-dichlorophenol, and 4-chlorophenol in soil after sorption on humified and nonhumified organic matter. J. Environ. Qual. 28, 1127-1135.
33. Bolan, N. S., Baskaran, S., 1996. Biodegradation of 2,4-D herbicide as affected by its adsorption-desorption behaviour and microbial activity of soils. Aust. J. Soil Res. 34, 1041-1053.

Developments in Soil Science, Volume 28A
Editors: A. Violante, P.M. Huang, J.-M. Bollag and L. Gianfreda

THE EFFECT OF SOIL MINERAL-ORGANIC MATTER INTERACTION ON SIMAZINE ADSORPTION AND DESORPTION

Á. Zsolnay[a], M. C. Hermosín[b], A. Piccolo[c] and L. Gianfreda[c]

[a]Institut für Bodenökologie, GSF; D-85764 Neuherberg bei München, Germany

[b]Instituto de Recursos Naturales y Agrobiologiá de Sevilla, CSIC, Avd. de Reina Mercedes 10, E-41080 Sevilla, Spain

[c]Dipartimento di Scienze Chimico-Agrarie, Via Universitá 100, I-80055 Portici, Italy

The desorption behaviour of simazine from a pure clay (montmorillonite), two "pure" humic substances, and from clay-humic complexes was investigated using a dynamic column system. The mobile and non-mobile fractions of simazine, the first order desorption coefficient, and the *in situ* distribution coefficient were measured. Generally, the desorption properties of simazine from the clay-humic complexes could not be predicted from those obtained from the studies done with a pure clay or humic substances alone.

1. INTRODUCTION

The regulation of pesticide usage requires a predictive capability, which is ideally based on a mathematical evaluation using indices [1] or deterministic and stochastic models [2-6]. To accomplish this not only are the soil's physical and hydrological properties needed but also the adsorptive/desorptive behaviour of the pesticide on the soil's matrix must be estimated. These processes are largely affected by two major soil components: the humic substances [*e.g.* 7-9] and the clay minerals [*e.g.* 10-13]. Under natural conditions only a very small portion of the humic substances are in a free state [14]. This puts into question the value of experiments done with extracted humics. By the same token, pure clays are most likely not present in the environment. They are complexed in some manner to the humic materials. This has been reflected in some studies dealing with the adsorption and desorption of pesticides in soils [15-17].

It is generally presumed that clay-humic complexes do not behave in an additive manner. That is to say the sorption properties of the complexes can not be predicated by the sorption attributes of their components. This, however, has not been sufficiently documented in experiments. In this study, the desorption properties of a well studied herbicide, simazine (2-chloro- N, N′-diethyl-1,3,5-triazine-2,4-diamine) from clay with and without the presence of two strongly differing humic substances was investigated. The ultimate goal was to see to what the degree the behaviour of the clay-humic complexes was a sum of its parts.

Experimentally a batch approach could have been used in which matrix bound simazine is sequentially extracted. However, the use of small columns, which are continually eluted, was felt to be superior [19]. With this approach very small amounts of samples could be investigated, and the results readily modelled to provide more detailed desorption parameters than could have been obtained with batch desorption studies.

2. MATERIALS AND METHODS

The matrices used in this study are given in Table 1. The humic substances selected were not obtained from soil, since it was felt that it would be more informative to investigate the behaviour of humic acids known to have significantly different attributes. The selected pesticide was simazine, which has been widely used and for which a great deal of scientific literature is available. The degree to which the humic substances could be complexed with the clays was determined by combusting the matrices in a Carlo Erba NA1500 and by multiplying the organic carbon contents by 1.8. Pesticide binding on the different matrices was obtained by (1) adding 50 mg of matrix to a 10 ml saturated solution of simazine (ca. 25 μ*M*); (2) shaking at room temperature for 24 h; and (3) filtering and washing once with 10 ml of 10m*M* $CaCl_2$. The resulting concentration of simazine on the different matrices can be seen in Figure 1. In both cases the complexing of humic substances to the clay resulted in less simazine adsorption than on the clay alone. This basically means that humic substances, despite their high affinity for simazine, interfere with the ability of clay to adsorb this pesticide.

Table 1
Matrices used in this study

AM18	Montmorillonite complexed with 18 meq of Alx(OH)y g^{-1}
COX	Humic substance obtained from oxidised coal (relatively rich in aromatic groups)
LIG	Humic substance obtained from lignite (relatively rich in aliphatic and carboxylic groups)
AM18-COX	1.93% COX co-precipitated with AM18
AM18-LIG	1.85% LIG co-precipitated with AM18

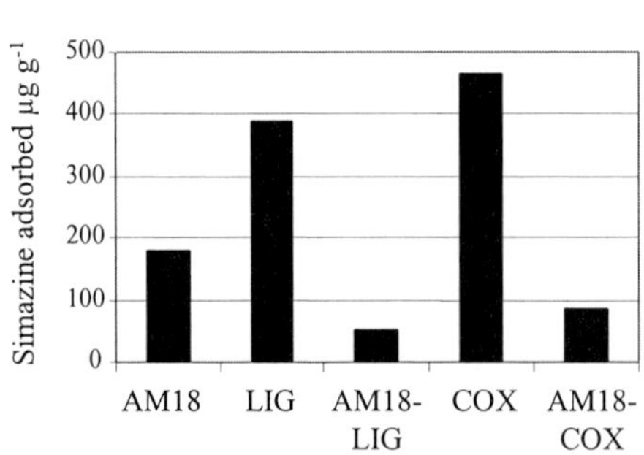

Figure 1. The amount of simazine adsorbed on the various matrices used.

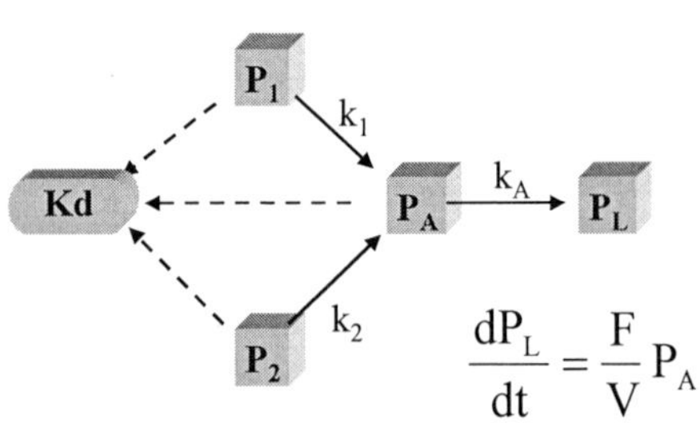

Figure 2. Schematic illustration of the model approach used. Cf. text for details.

The desorption study approach used is described in greater detail elsewhere [18]. A total of 70 mg of matrix with bound simazine was mixed with 15g of quartz sand and placed in small columns, which were continually eluted at room temperature in the dark with a pH 7 phosphate buffer. The flow rate was 65 mm h^{-1}, which was equivalent to 1.6 pore volumes h^{-1}. The leachate was collected at regular intervals and analysed without pre-treatment in an HPLC with a C-8 reverse phase column and UV detector. Simazine was found not to adsorb on the quartz sand. All runs were done in triplicate.

It was assumed that the pesticide was in two pools, a readily desorbable one (P_1 in Figure 1) and a poorly desorbable one (P_2). The desorption into the aqueous phase (P_A) is described by the first order desorption constants k_1 and k_2. Finally the pesticide is leached out of the column (P_L) at the rate k_A. In Figure 1 F is the flow rate and V the volume of the aqueous phase. Kd is the distribution coefficient.

$$Kd = \frac{(P_1 + P_2)\,V}{P_A M} \tag{1}$$

where M is the mass of the solid phase. An example of the results and the calculated curve are given in Figure 3. Simazine, which leached out within 70 h was considered to be relatively mobile. The rest was the relatively immobile fraction. Under the conditions used here and because of the short elution times, microbial metabolism of the pesticide was considered to be negligible.

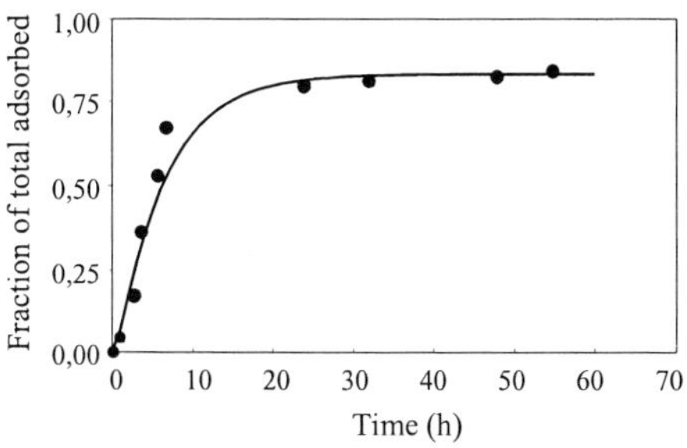

Figure 3. Cumulative breakthrough curve for AM18-COX.

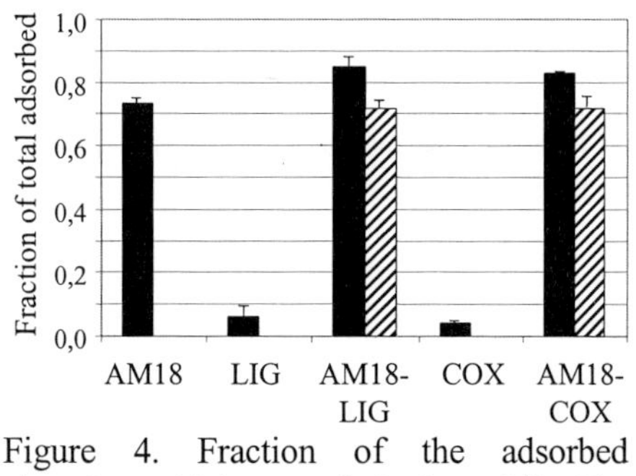

Figure 4. Fraction of the adsorbed simazine, which was relatively mobile. The solid bars are measured values, the hatched ones estimated. The lines above the bars are the standard errors.

3. RESULTS

The fraction of the adsorbed simazine, which was relatively mobile can be seen in Figure 4, while the relatively immobile amount is illustrated in Figure 5. The estimated effects (E_E) were obtained by assuming that there components were additive:

$$E_E = (F_C\, E_C) + (F_H\, E_H) \tag{2}$$

where F_C and F_H are the clay and humic substance fractions, respectively and E_C and E_H the effects of the individual components. Since in this study the masses of the complexes were about 98% clay, the estimated effects were largely the same as the measured effect of the clay alone. Although simazine was far more desorbable from clay alone than from the humic substances, the fraction of relatively mobile simazine was not altered to a large degree when humic substances were complexed to the clays. If anything the addition of humic substances to the clay tended to decrease its ability to bind with simazine. Therefore an additive estimate would over predict simazine binding by a factor of about two (Figure 5).

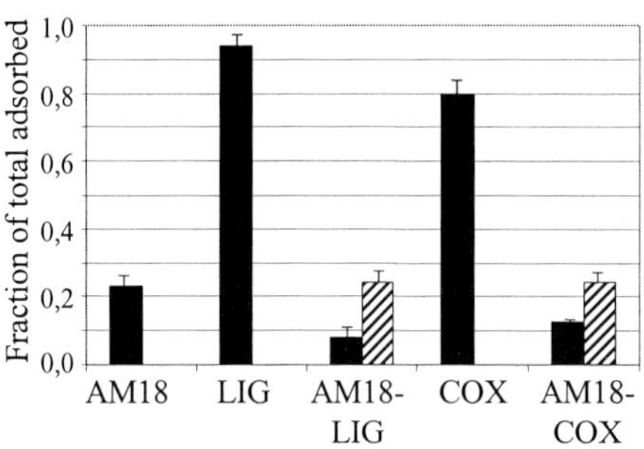

Figure 5. Fraction of the adsorbed simazine, which was relatively immobile. The solid bars are measured values, the hatched ones estimated. The lines above the bars are the standard errors.

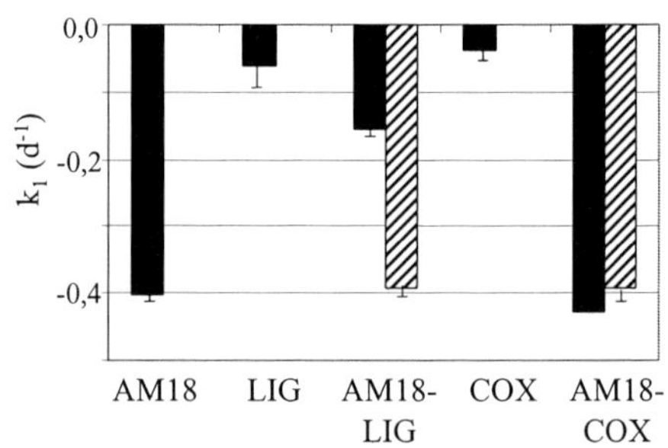

Figure 6. First order desorption constants for the mobile fraction. The solid bars are measured values, the hatched ones estimated. The lines below the bars are the standard errors.

Figure 6 shows that qualitative aspects of the desorption can be quite different than the quantitative ones. The addition of lignite humic substances to the clay resulted in a significantly lower desorption rate then from the clay alone, even though the total amount of mobile simazine had been the same for both matrices (Figure 4). The addition of LIG resulted in a desorption, which was more spread out and did not result in a sharp peak. This is also reflected in the maximum *in situ* concentrations (Figure 7) and could have ecological applications, since many processes are dependent on the concentration and not necessarily on the total amount of a given compound.

Simazine, which had been adsorbed on the more aromatic humic material from oxidised coal (COX), behaved similarly to that, which had been adsorbed on LIG, in regards to the rate of desorption and the *in situ* concentration. However contrary to its desorption behaviour from AM18-LIG, simazine, which had been adsorbed on AM18-COX, desorbed somewhat more rapidly and had a significantly higher *in situ* concentration than the simazine, which desorbed from the pure clay (Figures 6 and 7). The reason for these results can only be conjectured upon, but the point should be made that the effect of humic substances on the physical properties of clays may differ dramatically, depending on the chemical composition of the humics.

This contrary behaviour of LIG and COX, can also be seen in their effect on the distribution coefficient (Equation 1, Figure 8). All values are much higher than what one would expect from batch studies, but it must be kept in mind that the physical environment within a flow-through column with a very small amount of active matrix is quite different than from that in the shaken batch environment. Kd for pure COX is considerable higher than for LIG, but when complexed to the clay, the more aromatic humic substance suppresses the Kd value obtained with the pure clay matrix, while LIG tends to enhance it. In both cases the behaviour of the complex can not be predicted from the individual components.

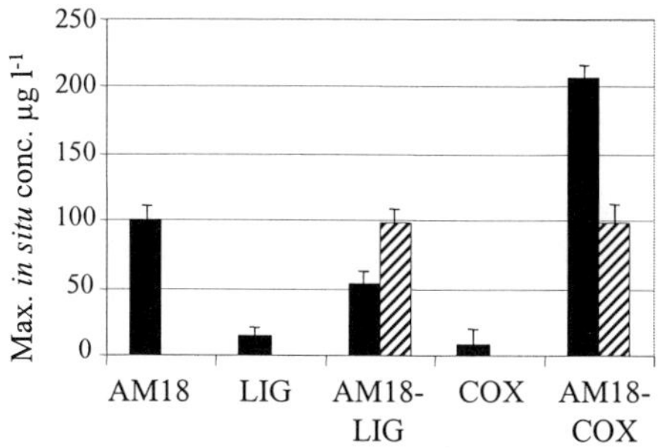

Figure 7. The maximum *in situ* concentrations of simazine in the aqueous phase. The solid bars are measured values, the hatched ones estimated. The lines above the bars are the standard errors.

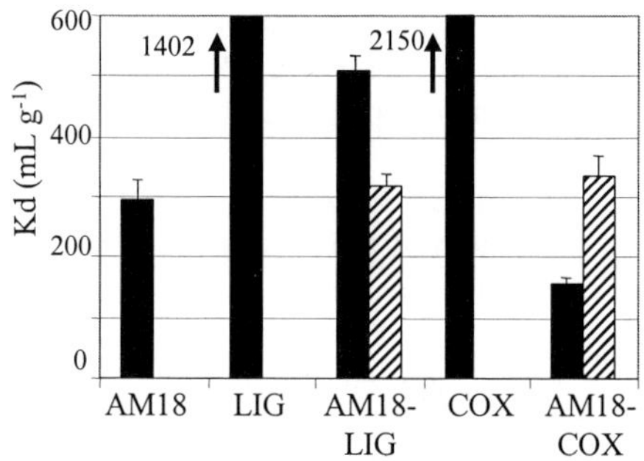

Figure 8. The distribution coefficient of simazine. The solid bars are measured values, the hatched ones estimated. The lines above the bars are the standard errors.

4. CONCLUSIONS

- The addition of humic substances to a clay tended to decrease the clay's ability to adsorb simazine (Figure 1).
- Humic substances only had a small positive effect on the amount of simazine, which was readily desorbable from a clay (Figure 4).
- The addition of a more aliphatic humic substance to the clay strongly decreased the rate at which the mobile simazine was desorbed. The effect of a more aromatic humic substance was considerably smaller and tended to increase the desorption rate (Figure 6). These effects were also reflected in the maximum *in situ* concentration of the simazine (Figure 7).
- Similarly, the addition of the more aliphatic humic substance to the clay increased the distribution coefficient for simazine, while the more aromatic humic substance decreased it.
- The desorption properties of simazine from clay-humic complexes could usually not be predicted from its desorption properties from the pure clay or humic substances alone.
- The use of small dynamic columns has several advantages over batch desorption studies.

REFERENCES

1. Khan, M.A., Tiang, T., 1989. Mapping pesticide contamination potential. Environ. Management 13, 233-242.

2. Leistra, M., 1973. Computation models for the transport of pesticides in soil. Residue Rev. 49, 87-130.
3. van Genuchten, M.T., Wagenet, R.J., 1989 . Two-site/two-region models for pesticide transport and degradation: Theoretical development and analytical solutions. Soil Sci. Soc. Am. J. 52, 1303-1310.
4. Boesten, J.J.T.I., van der Linden, A.M.A., 1991. Modeling the influence of sorption and transformation on pesticide leaching and persistence. J. Environ. Qual. 20, 425-435.
5. Banton, O., Lafrance, P., Martel, R., Villeneuve, J.P., 1992. Planning of soil-pore water sampling campaigns using pesticide transport modeling. Ground Water Monit. Rev. 1, 195-202.
6. Wagner, B., Schewes, R., Maidl, F.X., Fischbeck, G., 1995 . Measurement of residues and simulation of the movement of atrazine in deeper soil layers after long-term application in agriculture. Z. Pflanz. Bodenk. 158, 333-338.
7. Barriuso, E., Schiavon, M., Andreux, F., Portal, J.M., 1991. Localization of atrazine non-extractable (bound) residues in soil size fractions. Chemosphere 22, 1131-1140.
8. Senesi, N., 1992. Sci. Binding mechanisms of pesticides to soil humic substances. Total Environ. 123/124, 63-76.
9. Wang, Z., Gamble, D.S., Langford, C.H., 1992. Interaction of atrazine with Laurentian soil. Environ. Sci. Technol. 26, 560-565.
10. Gilchrist, G.F.R., Gamble, D.S., Kodama, H., Khan, S.U., 1993. Atrazine interactions with clay minerals: Kinetics and equilibria of sorption. J. Agr. Food Chem. 41, 1748-1755.
11. Laird, D.A., Yen, P.Y., Koskinen,. W.C., Teinheimer, T.R., Dowdy, R.H., 1994. Sorption of atrazine on soil clay components. Environ. Sci. Technol. 28, 1054-1061.
12. Barriuso, E., Laird, D.A., Koskinen, W.C., Dowdy, R.H., 1994. Atrazine desorption from smectites. Soil Sci. Soc. Amer. J. 58, 1632-1638.
13. Worrall, F., Parker, A., Rae, J.E., Johnson, A.C., 1996. Eur. Equilibrium adsorption of isoproturon on soil and pure clays. J. Soil Sci. 47, 265-272.
14. Zsolnay, A., 1996. Dissolved humus in soil waters. *In*: Piccolo, A. (Ed.). Humic Substances in Terrestrial ecosystems. Elsevier, Amsterdam, pp. 171-223.
15. Khan, S.U., 1974. Adsorption of 2,4-D from aqueous solutions by fulvic acid-clay complex. Environ. Sci. Technol. 8, 236-238.
16. Payaperez, A., Cortes, B. A., Sala, M.N., Larsen, B., 1992. Organic matter fractions controlling the sorption of atrazine in sandy soils. Chemosphere 25, 887-898.
17. Celis, R., Cox, L., Hermosin, M.C., Cornejo, J., 1996. Retention of metamtron by model and natural particulate matter. Intern. J. Environ. Anal. Chem. 65, 245-260.
18. Zsolnay A., The use of small columns to study pesticide distribution and desorption behaviour. 2000, *In*: Cornejo, J., Jamet, P. (Eds.). Pesticide/Soil Interactions: Some Current Research Methods. Elsevier, Amsterdam, pp. 89-95.

Developments in Soil Science, Volume 28A
Editors: A. Violante, P.M. Huang, J.-M. Bollag and L. Gianfreda

SORPTION AND RELEASE OF ENDOCRINE DISRUPTOR COMPOUNDS ONTO/FROM SURFACE AND DEEP HORIZONS OF TWO SANDY SOILS

E. Loffredo and N. Senesi

Dipartimento di Biologia e Chimica Agroforestale e Ambientale, Università di Bari, Via Amendola 165/A, Bari 70126, Italy

Adsorption and desorption processes of environmental endocrine disruptor compounds (EDCs) bisphenol A (BPA), octylphenol (OP), 17-alpha-ethynilestradiol (EED) and 17-beta-estradiol (17ED) onto/from a surface and a deep horizon of two sandy soils have been investigated. Adsorption kinetics of EDCs onto soils are generally very fast, mainly occurring in the first few hours of contact. Experimental adsorption data were best fitted in a linear isotherm for BPA, in a nonlinear Freundlich isotherm for EED, either in a linear or a nonlinear Freundlich model for OP, and in a Langmuir isotherm for 17ED. The values of the Freundlich constant, K, and of the distribution coefficient, K_d, calculated from experimental isotherms showed that these parameters are positively correlated with the organic carbon (OC) content of the soils and that surface horizon soils exhibit a much higher adsorption capacity than deep horizon soils for any EDC. The small differences of the organic carbon partition coefficient (K_{OC}) values measured among the various soils suggest that not only the content but also the nature and properties of OC affect the extent of adsorption, and that other soil components, e.g., clay minerals, might be involved in the adsorption process of EDCs. Among the various EDCs examined, OP appears to be the most adsorbed, and EED and 17ED show a similar extent of adsorption onto any soil. Adsorption of BPA is generally reversible and its desorption occurs quickly and is completed after few desorption steps. On the contrary, adsorption of OP and EED is mostly irreversible, partial desorption occurs slowly and most soil samples retain high amounts of adsorbate at the end of the experiment.

1. INTRODUCTION

Several man-made chemicals occurring in the environment are known to interact with the development and functioning of endocrine systems in wildlife and humans by acting as hormone-like substances [1]. These compounds are commonly called "endocrine disruptors" and can imitate or block or interfere with actions of natural hormones in the organism [2, 3]. Many of these endocrine disruptor compounds (EDCs) have estrogenic (feminizing) activity that may compromise reproductive fitness by causing disorders of the reproductive tract and possibly induce the development of steroid-hormone-dependent cancers [4, 5]. Some xenoestrogens can interact with estrogenic receptors and stimulate estrogenic activity of cells in the liver or gonadal tissues of organisms [6].

Compounds proven or suspected to act as EDCs include natural and synthetic estrogens of human origin; common agricultural products, such as pesticides; industrial chemicals; and paper, paint, plastic and pharmaceutical products [7, 8]. These compounds may enter the soil through current agricultural practices and by application, discharge and/or disposal of urban and industrial effluents, sludges and wastes. Although research needs for risk assessment of health and environmental effects of EDCs have been recently stressed [2], relatively little is known about the impact and fate of EDCs introduced from various sources into soil.

Generally, the response of soil to estrogenic risk of EDCs is related to the distribution and speciation of EDCs in the various soil phases. Depending on their molecular structure and physical and chemical properties in the various soil phases, EDCs can either be strongly bound to the solid soil fractions and accumulate in the top soil layer or be moved down to deeper soil horizons and groundwater. Adsorption onto soil solid phases is generally considered the most important process that controls the mobility, transport, accumulation, bioavailability and toxicity of organic xenobiotics in soil [9, 10]. Thus the evaluation of the kinetics and extent of adsorption and desorption processes of EDCs onto/from different soil horizons is very important for understanding their behavior and performances in soil.

The objective of this study is to investigate adsorption and desorption processes of four ascertained EDCs, bisphenol A (BPA), 17-alpha-ethynilestradiol (EED), 17-beta-estradiol (17ED) and octylphenol (OP) onto/from a surface and a deep horizon of two sandy soils.

2. MATERIALS AND METHODS

2.1. Soils

The soil samples used in this work were collected from the surface (0-30 cm) and deep (30-90 cm) horizons of two acidic sandy soils originating from Portugal (P) and Germany (G), and are labeled, respectively, as P30 and G30 and P90 and G90. Before analysis and experiments with EDCs, each soil was air-dried, crushed and passed through a 2-mm sieve.

Soil samples were analyzed for their most relevant physical and chemical properties, and results obtained are shown in Table 1. In particular, the surface horizons of both soils (samples P30 and G30) have much higher organic carbon (OC) and related organic matter (OM) contents than the corresponding deeper horizons (P90 and G90). Further, the surface horizon of the G soil (G30 sample) is much richer in OC and OM than the corresponding P horizon (P30 sample), whereas the opposite is true for the G90 sample with respect to the P90 sample.

2.2. Endocrine disruptor compounds

The EDCs examined in this work are: (a) bisphenol A (BPA) [2,2-(4,4-dihydroxydiphenyl) propane]; (b) octylphenol (OP) [4-(1,1,3,3,-tetramethylbutyl)phenol]; (c) 17-alpha-ethynilestradiol (EED) [17α-ethynil-1,3,5(10)-estratriene-3,17β-diol]; and (d) 17-beta-estradiol (17ED) [1,3,5(10)-estratriene-3,17β-diol]. All compounds, 99% purity, were obtained from Sigma-Aldrich Chemie GmbH, Steinheim, Germany. Their molecular formulas are shown in Figure 1.

BPA is a xenoestrogen formed as an intermediate compound in the preparation of epoxy resins and polycarbonates; it is also used in manufacturing adhesives, building materials, compact disks and electrical and electronic parts and in agriculture as a fungicide. OP is a

Table 1
Some physical and chemical properties of soils examined

Parameter	P30	P90	G30	G90
Sand *%*	93.0	93.9	94.9	94.8
Silt *%*	4.4	3.5	3.6	3.2
Clay *%*	2.6	2.6	1.5	2.0
pH *(H_2O)*	4.8	5.5	5.9	5.4
pH *(KCl 1N)*	4.5	4.8	5.5	4.6
Electrical conductivity at 25°C (1:2) *dS/m*	0.050	0.040	0.095	0.092
Organic carbon *g/kg*	3.6	1.8	9.3	1.1
Organic matter *g/kg*	6.3	3.3	16.0	1.8
Total N *g/kg*	0.4	0.3	0.7	0.2
C/N ratio	9.1	6.0	12.6	5.3
Available P *mg/kg*	36	14	16	7
K *mmol/kg*	0.887	0.788	2.049	1.893
Ca *mmol/kg*	4.005	3.262	14.051	2.830
Na *mmol/kg*	0.962	1.024	1.081	0.796
Mg *mmol/kg*	0.531	0.429	1.997	0.456

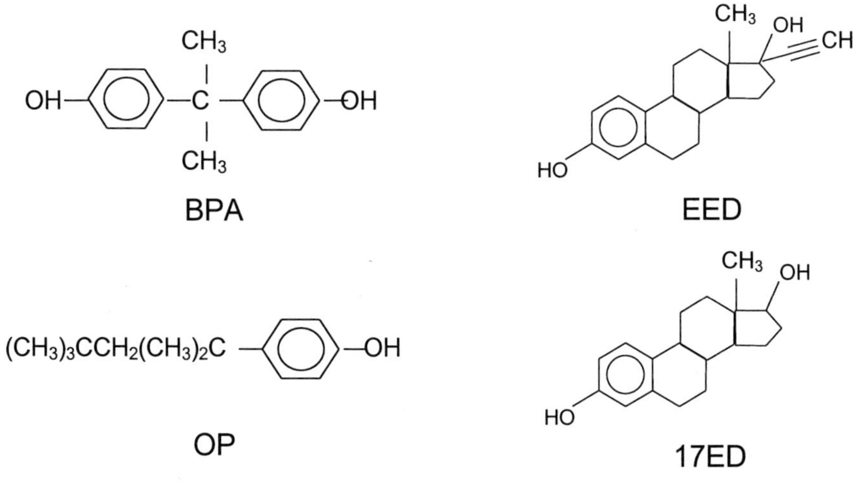

Figure 1. Molecular formulas of EDCs examined.

xenoestrogen formed as a stable biodegradation metabolite from octylphenol polyethoxylates that are widely used in the formulation and production of plastics, paints, pesticides and detergents. EED is a synthetic estrogen used for medical purposes, often in combination with progestogen as an oral contraceptive. 17ED is the most potent mammalian estrogenic hormone.

2.3. Adsorption kinetics

Adsorption kinetics were measured for each EDC to evaluate the adsorption rates onto soil samples examined and to establish an adequate equilibration time to be used for measuring adsorption isotherms.

Ten aliquots of 5 g of soil (air-dried and 2-mm sieved) were each suspended in 20 mL of either a water solution of BPA at a concentration of 10 mg L^{-1} or a 20% (v/v) ethanol/water solution of EED or 17ED at a concentration of 20 mg L^{-1}. For OP, ten aliquots of 1 g of soil were each suspended in 25 mL of a 6% (v/v) ethanol/water solution of OP at a concentration of 1 mg L^{-1}. Addition of ethanol to water was necessary to increase the solubility of EED, 17ED and OP.

The mixtures were then mechanically shaken for one of ten different time periods: 0.25, 0.5, 1, 2, 4, 8, 16, 24, 48, and 72 h. At the end of each time period the suspensions were centrifuged, and the supernatants were analyzed by high performance liquid chromatography (HPLC) to determine the residual concentration of each EDC in solution, using the procedure adopted for obtaining adsorption isotherms as described in section 2.4. All experiments were conducted in triplicate at a temperature of 20 ± 2°C.

2.4. Adsorption isotherms

Adsorption isotherms of each EDC onto each soil sample were obtained using a batch equilibrium (slurry-type) method. Aliquots of 5 g of soil were added with 20 mL of either water solutions of BPA at concentrations of 1, 2, 4, 8, 12, 20, and 40 mg L^{-1} or 20% (v/v) ethanol/water solutions of EED or 17ED at concentrations of 1, 2, 4, 8, 12, and 20 mg L^{-1}. For OP, aliquots of 2 g of soil were added to 15 mL of 10% (v/v) ethanol/water solutions of OP at concentrations of 0.1, 0.2, 0.5, 1, 2, 4, and 5 mg L^{-1}. All experiments were conducted in triplicate.

Equilibration was achieved by mechanical shaking of mixtures for 24 h at 20 ± 2°C in the dark. Suspensions were then centrifuged at 17,400 g for 15 min, and the supernatant solutions were removed and analyzed for the equilibrium concentrations of each EDC by HPLC using a Thermo Separation Products Liquid Chromatograph. For the determination of BPA, EED and 17ED, a 15-cm Merck LiChrospher® 60 RP-Select B column and an ultraviolet detector operating at 280 nm were used. For the determination of OP, a SUPELCOSIL™ LC-18 column and a fluorescence detector operating at 230 nm excitation and 310 nm emission were used. In all cases, the mobile phase used was a solution of acetonitrile/water at a ratio of 40/60 (v/v) for BPA, 50/50 (v/v) for EED and 17ED, and 75/25 (v/v) for OP. The amounts of each EDC adsorbed were derived from the difference between the initial concentration and the equilibrium concentration of EDC in solution.

To construct adsorption isotherms, experimental adsorption data obtained for each EDC on each soil were tentatively fitted to both a linear and nonlinear Freundlich equation:

$$x/m = K\,C^{1/n} \qquad (1)$$

and the Langmuir equation:

$$x/m = (KCb)/(1+KC) \quad (2)$$

where x/m is the amount of each EDC adsorbed in μg g^{-1}, C is the equilibrium concentration of EDC in solution in μg mL^{-1}, 1/n indicates the degree of nonlinearity between solution concentration and amount adsorbed, and b is the Langmuir adsorption maximum.

The magnitude of adsorption, i.e., the adsorption capacity of the substrate, was estimated by the values of both the Freundlich constant, K, and the distribution coefficient, K_d, which is defined as the mean value of the ratios of the amount of EDC adsorbed at each equilibrium concentration, and can be calculated according to the equation:

$$K_d = [(x/m)/C]_{mean} \quad (3)$$

Further, the organic carbon partition coefficient, K_{OC}, which provides the amount of EDC adsorbed per unit of OC present in the substrate, was calculated for all soil samples according to the equation:

$$K_{OC} = (K \times 100)/OC\% \quad (4)$$

2.5. Desorption isotherms

Desorption isotherms of BPA, OP, and EED were obtained by measuring the sequential release of each EDC immediately after its adsorption onto the various soil substrates. Attempts made to measure desorption of 17ED from soils examined were unsuccessful. This was because 17ED apparently degraded during the experiments, as shown by the appearance of unquantifiable degradation products in the HPLC analysis.

To obtain previous adsorption of EDC, the substrates were equilibrated for 24 h with either an aqueous solution of BPA or a 20% (v/v) ethanol/water solution of EED at concentrations of 20 mg L^{-1} or a 10% (v/v) ethanol/water solution of OP at a concentration of 5 mg L^{-1}, using in all cases the ratios of solution/substrate referred to above in the description of adsorption experiment (section 2.4).

After adsorption was obtained, the mixtures were centrifuged, and the equilibrium solution was carefully removed and replaced with the same volume of bidistilled water or the appropriate ethanol/water solution. After each desorption step, the amount of dissolved EDC present in the equilibrium solution that remained entrapped in the substrate was duly calculated and subtracted from the total amount of EDC measured in the supernatant solution. The suspensions were then shaken mechanically for 24 h to obtain a new equilibrium condition and centrifuged. For all substrates, the desorption procedure was repeated a maximum of five times for BPA and OP and four times for EED or until the concentration of the EDC in the supernatant solution fell below the lower limit of the amount detectable in the conditions used. After each desorption step, the concentrations of BPA, OP and EED in the supernatant solutions were measured by HPLC under the same conditions used for the adsorption studies and described above in section 2.4. The amount of compound that remained adsorbed was calculated by difference. All desorption experiments were conducted in triplicate.

For comparative purposes, in all cases both the adsorption parameters, K_{ads} and $1/n_{ads}$, and desorption parameters, K_{des} and $1/n_{des}$, were calculated by using the nonlinear Freundlich equation (eq. 1) described above in section 2.4. The magnitude of K_{des} vs. K_{ads} is considered to provide an indication of the degree of reversibility/irreversibility of the adsorption process [11], whereas the magnitude of $1/n_{des}$ vs. $1/n_{ads}$ may be related to the rate of the desorption process [12].

3. RESULTS AND DISCUSSION

3.1. Adsorption kinetics

Representative examples of adsorption kinetics curves measured for BPA, OP, EED and 17ED are shown in Figure 2. In all cases, adsorption of EDC onto soils examined appears to occur in two phases, a rapid one occurring in the first few hours of contact (generally less than 10 h), which generally corresponds to more than 90% of total adsorption, and a slow one that may need several hours until the attainment of the equilibrium. BPA appears to be the most quickly adsorbed EDC onto all substrates.

The rapid adsorption phase would occur on the most reactive and/or accessible sites of the substrate, whereas the slower adsorption may reflect the involvement of less reactive and/or more sterically hindered sites.

In all cases, an equilibration time of 24 h has been considered adequate and has been used for the measurements of adsorption isotherms.

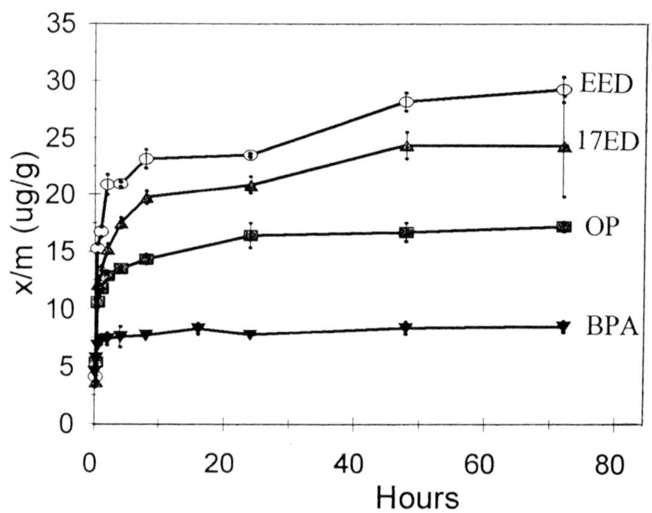

Figure 2. Representative adsorption kinetics curves of EDCs onto some soil samples.

3.2. Adsorption isotherms

Experimental adsorption data for BPA onto all soils fit better in linear, C-type isotherms (Figure 3) and for EED in nonlinear, L-shaped ($1/n<1$) Freundlich isotherms (Figure 4).

Adsorption data of OP onto G-soils fit better in linear isotherms, and onto P-soils in nonlinear, Freundlich isotherms (Figure 5). A linear, C-type isotherm indicates that a constant partition of the EDC occurs between the solution and the substrate, that is, adsorption is directly proportional to the solution concentration. A nonlinear, L-shaped isotherm indicates that the EDC has a moderately high affinity for the substrate in the initial stages of adsorption, which occurs with increasing difficulty as adsorption sites are filled. In both cases, no limiting adsorption (saturation) occurs over the whole concentration range examined for any EDC onto any substrate examined.

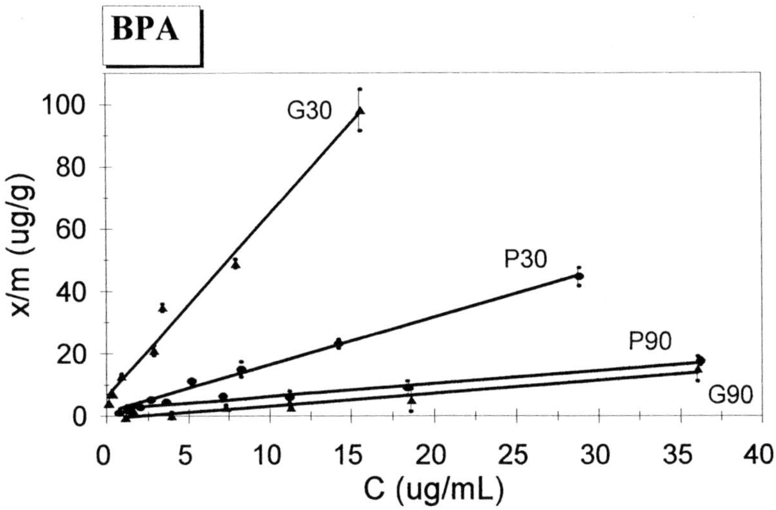

Figure 3. Linear adsorption isotherms of BPA onto soils examined.

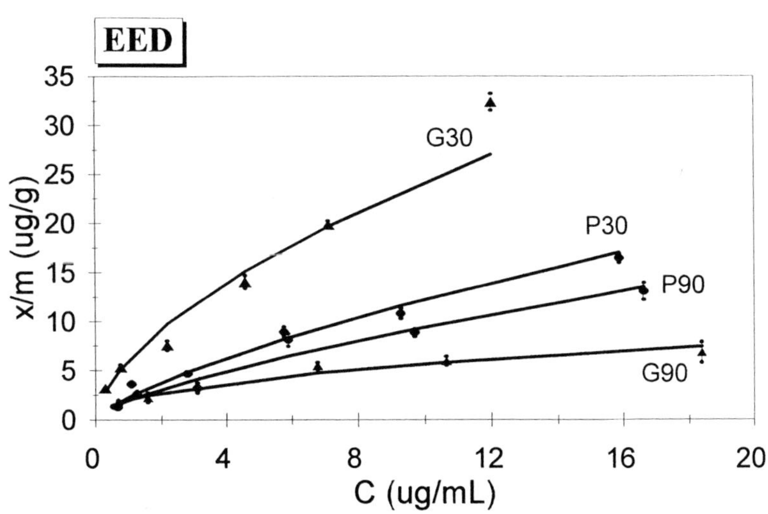

Figure 4. Freundlich nonlinear adsorption isotherms of EED onto soils examined.

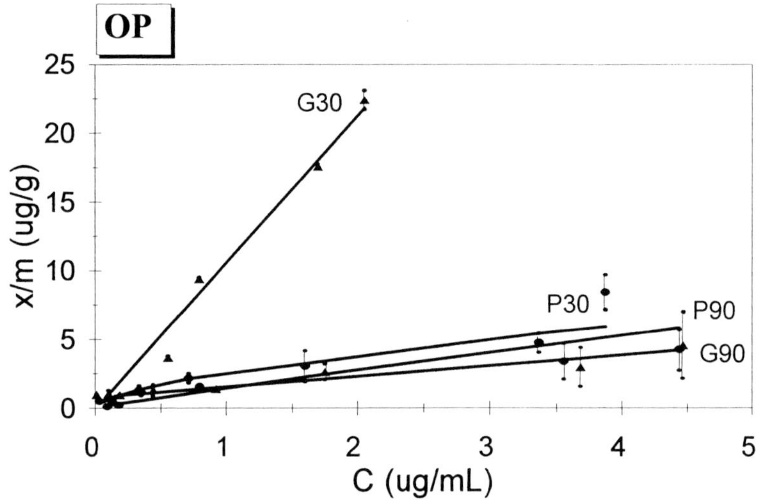

Figure 5. Freundlich nonlinear (P30 and P90) and linear (G30 and G90) adsorption isotherms of OP onto soils examined.

For 17ED the best fits of experimental adsorption data onto all soils examined are in a Langmuir-type isotherm (Figure 6). This type of isotherm suggests that 17ED has a moderately high affinity for the substrate in the initial stages of adsorption, whereas it has increasing difficulty in finding vacant sites as they are filled, finally reaching a maximum of adsorption, i.e., saturation.

The correlation coefficients, r, the Freundlich parameters, K and 1/n where applicable, the distribution coefficients, K_d, and the organic carbon partition coefficient, K_{OC}, calculated from experimental data for BPA, OP, EED and 17ED are shown, respectively, in Tables 2, 3, 4 and 5. The K and K_d values follow the same trend for adsorption of any EDC onto soils examined. In general, surface horizon soils, and especially soil G30, appear to have a much higher adsorption capacity for any EDC examined than the corresponding deep horizon soils. This effect appears to be related to the OC content of soil samples examined, as it is shown by the direct correlations ($P < 0.01$ and $P < 0.05$) found between K and K_d values of any EDC and soil OC% (Figures 7, 8, 9 and 10).

When K values for soil samples are normalized to OC, which means that the content of OC is assumed as the only factor that determines the extent of adsorption of each EDC by soil, a slight difference of K_{OC} values is measured between various soils. These results suggest that the OC nature and properties, and not only its content, affect the adsorption capacity of OC for any EDC. Further, the different trends obtained for K_{OC} values, with respect to K and K_d values, would suggest that soil components other than OC, such as soil clay minerals, could be involved in the adsorption of EDCs. For BPA the K_{OC} values follow the order: G30 > P30 > G90 > P90, whereas for OP the K_{OC} values are similar for the surface and deep horizons of the same soil but differ between the P-soils and G-soils. For EED and 17ED similar K_{OC} values are obtained for surface soils, but they differ from K_{OC} of deep soils.

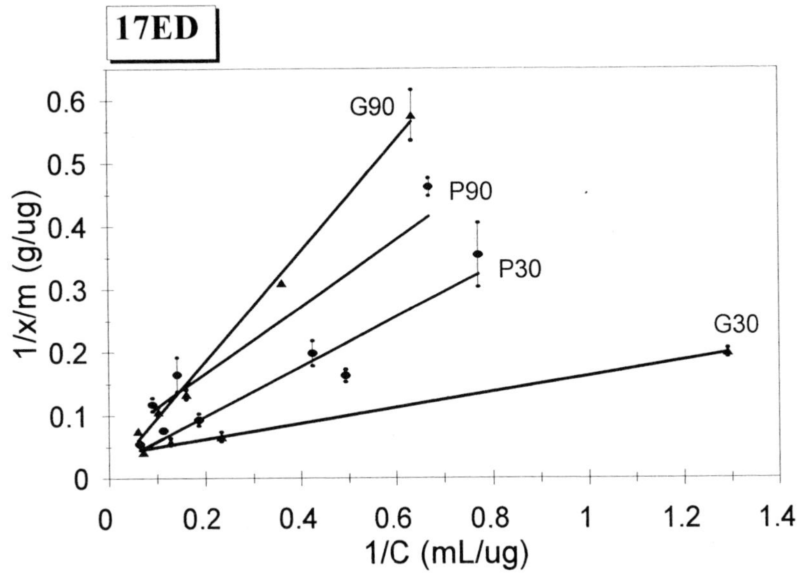

Figure 6. Langmuir adsorption isotherms of 17ED onto soils examined.

Table 2
Correlation coefficients, r, Freundlich K values, distribution coefficients, K_d, and organic carbon partition coefficients, K_{OC}, for BPA adsorption onto soils examined

Soil	r	K (L Kg^{-1})	K_d (L Kg^{-1})	K_{OC} (L Kg^{-1})
P30	0.995	1.59	1.74	436
P90	0.980	0.61	0.53	335
G30	0.977	6.54	12.91	703
G90	0.974	0.45	0.71	424

Table 3
Correlation coefficients, r, Freundlich parameters, K, 1/n, distribution coefficients, K_d, and organic carbon partition coefficients, K_{OC}, for OP adsorption onto soils examined

Soil	r	K (L Kg^{-1})	1/n	K_d (L Kg^{-1})	K_{OC} (L Kg^{-1})
P30	0.960[a]	2.60	0.60	4.76	713
P90	0.939[a]	1.45	0.93	1.69	797
G30	0.988[b]	10.03	-	8.68	903
G90	0.955[b]	0.94	-	1.39	1023

[a] Freundlich nonlinear isotherm
[b] Linear isotherm

Table 4
Correlation coefficients, r, Freundlich parameters, K, 1/n, distribution coefficients, K_d, and organic carbon partition coefficients, K_{OC}, for EED adsorption onto soils examined

Soil	r	K (L Kg^{-1})	1/n	K_d (L Kg^{-1})	K_{OC} (L Kg^{-1})
P30	0.969	2.30	0.72	1.73	631
P90	0.982	1.73	0.73	1.17	953
G30	0.985	5.92	0.61	3.77	637
G90	0.987	1.66	0.51	0.70	1562

Table 5
Correlation coefficients, r, distribution coefficients, K_d, and organic carbon partition coefficients, K_{OC} for 17ED adsorption onto soils examined

Soil	r	K_d (L Kg^{-1})	K_{OC} (L Kg^{-1})
P30	0.967	1.98	543
P90	0.922	1.40	768
G30	0.999	3.47	373
G90	0.997	1.02	960

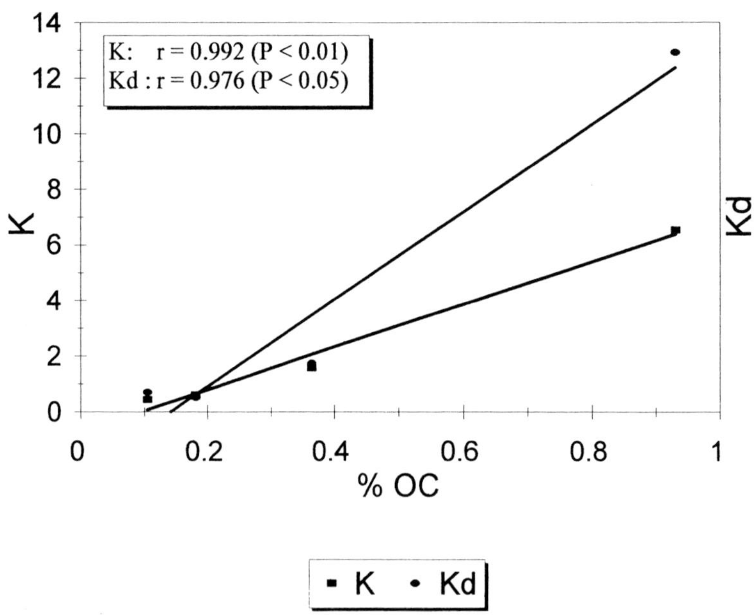

Figure 7. Relationship of Freundlich K values and distribution coefficients, K_d, for BPA with OC % of soils examined.

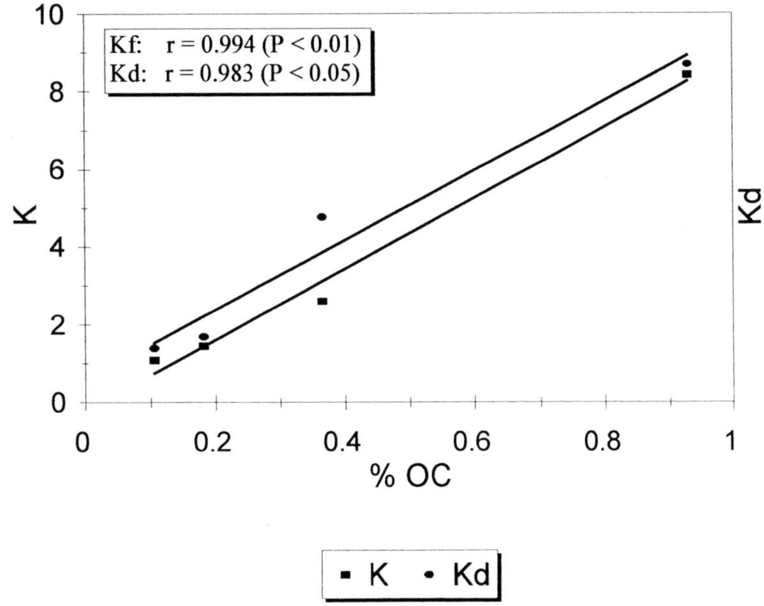

Figure 8. Relationship of Freundlich K values and distribution coefficients, K_d, for OP with OC % of soils examined.

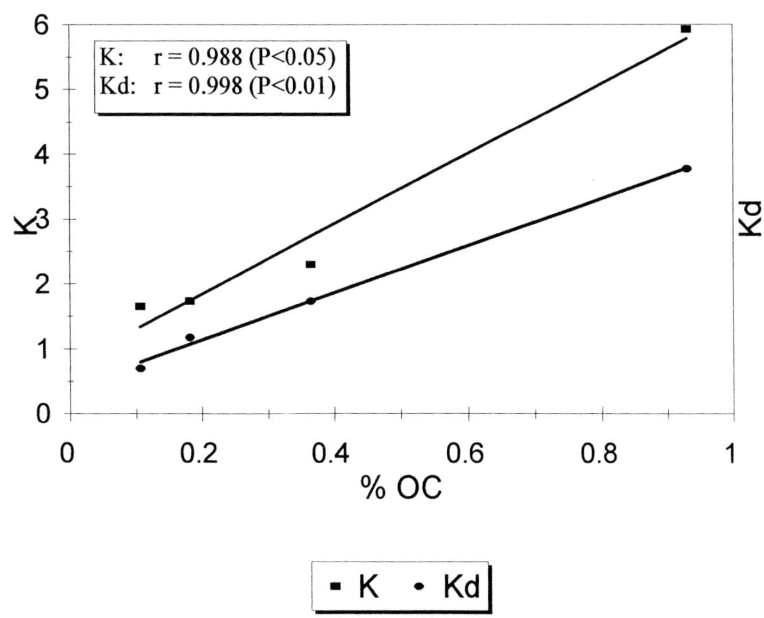

Figure 9. Relationship of Freundlich K values and distribution coefficients, K_d, for EED with OC % of soils examined.

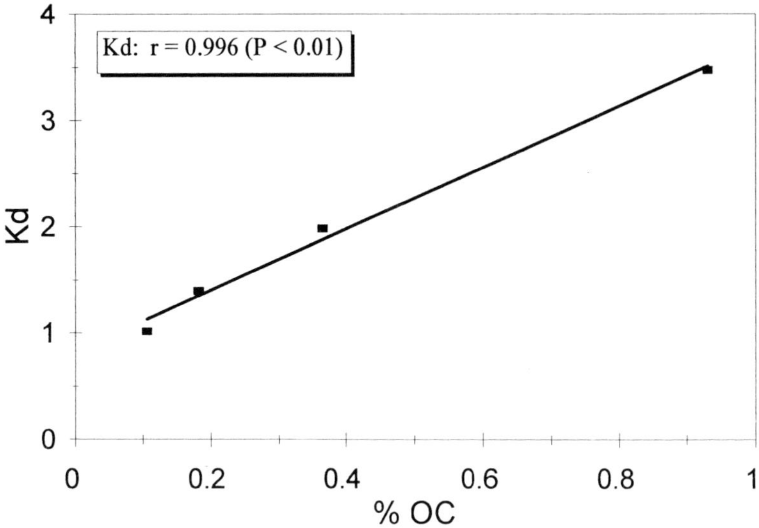

Figure 10. Relationship of distribution coefficients, K_d, for 17ED with OC % of soils examined.

The comparison of either K or K_d values of the various EDCs examined shows that OP is generally the most adsorbed among the EDCs; whereas EED and 17ED show a similar extent of adsorption onto any soil examined.

3.3. Adsorption-desorption studies

Even though adsorption data of each EDC onto various substrates could be best fitted in one of the equations described in section 2.4, a nonlinear Freundlich equation was used in all cases to fit both experimental adsorption data and desorption data to obtain adsorption and desorption parameters comparable for any EDC and any substrate examined. This procedure was allowed by the generally good correlation coefficients obtained in all cases for a nonlinear Freundlich fitting of both adsorption data and desorption data.

Correlation coefficients, r, Freundlich nonlinear parameters for adsorption and desorption, and total amounts (averages of three replicates) of BPA, OP and EED that are desorbed during the experiment from each substrate, expressed as the percentage of the initially adsorbed amount (100%), are shown, respectively, in Tables 6, 7 and 8.

For P-soils the K_{des} values of BPA are much lower than the corresponding K_{ads} values, which suggests that a highly reversible adsorption of BPA occurs on these substrates. On the contrary, G-soils show a K_{des} value slightly larger than the corresponding K_{ads} value, thus suggesting a moderate irreversibility for adsorption of BPA on these substrates. The values of $1/n_{des}$ are generally higher than those of $1/n_{ads}$, thus indicating a fast rate of desorption of BPA from most substrates. However, $1/n_{des}$ is smaller than $1/n_{ads}$ for soil G30, which indicates that for this substrate the rate of desorption of BPA is slower than the rate of adsorption.

Table 6
Correlation coefficients, r, for Freundlich nonlinear isotherms, Freundlich adsorption and desorption parameters and total amounts desorbed after a number of desorption steps (indicated between parentheses in last column) for BPA from soils examined

Soil	*ADSORPTION*			*DESORPTION*			
	r	K_{ads} (L Kg^{-1})	$1/n_{ads}$	r	K_{des} (L Kg^{-1})	$1/n_{des}$	% desorbed
P30	0.988	1.65	1.02	0.947	0.21	1.93	100 (4)
P90	0.937	0.35	1.22	0.978	0.06	1.94	100 (3)
G30	0.982	15.2	0.61	0.963	23.7	0.23	49.6 (5)
G90	0.893	1.00	0.69	0.996	1.19	0.84	100 (4)

Table 7
Correlation coefficients, r, for Freundlich nonlinear isotherms, Freundlich adsorption and desorption parameters and total amounts desorbed after a number of desorption steps (indicated between parentheses in last column) for OP from soils examined

Soil	*ADSORPTION*			*DESORPTION*			
	r	K_{ads} (L Kg^{-1})	$1/n_{ads}$	r	K_{des} (L Kg^{-1})	$1/n_{des}$	% desorbed
P30	0.960	2.60	0.60	0.923	8.36	0.006	4.5 (5)
P90	0.939	1.45	0.93	0.930	2.32	0.36	84 (5)
G30	0.965	8.40	1.05	0.992	26.96	0.003	2.3 (5)
G90	0.879	1.08	1.02	0.962	2.56	0.30	79 (5)

Table 8
Correlation coefficients, r, for Freundlich nonlinear isotherms, Freundlich adsorption and desorption parameters and total amounts desorbed after a number of desorption steps (indicated between parentheses in last column) for EED from soils examined

Soil	*ADSORPTION*			*DESORPTION*			
	r	K_{ads} (L Kg^{-1})	$1/n_{ads}$	r	K_{des} (L Kg^{-1})	$1/n_{des}$	% desorbed
P30	0.969	2.30	0.72	0.940	11.03	0.15	41.3 (4)
P90	0.982	1.73	0.73	0.966	10.18	0.08	32.3 (4)
G30	0.985	5.92	0.61	0.886	30.26	0.03	11.4 (4)
G90	0.987	1.66	0.51	0.915	4.91	0.08	37.3 (4)

For all samples examined, K_{des} values of OP and EED are larger (up to five times) than the corresponding K_{ads} values, and $1/n_{des}$ values are much smaller (up to three orders of magnitude) than the corresponding $1/n_{ads}$ values. These results suggest a marked irreversibility of adsorption and a rate of desorption much slower than the rate of adsorption of OP and EED for all the substrates.

The percentages of initially adsorbed (100%) BPA, OP and EED that remain adsorbed onto each substrate after each desorption step are shown, respectively, in Tables 9, 10 and 11 and summarized graphically in Figure 11. Desorption of BPA for most soils is completed (100%) after three or four desorption steps; the only exception was for sample G30, which still retained half of the amount of adsorbed BPA at the end of the desorption experiment. A total (100%) desorption of initially adsorbed OP and EED has never been measured for any substrate examined. However, amounts of OP desorbed from soil samples P90 and G90 reach values of 84% and 79%, respectively, after five desorption steps, whereas only a few percent is desorbed from samples P30 and G30. In the case of EED, all samples still retain a high amount of EED at the end of the experiment.

Table 9
Percentages (%) of initially adsorbed (100%) BPA that remain adsorbed onto each soil examined after each desorption step

Soil	Percentages of BPA remaining adsorbed after step				
	1	2	3	4	5
P30	41.8	17.9	1.6	0	-
P90	28.6	2.0	0	-	-
G30	79.4	67.2	56.1	53.5	50.4
G90	43.0	21.2	8.3	0	-

Table 10
Percentages (%) of initially adsorbed (100%) OP that remain adsorbed onto each soil examined after each desorption step

Soil	Percentages of OP remaining adsorbed after step				
	1	2	3	4	5
P30	97.8	97.2	96.6	96.0	95.5
P90	48	28	19	15	16
G30	98.3	98.2	98.0	97.8	97.7
G90	42	35	25	21	21

Table 11
Percentages (%) of initially adsorbed (100%) EED that remain adsorbed onto each soil examined after each desorption step

Soil	Percentages of EED remaining adsorbed after step			
	1	2	3	4
P30	92.5	66.5	64.0	58.7
P90	81.4	72.1	71.8	67.7
G30	99.4	94.0	90.4	88.6
G90	75.7	73.1	71.6	62.7

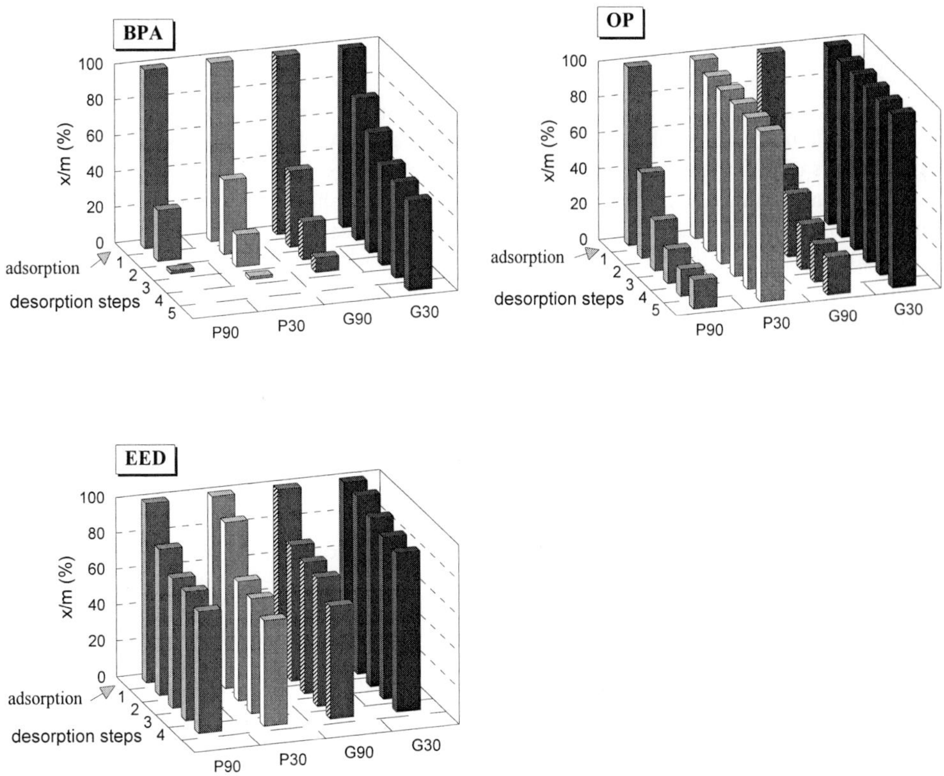

Figure 11. Percentages (%) of initially adsorbed (100%) EDC that remain adsorbed onto each soil examined after each desorption step.

4. CONCLUSIONS

Adsorption kinetics of all EDCs onto all soils examined are generally very fast, mainly occurring in the first few hours of contact. Experimental adsorption data were best fitted in different isotherms for the various EDCs: a linear isotherm for BPA, a nonlinear Freundlich isotherm for EED, either a linear or a nonlinear Freundlich model for OP, and a Langmuir isotherm for 17ED.

The K and K_d coefficients follow the same trend for any EDC, thus suggesting that either one of them is useful for evaluating soil adsorption capacity for the various EDCs. Surface horizon soils, which are richer in OC content, exhibit a much higher adsorption capacity than deep horizon soils. The small differences in K_{OC} values measured among the various soils suggest that not only the content but also the nature and properties of OC affect the extent of adsorption, and that other soil components, e.g., clay minerals, might be involved in the adsorption process of EDCs. Among the various EDCs examined, OP appears to be the most adsorbed, and EED and 17ED show a similar extent of adsorption onto any soil.

With the exception of one soil sample, adsorption of BPA is reversible, and desorption occurs quickly and is completed after few desorption steps. On the contrary, adsorption of OP and EED is mostly irreversible, partial desorption occurs slowly and most soil samples retain high amounts of adsorbate at the end of the experiment.

In conclusion, the sandy soils examined in this work are shown to be able to adsorb variable amounts of EDCs that tend to accumulate in soil surface horizons that are richer in OC. Adsorbed EDCs may either be desorbed quickly and completely, and are thus expected to move down the soil profile to contaminate the groundwater table, or they may be slowly and only partially desorbed, thus resulting in retention of most of the compound by soil solid phases, especially on the surface layer, with corresponding soil contamination.

ACKNOWLEDGMENT

This research is part of the European Commission Project "Prendisensor" and has been supported by Grant n° ENV4-CT97/0473.

REFERENCES

1. EDSTAC. Endocrine Disruptors Screening and Testing Advisory Committee. 1998. Final Report. August 1998. Available on line: www.epa.gov/opptintr/opptendo/whatsnew.htm.
2. Kavlock, R.J., Daston, G.P., Derosa, C., Fenner-Crisp, P., Earl Gray, L., Kaattari, S., Lucier, G., Lustre, M., Mac, J.M., Maczka, C., Miller, R., Moore, J. Rolland, R., Scott, G., Sheehan, M., Sinks, T., Tilson, H.A., 1996. Research needs for risk assessment of health and environmental effects of endocrine disrupters: a report of U.S. EPA-sponsored workshop. Environ. Health Perspect. 104, 715-740.
3. Keith, L.H., 1997. Environmental Endocrine Disruptors. A Handbook of Property Data. J. Wiley and Sons, NY.
4. Colborn, T. Clement, C., 1992. Chemically-Induced Alterations in Sexual and Functional Development: The Wildlife/Human Connection. Princeton Sci. Publ., Princeton, NJ.
5. Nolan, C. (Ed.), 1998. Ecosystem Research Reports Series No. 29: Endocrine-Disrupters Research in the EU. Report EUR 18345, Environ. Clim. Res. Progr., Office for Official Publication of the EC, Brussels-Luxembourg.
6. Anonymous, 1996. Chemicals with estrogen-like effects. Tema Nord Environment, Nordic Council of Ministers, Copenhagen.
7. Colborn, T., vom Saal, F.S., Soto, A.M., 1993. Developmental effects of endocrine-disrupting chemicals in wildlife and humans. Environ. Health Perspect., 101, 378-384.
8. Keith, L.H., 1997. Environmental endocrine disruptors: An overview of the analytical challenge. Annual Symposium on Waste Testing & Quality Assurance, DATA, Alexandria, VA.
9. Stevenson, F.J., 1982. Humus Chemistry: Genesis, Composition, Reactions. J. Wiley and Sons, NY.
10. Senesi, N., Miano, T.M., 1995. The role of abiotic interactions with humic substances on the environmental impact of organic pollutants. *In*: Huang, P.M., Berthelin, J., Bollag,

J.-M., McGill, W.B., Page, A.L. (Eds.), Environmental Impact of Soil Component Interactions: Natural and Anthropogenic Organics, Vol. I. CRC-Lewis, Boca Raton, FL, 311-335.

11. McCall, P.J., Laskowski, D.A., Swann, R.L., Dishburger, H.J., 1981. Test protocols for environmental fate and movement of chemicals. *In*: Proc. AOAC 94th Annual Meeting, Washington, D.C., Assoc. Official Anal. Chem., Arlington, VA, pp. 89-109.
12. Pignatello, J.J., Huang, L.Q., 1991. Sorptive reversibility of atrazine and metolachlor residues in field soil samples. J. Environ. Qual. 20, 222-228.

Developments in Soil Science, Volume 28A
Editors: A. Violante, P.M. Huang, J.-M. Bollag and L. Gianfreda

DISTRIBUTION OF TRINITROTOLUENE BETWEEN AQUEOUS AND SOLID PHASE SOIL ORGANIC MATTER

J. Eriksson[a, b] and U. Skyllberg[a]

[a]Department of Forest Ecology, Swedish University of Agricultural Sciences, S-901 83 Umeå, Sweden*

[b]Defence Research Establishment, FOA, Division of NBC-Protection, S-901 82 Umeå, Sweden

Soil organic matter (SOM) pertaining to solid and aqueous phases plays an important role in retention and transport of chemicals in soil. In this study the binding of trinitrotoluene (TNT), and its degradation products to SOM was determined. The main objective was to differentiate between TNT adsorbed to particulate organic matter (POM) and TNT bound to dissolved organic matter (DOM). We used a batch soil-solution system equilibrating POM and DOM from a Spodosol humus layer with TNT in 0.05 M NaCl as background electrolyte. The system was adjusted to give pH values and DOM concentrations typical for acidic forest organic soils. An HPLC C-18 column was used to separate TNT bound to DOM from TNT in solution. In order to quantify TNT bound to DOM, we used ^{14}C-labelled TNT. Typically, approximately 80% of the added TNT was free, 3% was associated with DOM and approximately 15% was associated with POM at equilibrium. Linear, Freundlich and Langmuir isotherms were fitted to the data. The bonding of TNT to DOM was clearly non-linear (linear K_d, $R^2<0.34$) and increasingly better fitted by the Langmuir isotherm with increasing pH ($0.93<R^2<0.99$). The bonding of TNT to POM was more linear than the bonding to DOM, but the Langmuir and Freundlich isotherms showed the best fit irrespective of pH ($0.993<R^2<0.999$). The adsorption capacity of TNT modeled by the Langmuir isotherm (q_{max}) was 6 to 22 times higher for POM than for DOM, depending on pH. Values of the Freundlich and Langmuir association constants furthermore indicate that the bonding of TNT was stronger to DOM than to POM and that the strength of bonding to organic carbon increased with increasing pH for both organic phases. These results could be explained by a predominant bonding of TNT to specific pH-dependent sites in DOM and by a combination of predominantly hydrophobic partitioning and some specific bonding of TNT in POM.

1. INTRODUCTION

Trinitrotoluene (TNT) is an explosive first synthesized by Wilbrand in 1863. Its industrial use was initiated at the beginning of the twentieth century and during World Wars I and II, TNT was produced in millions of tons. TNT and its degradation products have toxic and mutagenic effects on several organisms, including humans [1].

TNT is not stabile under natural conditions in soils and degrades within hours to nitrosodinitrotoluene (NDNT) and hydroxylaminedinitrotoluene (HADNT). The latter transforms further into aminodinitrotoluene (ADNT) and diaminonitrotoluene (DANT) [2]. Recently, it has been shown that NDNT and HADNT can react to form azoxy compounds, which, like NDNT and HADNT, are very reactive and bind strongly to organic substances [3].

The mobility and retention of TNT and its degradation products are determined by their adsorption to soil particles and potential association with mobile (dissolved) organic matter. Of active soil particles, the binding of TNT (as well as other nitro aromatic compounds, NAC) to clays has been most intensively studied [4-7]. An electron-donor-acceptor complex (EDA) between the electron-excessive clay surface and the electron-lacking NAC has been proposed as an explanation for the very strong bonding of NAC to 2:1 clays in the presence of weakly hydrated cations such as K^+ [5, 6]. Extracted humic acids (HA) have been used in most studies of sorption of TNT to organic matter [8-10]. Reported results vary from non-significant bonding of TNT to HA in a sterile system [10] to extensive bonding of TNT, ADNT and DANT to Aldrich HA [8]. In several studies, adsorption of TNT and its decomposition products have been monitored as a function of time [3, 11, 12].

In soils, SOM can be partitioned into a two-phase system: DOM in the aqueous phase and POM in the solid phase. In this study, the bonding of TNT and its degradation products to DOM and POM is determined in the two-phase organic system. To our knowledge, there are no reported studies of the bonding of TNT to DOM. Nor has the bonding of TNT to DOM and POM been determined simultaneously.

2. MATERIALS AND METHODS

2.1. Soil material

The SOM used in this investigation was collected from the organic horizon of a Spodosol [13], located near Umeå in northern Sweden. The upper 2 cm of the organic horizon, i.e., the less decomposed fermentation layer, was sampled and homogenized through a 4 mm cutting sieve. The soil material was stored in darkness at 4°C. General soil chemistry data are presented in Table 1.

Table 1
Selected chemistry measurements of the organic soil, including adsorbed metal cations

pH[¤]	org-C	LOI[†]	$CEC_{pH\ 8.2}$	Adsorbed cations[§]
	%		cmol$_C$ kg^{-1}	
3.46	42	86	366	42.3

[¤]The pH was measured in 10mM $CaCl_2$. [†]Loss on ignition. The residual, i.e., non-oxidizable material (14% at 570°C) consists mainly of primary minerals, e.g., quartz. [§]$K^+ + Na^+ + 2Mg^{2+} + 2Ca^{2+} + 2Mn^{2+} + 3Al^{3+} + 3Fe^{3+}$

Dissolved organic matter (DOM) used in the experiments was extracted from the organic horizon by a modification of the method of Adams and Byrne [14]. Thirty grams of moist soil was added to 200 mL of Millipore water in two 250 mL polycarbonate (Nalgene) centrifuge bottles. To each bottle, 2.4 g of Na^+-saturated Chelex 20 resin (BioRad, CA) was added to remove Al^{3+} and other polyvalent cations bound to SOM. After gentle shaking for 48 hours, the bottles were centrifuged for 10 min at 6000 rpm with a Beckman J2-21M/E centrifuge. The supernatant, now with a high concentration of DOM, was decanted, and the complete procedure was repeated once. The combined DOM extract was kept for the experimental part, while the soil material with the resin was discarded. In the two-phase organic system, the DOM extract and the original soil organic matter were mixed in various proportions.

2.1.1. Organic matter

Before each adsorption experiment, the extracted DOM solution was filtered through a 0.45 μm polypropylene filter and analyzed with respect to dissolved organic carbon (absorbance at 254 nm). The absorbance was related to DOC of known standards. The concentration of SOM in the experiments was expressed as total soil organic carbon and determined by dry combustion (LECO).

2.1.2. General soil chemical analysis

Adsorbed metals (Na, K, Mg, Ca, Mn, Al, Fe) were extracted with 0.5M $CuCl_2$ using the method of Skyllberg and Borggaard [15]. The cation-exchange capacity (CEC) of the organic soil was calculated as the sum of total acidity determined at pH 8.2 [16] and charges pertaining to non-acidic cations (K^+, Na^+, Ca^{2+}, Mg^{2+}). Water and inorganic content was measured as loss of weight after drying (105°C) and ash content as loss on ignition (LOI) at 540°C. Based on the acid pH and the low mineral content, the content of clay minerals with permanent charge was likely negligible (not determined).

2.2. Chemicals

2.2.1. Trinitrotoluene stock-solution

The ^{14}C-labeled TNT was synthesized from labeled toluene at the Defence Research Establishment, FOA in Grindsjön, Sweden. The TNT was dissolved in water. Before use, the solution was filtered through a 0.45 μm filter and analyzed by HPLC.

2.3. Analysis

2.3.1. HPLC separation and UV detection

Free concentrations of TNT, ADNT and DANT in filtered samples from the various adsorption experiments were separated by reverse phase HPLC. This was done immediately, without storage delay, to prevent unwanted reactions. Chromatogram peaks were identified based on retention times combined with the UV spectra of known standards. Analysis was performed with a system consisting of a Waters' autosampler 717, HPLC pump 510 and a photodiode array detector PAD 996. The column used was a Merck Puraspher RP-18, with 5 μM spheres. The mobile phase was methanol (HPLC grade) and 10 mM phosphate buffer (pH 7) in a proportion of 50/50 (V/V) for TNT and ADNT and 30/70 when analyzing DANT. The flow rate was 1 mL min^{-1} and the injection volume was between 10 μL and 250 μL, depending on the analyte concentration. Concentrations of TNT, ADNT and DANT were measured at 254 nm. The hardware was controlled and monitored from a computer with Millennium 32 software.

2.3.2. ^{14}C determination

TNT labeled with ^{14}C was used to determine TNT associated with DOM. The HPLC analysis effluent was fractionated, collected and subsequently analyzed with respect to ^{14}C with a Beckman LS 5000CE liquid scintillation system. The first 15 mL in one HPLC run was collected in five aliquots. The 1st fraction contained DOM. The 2nd and 5th fractions represented unknown activity. The 3rd and 4th fractions contained TNT and ADNT, respectively. Each fraction was mixed with Beckman scintillation cocktail before analysis. The ^{14}C radioactivity associated with POM was calculated as the difference between total ^{14}C radioactivity and the radioactivity associated with the five HPLC fractions. Because TNT decomposes rapidly in soils, many of its decomposition products likely are involved in the sorption mechanisms. Since degradation products in the TNT family associated with POM and DOM were not measured, we define TNT* as TNT and these unknown compounds associated with POM.

2.3.3. Adsorption isotherms

The adsorption isotherms were modeled using linear (eq. 1), Freundlich (eq. 2) and Langmuir (eq. 3) adsorption functions. The isotherms were fitted using non-linear regression by minimizing the sum of squared differences of observed and fitted values (Marquardt-Levenberg algorithm).

Linear:
$$C_s = K_d \cdot C_w \tag{1}$$

Freundlich:
$$C_s = K_f \cdot C_w^N \tag{2}$$

Langmuir:
$$C_s = \frac{q_{max} \cdot K_L \cdot C_w}{1 + K_L \cdot C_w} \tag{3}$$

C_s is the concentration of adsorbed TNT* expressed in relation to mass of carbon (mol g^{-1} C); C_w is the equilibrium solute concentration in the solution (M); K_d and K_f are partitioning coefficients for the linear and Freundlich equation, respectively, (L g^{-1} C); N is the power of C_w and is generally ≤ 1; q_{max} is the maximum adsorption capacity of the surface with the assumption that the adsorbate is arranged in a monolayer (mol g^{-1} C); and K_L is the Langmuir constant.

2.4. Adsorption experiments

The soil organic matter was separated into DOM (organic carbon passing a 0.45 μm filter) and POM, which did not pass the filter. Experiments encompassing only DOM were defined as one organic phase (one-phase) systems, and experiments encompassing DOM + POM were defined as two organic phase (two-phase) systems. Thus, in the two-phase system SOM = POM + DOM. Note that concentrations of POM and DOM were expressed as organic carbon.

Solutions of TNT, soil-extracted DOM and moist organic soil were mixed in various proportions in 50 mL borosilicate glass Erlenmeyer flasks. In all experiments, the flasks were shaken on a reciprocal shaker (150 rpm) for 15 min every second hour, although the total time of equilibration varied. In all experiments, a TNT solution without organic matter but otherwise treated the same was analyzed as a control of unwanted side-reactions, e.g., adsorption to glass walls. Diluted solutions of HCl and $Ca(OH)_2$ were used to adjust pH, and 0.05 M NaCl was used as ionic medium in all experiments. All experiments were carried out in darkness and the temperature was kept at 22 ± 1°C.

2.4.1. Kinetic experiments

The time-dependent adsorption of TNT in the two-phase system was studied with three POM concentrations: 0, 160 and 640 mg C L^{-1}; the concentration of DOM was approximately 75 mg C L^{-1}. The pH was not adjusted and was found to vary between 5 and 6. Suspension aliquots were sampled after 0.01, 0.08, 0.2, 1.0 and 7.0 days.

2.4.2. Equilibrium experiments, one-phase system

Initial concentrations of TNT ranged from 0.2 to 290 μM. Adsorption was studied at pH 5.9 (non-adjusted) and 4.6. The DOM concentrations were 97-99 mg C L^{-1}. The equilibrium time was 22 h, based on the results from the kinetic experiments.

2.4.3.Equilibrium experiments, two-phase system

The initial concentrations of TNT ranged from 0.1 to 300 μM. The concentrations of POM and DOM were determined afterwards and were approximately 650 and 100 mg C L^{-1}, respectively. Adsorption was examined at pH 4.4, 5.2, 5.6 and 6.2. The equilibrium time was 20 h.

3. RESULTS AND DISCUSSION

3.1. Kinetic adsorption experiments

In presence of SOM, the free concentration of TNT decreased over time (Figure 1). The decrease can be divided into two phases: a rapid, exponential decrease that grades into a slower, more linear phase after less than 24 h. This was most obvious at the two highest concentrations of POM (160 and 640 mg C L^{-1}).

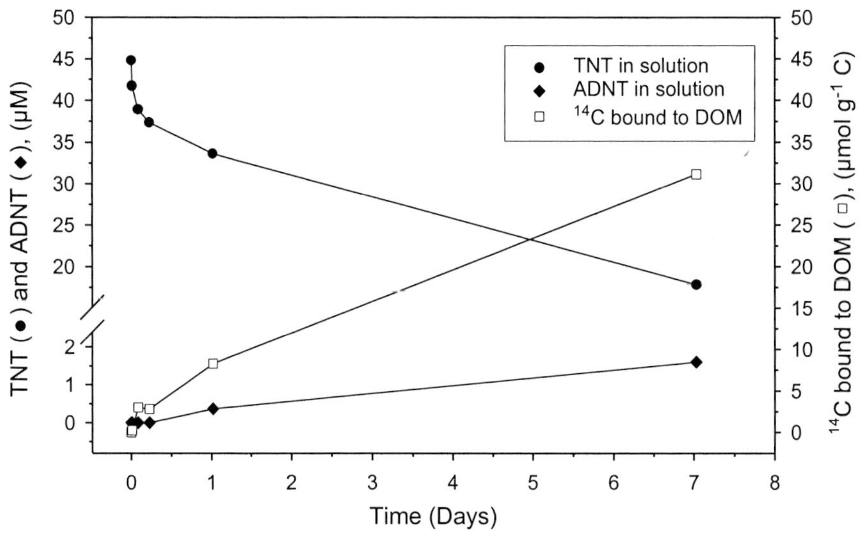

Figure 1. Decrease of TNT, formation of ADNT and sorption of ^{14}C to DOM in presence of DOM and POM. Concentrations were 75 and 640 mg C L^{-1} of DOM and POM, respectively.

For clarity, Figure 1 shows only the data for 640 mg C L^{-1}. Free ADNT was detected after 24 h and then increased with time. We interpret the first, exponential decrease in TNT concentration as mainly an effect of adsorption kinetics of TNT*, and the linear phase, extending up to 7 days, as mainly an effect of TNT degradation. In order to avoid extensive degradation, but reaching as close to chemical equilibrium as possible, we chose an equilibration time of 20-22 h in our equilibrium adsorption experiments.

The concentration of adsorbed TNT* and the concentration of free ADNT increased with increasing total concentration of POM + DOM. Recent findings based on ^{15}N-NMR show that HADNT, the precursor to ADNT in the decomposition pathway of TNT, and azoxy compounds are largely responsible for the strong bonding of TNT* to organic substances under reducing conditions [3, 9]. Therefore, we suggest that an increased concentration of SOM results in an increased rate of formation of reactive TNT derivatives and that these compounds associate strongly with organic substances.

3.2. Equilibrium adsorption experiments

3.2.1. Binding of TNT to dissolved organic matter (DOM)*

Figure 2 shows the concentration of TNT* associated with DOM as a function of free concentration of TNT and pH. Curves for pH 4.8 and 5.9 in the two-phase system represent an average for pH 4.4 & 5.2 and 5.6 & 6.2, respectively. First, it can be noted that the concentration of TNT* bound to DOM (per mass of carbon atom, μmol g^{-1} C) is higher in the two-phase system than in the one-phase system. This is in line with the observation from the kinetic study. Again, this effect most likely is explained by enhanced degradation of TNT resulting in an increased bonding of TNT* with increased concentration of organic matter.

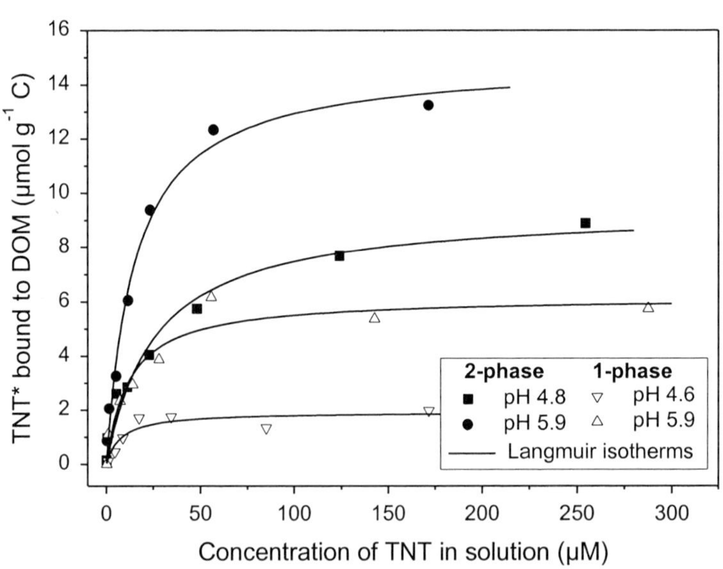

Figure 2. Association of TNT* with dissolved organic matter at different pH and solution concentrations of TNT. Organically bound TNT* was determined by ^{14}C.

The bonding of TNT* to DOM was clearly non-linear and could be satisfactorily modeled by either Langmuir or Freundlich isotherms. The Langmuir equation (3) gave the best fit both in one-phase and two-phase systems ($0.915<R^2<0.980$). The non-linear shape indicates that DOM reaches saturation with respect to TNT*, which could be explained by a reaction involving a limited number of specific sites in DOM. This type of bonding is often referred to as specific bonding, and it is assumed to have an ionic character. The general observation that the TNT* binding increased with increasing pH might indicate that negatively charged weak acid groups in DOM are involved in the bonding. Carboxyl groups are the most numerous weak acid groups in DOM and POM. Our observations are in line with recent findings from a long-term soil-slurry experiment of an organic soil with added TNT. Using ^{14}C-TNT in combination with ^{15}N-NMR to characterize the bonding of TNT and its degradation products (TNT* using our definition) in fulvic acid (FA) and humic acid (HA) extracts, Achtnich and co-workers found that the binding to HA had an ionic character after four days of reaction and that the capacity of FA + HA to bind TNT* was limited. TNT* bound to HA mainly represented azoxy compounds [3].

3.2.2. Binding of TNT to particulate organic matter (POM)*

Typically at equilibrium, approximately 80% of the added TNT was free, 3% was associated with DOM and approximately 15% was associated with POM. In contrast to DOM, the binding of TNT* to POM did not vary significantly with pH (data not shown). In Figure 3, the binding of TNT* to DOM and POM is illustrated for the two-phase system at pH 5.6. For POM, Freundlich and Langmuir equations gave similar fits ($0.993<R^2<0.999$) at all pH values. The linear equation (1) gave slightly worse fits ($0.932<R^2<0.984$), with lower R^2 values at higher pH. Based on the Langmuir equation, the binding capacity was 15, 22, 7 and 6 times higher for POM than for DOM at pH 4.4, 5.2, 5.6 and 6.2, respectively.

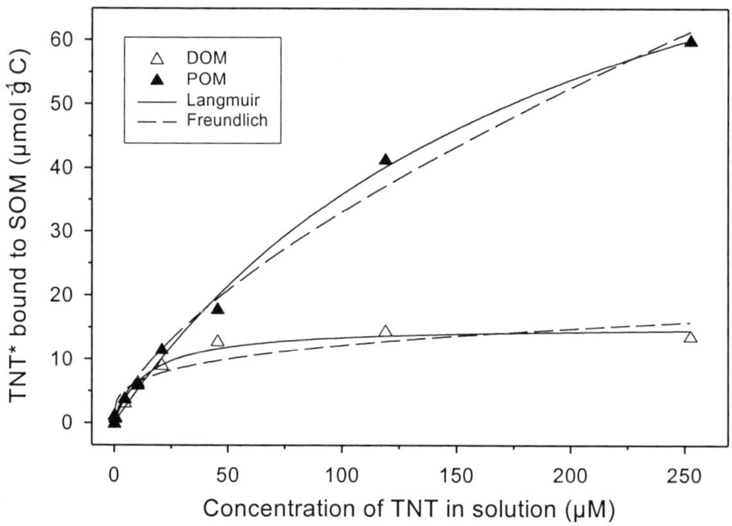

Figure 3. Adsorption isotherms for TNT* bound to POM and DOM at pH=5.6 and an ionic strength of 50 mM NaCl.

Considering the hydrophobic character of the benzene ring and the linear type of adsorption of non-polar compounds by SOM [17], hydrophobic partitioning might explain the discrepancy between POM and DOM isotherms in Figure 3. We conclude that at low concentration of free TNT, bonding seems to take place at specific sites in both POM and DOM. When these sites are saturated, little additional TNT* binds to DOM; hydrophobic partitioning becomes the predominant adsorption mechanism in POM.

ACKNOWLEDGMENTS

We are grateful for the help and support of the personnel at the Defence Research Establishment (FOA), especially Mats Ahlberg, Åsa Fällman and Lars Hägglund. The Swedish Armed Forces provided financial support for this study.

REFERENCES

1. Kaplan, D.L., Kaplan, A.M., 1982. 2,4,6-Trinitrotoluene-surfactant complexes: decomposition, mutagenicity, and soil leaching studies. Environ. Sci. Technol. 16, 566-571.
2. Rieger, P.-G., Knackmuss, H.-J., 1995. Basic knowledge and perspectives on biodegradation of 2,4,6-trinitrotoluene and related nitroaromatic compounds in contaminated soil. *In*: Spain J.C. (Ed.), Biodegradation of Nitroaromatic Compounds. Plenum Press, New York, pp. 1-18.
3. Achtnich, C., Fernandes, E., Bollag, J.-M., Knackmuss, H.J., Lenke, H., 1999. Covalent binding of reduced metabolites of [$^{15}N_3$]TNT to soil organic matter during a bioremediation process analyzed by ^{15}N NMR spectroscopy. Environ. Sci. Technol. 33, 4448-4456.
4. Haderlein, S.B., Weissmahr, K.W., Schwarzenbach, R.P., 1996. Specific adsorption of nitroaromatic explosives and pesticides to clay minerals. Environ. Sci. Technol. 30, 612-622.
5. Haderlein, S.B., Schwarzenbach, R.P., 1993. Adsorption of substituted nitrobenzenes and nitrophenols to mineral surfaces. Environ. Sci. Technol. 27, 316-326.
6. Weissmahr, K.W., Haderlein, S.B., Schwarzenbach, R.P., Hany, R., Nuesch, R., 1997. In situ spectroscopic investigations of adsorption mechanisms of nitroaromatic compounds at clay minerals. Environ. Sci. Technol. 31, 240-247.
7. Selim, H.M., Iskandar, I.K., 1994. CRREL Report 94-7. NTIS, Springfield.
8. Li, A.Z., Marx, K.A., Walker, J., Kaplan, D.L., 1997. Trinitrotoluene and metabolites binding to humic acid. Environ. Sci. Technol. 31, 584-589.
9. Daun, G., Lenke, H., Reuss, M., Knackmuss, H.J., 1998. Biological treatment of TNT-contaminated soil. 1. Anaerobic cometabolic reduction and interaction of TNT and metabolites with soil components. Environ. Sci. Technol. 32, 1956-1963.
10. Held, T., Draude, G., Schmidt, F.R.J., Brokamp, A., Reis, K.H., 1997. Enhanced humification as an in-situ bioremediation technique for 2,4,6-trinitrotoluene (TNT) contaminated soils. Environ. Technol. 18, 479-487.
11. Hundal, L.S., Shea, P.J., Comfort, S.D., Powers, W.L., Singh, J., 1997. Long-term TNT sorption and bound residue formation in soil. J. Environ. Qual. 26, 896-904.

12. Drzyzga, O., BrunsNagel, D., Gorontzy, T., Blotevogel, K.H., Gemsa, D., VonLow, E., 1998. Incorporation of ^{14}C-labeled 2,4,6-trinitrotoluene metabolites into different soil fractions after anaerobic and anaerobic-aerobic treatment of soil/molasses mixtures. Environ. Sci. Technol. 32, 3529-3535.
13. Soil Survey Staff, 1992. SMSS technical monograph No.19. Pocahontas Press, Inc., Blacksburg, VA.
14. Adams, M.A., Byrne, L.T., 1989. ^{31}P-NMR analysis of phosphorus-compounds in extracts of surface soils from selected Karri (*Eucalyptus-diversicolor* F Muell) forests. Soil Biol. Biochem. 21, 523-528.
15. Skyllberg, U., Borggaard, O.K., 1998. Proton surface charge determination in Spodosol horizons with organically bound aluminum. Geochim. Cosmichim. Acta 62, 1677-1689.
16. Thomas, G.W. 1982. Exchangeable cations. *In*: Page, A.L. (Ed.) Methods of soil analysis. Part 2. Agronomy monograph. No. 9, 2nd Edition. American Society of Agronomy, Soil Science Society of America, Madison, WI, pp. 159-165.
17. Chiou, C.T., 1989. Theoretical considerations of partition uptake of nonionic organic compounds by soil organic matter. *In*: Sawhney B.L., Brown, K. (Eds.), Reactions and Movement of Organic Chemicals in Soils. Soil Science Society of America, Madison, WI, pp. 1-29.

Developments in Soil Science, Volume 28A
Editors: A. Violante, P.M. Huang, J.-M. Bollag and L. Gianfreda

RETENTION AND MOBILITY OF CHEMICALS IN SOIL

M. De Nobili[a], R. Francaviglia[b] and P. Sequi[b]

[a]Dipartimento di Produzione Vegetale, Via delle Scienze 208, 33100 Udine, Italy

[b]Istituto Sperimentale per la Nutrizione delle Piante, Via della Navicella 2, 00184 Rome, Italy

Problems related to prediction of mobility and fate of a chemical in soil are intriguing and complex issues for soil chemists. It should not be surprising that so many mistakes are currently made, when dealing with problems of this kind, by professionals who are not familiar with the complexity of the soil environment. A chemical that enters the soil either intentionally or accidentally may (i) react with the soil surfaces, (ii) precipitate as insoluble compound, (iii) be taken up, and eventually released, by soil microorganisms, or (iv) be taken up by plant roots, with the further possibility of being delivered outside the soil and to other environmental compartments through the food chain. Of course, while dissolved in the soil solution, the chemical may also (v) move to other soil horizons or even to surface or ground waters. The soil solution is, in other words, the general crossroads of possible fates for any chemical in soil. Developing an understanding of how a chemical dissolved in the soil solution may undergo reactions that influence its retention and mobility leads to an understanding of its possible fate. Knowledge of the soil solution does not simply mean knowledge of the solutes' concentrations in soil water, because the composition of the soil solution differs in space and time. Only a definite proportion of the soil solution volume moves within the soil itself. Moreover, the soil solution is crowded not only with solutes, but also with inorganic and organic particles or even living components. So the prediction and modeling of retention and mobility of chemicals require a thorough understanding, not only of hydrological properties, but also of time and space factors, together with a holistic approach that combines contributions by physicochemical interactions, plant uptake and microbial decomposition.

1. CHARACTERISTICS AND MOBILITY OF THE SOIL SOLUTION

The soil solution is located in a complex system of pores which, according to a classification proposed by Greenland [1], may be named, depending on their diameters, as fissures (equivalent cylindrical diameter >500 μm), transmission pores (50-500 μm), storage pores (0.5-50 μm), residual pores (0.005-0.5) and bonding spaces (<0.005 μm). Pores larger than 50 μm normally drain under gravity, and allow free water and air movement. The water content of the soil, when these pores have drained, corresponds to the so-called field capacity of soil. The whole space composed of pores larger than 50 μm is also named aeration porosity. In order to meet adequate aeration requirements for arable crops, its volume should represent at least 10% of the soil volume. Another 10% of the soil volume may be made of pores smaller than 5 μm; these are particularly important because most interactions at the molecular level take place inside them. Between the two size limits of 5 and 50 μm, are two groups of pores: the first holds water retained by capillary forces against gravitational forces (storage pores), and the second, i.e., residual pores, acts as an effective reservoir of nutrients (Figure 1).

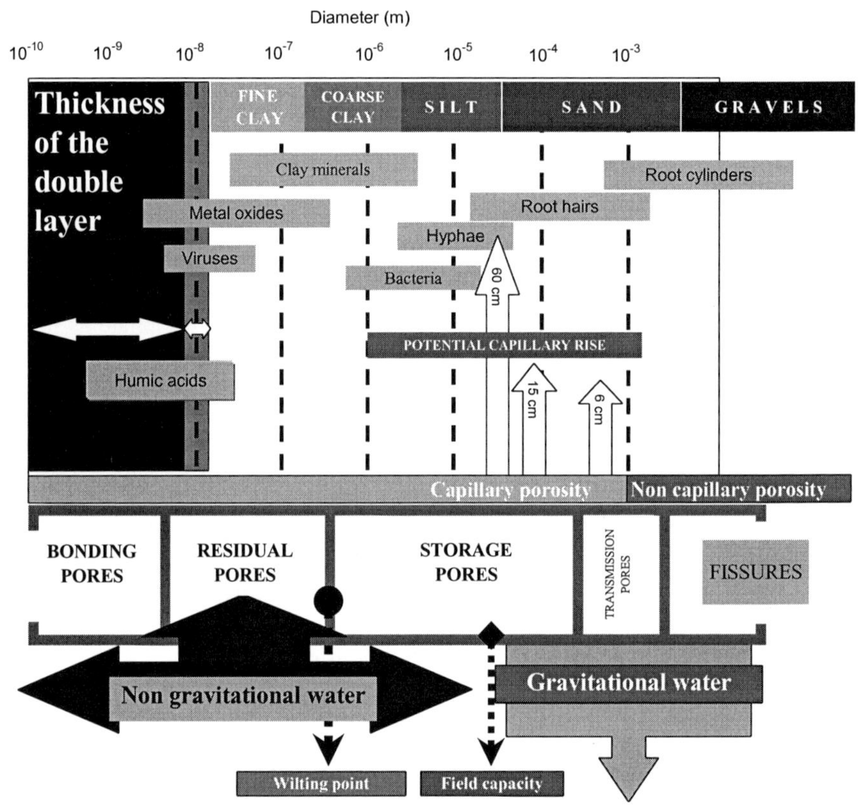

Figure 1. Size range of some living organisms and organs, mineral and organic colloidal particles, and soil pores commonly referred to in soil science (modified from Kretzschmar et al. [2]).

1.1. Forces acting on the soil solution

The movement of the soil solution in storage pores depends on a gradient in the matric potential ψ_m from one soil zone to another. Matric potential is the result of two phenomena: attraction (or adsorption) of soil solids for water, and capillarity. The direction of flow is from a zone of higher ψ_m to one of lower moisture potential. Although this movement may be slow, it is extremely important, especially in supplying water to plant roots.

The other components of total soil water potential Ψ are the osmotic potential Ψo and the gravitational potential Ψg. The osmotic potential originates from the presence of solutes (inorganic salts or organic compounds) in the soil solution and has little effect on the mass movement of water in soils, its major effects being on the uptake of water by plant roots. The osmotic potential becomes significant in soils with high soluble salt levels that can constraint the normal uptake of water by roots. The gravitational potential Ψg depends on the gravity force that tends to pull the water downward. Gravity plays an important role in removing excess water from the upper rooting zones through the widest transmission pores and fissures following heavy precipitation or irrigation.

The general relationship of soil water potential to free energy, i.e., the energy status of water depending on all the forms of energy available to do work other than potential, kinetic and electrical, is shown in Figure 2.

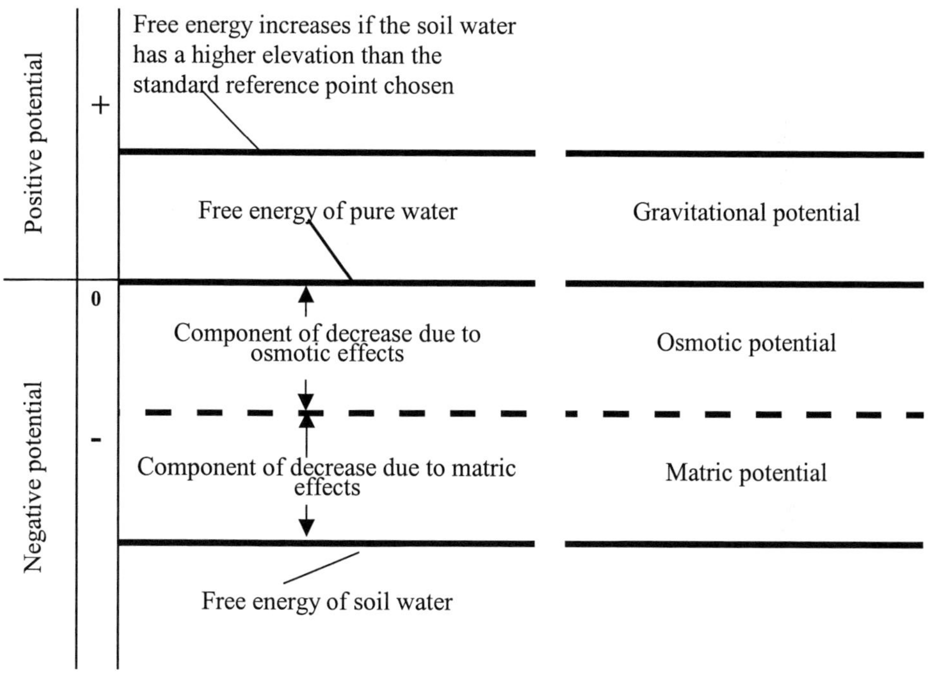

Figure 2. Relationship between the free energies of pure water and of soil water and the effect of elevation on free energy to illustrate the gravitational potential (after Brady [3]).

Another important driving force of the upward movement of water in soil is absorption by plant roots. The importance of this force depends also on the fact that ions are selectively associated with this movement.

1.2. Flow patterns

For modeling purposes, the movements of the soil solution are often distinguished into three main possible categories: saturated flow, unsaturated flow, and vapor movement.

Saturated flow (i.e., the flow of soil solution when all pores are filled with water) occurs, in general, during and/or immediately following a heavy rain or irrigation application, when pores in the upper zones are often filled entirely with water.

The flow of water under saturated conditions is determined by two major factors: the hydraulic force driving the water through the soil (gravity) and the hydraulic conductivity, or the ease with which the soil pores permit water movement.

Hydraulic conductivity can be expressed mathematically as

$$K = V/f \tag{1}$$

where V is the volume of water moved per unit of time and unit area section of the soil, f is the driving force of water movement, and K is the hydraulic conductivity of the soil. The hydraulic conductivity of a uniformly saturated soil is essentially constant and is dependent on the size and configuration of the soil pores. However swelling soils, i.e., soils that contain swelling clays such as smectites, are affected by changes in the salt concentrations in the pore water and may display a much larger conductivity when permeated by leacheates of high salt concentrations [4]. All clays are sensitive to changes in pH: a low pH of leacheates promotes positive edge-to-negative surface interactions of clay particles, enhancing aggregation and increasing soil permeability. On the contrary, a high pH causes dissociation of hydroxyl groups on edges and, particularly at low salt concentrations, causes the dispersion of clay particles, which decreases the conductivity of the soil.

The driving force of the saturated flow, known as hydraulic gradient, is the height of the saturated soil column, and the volume of water moving down the column will depend on this force as well as on the hydraulic conductivity of the soil. The same hydraulic force will cause horizontal and upward flow, if downward movement is impeded. The rate of such flow is usually not quite as rapid, since the force of gravity does not assist horizontal flow and hinders upward flow.

Any factor affecting the size and configuration of soil pores will influence hydraulic conductivity. The total flow rate in soil pores is proportional to the fourth power of the radius, whereas the cross-sectional area is proportional to the second power of the radius. Thus, flow through a pore 1 mm in radius is equivalent to that in 10,000 pores with a radius of 0.1 mm, even though it takes only 100 pores of radius 0.1 mm to give the same cross-sectional area as a 1-mm pore. Obviously, the macropore spaces will account for most of the saturated water movement.

The particle size and structural characteristics of soils are the properties that are more closely related to hydraulic conductivity. Under saturated conditions, sandy soils generally have higher conductivities than fine textured soils. Likewise, soils of stable granular structure conduct water much more rapidly than those with unstable structural units that breaking down upon wetting. Fine clay and silt can clog small connecting channels of even the largest pores. Fine-textured soils that crack during dry weather allow, at first, rapid water movement; but later, the cracks swell and close, thereby drastically reducing water movement.

In saturated soils, the relatively rapid water movement occurs through the widest and most continuous pores, whereas, in unsaturated soils, these pores are filled with air. Water movement can therefore occur only through the finer pores and is much slower. Figure 3 shows the general relationship between matric potential (Ψm) and hydraulic conductivity. Note that at or near zero water potential (saturated flow region), the hydraulic conductivity is thousands of times greater than at potentials that are typical of unsaturated flow (–0.1 bar and below). At high levels of the water potential (high moisture content), hydraulic conductivity of unsaturated soils is larger in the sandy soils than in the clayey soils, whereas the opposite is true at low potential values. In fact, the dominance of large pores in the coarse-textured soils encourages saturated flow, whereas the prominence of thinner (capillary) pores in clayey soils encourages unsaturated flow. Unsaturated flow is primarily governed by the matric potential gradient between moister and drier soil areas. Movement will be from a zone of thick moisture films (higher matric potential, e.g., –0.001 MPa) to one of thin films (lower matric potential, e.g., –0.1 MPa).

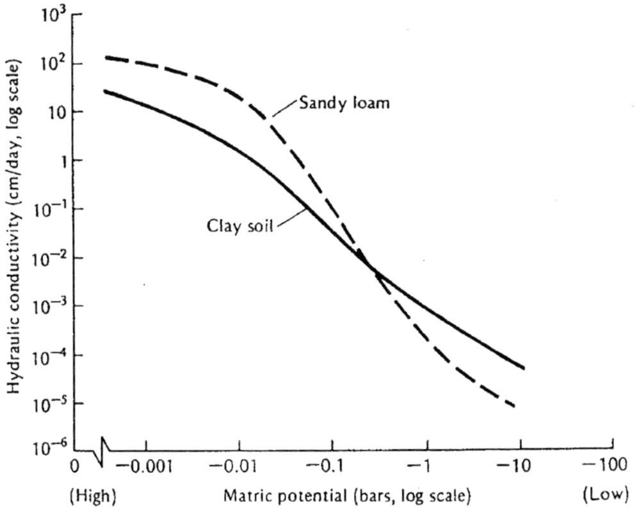

Figure 3. Generalized relationship between matric potential and hydraulic conductivity for a sandy soil and a clay soil (after Brady [3]).

The influence of potential gradient on water movement is illustrated by the moisture curves shown in Figure 4.

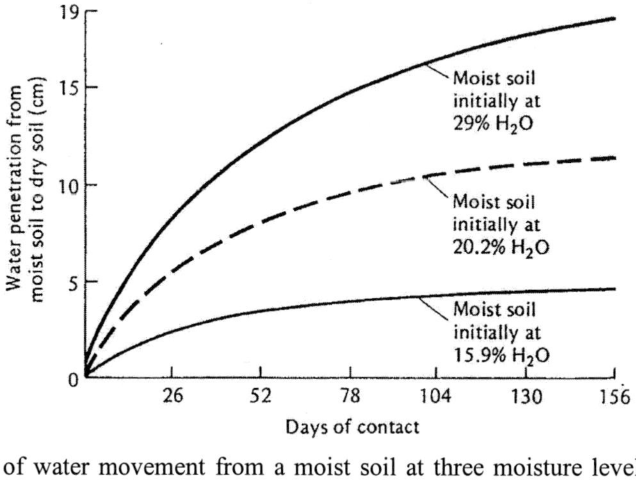

Figure 4. Rate of water movement from a moist soil at three moisture levels to a drier one (after Gardner and Widtsoe [5]).

Water moves more rapidly from the areas with higher moisture content. The higher the water content in the moist soils, the greater the matric potential gradient between the moist and dry soil and, in turn, the more rapid the flow. In this case, the rate of movement obviously is a function of the matric potential gradient.

The third flow pattern occurs in the gas phase as vapor flow. Two types of water vapor movement occur in soils, internal and external. Internal movement takes place within the soil, that is, in the soil pores. External movement occurs at the land surface, where water is lost by surface evaporation.

Water vapor moves from one point to another within the soil in response to differences in vapor pressure. Thus, water vapor will move from a moist soil area where the soil air is nearly 100% saturated with water to a drier soil area where the vapor pressure is somewhat lower. Likewise, if the temperature of one part of a uniformly moist soil is lowered, the vapor pressure will decrease and water vapor will move toward this cooler part. Heating will have the opposite effect. Figure 5 illustrates the relationship.

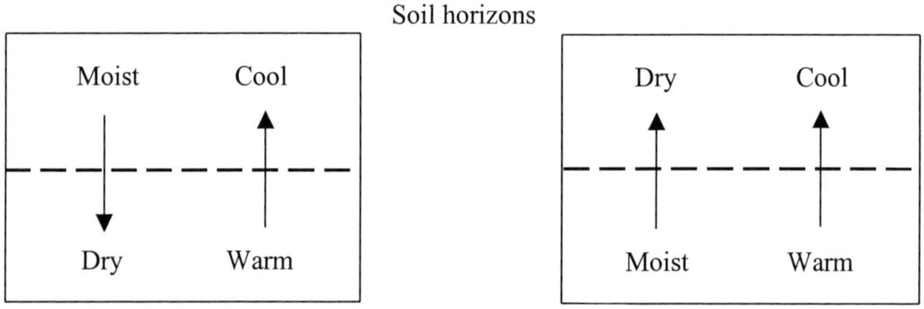

Figure 5. Vapor movement between soil horizons differing in temperature and moisture (after Brady [3]).

The actual amount of water vapor in a soil at optimum moisture for plant growth is surprisingly small, being at one time perhaps no more than 10 kg in the upper 15 cm of a hectare of soil. This amount is negligible compared with the 375,000 kg of liquid water contained in the same soil volume. Because the amount of water vapor is small, its movement in soils is of limited practical importance if the soil moisture is kept near optimum for plant growth. In dry soils, however, water vapor movement may be of considerable significance, especially in supplying moisture to drought-resistant desert plants, many of which can exist at extremely low moisture levels.

Following the approach of van Genuchten and Wierenga [6], a simplified soil system consists of the following five regions (Figure 6):

1 air spaces (unsaturated soil);
2 mobile water located inside the large (inter-aggregate) pores. Water flow is assumed to occur in this region only. Solute transfer in this region occurs by both convection and longitudinal diffusion;
3 immobile water located inside the small (intra-aggregate) pores and at the contact points of aggregates. Solute exchange between mobile and immobile waters is by diffusion only;

4 a dynamic soil region located sufficiently close to the mobile water phase. Solute adsorbed by this portion of soil is in equilibrium with that in the mobile water phase;

5 a stagnant soil region that is in direct contact with the immobile water phase. Solute adsorption in this region is limited by solute diffusion from the mobile water phase into the immobile water phase.

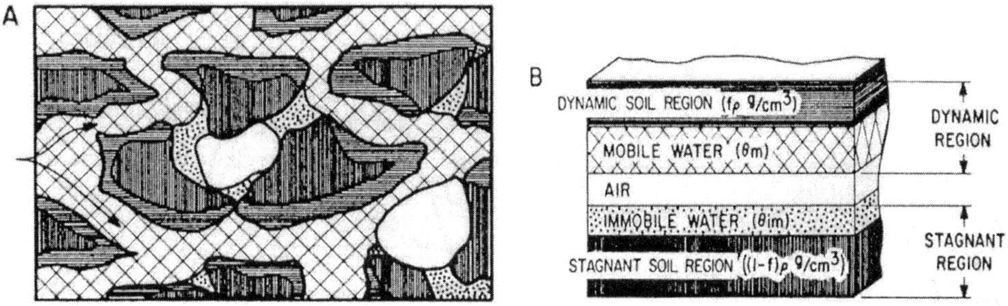

Figure 6. A schematic diagram of an unsaturated aggregated porous medium. A: actual model. B: simplified model. The shading patterns in A and B represent the same region (after van Genuchten and Wierenga [6]).

Soils are in most cases non-uniform; this causes deviations from model flows and deserves specific attention.

A first important case is the occurrence of irregular horizons: uniform profiles are very rare, whereas layered conditions in the profile are very common.

Impervious silt or clay pans are not rare, nor are sand and gravel lenses or other subsurface layers and in all cases downward movement of water is impeded. Stratification influences the amount of water held by the upper part of the soil and causes a much higher field moisture level than would normally be found in freely draining soils. Another condition of the soil that affects the flow regime is the configuration of the soil layers [7]. Horizontal layers within concave troughs or depressions are more likely to detain perched bodies of water and solutes, whereas slanted or sloping layers may direct the flow towards water supplies.

Depending on irregular features in the soil profile, preferential and by-pass flow patterns can easily occur. Preferential flow is a useful generic term to describe the process whereby water and chemicals move through a porous medium, e.g., a soil, following favorable routes and bypassing other parts of the medium. Preferential pathways may be of various types. The most evident follow pre-existing features that are either formed physically by shrinking and cracking of clay, or biologically by burrowing animals or decaying roots. However, there is a quite different sort of concentrated flow that cannot be so readily discerned, as it takes place in soil layers without preexisting features of this kind. Such a flow may occur spontaneously and sporadically at seemingly random sites within the soil profile. This is associated with a still poorly understood phenomenon, termed unstable flow, wetting-front instability or channeling [8]. This phenomenon occurs most notably during transition of infiltrating water from a fine-textured to a coarse-textured layer. Instead of advancing as a smooth front, the percolating water may concentrate at specific locations to break into the coarse sublayer in the form of finger-like or tongue-like protrusions that eventually grow to become pipes. Thereafter, flow through the coarse-textured layer takes place through these pipes rather than uniformly through the entire layer. In aggregated and cracked soils, preferential flow pathways are developed in the wide seasonal fissures. In these soils, a large fraction of the flow occurs via inter-aggregates.

Another well-known approach for describing physical non-equilibrium in soils is the two-flow domain model [9], in which water is assumed to flow through two different-sized pores with distinct water velocity, water content, solute concentration, and dispersion coefficient: the fast flow region and the slow flow region.

A last very common deviation is caused by the presence of repellent surfaces, which influence several soil properties. Water repellent soils have been reported in many countries and can lead to erosion [10] and to unstable wetting fronts [11], which is in turn related to increased risk of groundwater contamination. Water repellency can originate from organic coatings produced by the growing microorganisms, e.g., mycelia from the Basidiomycete fungi [12]. Many plants produce waxes in order to make their leaves water repellent because this facilitates the removal of particulate depositions (dust, spores, etc.) and results in a cleaning of plant surfaces through rain, fog or dew [13]. These waxes are introduced to the soil after plants die and decompose. In California, De Bano et al. [14] reported that wild fires on chaparral watersheds caused water repellency in soil. The burning of the litter produced volatile organic substances that moved into the underlying soil, causing the repellency.

Water repellency has typically been related to dry soils, and significantly affects infiltration, evaporation and other water-soil interactions. Of course, water repellent surfaces may completely change the mobility of hydrophobic solutes.

1.3. Driving agents

Precipitation water is the main agent of solute movement in soil. Leaching, i.e., displacement of the soil solution with consequent downward transport of dissolved chemicals out of the soil profile, occurs whenever

$$P - SR > WHC + ET \tag{2}$$

i.e., where precipitation (P) less the fraction of rain water flowing through surface runoff (SR) overcomes the sum of soil water-holding capacity (WHC) plus evapotranspiration water (ET) lost to the atmosphere during the period of time considered.

Often SR water is underestimated, but can be an impressive amount in Mediterranean climates; the average figure for Italy is reported to be 51% of precipitation water [15]. Soil WHC may range from 0.2 mm in sandy soils, to ten thousand times as much in clay soils. ET is one of the driving agents governing the upward movement of water within soils, the other being absorption by roots. Under vapor form, water losses from soils occur in two ways: a) by evaporation of water at the soil surface, and b) by transpiration from leaf surfaces. The combined loss resulting from these two processes, termed evapotranspiration, is responsible for most of the water lost from soils under normal field conditions. The water loss rate is basically determined by differences in moisture potential identified as the vapor pressure gradient, i.e., the difference between the vapor pressure at the leaf or soil surface and that of the atmosphere. The vapor pressure gradient in turn is related to a number of climatic and soil factors, e.g., radiant energy, atmospheric vapor pressure, temperature, wind, and soil moisture supply. By considering the highest potential ET values occurring in semi-arid climates, it is not surprising to find that leaching processes cannot actually take place in many areas [16].

2. MOBILITY OF DISSOLVED CHEMICALS

Soil particles normally form aggregates (5-250 μm and more) consisting of mineral grains from parent material that are bound together by natural organic matter and amorphous materials such as iron oxides. A solute driven through the soil by advective transport in transmission pores (Figure 7), should, with time, reach equilibrium with the more or less stagnant volume of the soil solution that surrounds the outside of the aggregates and with the water filling the aggregate's pores. Several transport steps are required before a solute reaches equilibrium with the solid phase. The first step is diffusion across the stagnant layer around aggregates. The time required for this step is, in most cases, less than 20 s and depends on the thickness of the stagnant film. The film is about 200 μm thick for large aggregates (>200 μm), but shrinks proportionally with size, because small aggregates cannot impart a significant friction to the fluid.

For a molecule with a diffusion coefficient Dw_0 10^{-5} cm^2 s^{-1}, the flow velocity that would allow a pulse of solute to reach equilibrium with a 200 mm film of immobile water is of the order of magnitude of 10^{-3} cm^2 s^{-1}. As reported before, the velocity with which water percolates through the soil depends on the soil porosity and particle-size composition. To reach the sorption sites on sorbent surfaces, a solute, carried by the mass flow, must then travel further inside, within the immobile water filling the interstices between particles.

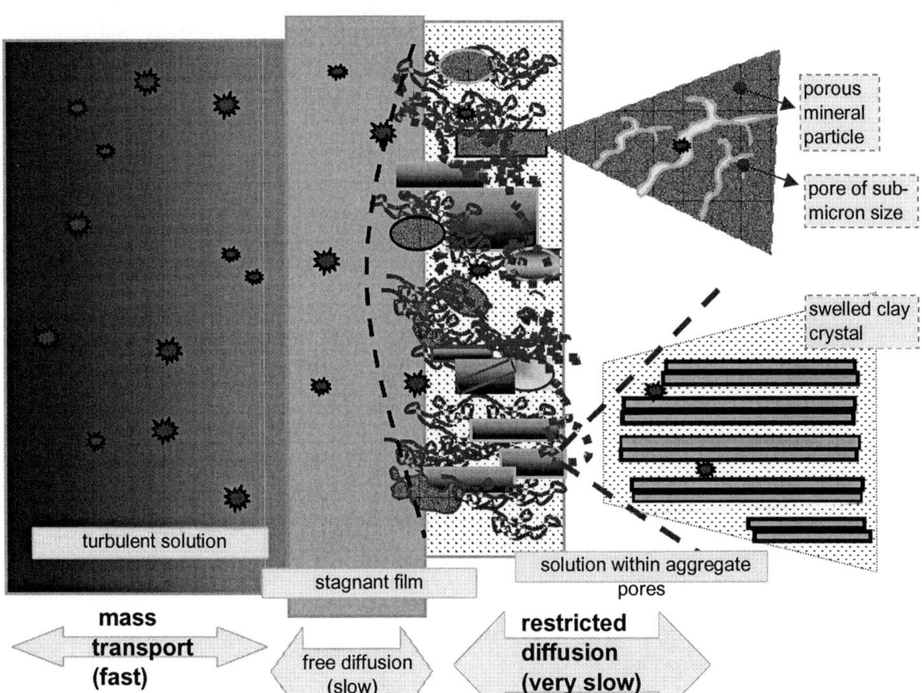

Figure 7. Movement of solutes in the different soil solution regions and the prevailing transport mechanisms therein.

In the larger pores, the solutes diffuse as if they were in free solution, but in many soils a considerable fraction of the total porosity has openings that are comparable in size to the size of solute molecules. Movement of solutes within aggregates is affected by strong steric limitations, which are enhanced by the existence of dead end pores and by the need to diffuse through the net of organic polymers (polysaccharides, humic substances) protruding into the solution and partially blocking larger pores.

Inside aggregates, diffusion coefficients are therefore greatly reduced and this affects not only adsorption, but also desorption kinetics. Decades long desorption kinetics of this kind are observed for soils contaminated with volatile organic substances (VOCS) such as trichloroethylene, tetrachloroethylene, toluene and xylene, and are probably caused by adsorption in nanometric-sized pores [17, 18]. A quicker release of 1,2-dibromoethane and of its residues from soils treated in the laboratory for short periods, and a slower two-stage kinetic desorption observed in field-contaminated soils where the compound was used as fumigant for many years, confirms that the rate-determining step is governed by diffusion.

2.1. Influence of soil surfaces

When the velocity of the flow does not allow the reaching of an equilibrium with the solid phase, the attenuation of advective driven substances can, eventually, only be due to physical constraints (i.e., layers of low hydraulic conductivity) and is only marginally or indirectly affected by chemical interactions. On the contrary, if the mass flow is sufficiently slow to allow equilibration with the immobile films, diffusion-driven dispersion of substances is strongly influenced by interactions with soil components. Both, however, are controlled, at the macroscopic or microscopic level, by the manner by which water is held in the soil particle system.

Clay particles interact through layers of adsorbed water, through the diffuse layer of exchangeable cations and, in only some cases, by direct particle contact. Inter-penetration of ionic layers results in repulsion and particle dispersion and ultimately in reduction of the hydraulic conductivity of the soil. Water molecules make hydrogen bonds with the hydroxy groups and oxygen atoms of soil minerals, forming a layer of adsorbed water molecules followed by several layers of highly structured water. To become adsorbed, a solute would need to break these interactions.

The interactions of most substances with soil components occur at the solid-water or solid-gas interfaces; attenuation processes are therefore ruled by soil components that possess larger specific surface areas. If the adsorption isotherm of a dipositive metal, e.g., lead, is experimentally obtained as it normally is, in batch experiments on a soil suspension, it will measure the maximum potential capability of the soil to adsorb the substance. Leaching experiments show a deviation from the Langmuir type isotherm obtained in batch experiments (Figure 8) towards a constant adsorption isotherm, because more surface is exposed when soil particles are suspended. As the equilibrium concentration of Pb increases, the concentrations of adsorbed Pb in the two soil sections coincide and the isotherms become parallel. Under these conditions, the difference between adsorption isotherms can be accounted for by the difference in exposed surface area.

The clay-size fraction, that is the size fraction of inorganic soil components with an effective diameter of less than 2 μm, and the colloidal organic materials usually play the most important roles in solute–soil interactions. The clay-size fraction includes both clay minerals and small particles of amorphous components such as iron oxides, together with other minerals that are the result of either chemical or physical weathering. This fraction possesses two predominant kinds of surfaces:

1) siloxane surfaces, which essentially consist of the upper layer of oxygen atoms in Si-coordinated oxygen tetrahedrons;

2) oxide-hydrous oxide surfaces, which consist of broken bonds that are saturated by hydroxyls when in contact with an aqueous phase.

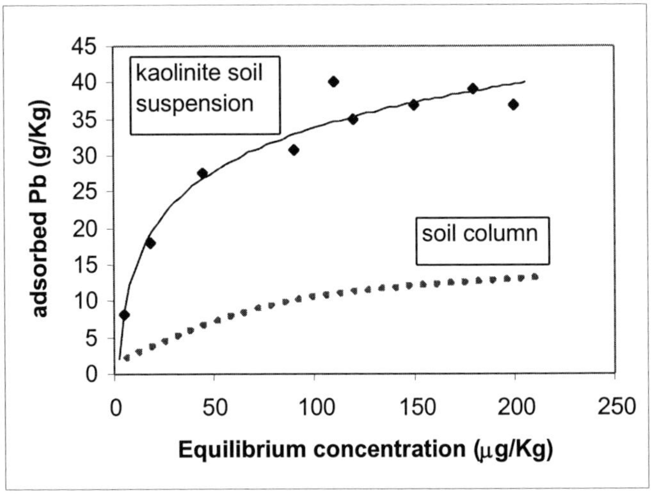

Figure 8. Adsorption isotherm (soil suspension) and adsorption measured by leaching tests in a soil column. The lower adsorption in the soil column is caused by the lower surface exposed as compared to suspended soil particles.

The unsatisfied charges on this second kind of surface are associated with a variable charge originating from proton association-dissociation equilibria. Weathered soils, high in oxide and oxyhydroxide minerals, non-crystalline forms and organics, possess a variable charge density that can even switch its overall sign from plus to minus as a function of pH. The surface of organic materials has a variable negative charge that depends on the presence of functional groups such as hydroxyls, carboxyls, phenolics and amines. Similar to the hydroxyl functional groups in clay minerals, these groups can protonate or deprotonate according to the pH of the soil solution. The surface can therefore acquire an overall positive or, more frequently, negative charge, according to the soil pH. The parameter that expresses these properties of surfaces is the point of zero charge (PZC), which is the pH at which the overall surface charge becomes zero. On the contrary, soils containing 2:1 clay minerals, such as micas, smectites and vermiculites, have a prevalence of siloxane surfaces and a predominant fixed negative charge.

The surplus surface negative charge of most soil components affects the behavior of both cations and anions in solution because of the coulombic interactions between ionic charges and charged particle surfaces. The distribution of ions (n) at any distance (x) from the surface is

$$n = n_i \, [cotan \; x/2 \; (8pz_i^2 e^2 n_i/\varepsilon kT)^{1/2}]^2 \tag{3}$$

where n_i is the number of ions in the bulk soil solution, z_i is the valence of the ions, e is the charge of the electron, ε is the dielectric constant, k is the Boltzmann constant and T is the temperature.

The distance at which the surface potential alters the composition of the solution depends not only on the surface charge density, but also on the concentration and charges of the ions in solution. From all this, it follows that anions will be repelled from the space surrounding soil particles. Repulsion increases with increasing solution concentration, and at low water contents, anions will be effectively excluded from the space between particles (domains and micro-domains of clay particles). Anions will have a smaller effective available water volume than non-reactive solutes (10-20%) and, because anion concentration is highest in the center of pores, where water velocity is maximum, net anion flux will exceed water flux.

However, aluminum and iron oxides have rather high PZC values, and between pH 4 and 7, soil surfaces also possess positively charged sites that can electrostatically attract anions. Chloride and nitrate, when attracted by positive charges, are subject to simple mass action equilibria and behave similarly to cations, but most anions, including organic anions, have a greater affinity for soil surfaces than their concentration would suggest. Although the pH of the soil solution influences adsorption, it acts more by protonating the conjugate acid of the anion than by its action on the surface potential. The mechanism in this case is not a simple electrostatic attraction, but involves ligand exchange: the anions penetrate into the coordination shells of iron or aluminum atoms on the surface of the hydroxide. Anions that are simply electrostatically attracted are reversibly adsorbed, whereas when ligand exchange occurs, it is unlikely that ions can be desorbed using the same amount of energy that was released by their sorption, and large hysteresis phenomena can occur. Sorption can become reversible again by changing the pH: adsorbed phosphate may be desorbed following an increase in pH of the soil before reaching a new equilibrium. This is not an unlikely occurrence: many agricultural soils of the cold temperate climate are acidic and become increasingly acid with cultivation because of fertilizer addition and crop harvest. When the pH decreases below the pH at which yield reductions become visible (which may take five to ten years depending on conditions), the soil is usually limed and the pH brought back to near neutrality levels.

The PZC of a soil generally increases with depth, as clay and oxide content increase and organic matter decreases. Subsurface horizons of accumulation of clay and oxides can represent effective barriers for the downward migration of anions.

The time needed for a chemical reaction to occur in soil can vary from microseconds for ion pairing, complexation, and ion exchange, to milliseconds for many sorption reactions, and to several years for precipitation, dissolution, and mineral crystallization reactions.

The soil solution is in rapid dynamic equilibrium with the composition of the liquid phase held within the double layer and determines not only the relative concentration of the different ionic species, but even the extension of the double layer itself.

The maximum distance at which ionic species feel the action of the surface charge depends on the total ionic concentration and on the prevalent charge of the ions present in the liquid phase. High ionic concentrations and predominance of dipositive over monopositive cations cause a compression of the double layer, because of the enhanced shielding of electrostatic charges. The extension of the double layer from the solid surface is given by

$$d = \frac{1}{n e}\sqrt{\frac{\varepsilon k T}{8 \pi C_0}} \tag{4}$$

where n is the number of charges on the ion, e is the charge of the electron, ε is the dielectric constant, k is the Boltzmann constant and C_o is the concentration of the ion in the bulk solution.

The volume from which anions are excluded can therefore be reduced by half if the solution contains dipositive cations. If we consider that the amount of solute that is retained in a soil layer after a leaching event corresponds to the sum of the fraction adsorbed on solids and of that dissolved in the solution retained by the reserve pores, we can easily foresee that the mobility of negatively charged chemicals will be greatly increased if dipositive ions prevail in the leachate. Cations will be affected as well, not only because the volume where they are retained by electrostatic attraction becomes smaller, but also by the new equilibrium of the exchange mechanism itself.

Many attempts have been made to try to generalize the adsorption behavior of organic compounds in soils [4, 19], but in many cases, application of these models does not allow reliable predictions. The distribution coefficient *Kd*, which is calculated as the ratio of the total equilibrium concentration of the substance in the sorbed phase to that in solution, can be seen as the sum of the contributions of all the different possible forms of adsorption for all the different chemical species of the chemical in solution:

$$K_d = \frac{C_{om} \cdot f_{om} + C_{min} \cdot A + C_{ie} \cdot \sigma_{ie} \cdot A + C_{xn} \cdot \sigma_{xn} \cdot A}{C_{neutr.} + C_{ion}} \qquad (5)$$

where:
C_{om} is the concentration of the chemical adsorbed on organic matter;
f_{om} is the weight fraction of organic matter in soil;
C_{min} is the concentration of the chemical adsorbed on minerals;
A is the area of mineral surfaces per unit mass of soil;
C_{ie} is the concentration of ionized sorbate electrostatically attracted by surface charges;
σ_{ie} is the net concentration of surface charges;
C_{xn} is the concentration of neutral sorbate bonded by reversible reaction to solid surfaces;
σ_{xn} is the concentration of bonding sites;
$C_{neutr.} + C_{ion}$ is the sum of the concentrations of the neutral and charged species of the chemical in solution.

The mathematical evaluation of these parameters is very complex and cannot be precise: values need to be determined experimentally for every specific soil and chemical.

The retardation of the migration of chemicals in soil depends on adsorption and can be calculated on the basis of the distribution coefficient and of the characteristics of soil porosity:

$$R = \left[1 + Kd * \frac{\rho d}{n} \right] \qquad (6)$$

where R is the retardation coefficient, ρ_d the bulk density and n the total porosity [20].

2.2. Influence of dissolved chemicals

The composition of the soil solution can have a fundamental influence on the potential mobility of a solute. The mobility of nonpolar compounds is modified by the amount and quality of solvated inorganic ions in the soil solution. This is indirectly caused by the impact of dissolved salts on the aqueous solubilities; this effect is rather small for relatively soluble polar non-ionic compounds, but may approach a factor of 2 for large nonpolar molecules [19].

Polynitroaromatic compounds, such as trinitrotoluene, can be adsorbed on the surface of clay minerals due to specific interactions that involve formation of an electron donor-acceptor complex with oxygen atoms on the siloxane surfaces. When the cations retained near these surfaces by electrostatic attraction are weakly hydrated large monopositive cations such as K^+ or NH_4^+, a strong electron donor-acceptor complex forms between the retained cation and the nitro-compound, but only very weak interactions occur with the strongly hydrated cations such as Na^+, Ca^{2+}, Mg^{2+} and Al^{3+} [21].

Dipositive cations also affect the distribution coefficient of organic anions owing to the formation of ion pairs that reduce repulsion from negatively charged surfaces. It is therefore obvious that during movement within the soil, the potential mobility of both organic and inorganic solutes will be affected by the ionic strength and composition of the solution. However, the soil solution itself can be subject to changes that can be either the consequence of common agricultural practices, such as addition of fertilizers or pH amendments, or the result of natural processes, such as those occurring in submerged soils.

The pH and buffering strength of the soil solution are very important parameters in determining the behavior and fate of ionizable molecules.

2.3. Reactions causing differences in chemical adsorption

In the soil environment, many chemicals can undergo rapid biological or abiotic transformations that alter their concentration in the bulk solution. This affects not only the overall rate of diffusion, which depends on the gradient concentration, but also the various mechanisms of interaction that are at the base of the adsorption process. The dissociation of ionizable functional groups, such as carboxyls and phenolic OH, and the protonation of bases are the most common examples of reactions that have strong effects on the solid/solution partition coefficient and therefore on the potential mobility of chemicals. It is obvious that the pH of the soil is one of the main factors that determine the extent of these kinds of reactions, but not the only one: the actual behavior of a chemical in soil depends on the chemico-physical properties and composition of the soil components. Again, we must consider that, in the proximity of charged surfaces, the composition of the liquid phase differs from that of the bulk soil solution. Hydrogen ions are attracted by negative charges and their concentration is therefore larger near the solid's surfaces than in the outer part of the double layer; the same happens to aluminum ions that contribute to soil acidity through the dissociation of the water molecules of the first hydration shell. Both ions are retained as exchangeable cations and make up the exchangeable acidity of the soil. This exchangeable acidity can often be a few orders of magnitude larger than the acidity of the soil solution and substantially contributes to the buffering capacity of the soil.

A neutral molecule such as aniline (pKa: 4.63) can freely diffuse inside the double layer, and, if the pH is sufficiently lower than in the bulk solution, become protonated. The protonated aniline produced by the reaction with a hydrogen ion is electrostatically attracted by the negatively charged surface and remains within the double layer, where it acts as a substitute of the proton consumed in the reaction. A protonated aniline molecule diffusing from the bulk solution to the double layer region is in principle attracted by the negatively charged surfaces, but may not actually be retained in a very acid soil due to the fact that the monopositive protonated aniline molecules cannot displace strongly held tripositive aluminum cations. So, contrary to expectations, protonation can result in a much stronger adsorption of weak bases in a mildly acid soil than in a very acid one (Figure 9).

Compounds that acquire a negative charge from the dissociation of an acid group will experience, on the contrary, a dramatic decrease in adsorption. Trichlorophenol, which has a pKa of 6.13, will only be about 10% in this neutral form at pH 7; its adsorption coefficient on organic matter (Koc), which is about 2230 for the neutral molecule, drops to about zero for the ionized form [22]. Negatively charged, ionized species are, in fact, not only prevented by electrostatic repulsion from direct contact with surfaces, but the ionization itself increases the solubility of the molecule and therefore reduces the energy gain derived from hydrophobic adsorption.

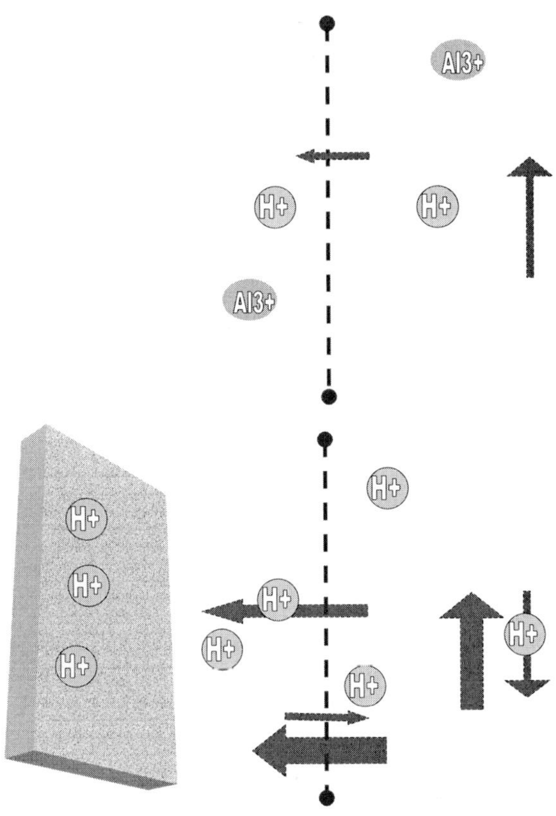

Figure 9. Effect of pH on protonation of aniline, and of exchangeable aluminum on the retention of aniline in two acidic soils. Although a larger fraction of aniline is protonated in the more acid soil, retention of the cationic species is diminished because exchange sites are occupied by strongly held Al^{3+} ions.

All this results in a decreased affinity for the solid phase and in an increased hazard of groundwater contamination. To evaluate the actual risk, we must, however, consider the interactions with the double layer. The lower pH that is expected in the proximity of solid surfaces favors the association of a proton to ionized weak acid groups, shifting the equilibrium to the neutral form and increasing the concentration gradient-driven diffusion toward the surface. When double layers are fully expanded, double layers from different particles come to overlap in the bonding pores and in the smaller residual pores, and negatively charged molecules are virtually excluded from the whole micropore volume. Even diffusion into storage pores can be prevented when these pores are not completely filled and the double layer extends for a large fraction of the water films (Figure 10). This means that molecules will be confined to that fraction of soil porosity where water is not held by matrical forces and can freely move under the effect of gravity (gravitational water).

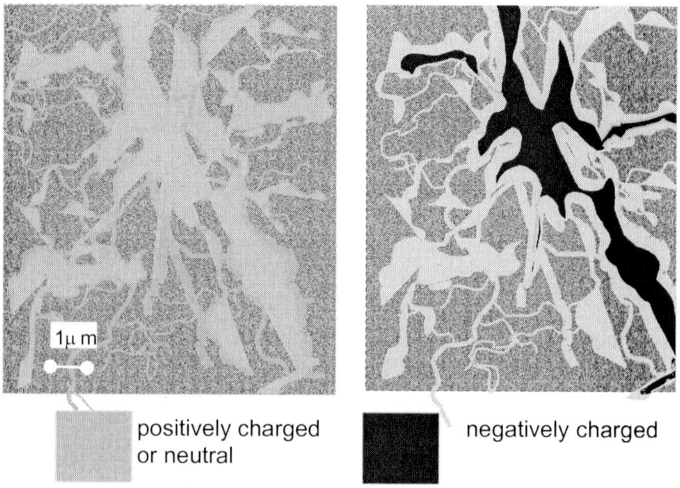

Figure 10. Volume available for the diffusion of charged solutes in water-filled soil pores. Note that negatively charged solutes are excluded from a large fraction of the pore space.

2.4. Biodegradation

Soil is not an aseptic environment, but is the home of a large and well-differentiated microbial community [23]: soil microbial biomass represents from 1 to 3% of the soil organic carbon and is mostly concentrated in the upper soil horizons. Besides being the main agent for the turnover of organic carbon and nitrogen in soil, microbial biomass can actively decompose xenobiotic molecules. This occurs either as the result of a detoxification mechanism or from microbial utilization of xenobiotics as an energy source, alone or together with natural organic molecules (commensalism). In most cases, complete biodegradation to carbon dioxide is not achieved and partial degradation generates a variety of degradation products. Typical biodegradation reactions include: decarboxylation, oxidation of amino groups, reductive dehalogenation and hydrolysis. Both aerobic and anaerobic bacteria carry out some of these reactions, whereas others, such as dehalogenation, occur only under anaerobic conditions.

The biodegradation of an organic chemical in soil is catalyzed by enzymes that are located inside living microorganisms, released from cells as extracellular enzymes, or occur as stabilized enzymes in soil organic matter. Biodegradation reactions are generally first-order with respect to the organic chemical's concentration; models of migration of chemicals in soil must, therefore, include a first-order degradation term:

$$-\frac{dC}{dt}=kC \tag{7}$$

where C is the concentration of the chemical and k is the rate constant. Biodegradation rates constants are often calculated, but are obviously influenced by all those factors that are known to affect soil microbial activity, such as pH, water content, substrate availability, organic carbon and clay content, etc. Their values should be used only for rough estimates or to compare the resistance to biodegradation of various chemicals.

By comparing microbial activity levels in soils of the same kind, it is possible to note the fundamental importance of soil porosity for soil microorganisms. This factor is often overlooked, but its importance should not be surprising: soil microorganisms not only use soil pores as their living habitat, but they also need water to survive. The distribution of the dimensions of soil pores determines which fraction of soil pores remains water-filled at a given water potential (Figure 11). The degradation rate actually reaches a maximum at water potentials where those pores that host most bacterial species, that is between 0.8 and 30 μm in diameter, are filled with water.

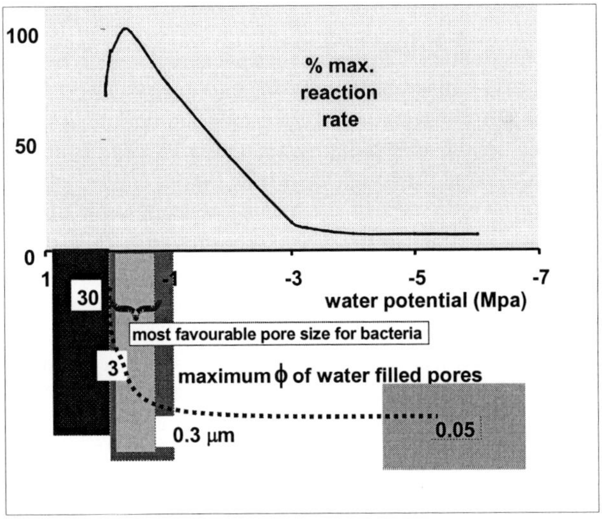

Figure 11. Relationships between biodegradation rate, water potential and diameter of water-filled pores.

Soil management practices, such as no-tillage [24], that cause an increase of soil porosity in this size range increase microbiological activity and will increase the biodegradation potential of the soil.

2.5. Consecutive reactions

Many transformations often occurring in soils are chained reactions due to the action of microbial associations; they result in consecutive reactions that may or may not lead to the accumulation of intermediate products, but strongly affect the overall fate of the substance. A simple, but environmentally important example, is the sequence of reactions that follow the application of the most widely used nitrogen fertilizer: urea. This neutral organic compound is readily hydrolyzed by soil urease activity into ammonium and carbon dioxide. Contrary to urea that would interact only weakly with soil surfaces and could be easily leached into the subsoil, ammonium is a cation and therefore can be strongly held by cation exchange near the negatively charged surfaces of soil colloids. Ammonium, however, can be oxidized by different genera of chemoautotrophic bacteria, first to nitrite and then to nitrate. Nitrite does not normally accumulate in the soil and is readily converted to nitrate; the overall result is a chain reaction kinetic mechanism [25, 26] consisting of two consecutive irreversible first-order reactions where

$$A \longrightarrow B \longrightarrow P$$

$$dA/dt = -k_1A \quad dB/dt = k_1A - k_2B \quad dP/dt = k_2B, \quad \text{with } K_1 \ll K_2 \tag{8}$$

Nitrate, unlike other anions such as phosphate, is not retained in most soils unless they are below the PZC. Therefore, it can be easily leached away; any downward movement of water will produce a loss proportional to the concentration of nitrate in the soil solution and the mass of percolated water. Nitrate is subject to anion exclusion and the rate at which it diffuses into the soil is larger than the rate of oxygen diffusion [27]. For this reason, the ion can easily reach anaerobic soil layers or sites within aggregates where it can be reduced.

Nitrate is converted under anaerobic conditions to reduced gaseous forms such as N_2O and N_2. Therefore, in soil, the consecutive reactions of ammonium oxidation can be linked through a diffusion step to the dissimilatory reduction of nitrate, in spite of the apparent incongruency of such a system. Although denitrification is carried out by chemoautotrophic bacteria, the process is favored in soil by a high concentration of readily decomposable organic carbon because the increased oxygen consumption that derives from microbial activity can induce the onset of anaerobic conditions. The impact of these transformations, which can take place within the time employed by the soil solution to move downward along the soil profile, may be of primary importance. A substance that is adsorbed onto a soil surface, can subsequently be ejected after a biochemical "one-way" reaction from the surface, and repulsed towards the outer and larger pores where the soil solution may easily move and leach it away.

2.6. Ion movement and plant nutrient uptake

A prerequisite of uptake is contact between plant roots and the nutrients in soil. This contact occurs by two different, but complementary, processes: growth of roots to places where nutrients are located (root interception) and movement (mass flow and diffusion) of nutrients through the soil to the root surface. Root growth is measured in meters, and depends on species and environmental conditions. By contrast, nutrient transport from soil to roots ranges from fractions of a millimeter to a few centimeters, but interception contributes only a small part to the total nutrient uptake (Table 1).

Movement of nutrients from soil to root is brought about by two mechanisms: mass flow and diffusion [29]. Mass flow is the convective transport of dissolved nutrients in the soil solution: the nutrients move towards plant roots as a result of shoot transpiration. Diffusion is the movement of a substance under the effect of a concentration gradient, which in this case is caused by depletion of nutrients near the root surface. The total flux of nutrients from soil to roots, F_T, is the sum of mass flow F_M and diffusion F_D. The contribution of mass flow is given by the product of the volume of water absorbed v and the concentration C_L of the nutrient in the soil solution. Only small fractions of the total P and K taken up by crop plants reach the roots by mass flow: the major part, about 95%, is ascribed to diffusion from soil to root. Only nitrate is often assumed to arrive at the root surface mainly by mass flow (Table 1). The plant requirements for other nutrients, e.g., Ca^{2+} and Mg^{2+}, in calcareous soils may be supplied entirely by mass flow. Diffusion in soil follows Fick's first law, but the effective diffusion coefficient in soil is much smaller than in free solution, due to the complex three-dimensional structure of the pores, and to changes in water content.

Table 1
Estimated contribution of root interception, mass flow, and diffusion to the mineral nutrition of a field-grown maize crop (after Barber [28])

Mineral nutrient	Process			
	Uptake	Interception	Mass flow	Diffusion
K	195	4	35	156
N	190	2	150	38
P	40	1	2	37

Nutrient uptake is also influenced by the presence of other ions. Addition of a cation (e.g., K^+) or an anion may decrease the absorption of other ions of the same type (e.g., Mg^{2+}) and increase the absorption of oppositely charged ions. Other more specific interactions also occur. One example is the antagonism between NO_3^- and Cl^- (Table 2). Ammonium ions can also inhibit the uptake of nitrate ions through the suppression of the symport action of NO_3^-.

The mechanism of nutrient uptake by roots is very complex and is regulated by the functioning of the root cell membrane. The plasma membrane of the cell actually represents a barrier to the uptake of nutrients into the root: the plasma membrane is hydrophobic and charged hydrated ions do not cross it easily. The physiological reason for this is that the membrane must act as a barrier, preventing not only unwanted ions from getting in, but also nutrients from getting out of the cell. The membrane regulates the uptake of nutrients and the release of unwanted solutes, acting even against diffusion gradients by means of an energy-dependent transport (active transport) based on the energy released by the hydrolysis of ATP.

This energy-dependent transport is performed by the so-called proton pump, which maintains the pH of the cytoplasm near neutrality by releasing protons in the soil solution. This allows the uptake of nutrients by means of a passive transport system, which, on the contrary, does not use energy, but works along the diffusion gradient (e.g., Ca^{2+} in soil solution = 1000 μM; cytoplasm = 1 μM). It is important to note that the proton pump (Figure 12) transfers H^+ ions out of the cell into the apoplast (soil solution) and consumes energy in order to operate, not only against the concentration gradient, but also against the build up of charge on the outside of the membrane (electrical charge gradient).

Table 2
Effects of nitrate concentration in the rooting medium on chloride uptake (after Glass and Siddiqi [30])

Concentration in nutrient solution (mmol L^{-1})		Cl^- concentration (μmol g^{-1} fresh weight)	
Cl^-	NO_3^-	In roots	In shoots
1	0	52	93
1	0.2	26	73
1	1	13	54
1	5	9	46

The proton pump has several functions for plant nutrition: it acidifies the soil (making P and Fe more soluble) and the H^+ activates the other transporters (symporters and uniporters). These transporters counteract the electrical charge gradient created by the proton pump by letting the nutrients in. The electrical charge gradient would be sufficient in itself, in terms of energy to drive positive ions in or move negative ions out of the cell to balance the charge on the outside, but since the plasma membrane is hydrophobic, this cannot happen. To allow a selective compensation of the charge gradient, the membrane contains proteins (hydrophilic holes), which let the positively charged ions in. These are termed channels (fast passages for a single ion at a time), uniporters (proteins in the membrane letting in a single ion at a time), or symporters (proteins in the membrane letting two ions pass e.g., anions NO_3^- or PO_4^{2-} and H^+ at the same time). Depending on how they work, they are selective for a particular ion (e.g., K^+-channel, Ca^{2+}-channel) and they (normally) only let ions (K^+, H^+, Ca^{2+}, Mg^{2+}) go one way (Figure 13).

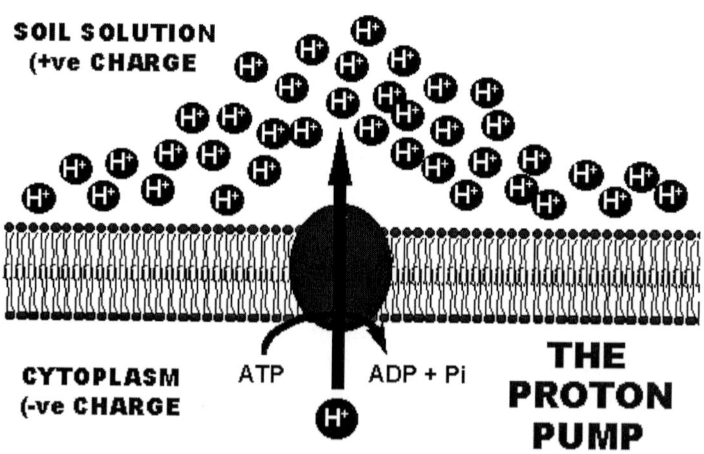

Figure 12. Active transport through the H+ pump (adapted from Marschner [31]).

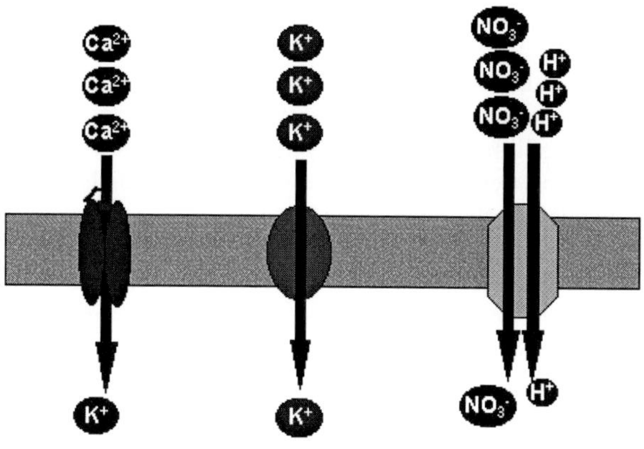

Figure 13. Schematic representation of channels, uniporters and symporters. Ions cross the plasma membrane (the grey line) through proteins (the grey bits) to enter the cell. Note the little ball gate at the top of the channel that regulates uptake (it's like a plug that can move in and out of the hole). Also note that the channel is a pore, whereas the others are not (adapted from Marschner [31]).

3. IMPORTANT DEVIATIONS

The more important deviations from the model flows outlined above do not happen in anomalous situations, but are very common in particular soils or topographic environments. We will consider the case of mobility at the solid state and that of anaerobic environments.

3.1. Mobility at the solid state

If soil particles are either present as discrete entities or soil aggregates are relatively unstable, small soil crumbs and simple particles may detach from the pores' surfaces and migrate down the soil profile carried by gravitational water. Of course, the behavior of tightly adsorbed or loosely retained chemicals may not change during or after migration. Nevertheless, the chemicals will be carried away together with the particles they are sorbed to and eventually reach zones where the composition of the soil solution is different from that of the upper horizons, causing their desorption.

Soil particles can be mobilized by leachates of high pH, high Na^+ and low ionic strength and/or by dissolution of cementing agents, such as organic matter, by increased mineralization or reduced C inputs or, in the case of iron oxides, from dissolution caused by anoxic conditions [2]. Mobilization of soil particles affects the downward migration of cationic species that, being strongly retained by negatively charged surfaces, would otherwise be fairly immobile [32]. In the same way, particle-mediated transport is likely to increase the potential mobility of non-soluble or of sparingly soluble organic compounds that are mostly partitioned in the solid phase. Together with preferential flow, particle-mediated transport can probably explain the large difference between predicted and actual mobilities often found in field studies.

The behavior of soil particles moving in vertic soils shows peculiar features with respect to other kinds of soils, as it is not associated with drainage water. In these soils during the dry season, the presence of consistent amounts of shrinking clays forms large cracks; soil crumbs may fall in these cracks, refilling them partially or totally. When the season turns to wetness, clays swell enormously, and tremendous pressures build up, which crush and mix thoroughly soil components in such a way that lateral and upward movement of solid components occurs regularly in a large part of the soil profile. Organic matter is smashed so intimately with mineral soil components that the soil profile becomes dark black even if the organic matter content does not reach values higher than 2-3%. Adsorbed chemicals would probably behave analogously, but information available on this peculiar contribution to the mobility of chemicals is scarce. The behavior of organic chemicals adsorbed outside and within the clay platelets during the swelling-shrinking cycles is also affected. The huge pressures involved during shrinking of clay particles, coupled to the catalytic effect of surfaces, can in fact favor condensation and coupling reactions as well as hydrolysis, making the prediction of the movement of chemicals in vertic soils a case of its own.

3.2. Anoxic environments

Soil is a living world, and is so rich in life that O_2 must be continuously renewed in order to allow optimal rates of essential metabolic processes of soil organisms and plant roots. The exchange of gases between the soil and the atmosphere is facilitated by two mechanisms, mass flow and diffusion. Mass flow of air is apparently due to temperature or pressure differences between the atmosphere and the soil air, and is relatively unimportant in determining the total exchange that occurs. However, in the upper few centimeters of soil, diurnal changes in soil temperature may result in mass flow of some significance. The extent of mass flow is determined by factors such as soil and air temperature, barometric pressure, and wind movements. Most of the gaseous interchange in soils occurs by diffusion, and through this process each gas tends to move in a direction determined by its own partial pressure. Since the rate of diffusion of O_2 in air is about 8,000 times higher than that in water, 1-2 hours are enough to establish anaerobic conditions in the soil body when the soil is saturated with water.

This behavior may also be taken as an example of the great differences between soil and water environments. A water environment is aerobic to its maximum depth, and normally even its sediment may be fully aerobic. A submerged soil becomes completely anaerobic soon after waterlogging, with the exception of the upper layer (about 1 cm) in contact with the water layer. A further consequence is that the entire soil body should be considered aerobic, provided that non-capillary porosity will be refilled with some air in a solution of continuity with the atmosphere. This is because the rate of diffusion of O_2 in the soil solution will normally be sufficient to reach any pores of smaller size, since they are rarely more than 1 cm apart from the wider channels. A last consideration is that saturated flow models must be assumed as reference models only on a temporary basis, because soil saturation leads to sudden anaerobic conditions, which cannot be ideal for a reference soil.

As is well known, in water environments accumulation of organic matter by natural processes (e.g. bogs, peats) or anthropogenic pollution (discharge of wastes with high BOD content) may lead to the establishment of anaerobic conditions. Similarly, an aerobic soil may easily become anoxic if treated with fresh organic amendments. This is the reason why organic matter of fertilizers must be stabilized to prevent undesirable consequences, which may include formation of an environment unfavorable to growth of crop roots, loss of nitrogen by nitrification-denitrification processes (Figure 14), and production of large amounts of organic intermediates and acids. Such substances can also act as complexing agents of metals in soil, making them soluble and leachable.

Most of the redox reactions that occur in the soil environment are microbially mediated. Even in those cases when it is possible to talk of abiotic oxidation or reduction, because metabolically active microorganisms are not directly involved, biological activity has a predominant influence, because it determines the abundance of electron donors and acceptors in the medium [33].

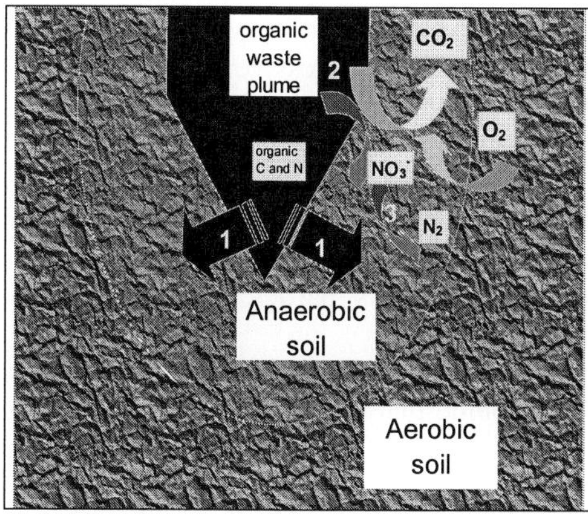

1. Organic matter is decomposed aerobically, consuming all oxygen
2. Nitrate produced from ammonium diffuses in the anaerobic zone
3. Denitrification occurs in the anaerobic soil zone in proximity of the organic waste plume

Figure 14. Decomposition of an organic waste can cause the onset of anoxic conditions in the soil zone affected by the organic plume.

The redox potential (Eh), measured by a platinum electrode immersed in a soil suspension, is the expression of the redox status of the soil and therefore of the tendency of a substance to be oxidized or reduced. Knowledge of the soil pH is also essential to predict if the reduction of a certain oxidant is energetically favored ($\Delta G^0 < 0$). Redox reactions in soil, in fact, involve, in most cases, a combination of proton and electron transfer. Electrons do not exist as a free species, and it is therefore necessary, for the reaction to occur, that the oxidant and reductant come into close contact and an oxidation reaction must balance the reduction reaction or *vice versa*.

Mangancse oxides are widespread oxidizers that can carry out the oxidation of inorganic ions such as chromium and plutonium, increasing their toxicity and potential mobility. These elements can exist as both cationic and anionic species, depending on their oxidation state. In solution, Cr(III) is in the form of a tripositive metal cation or its hydrolysis products, $Cr(OH)^{2+}$ and $Cr(OH)_2^+$. As a tripositive cation, Cr is strongly held by electrostatic attraction within the double layer, but Cr(VI), which is far more dangerous, being a suspected carcinogen, occurs as dichromate ($Cr_2O_7^{2-}$) or chromate ions ($Cr_2O_4^{2-}$). Manganese oxides can catalyze the oxidation of Cr(III) in soils above pH 5; the amount of Cr oxidized depends on the amount of Mn oxides present, but a fresh amorphous surface is necessary. A fresh surface can be generated within the soil from dissolution of MnO_2 by root exudates. Chelating substances, contained in root exudates, complex dissolved Mn^{2+} and thermodynamically favor the oxidation of Mn^{2+} from MnO_2 to Mn^{3+}. This is followed by dismutation to Mn^{2+} and formation of fresh amorphous MnO_2. The sudden switch of oxidized Cr from the cationic to the anionic form causes the immediate repulsion of the oxidized species from the negatively charged surfaces into the bulk solution. Depending on the redox status of the soil, the newly formed dipositive anion can either be quickly reduced again to Cr(III) or be forced to remain in that space of the soil solution that is outside the action of electrostatic repulsion [34]. The result is a dramatic change in mobility of the element in the soil.

Reducing substances are naturally produced in soil by microbial activity. When normal gaseous exchanges with the atmosphere are blocked by a layer of flooding water, or more simply, whenever impeded drainage causes the onset of anaerobic conditions in all or part of the soil profile, facultative anaerobes are forced to utilize electron acceptors besides oxygen. This process is called anaerobic respiration and causes the progressive reduction of soil components following the order of maximum energy yield: nitrates are reduced first, followed by manganese oxides and iron oxides. The intensity of reduction depends on the availability of substrate and is stronger in soils of large microbial biomass and high organic matter content. Anaerobic conditions are therefore accompanied by a general increase in the ionic strength of the soil solution and by the release of large amounts of soluble Fe^{2+} and Mn^{2+}. Depending on the soil position in the landscape and on its particle-size composition, anaerobic conditions can be a frequent feature for many soils.

Under strongly reducing conditions, obligate anaerobes, such as sulfate-reducing bacteria, become active and release into solution hydrogen sulphide that normally precipitates as iron sulfide, which eventually reacts in turn with alkylbromides to yield alkyl mercaptans or thiols. The mercaptans are even better nucleophiles than hydrogen sulfide and react further with other alkyl bromide molecules, producing dangerous dialkyl sulfides.

Soil humic substances can be reduced by many Fe(III)-reducing bacteria of the family of *Geobacteriaceae*, which act as electron carriers for the reduction of substances such as mercury, nitroaromatic compounds and chlorinated solvents [35]. The reduction of humic molecules, which probably involves quinone moieties as the direct electron acceptors and donor groups, is environmentally important, not only for their direct reducing capability, but also because it enhances the capacity of microorganisms to reduce insoluble Fe(III) oxides and consequently increases the rate of anaerobic degradation of aromatic hydrocarbons [36, 37].

4. CONCLUSIONS

The prediction and modeling of the fate and transport of chemicals in soil requires a sound holistic approach. Factors affecting the movement of water are of primary importance, but should be considered in a dynamic and not in a static way. To this purpose it is very important to understand the actual time scale of soil processes, particularly those related to diffusion, in order to predict whether a solute will reach or not equilibrium with the solid phase. Biological activity, which is the cause of many transformations of organic and inorganic chemicals that enter and migrate through the soil, is in many cases the key factor that determines both the behavior and the ultimate fate of a chemical in the soil environment.

REFERENCES

1. Greenland, D.J., 1977. Soil damage by intensive arable cultivation: temporary or permanent? Phyl. Trans. R. Soc. Lond. B. 281, 193-208.
2. Kretzschmar, R., Borkovec, M., Grolimund, D., Elimelech, M., 1999. Mobile subsurface colloids and their role in contaminant transport. Adv. Agron. 66, 121-193.
3. Brady, N.C., 1984. The Nature and Properties of Soils. Macmillan Publishing, New York.
4. Yong, R.N., Mohamed, A.M.O., Warkentin, B.P., 1992. Principles of Contaminant Transport in Soils. Elsevier, Amsterdam, Holland.
5. Gardner, W., Widtsoe, J.A., 1921. The movement of soil moisture. Soil Sci. 11, 230.
6. van Genuchten, M. Th., Wierenga, P.J., 1976. Mass transfer studies in sorbing porous media. I. Analytical solutions. Soil Sci. Soc. Am. J. 40, 473-480.
7. Hillel, D., 1989. Movement and retention of organics in soil: a review and a critique of modeling. *In*: Kostecki, P., Calabrese, E.J. (Eds.), Petroleum Contaminated Soils. Volume I: Remediation Techniques, Environmental Fate, Risk Assessment. Lewis Publishers, London, pp. 81-86.
8. Hillel, D., 1993. Unstable flow: a potentially significant mechanism of water and solute transport to groundwater. *In*: Russo, D., Dagan, G. (Eds.), Water Flow and Solute Transport in Soils. Springer Verlag, Heidelberg, pp. 123-135.
9. Skopp, J., Gardner, W.R., Tyler, E.J., 1981. Solute movement in structured soils: Two-region model with small interaction. Soil Sci. Soc. Am. J. 45, 837-842.
10. Wallis, M.G., Horne, D.J., 1992. Soil water repellency. Adv. Soil Sci. 20, 91-140.
11. Ritsema, C.J., Dekker, L.W., 1996. Water repellency and its role in forming preferred flow paths in soils. Aust. J. Soil Res. 34, 475-487.
12. Bond, R.D., Harris, J.R., 1964. The influence of the microflora on the physical properties of soils. 1. Effects associated with filamentous algae and fungi. Aust. J. Soil Res. 2, 111-122.
13. Neinhuis, C., Barthlott, W., 1997. Characterization and distribution of water-repellent, self-cleaning plant surfaces. Ann. Bot. (London) 79, 667-677.
14. De Bano, L.F., Savage, S.M., Hamilton, D.A., 1976. The transfer of heat and hydrophobic substances during burning. Soil Sci. Soc. Am. J. 40, 779-782.
15. ISTAT, 1991. Statistiche forestali. Istituto Poligrafico dello Stato, Roma.
16. Sequi, P., Indiati, R., 1997. Minimizing surface and ground water pollution from fertilizer application. *In*: Rosen, D., Tel-Or, E., Hadar, Y., Chen, Y. (Eds.), Modern Agriculture and the Environment. Kluwer Academic Publishers, Great Britain, pp. 147-158.
17. Ball, W.P., Roberts, P.V., 1991. Long term sorption of halogenated organic chemicals by aquifer materials. Part 2. Intraparticle diffusion. Environ. Sci. Technol. 25, 1237-1249.
18. Pavlostathis, S.G., Jaglal, K., 1991. Desorptive behaviour of trichloroethylene in contaminated soil. Environ. Sci. Technol. 25, 274-279.
19. Schwarzenbach, R.P., Gschwend, P.M., Imboden, D.M., 1993. Environmental Organic Chemistry, Illustrative Examples, Problems and Case Studies. John Wiley and Sons, Inc., New York.
20. Liptbak, B.G., Liu, D.H., 2000. Groundwater and Surface Water Pollution. Lewis Publishers, Boca Raton.
21. Haderlein, S.B., Schwarzenbach, R.P., 1993. Adsorption of substituted nitrobenzenes and nitrophenols to mineral surfaces. Environ. Sci. Technol., 27, 316-326.
22. Schwarzenbach, R.P., Gschwend, P.M., Imboden, D.M., 1995. Environmental Organic

Chemistry, Illustrative Examples, Problems and Case Studies. John Wiley and Sons, Inc., New York.
23. Paul, E.A., Clark, F.E., 1996. Soil Microbiology and Biochemistry. Academic Press, San Diego, CA.
24. Pagliai, M., De Nobili, M., 1993. Relationship between soil porosity, root development and soil enzyme activity in cultivated soils. Geoderma 56, 243-256.
25. Richter, J., 1987. The Soil as a Reactor: Modeling Processes in the Soil. Catena Verlag, Cremlingen, Germany.
26. Day, P.R., Doner, H.E., McLaren, A.D., 1978. Relationships among microbial populations and rates of nitrification and denitrification in a Hanford soil. *In*: Nielsen, D.R., McDonald, J.G. (Eds.), Nitrogen in the Environment, Vol. 2. Academic Press, New York, pp. 305-364.
27. Reddy, K.R., Patrick, W.H., 1980. Evaluation of selected processes controlling nitrogen loss in paddy soils. Soil Sci Am. J. 44, 1241-1243.
28. Barber, S.A., 1995. Soil Nutrient Bioavailability. A Mechanistic Approach. John Wiley and Sons, New York.
29. Barber, S.A., 1962. A diffusion and mass-flow concept of soil nutrient availability. Soil Sci. 93, 39-49.
30. Glass, A.D.M., Siddiqi, M.Y., 1985. Nitrate inhibition of chloride influx in barley: implications for a proposed chloride homeostat. J. Exp. Bot. 36, 79-90.
31. Marschner, H., 1995. Mineral Nutrition of Higher Plants. Academic Press, London.
32. Faurè, M. H., Sardin, M., Vitorge, P., 1996. Transport of clay particles and radioelements in a salinity gradient. J. Contamin. Hydrol. 21, 255-267.
33. Sparks, D.L., 1995. Redox chemistry of soils. *In*: Environmental Soil Chemistry. Academic Press, San Diego, CA, pp. 187-202.
34. Bartlett, R.J., 1997. Chromium redox mechanism in soils: should we worry about Cr(VI)? *In*: Canali, S., Tittarelli, F., Sequi, P. (Eds.), Chromium Environmental Issues. Franco Angeli, Milano, Italy, pp. 2-20.
35. Coates, J.D., Ellis, D.E., Blunt-Harris, E.L., Gaw, C.V., Roden, E.E., Lovley, D.R., 1998. Recovery of humic reducing bacteria from a diversity of environments. Appl. Env. Micr. 64, 1504-1509.
36. Lovley, D.R., Coates, J.D., Blunt, E.L., Phillips, E.J.P., Woodward, J.C., 1996. Humic substances as electron acceptors for microbial respiration. Nature 382, 445-448.
37. Lovley, D.R., Woodward, J.C., Chapelle, F.H., 1996. Rapid anaerobic benzene oxidation with a variety of chelated Fe(III) forms. Appl. Env. Microb. 62, 288-291.

Developments in Soil Science, Volume 28A
Editors: A. Violante, P.M. Huang, J.-M. Bollag and L. Gianfreda

SOIL AGGREGATE HIERARCHY IN A BRAZILIAN OXISOL

G. Vrdoljak[a] and G. Sposito[b]

[a]Electron Microscope Lab, 26 Giannini Hall, University of California Berkeley, California 94720-3330

[b]Division of Ecosystem Sciences, Hilgard Hall #3110, University of California Berkeley, California 94720-3110 USA

Light and electron microscopy were used to investigate the applicability of the aggregate hierarchy model of Tisdal and Oades [1] to the micromorphology of the A horizon in a Brazilian Xanthic Hapludox soil under both forest and agricultural crop cover. The arrangement of minerals, amorphous material, organic matter, and biota in aggregates of diameter <20, <53, 100-250, and >2000 μm was highly dependent on aggregate size, consistent with the hierarchical model. Detailed examination of the levels of structural organization in the soil led to an improved model of aggregate hierarchy for Oxisols. Face-to-face orientation of kaolinite in domains 200 nm to 2 μm diameter were bound together with goethite and organic matter. Domains combined with bacteria, fungal hyphae, and polysaccharides to form larger clusters (2-20 μm). These clusters combined with larger, less decomposed, organic materials and silt-sized minerals to form microaggregates (20-250 μm). Soil aggregates > 250 μm diameter were formed by the combination of microaggregates and parts of plant roots. Oxisol aggregation thus follows a process of stepwise amalgamation and disruption of hierarchical order, with the content of organic matter being a key factor in the stabilization of soil structure. Organic matter was abundant in the forested soil, but was nearly absent from the cultivated soil.

1. INTRODUCTION

Population and cultural changes within tropical regions are placing increasing demands on the agricultural use of fragile highly-weathered soils [2]. Current methods of slash and burn agriculture allow the soils of the humid tropics to remain productive for only a few years, but contribute significant amounts of greenhouse gases to the atmosphere while destroying 2 million ha of tropical forest annually in the Amazon region alone [3]. The loss in agricultural productivity and the difficulty in vegetation reestablishment [4] are thought to be from degradative changes in soil structure [5]. A better understanding of aggregate structure in tropical soils thus may aid in the development of more efficient agricultural practices.

Soil structure, the result of the process by which primary soil particles (sand, silt and clay) are aggregated with organic and amorphous materials, controls aeration; water infiltration; water, gas, and solute transport; drainage; soil fertility; and the ease of soil tillage [6]. The Oxisol order has the least understood soil structure of all twelve orders [7], although 25% of the

Earth surface has a humid tropical climate, and Oxisols cover the largest area of any of the soil orders found in the tropics [7].

Aggregate hierarchy, a model of soil structure proposed by Tisdall and Oades [1], describes the levels of structural organization exhibited in soil aggregates. At the smallest scale, individual clay particles bind together in packets to create a floccule or domain <20 μm in diameter. Floccules or domains combine together to form larger microaggregates, 20-250 μm. Microaggregates bind together to form stable macroaggregates, >250 μm [8].

Using transmission electron microscopy (TEM), Cambier and Prost [9] found evidence for levels of hierarchical organization in a ferralitic soil (clay rich in oxides and hydroxides of iron and aluminum) from Senegal. However, Oades and Waters [10] used destructive slaking experiments to show that an Alfisol and a Mollisol had an aggregate hierarchy, whereas an Oxisol did not. But, some evidence supporting a hierarchical organization of soil components in an Oxisol was found through scanning electron microscopy (SEM) of selected size fractions of soil aggregates by Waters and Oades [11]. Golchin et al. [12] have recently proposed that hierarchical levels of soil structure are reflected in the type of organic matter present at each stage of the aggregation process. To characterize decisively the presence or lack of aggregate hierarchy in an Oxisol, therefore, the levels of soil structure should be observed directly.

Traditionally soil structure has been studied by a combination of physical, chemical, and microscopic techniques to cover the extremely wide range of size scales in soil aggregates. The fundamental interactions of clay particles (< 2 μm) up to large aggregates or peds (1-2 cm) can cover several orders of magnitude in scale. In this paper, the structure in a representative Oxisol was characterized by a range of advanced microscopic techniques. The soil investigated was a "benchmark" soil both in forested and agricultural ecosystems collected by Cheryl Palm and Pedro Sánchez as part of the Tropical Soils Program at North Carolina State University. The soil was chosen as representative of a typical kaolinitic Oxisol, abundant in the Amazon region.

2. MATERIALS AND METHODS

The soil under study was a Xanthic Hapludox collected in 1991 at an EMBRAPA research station outside of Manaus, Brazil. The Manaus soil is representative of those developed on the Barreiras Sediments and constitutes more than 10 % of the total Amazon basin area [13]. A soil sample (MF) from a tropical forest site was collected from the 0-8 cm depth (23.5 % sand, 1 % silt, and 75 % clay; 830 kg m^{-3} bulk density). Another sample (21.8 % sand, 3.6 % silt, and 75 % clay; 1200 kg m^{-3} bulk density) was collected from the 0-20 cm depth at a nearby site under continuous corn-cowpea cultivation (MC). The cultivation caused thickening of the 'A' horizon, necessitating deeper sampling to 20 cm. Three field replicates for each soil were sampled from pits located 50 m apart. The soil was placed in airtight plastic bags, kept at field moisture content, and stored at 4 °C.

To prepare the soils for examination by petrographic microscope, eight 2.7 x 2.5 cm thin sections of the MF and MC samples were cut. The samples were first air dried at 20 °C to constant mass and then dried at 40 °C for 48 h. The samples were impregnated with LR White Acrylic Resin™ under vacuum. The soil samples and resin were covered with foil and left at 60 °C for 12 h to cure into a hard block, after which the block was cut with a diamond saw and mounted onto a glass slide with UV Cement™. Sample thickness was reduced to 30 μm by polishing on rotating lapping plates, first with silicon carbide and then with alumina grit.

The MF and MC samples were analyzed with a petrographic microscope according to the technique of FitzPatrick [14]. The volume percentage of selected minerals in the soil was inferred by point counting [15].

Size fractions of aggregates > 53 μm and < 53 μm diameter were obtained by sieving the MF and MC soils through a 270 (53 μm) USA standard sieve. The soil was then air dried at 20 °C until constant mass was reached. Aggregates were placed onto a scanning electron microscope sample puck coated with adhesive carbon conductive tape softened by warming under a 100 W lamp for 2 min. The samples were plasma sputter coated with 15 nm of Au/Pd alloy on a Balzers Med 010 sample coater and analyzed in a Jeol JSM-35CF scanning electron microscope. Operating voltages of 15-25 kV were used in imaging depending on the magnification used. Standard gamma filtering ($\gamma = 1/3$) was done during imaging to give selective contrast expansion in images [16]. For stereoscopic images, the samples were photographed at 0° and at 9-10° tilt along an axis perpendicular to the column of the microscope. Anaglyphs, or red and green color separations of images, were obtained by combining the red and green channels of the stereoscopic pair into a final image.

Ultrathin sections for aggregates in the size ranges of 2-20 μm, < 53 μm, 100-250 μm, and <2000 μm diameter were prepared. The aggregates were imbedded into a 2 % agar solution, solidified and then trimmed. The samples were then fixed with 2 % glutaraldehyde in 0.1 M Na-cacodylate buffer (CACO) at pH 7.2 for 1-2 h. After rinsing with CACO three times for 15 min each, the aggregates were stained with 1 % OsO_3 in CACO for 1-2 h. The samples were rinsed with CACO three times for 5 min each and then rinsed with distilled water three times for 10 min each. Final staining was done with 0.5 % uranyl acetate for 1 h at room temperature or at 4 °C for longer periods. The samples were then rinsed with CACO three times for 15 min periods. Dehydration of the sample was accomplished by washing it with increasing concentrations of acetone:water mixtures until 100 % acetone was reached (steps are 35 %, 50 %, 70 %, 80 %, 95 %,100 %, and 100 % acetone:water). The aggregates were then infiltrated with Spurrs' [17] epoxy resin:acetone mixtures (2:1, 1:1, and 1:2) and gently mixed for 1 h at each stage. The aggregates were then infiltrated with pure resin for 1 h, and then again overnight. Finally, the samples were embedded into molds and left to cure at 60 °C for two days. After curing, the samples were trimmed and ultrathin sections (~60 nm) were cut on a Sorval MT-6000-XL microtome utilizing glass or diamond knives with a cutting speed of 0.1 mm/s. Sections were deposited onto copper transmission electron microscope grids coated with a support film of formvar and carbon for ultimate use in the transmission electron microscope.

Analysis of the sections was done on a Philips 300 transmission electron microscope (operating at 100 kV), a Philips 400 transmission electron microscope (operating at 120 kV), or on a JEOL 200CX (operating at 200 kV) analytical transmission/scanning transmission electron microscope. Energy dispersive X-ray spectra and X-ray element maps were collected with a KEVEX system 8000 silicon detector attached to the JEOL instrument using a variety of electron probe diameters to localize chemical information.

3. RESULTS

3.1. Petrographic microscopy

The minerals in the MF and MC samples were previously determined by X-ray diffraction [18] to be quartz, rutile, anatase, kaolinite, goethite and hematite. The distribution of the minerals, determined by light microscopy, among the sand, silt, and clay size fractions is given in Table 1.

Table 1. Distribution of minerals among the size fractions of the MF and MC soils

MF soil			MC soil		
Size fraction	minerals present	volume % of soil	Size fraction	minerals present	volume % of soil
Sand 2000-50μm	quartz	10-15	Sand 2000-50μm	quartz	15-30
Silt 50-2μm	quartz, rutile, anatase, zircon	< 2	Silt 50-2μm	quartz, rutile, anatase, zircon	0.5-2
Clay < 2μm	kaolinite, goethite, hematite, gibbsite	> 50	Clay <2μm	kaolinite, goethite, hematite, gibbsite	> 50

Low-magnification photomicrographs of the MF and MC samples are shown in Figure 1. Weakly accordant (extent to which opposite faces of aggregates are moulds of each other) granular aggregates well separated from one another and appearing completely surrounded by pore space are defined as 'complete' [14]. The MF and MC soil samples have a complete structure. The abundant (> 50 % volume) matrix of the MF and MC samples has a brownish yellow color from goethite minerals. Reddish regions composed of humified organic matter also occur. Speckling was seen in the plasma, which has been described previously [14] to arise from secondary crystalline goethite. The fabric of the matrix is strongly isotropic with weak reticulate anisotropic zones having a broad size (~800 μm). Less than 5 % or 0.5-2 % of the aggregate area is composed of pore space in the MF or MC soil sample, respectively.

Both fresh and very decomposed organic material was seen in the MF samples. Mite fecal pellets of 20 μm diameter often surrounded decaying organic features in thin section (not shown). The fresh material which can be recognized is mainly roots, cellular debris, and some leaf fragments. Most of the organic matter is highly decomposed and extends 10 to 50 μm across the section. Charcoal was seen only once in thin section in the MF sample. It has a blocky shape with the characteristic cellular structure as seen in charcoal debris in soil [14].

Strongly decomposed isotropic organic residues with an elliptical, rounded, or bladed shape were seen in the MC samples. They occur very occasionally (< 0.5 volume %), are small (10-100 μm) in size, and occur with both clustered and random distribution patterns. No root cross sections were visible in the MC sections. Mite fecal pellets are similar to those in the MF soil, but occur with much less frequency (<0.5 volume%). Prominent black charcoal is seen in the MC soil with a frequency of about 0.5-2 %.

Clay coatings on sand grains and pores were observed in the MC soil sections (not shown). The coatings are translucent, have distinct laminations, and conform to the surfaces where formed. They have strong, thin black extinction bands between crossed polarizing filters, which implies a strong continuous orientation of clay [14].

3.2. SEM

Numerous whole aggregates from >53 and <53 μm diameter size fractions of the MF and MC soils were visualized by SEM. Figure 2 shows the rounded, nodular shape of the aggregates, which are completely draped with clay. Energy dispersive X-ray (EDX) spectroscopy showed the composition of the aggregate surfaces to be predominantly Si, Al, and Fe in ratios of 1:0.8:0.6, which is typical for the composition of kaolinite from the Amazon region [19].

The nodular, clay draped features were common to aggregates from all size fractions of the soil studied. The MC soil aggregates (Figures 2b, 3b, 4b, and 5b) had surfaces with a 'fluffy' or rougher appearance than in the MF soil sample because the clays have a more random orientation on aggregate surfaces. Clays on the surfaces of aggregates from the MF soil tended to have a planar, flat orientation.

The > 53 μm diameter aggregates tended to have a more complete rounded shape than the < 53 μm diameter aggregates. Nodular features of the aggregates were more apparent in the less spherical, smaller sized aggregates. All of the aggregates have very complex, irregular surfaces, with great variations in the height from protrusions to canyons or valleys. Stereoscopic images of the size fractions which highlight the irregular surfaces are shown in Figures 4 and 5.

The MF soil aggregates had more visible organic features than in the MC soil. Plant and fungal components were identified by SEM through comparison of their observed morphology to those seen in other similar studies of the root - soil interface [20]. Structures very similar in morphology to actinomycete filaments, commonly seen in soils of other orders [21], were also abundant in the MF soil (Figure 5a), but were not seen in aggregates from the MC soil. Other organic features were seen in the MF soil, such as a root tip or fungal hypha. The actinomycete filament - like structures found in the MF soil appeared to act as bridges, binding parts of the aggregate surfaces together (Figure 5a).

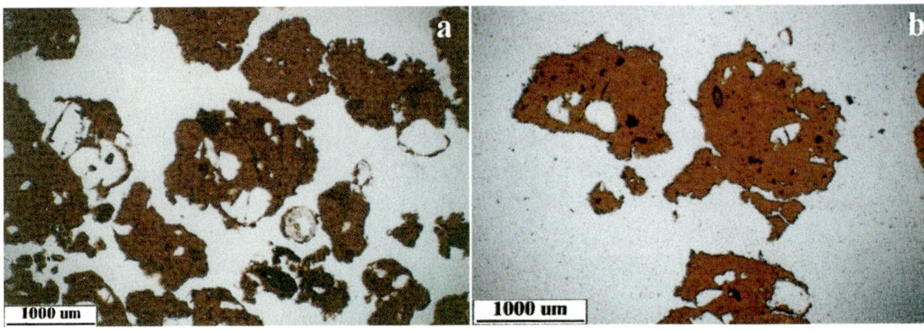

Figure 1

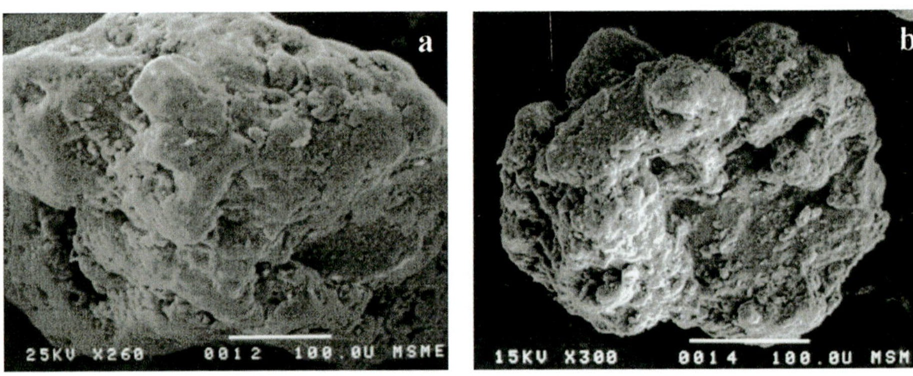

Figure 2

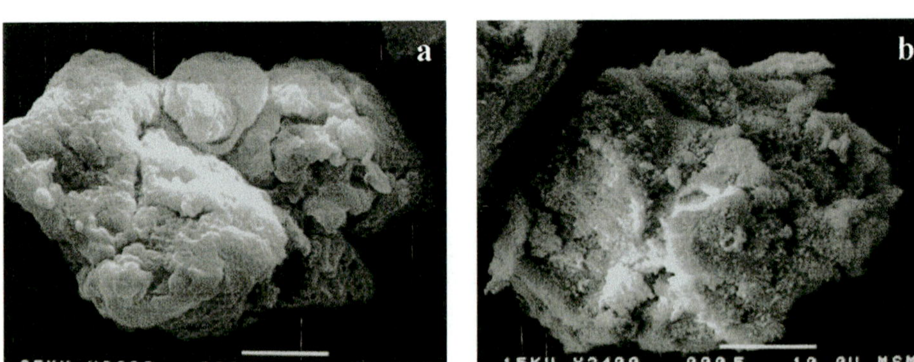

Figure 3

Figure 1. (a) Large scale image of several aggregates in the MF soil. The aggregates are complete (aggregates are well separated from one another and completely bound by pore space), with a characteristic crumb or granular shape. b) Large scale image of several aggregates in the MC soil. The aggregates are complete, with a characteristic crumb or granular shape.

Figure 2. (a) Micrograph of an aggregate from the MF soil, > 53 μm size fraction. Clay uniformly drapes the surface of the aggregate, masking internal structure. The aggregate has an elongated shape with nodular features draped by clay particles. The scale bar is 100 μm. (b) A complete aggregate from the > 53 μm fraction of the MC soil. The aggregate has an irregular rounded shape with various nodular features draped by clay. The scale bar is 100 μm.

Figure 3. (a) An aggregate from the < 53 μm size fraction of the MF soil. The aggregate has an irregular shape that consists of four smaller aggregates bound together and coated with clay. The scale bar in this micrograph is 10 μm. (b) Aggregate from the < 53 μm size fraction of the MC soil. It has a very 'fluffy' surface texture from the high amounts of clay which do not drape the surface of the aggregate evenly. This aggregate appears less well consolidated than that in Figure 3a. The scale bar in this micrograph is 10 μm.

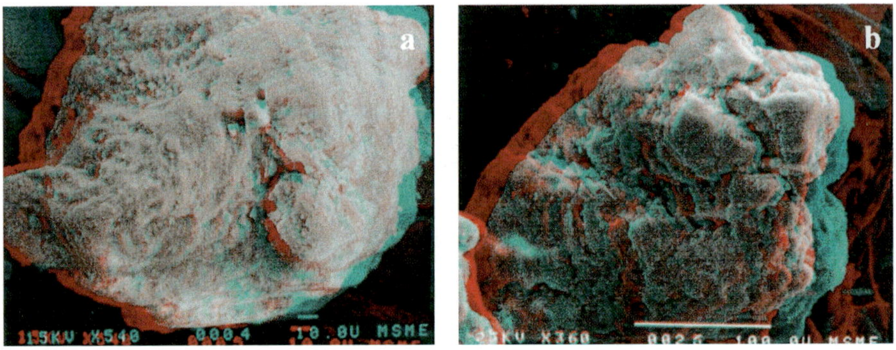

Figure 4. (a) Stereoscopic image of a > 53 μm diameter aggregate from the MF soil. The aggregate is roughly spherical with large protrusions on the left and right half of the aggregate which may be recently amalgamated smaller aggregates or primary minerals coated with clays. The scale bar in this micrograph is 10 μm. (b) Stereoscopic image of a > 53 μm diameter ellipsoidal shaped aggregate from the MC soil. It has a very nodular crenulated appearance. The larger protrusions in the tip of the aggregate appear to be primary minerals, such as quartz or rutile, coated by clay material and combined into the larger aggregate.

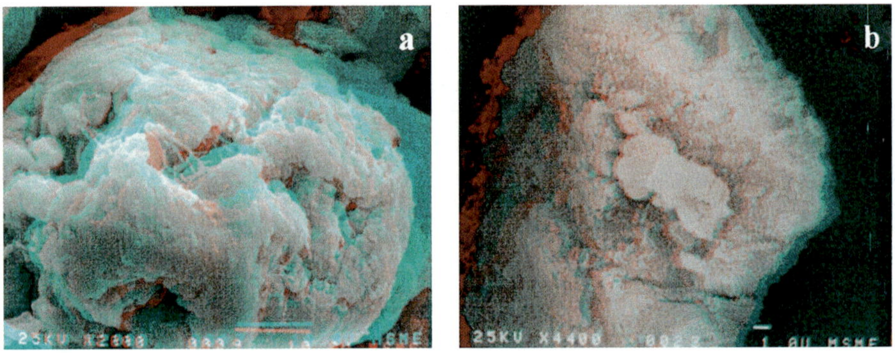

Figure 5. (a) Stereoscopic image of an aggregate from the MF soil with < 53 μm diameter. This aggregate has a spherical shape with an open porous structure facing the viewer. A large nodule is pointing out of the aggregate toward the viewer. Actinomycete filaments cover the surface of the aggregate. The scale bar in the micrograph is 10 μm. (b) Stereoscopic image of an aggregate from the MC soil with diameter < 53 μm. It has a subangular prismatic shape with a very 'fluffy' surface texture. Large flat particles appear to be in the process of peeling away from the aggregate. The scale bar in this micrograph is 1 μm.

3.3. TEM

Nearly 600 electron micrographs of aggregate sections from the MF and MC soil fractions were obtained. In total, 105 individual aggregates were sectioned and photographed by TEM.

Of these, 53 aggregates were < 20 μm; 44 were < 53 μm; 4 were 100-250 μm; and 4 were > 250 μm diameter. These size fractions were chosen to cover the steps of aggregate formation discussed by Tisdall and Oades [1]. The figures presented in this paper are representative of typical aggregates within each size - fraction. Measurements were made on each of the individual figures shown and from the total pool of images. Minerals were definitively identified by selected area diffraction or micro-micro diffraction, and the resulting patterns compared to those generated by electron diffraction computer simulation using the Cerius modeling package [22].

EDX spectroscopy was useful in locating iron oxides, titanium oxides, and uranium - or osmium - stained organic material. It was less useful in the quantification of chemical composition as high amounts of silicon and aluminum caused spurious X-ray emission, interfering with accurate quantitative analysis. A typical kaolinite observed had a relative atomic composition of 46 % Al, 50 % Si, and 4 % Fe. Other minerals not detected in the electron diffraction studies, or in previous X-ray diffraction studies of these soils [18], were located and identified by EDX spectroscopy. They were zircon, phosphate minerals, and barium sulfate minerals. These minerals, very uncommonly seen in the soils, were evident only in the larger aggregates (> 100 μm diameter). Zircon accumulated from the parent material due to its resistance to weathering [14]. Phosphate often adsorb to kaolinitic soils and precipitation of phosphate minerals is common in acid soils [23]. Barium sulfate was found unexpectedly in the soil, but may have been previously precipitated, and coated with iron oxides preventing dissolution.

Electron micrographs of selected aggregate sections from the < 20 μm fraction of the MF and MC soils are shown in Figure 6. The sections of the MF soil often had an organic matter core (Figure 6a), surrounded by randomly oriented kaolinite, goethite, and rutile minerals (identified by electron diffraction and EDX spectroscopy). Organic regions covered from 14-42 % of the aggregate section areas. Iron oxides (identified by EDX spectroscopy) were often seen along the periphery of organic masses in the aggregates (Figure 6a). The organic matter in the aggregates was highly decayed and often lacked structural features such as intact cell walls, membranes, or organelles. Occasional collapsed organelle or cell membranes, collapsed plant cell walls, and bacteria were identified as they are similar to structures found in studies of the root - soil interface [20].

Sections of the < 20 μm fraction of the MC soil showed very few organic features as compared to the MF soil (Figure 6b). They were more generally masses of soil mineral materials arranged in random orientation. The few organic matter regions identified were smaller and had much more decomposed material than those in the MF soil. From the areas represented in the micrographs, the < 20 μm fraction of MF had pores covering an average of 18 % of the section area and organic matter covering an average of 28 % of the section area. The < 20 μm fraction of MC had averages of 33 % pore space and 2 % organic matter areas.

The morphology of the section in Figure 7a is typical of aggregate sections found from the MF < 53 μm size fraction. The sections contained randomly oriented mineral matter surrounding decomposed organic matter. Pores extended randomly across the section area, and had a complex shape. In some cases (such that as shown in Figure 7a), the organic matter was not extensively decomposed and could be identified as a section of a cell from a plant root. The cell wall is surrounded by organic material or mucilage [20] which mixes and binds with the soil minerals. Amorphous organic regions are commonly seen in aggregates without identifiable large scale structural features such as a cytoskeleton, membranes, or organelles.

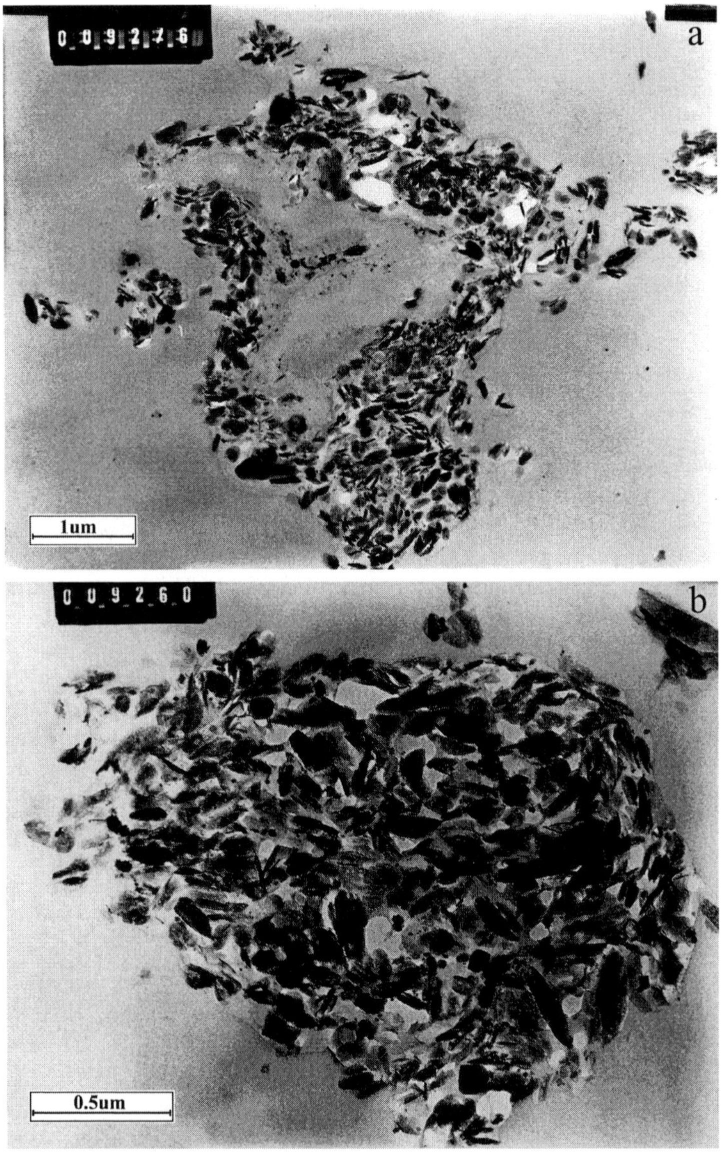

Figure 6. (a) Amorphous organic material covers 24 % of the area of this aggregate section. Some iron oxides (identified by EDX spectroscopy) are concentrated at the periphery of the organic mass. The minerals (mainly kaolinite, iron oxides, and rutile) are randomly oriented about the organic mass in the MF soil (< 20 μm diameter aggregate fraction). (b) Section of a typical MC soil aggregate, with no organic residues visible. Organic material would appear as amorphous areas of the section with darker contrast than the surrounding embedding medium around the aggregate. Pore space accounts for 35 % of the aggregate section area in this MC soil section (< 20 μm diameter aggregate fraction).

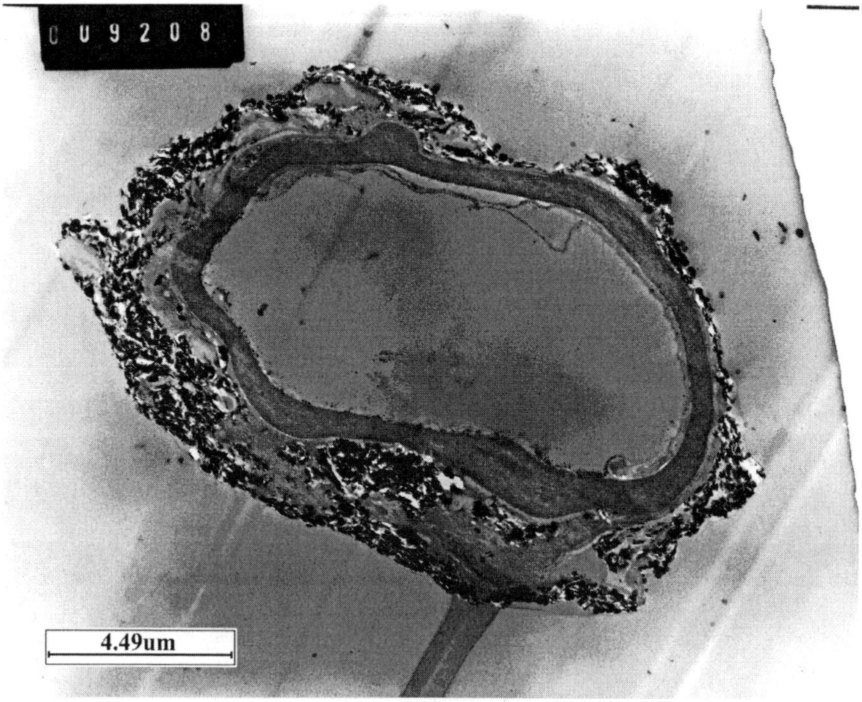

Figure 7a. This aggregate has a core of plant material, as evidenced by the thickened cell wall present. It is surrounded by mucilagenous excretions of the plant cell (polygonal shape suggestive of a root cell), and a thin layer of mineral material. Organic matter comprises 71 % of the section area of this MF < 53 μm diameter size fraction aggregate.

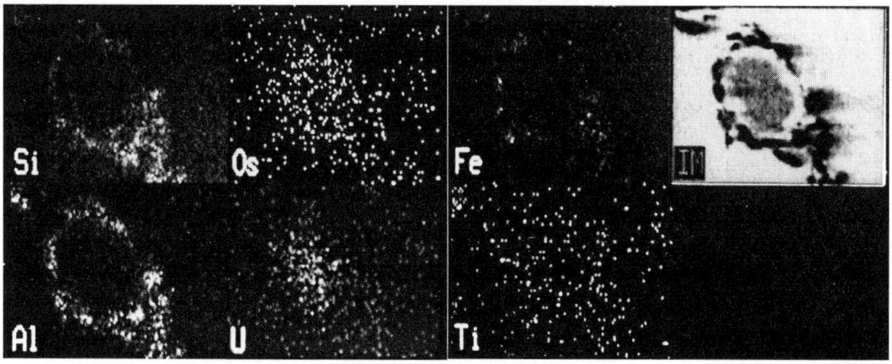

Figure 7b. X-ray element map from organic rich region of a MF < 53 μm diameter aggregate. Iron oxides (Fe) can be seen distributed sparsely, next to the organic mass. Higher amounts of Si and Al are visible due to kaolinite coating the organic material. The organic mass is enriched in both Os and U.

X-ray element maps of common soil elements (Si, Al, Fe, Ti, Na, and Ca) and selective stains for organic matter (U and Os) for aggregate sections from the < 53 μm size fraction of MF were collected (Figure 7b). They show that the chemical stains for organic matter were selective and did not significantly adhere to minerals. The element maps were useful in identification of organic regions in the soil and the location of clay minerals, titanium oxides, and iron oxides.

An electron micrograph of a < 53 μm MC aggregate sections is shown in Figure 7c. There tended to be less organic matter present in the MC soil aggregates than in the MF soil. When present, it was highly decayed, structureless and composed a maximum of 7 % of the aggregate section area. The aggregates were dominated by randomly - oriented kaolinite, goethite, and rutile minerals identified by pattern matching of electron diffraction patterns. Typically no recognizable organic materials were seen. The aggregates were a porous mass of randomly - oriented soil minerals. Occasionally, kaolinite crystals were observed (data not shown) to undergo rolling of basal (001) sheets along one of the crystallographic axes, indicative of early halloysite formation [24]. The < 53 μm fraction of MF had pores covering an average of 26 %, and organic matter covering an average of 19 % of the section area. The < 53 μm fraction of MC soil had 25 % pore area and 1 % organic matter area.

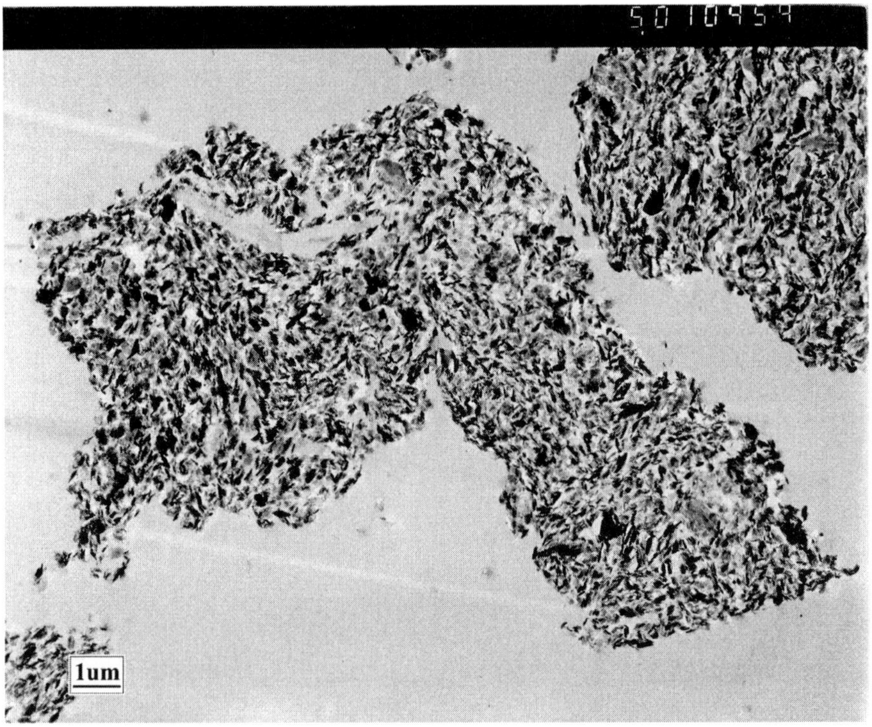

Figure 7c. This aggregate has a small humified organic region, surrounded by a kaolinite and goethite mineral matrix. Organic matter comprises 3 % of the section area of this MC < 53 μm diameter aggregate.

Aggregates 100-250 μm sized from the MF and MC soil were too large to observe entirely by TEM without excessive optical distortion. These aggregate sections were first viewed in small 2500 μm^2 areas, then translated to view other areas of the aggregate section. Aggregates from this size fraction of the MF soil had high amounts of organic material which were decayed, but could still be recognized in many cases as to their source (Figure 8a). Aggregates from the MC sample had very little organic matter and a matrix composed entirely of inorganic mineral material (Figure 8b). Electron diffraction patterns from this material indicated that it was composed predominantly of kaolinite. Any small amounts of organic material which were present in the MC aggregates were highly decomposed and structureless. Both the MF and MC soil aggregates in the 100-250 μm size fraction had significantly more quartz particles present than in smaller size fractions. During sectioning, quartz was shattered in the ultramicrotome by the diamond knife, leaving fragments in a hole where the quartz grain was.

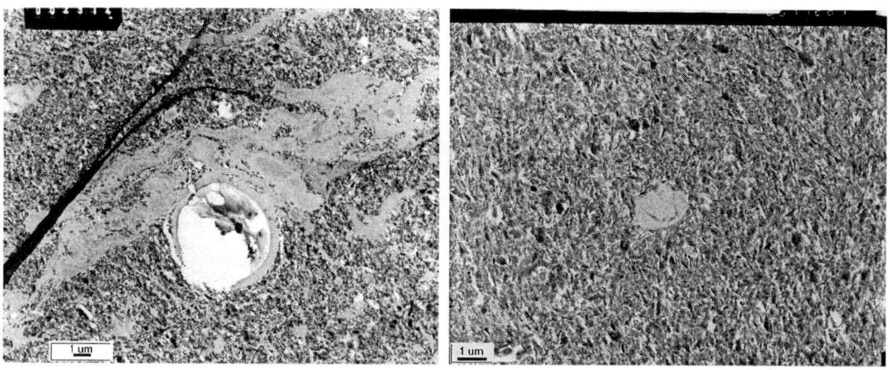

Figure 8. (a) Micrograph of a 100-250 μm MF aggregate section. Often, very large organic features, such as those shown here, were seen in this size class of MF aggregates. The organic material appears to be from a root which has undergone extensive decay. (b) The bulk of this MC 100-250 μm aggregate is randomly distributed kaolinite, with goethite, rutile, and very minor amounts of organic matter indicated.

Aggregates from the MF and MC soil samples with diameters greater than 250 μm were also too large to observe entirely by TEM without excessive optical distortions. These aggregate sections were viewed in the same fashion as the 100-250 μm diameter aggregates (2500 μm^2 areas). Representative regions of the MF and MC soil aggregates are shown in Figure 9. They appeared similar in morphology to that of the 100-250 μm size class.

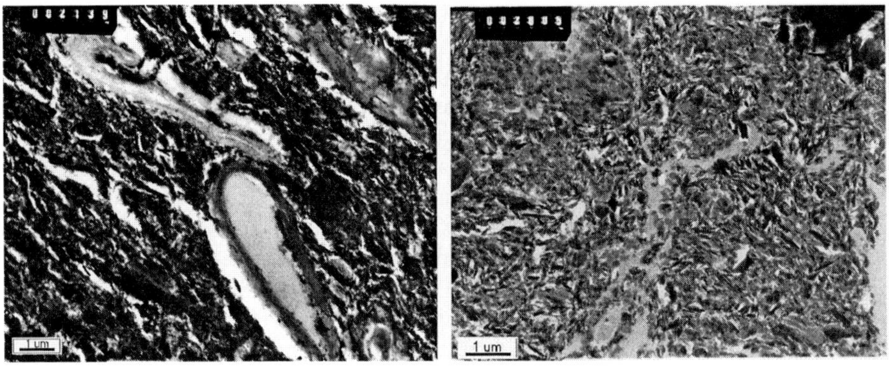

Figure 9. (a) Micrograph of a > 250 μm MF aggregate section. Abundant organic material is usually found in aggregates from this size class, e.g., collapsed plant cells (plant cell membranes lacking internal cytostructures or organelles) seen here. There are holes visible from tearing of the section during ultramicrotomy. This section was prepared using a glass instead of a diamond knife. (b) Micrograph of a region in a > 250 μm MC aggregate section. MC aggregates in this fraction have little organic material, and the organic material is usually highly decomposed.

4. DISCUSSION

Petrographic microscopy showed the MF and MC soil samples both have a micromorphology characteristic of highly weathered Oxisols [25]. The samples were very homogeneous in all thin sections, with a uniform distribution of coarse and fine constituents. The dominant mineral is kaolinite, as shown also by X-ray diffraction [18], which has been both altered from kaolinite present in the original parent material and produced from the weathering of other primary minerals [26].

Clay coatings, only seen in the MC soil sample, conformed to former pore surfaces where found. They most likely were formed by eluviation of clay from higher in the soil profile [25]. In the MF samples, higher organic matter content causes the clays to be more tightly flocculated or aggregated, preventing this downward movement from occurring.

The MF surface horizon studied showed an abundance of humified material, plant remains, some charcoal, and phytoliths similar to what was found by Verheye and Stoops [27] in an Oxisol. A dramatic decrease in organic content of the MC soil sample was seen, as expected for a cultivated soil [13]. As also shown by previous authors [13, 28], cultivation of Oxisols decreases aggregate porosity in the surface horizon.

Aggregates > 53 μm diameter in both soils had a rounded spherical shape with nodular features and a surface completely coated with clays. This result is in agreement with the morphology of Oxisol microaggregates as described by Waters and Oades [11]. Smaller diameter aggregates had a less rounded shape and the nodular appearance was more apparent. The smaller aggregates may represent younger aggregates that are in the process of forming in the soil and which are therefore less well consolidated by clay coatings. The kaolinite crystals have similar size and shape to those found by Stoops [29] in a laterite soil. Gibbsite, however, was not seen on the soil ped surfaces as found by Eswaran et al. [30] in Oxisols from Zaire.

Since the soils used in the present study are from the upper A horizon, gibbsite is unlikely to be seen. It is in the lower horizons, which undergo vigorous hydrolysis, that this material would be normally found [14]. The peds in the soils were very similar to the micronodules observed by Cambier and Prost [9]. Although much smaller than the aggregates of this study, the micronodules of Cambier and Prost [9] and the aggregates from the MF and MC soil samples are both compact nodulated aggregates covered by face-to-face oriented kaolinite. A more random orientation of clays at the surfaces of the MC aggregates, which gives these clays a 'fluffy' appearance, may be caused by rapid flocculation or deflocculation chemical conditions within this soil. Clays appear to be more mobile, as they are not bound well to the aggregate surface. The MF clays are arranged in more regular domain-like near-parallel orientations, characteristic of a chemical environment conducive to slower flocculation in which sufficient time for parallel attraction (face-to-face approach of clay plates) and aggregation of clay minerals is possible [31]. The pattern of clays coating the aggregates is very similar to that found with SEM by Waters and Oades [11] and by Cambier and Prost [9] in Oxisols.

Relatively few published studies of organic matter, biota, and mineral interactions within the soil have been done using SEM. Comparisons between the features found in the soil with other microbiological studies of plant roots and bacteria allowed identification of most soil organisms. Organic features were much more abundant in the MF samples than in the MC samples. This result indicates the higher organic content and biological activity in the soil of the forested ecosystem than in the soil under continuous cultivation. Actinomycetes were the prevalent organisms visible in MF by SEM. This observation has also been noted for a SEM study of the rhizosphere of a grassland soil by Campbell and Rovira [21]. Fungal hyphae similar to those found by Rovira and Campbell [32] were also seen, but not as often as were actinomycetes. Other organisms are likely present in the soil aggregates, but may be masked by minerals or by secretions. A scale-dependence on aggregate size and organic materials, as found by Waters and Oades [11] was not found in this SEM study, which, however, concentrated only on aggregate surfaces and did not investigate the cores of aggregates, as did Waters and Oades [11]. Thus, many organic features could have been masked by the outer clay-coated surfaces of the peds.

Electron diffraction patterns (data not shown) for the samples were dominated in all fractions by kaolinite. Other minerals, such as rutile and quartz, were identified by conventional selected-area electron diffraction techniques. These minerals did not section well and would often leave only fractured remnants. Micromicrodiffraction was found to be the best technique to identify small phases (< 60 nm diameter). The small crystallites observed were predominantly goethite. Hematite was not observed, possibly because of the small portion of the soil analyzed in this study. The preponderance of goethite and its close association with kaolinite (Figure 6b) is likely from Fe-substituted kaolinite domains acting as precipitation sites for goethite [33]. Furthermore, iron oxides promote aggregation of the clay minerals within Oxisols [34] and they may act as bridging agents.

The organic material in most of the aggregates was very decomposed, making its origin often indeterminable by electron microscopy (Figures 6a, 7b, and 8a). Some of the organic features, however, could be identified by comparison of their structure to results of previous studies done directly on the rhizoplane of plants [35]. Materials commonly observed in all size fractions of the MF samples were plant debris, bacteria, collapsed cell walls of various unidentifiable organisms (Figure 7b) and their organelles, and organic mucilage (Figures 6a, 7a, 8a, and 8b). Fine iron oxides were distributed at the contact between organic matter and the bulk mineral fabric of the soil material (Figures 6a, and 7b). This could be from chemical interactions between the iron oxides and functional groups in organic matter, or to organic matter serving as an effective precipitating agent.

In all the size fractions studied, the MC sample showed a clear lack of organic features as compared to the MF samples (Figures 6, 7, 8, and 9) The MC mineral fabric was identical in composition, individual crystal morphology, and overall orientation to the mineral fabric of the MF samples. The few organic materials seen in the MC samples were highly decomposed (Figures 7c and 9b), making it impossible to identify their origin. Cultivation of the soil thus had the effect of reducing the organic matter content of all size fractions of the soil aggregates. Organic matter surrounded by clay material is defined as occluded or protected [36]. The loss of occluded organic matter was previously thought to have been a minor factor in the total organic matter lost by cultivation [36] but occluded material has obviously been removed from the MC soil through cultivation. This was evident in the depletion of organic matter found in numerous aggregate sections of the MC samples. Golchin et al. [37] also found a reduction in quantity and a change in the composition of occluded organic matter in aggregates of tilled as compared to untilled soils of both Alfisols and Vertisols.

The pores of the MF < 20 μm diameter aggregates varied significantly. They were dendritic in shape, had diameters ranging from several nanometers to hundreds of nanometers, and covered from 9 to 36 % of the area in the aggregates shown. The MC aggregates in this size fraction were more porous (23-42 %). The difference in porosity found by TEM between the MF, MC soil and that obtained by light microscopy of the soil is likely due to the small sample size, or an artifact created by the algorithm used to differentiate pores from the soil matrix. The aggregates of the MC soil are a mass of soil minerals with pores that lack organic material to act as bridging agents or as infillings between pores.

The < 53 μm fraction, as well as the larger size fractions, of the MF sample had more identifiable organic components than the smaller fractions. Plant root sections were encrusted with kaolinite and iron oxides (Figure 7a), as well as with more decomposed cellular remnants with bacteria and decayed organelles. Phytoliths were also seen, but were relatively uncommon in sections. The more complete organic structures found indicate that these materials were recently deposited in the soil, or are better preserved from decomposition than that in the smaller size fractions. Thus, they represent a fresh input of organic material to the soil which is then utilized by microorganisms.

Iron oxides were randomly-distributed throughout the fabric of the soil samples, supporting the concept of their role as bridging agents between kaolinite plates and assemblages [34]. The distribution and disposition of minerals within the soil fabric is similar to that found by Santos et al. [38] in an Oxisol from Pernambuco, Brazil. Larger mineral grains were seen in the < 53 μm fraction. As mentioned above, rutile and quartz primary minerals would not section in the microtome and often left holes with occasional shards of the primary mineral left. Santos et al. [38] may have incorrectly assigned such holes in sections to intraaggregate pores.

The pores of the < 53 μm diameter aggregates covered the same areas of the aggregate sections in the MF and MC samples (26 and 25 % respectively). This similarity is from the increasing amount of inorganic material found in the larger MF soil aggregates. Organic matter covers less aggregate section area (19 %) in the < 53 μm size fraction of the MF sample than in its < 20 μm size fraction (28 %). Pores were highly complex in shape, consisting of small nanometer sized pores between individual clay particles and larger (several hundreds of nanometer) pores between larger assemblages of kaolinite. Bui et al. [39] found a similar size and arrangement of pores for an Oxisol from Brazil, but with a somewhat smaller porosity (17 %).

Larger aggregates (100-250, > 250 μm) appeared very similar in morphology to the < 53 μm diameter aggregates in the MF and MC samples. The only real difference was that the MF sample did have much larger and less - decayed organic masses present, but the surrounding mineral fabric was the same. Many larger primary minerals, such as quartz and rutile, were also seen in this size fraction for both the MF and MC samples than the smaller size fractions. Only eight aggregates were sectioned in this size class to minimize damaging the knife in the ultramicrotome from sectioning large quartz or rutile grains.

Quantitative EDX spectroscopy for the determination of Si or Al was unachievable with less than 15 % inaccuracy because the high amounts of these elements cause X-ray production from surrounding areas outside the electron beam to be detected [16]. This also made this technique unsuitable for locating Si and Al rich regions in the soils. EDX was useful, however, in locating Fe and Ti oxides and in identifying the composition of other minerals.

The use of TEM only for the study of sections of soil aggregates, limits information to two dimensions. Attempts were made to reconstruct an entire aggregate from sequential sectioning of a single < 53 μm diameter aggregate. Serial sections were collected, and regular structural features were seen extended through the aggregate, as expected. However, a complete reconstruction of the entire aggregate could not be made, as many sections of the aggregate were lost, became folded on collecting onto the TEM grid, or would not section cleanly. Complete reconstruction would take about 800 60-nm sections from a 50-μm aggregate, or 300 from a 20-μm aggregate — far too many to analyze efficiently by TEM.

5. CONCLUSIONS

The results of the techniques applied to the study of the MF soil can be generalized into a revised schematic model of aggregate hierarchy (Figure 10).

This model is similar to that proposed by Tisdall and Oades [1], but there are several enhancements and modifications that we propose. The majority of soil minerals was found to have very low crystallinity and may have in the past been confused with amorphous materials. The fundamental unit of structure for the MF soil is the face-to-face arrangement of kaolinite clay plates and goethite minerals. They come together in small domains and are bound by polysaccharides at the scale of approximately 200 nm to 2 μm. At the next stage of aggregation (2-20 μm), these domains combine together and with organic materials, such as bacteria, fungal hyphae, and polysaccharides, to form clusters. Clusters combine with silt sized mineral grains, larger organic materials (such as plant root cells, and decomposed plant and microbial materials) to form microaggregates at the 20-250 μm scale. Finally, aggregates are formed by the amalgamation of microaggregates bound together by organic materials, such as plant roots that are less decomposed than organic materials at other stages of aggregation. These organic materials act as strong binding agents, bridging separate clay minerals, domains, clusters, and microaggregates.

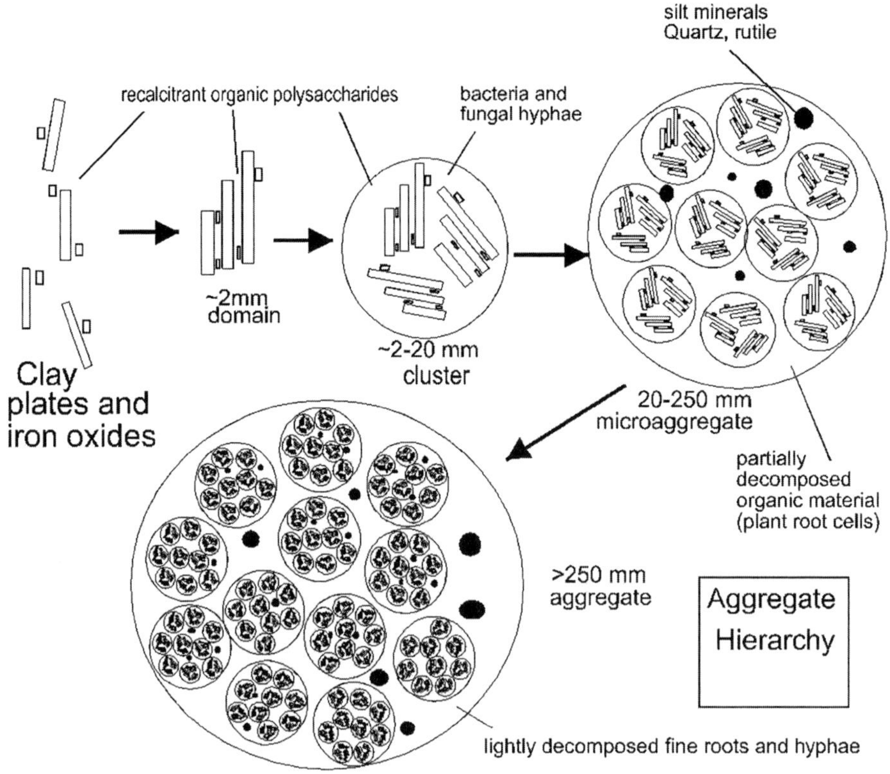

Figure 10. Enhanced model of aggregation hierarchy in uncultivated Oxisols. At the smallest scale of association, kaolinite clay (rectangular plates) and iron oxides (small dark rectangles) are bound together with polysaccharides into domains approximately 200 nm to 2 μm diameter. Domains combine with one another, along with bacteria and fungal hyphae to form clusters 2 μm m to 20 μm diameter. Clusters combine together with partially decomposed organic materials to form microaggregates 20 - 250 μm diameter. Clusters finally, form larger aggregates (> 250 μm) bound with lightly decomposed plant residue.

Thus, in our study, a benchmark Oxisol was found by direct examination to contain elements consistent with a hierarchical aggregate structure. The conceptual model shown in Figure 10 applies, but the for aggregate hierarchy in Oxisols because of the destructive methods (ultrasonic dispersion and fast wetting techniques) used in their experiments, which were insufficient to reveal the subtle stepwise formation of aggregates may not always follow the primary path shown (arrows), from smaller materials forming larger aggregates. Instead, materials at any stage of aggregation may slake and combine with other larger or smaller units. It is likely that Oades and Waters [10] did not find evidence differences in inorganic and organic bonding agents in Oxisols.

The MC sample showed very similar evidence for aggregate hierarchy. However, there was a nearly complete absence of visible organic material within this soil at all levels of aggregate hierarchy. This caused the soil sample to be structurally weaker and form more loosely associated aggregates that can disaggregate or aggregate dependent upon soil chemical conditions. Thus, cultivation of the MC soil greatly reduced the quantity of soil organic materials at all hierarchical levels in the soil structure.

These observations of the MC and MF soil samples imply that, for soil management in tropical ecosystems, organic materials play a vital role in maintaining soil structure. Organic materials also are very important in tropical ecosystems for maintaining the nutritive status of soils [40], but they are shown here also to be very important in maintaining their structure. For the sustainable use of soils in tropical regions, organic matter levels therefore should be controlled closely and, ideally, kept as nearly as possible to their physicochemical state under forested conditions. Recent progress toward this goal has been made in conjunction with continuous crop rotations with ground cover and no-tillage practices [41].

ACKNOWLEDGEMENT

This work was conducted under the auspices of the United States Department of Energy, supported in part by funds provided by the University of California for the conduct of discretionary research by Los Alamos National Laboratory. This work was also supported in part by the Director, Office of Energy Research, Office of Basic Energy Sciences, Materials Sciences Division of the U.S. Department of Energy under Contract No. DE-AC0376SFOO098.

REFERENCES

1. Tisdall, J.M., Oades, J.M., 1982. Organic matter and water - stable aggregates in soils. J. Soil Sci. 33, 141-163.
2. Fujisaka, S., Castilla, C., Escobar, G., Rodrigues, V., Veneklaas, E.J., Rhomas, R., Fisher, M., 1998. The effects of forest conversion on annual crops and pastures: Estimates of carbon emissions and plant species loss in a Brazilian Amazon colony. Agric. Ecosys. Environ. 69, 17-26.
3. Saatchi, S.S., Soares, J.V., Alves, D.S., 1997. Mapping deforestation and land use in Amazon rainforest by using SIR-C imagery. Remote Sens. Environ. 59, 191-202.
4. Buol, S.W., Sanchez, P.A., 1986. Red soils in the Americas: Morphology, classification, and management. *In*: Sinica, A. (Ed.), Proceedings of the International Symposium on red soils. Elsevier, Amsterdam, pp. 14-43.
5. Chauvel, A., Grimaldi, M., Tessier, D., 1991. Changes in soil pore – space distribution following deforestation and revegetation: An example from the Central Amazon Basin, Brazil. For. Ecol. Man. 38, 259-271.
6. Dexter, A.R., 1988. Advances in characterization of soil structure. Soil Tillage Res. 11, 199-238.
7. Wambeke, A.V., 1992. Soils of the Tropics: Properties and Appraisal. McGraw-Hill, New York.

8. Oades, J.M., 1993. The role of biology in the formation, stabilization, and degradation of soil structure. Geoderma 56, 377-400.
9. Cambier, P., Prost, R., 1981. Etude des associations argile-oxyde: organisation des constituants d'un matériau ferralitique. Agronomie 9, 713-722.
10. Oades, J.M., Waters, A.G., 1991. Aggregate hierarchy in soils. Aust. J. Soil Res. 29, 815-828.
11. Waters, A., Oades, J., 1991. Organic matter in water-stable aggregates. *In*: Wilson, W., (Ed.) Advances in Soil Organic Matter Research: The Impact on Agriculture and the Environment. The Royal Society of Chemistry, Cambridge, pp. 163-174.
12. Golchin, A., Baldock, J.A., and Oades, J.M. 1998. A model linking organic matter decomposition chemistry and aggregate dynamics. *In*: R. Lal, J.M. Kimball, R.F. Follet (Eds.), Soil Processes and the Carbon Cycle. CRC Press, Boca Ratan, Florida, pp. 245-266.
13. Cerri, C.C., Volkoff, B., Andreaux, F., 1991. Nature and behaviour of organic matter in soils under natural forest, and after deforestation, burning and cultivation, near Manaus. For. Ecol. Man. 38, 247-257.
14. FitzPatrick, E.A., 1993. Soil Microscopy and Micromorphology. John Wiley & Sons, New York.
15. Drees, L.R., Ransom, M.D., 1994. Light microscopic techniques in quantitative soil mineralogy. *In*: Luxmoore, R.J. (Ed.) Quantitative Methods in Soil Mineralogy. Proceedings of a symposium sponsored by Division S-9 of the Soil Science Society of America, SSSA Miscellaneous Publication. Soil Science Society of America, Madison, pp. 137-176.
16. Goldstein, J.I., Fiori, C.E., Echlin, P., Joy, D.C., Newbury, D.E., 1992. Scanning Electron microscopy and X-ray microanalysis: a text for biologists, materials scientists, and geologists. 2nd ed., Plenum Press, New York.
17. Spurr, A.R., 1969. A low-viscosity resin embedding medium for electron microscopy. J. Ultrastructure Res. 26, 31-43.
18. Malengreau, N.,Sposito, G., 1997. Short-time dissolution mechanisms of kaolinitic tropical soils. Geochim. Cosmochim. Acta 61, 4297-4307.
19. Costa, M.L., Moraes, E.L., 1998. Mineralogy, geochemistry and genesis of kaolins from the Amazon region. Mineralium Deposita 33, 283-297.
20. Foster, R.C., Rovira, A.D., Cock T.W., 1983. Ultrastructure of the Root-Soil Interface. The American Phytopathological Society, Minnesota.
21. Campbell, R., Rovira, A.D., 1973. The study of the rhizosphere by scanning electron microscopy. Soil Biol. Biochem. 5, 747-752.
22. Molecular Simulations Inc., 1998. Cerius2 User Guide.
23. Lindsay, W.L., Vlek, P.L.G. 1977. Phosphate Minerals. *In*: Dixon, J.B., and Weed, S.B. (Ed.) Minerals in Soil Environments, Soil Science Society of America, pp. 639-672.
24. Singh, B., Mackinnon, D.R., 1996. Experimental transformation of kaolinite to halloysite. Clays Clay Minerals 44, 825-834.
25. Stoops, G.J. and Buol S.W., 1985. Micromorphology of Oxisols. *In*: Douglas, L.A., Thompson, M.L., (Eds.) Soil Micromorphology and Soil Classification. Soil Science Society of America Special Publication 15, Madison, pp. 105-119.
26. Bravard, S., Righi, D., 1989. Geochemical differences in an Oxisol-Spodosol toposequence of Amazonia, Brazil. Geoderma 44, 29-42.

27. Verheye, W., Stoops, G., 1975. Nature and evolution of soils developed on the granite complex in the subhumid tropics (Ivory Coast). II. Micromorphology and mineralogy. Pedologie 25, 40-55.
28. Curmi, P., Kertzmann, F.F., Queiroz Neto, J.P., 1993. Degradation of structure and properties in an Oxisol under cultivation (Brazil). *In*: Ringrose-Voase, A.J., Humphreys, G.S. (Eds.), Soil Micromorphology: Studies in Management and Genesis. Developments in Soil Science, Proceedings of the Xth International Working-Meeting on Soil Micromorphology, Developments in Soil Science. Vol. 22, Elsevier, New York, pp. 569-579.
29. Stoops, G., 1970. Scanning electron microscopy applied to the micromorphological study of a laterite. Pedologie XX, 268-280.
30. Eswaran, H., Stoops, G., Sys, C., 1977. The micromorphology of gibbsite forms in soils. J. Soil Sci. 28, 136-143.
31. van Olphen, H., 1977. An Introduction to Clay Colloid Chemistry For Clay Technologists, Geologists, and Soil Scientists. 2nd ed, John Wiley & Sons, Toronto.
32. Rovira, A.D., Campbell, R., 1975. A scanning electron microscope study of interactions between micro-organisms and Gaumannomyces graminis (Syn. Ophiobolus graminis) on wheat roots. Microbial Ecol. 2, 177-185.
33. Muller, J., Manceau, A., Calas, G., Allard, P.I., Hazemann, J., 1995. Crystal chemistry of kaolinite and Fe-Mn oxides: relation with formation conditions of low temperature systems. Am. J. Sci. 295, 1115-1155.
34. Pinheiro-Dick, D., Schwertmann, U., 1996. Microaggregates from Oxisols and Inceptisols: dispersion through selective dissolutions and physicochemical treatments. Geoderma 74, 49-63.
35. Foster, R.C. and Martin, J.K., 1981. In situ analysis of soil components of biological origin. *In*: Paul, E.A., and Ladd, J.N. (Ed.) Soil Biochemistry. Vol. 5., Marcel Dekker, New York, pp. 75-110.
36. Roberts, W.P., Chan, K.Y., 1990. Tillage induced increases in carbon dioxide loss from soil. Soil Tillage Res. 17, 143-151.
37. Golchin, A., Clarke, P., Oades, J.M., Skjemstad, J.O., 1995. The effects of cultivation on the composition of organic matter and structural stability of soils. Aust. J. Soil Res. 33, 975-993.
38. Santos, M.C.D., Mermut, A.R., Ribeiro, M.R., 1989. Submicroscopy of clay microaggregates in an Oxisol from Pernambuco, Brazil. Soil Sci. Soc. Am. J. 53, 1895-1901.
39. Bui, N., Mermut, A.R, Santos, M.C.D., 1989. Microscopic and ultramicroscopic porosity of an Oxisol as determined by image analysis and water retention. Soil Sci. Soc. Am. J. 53, 661-665.
40. Fox, R.L., 1980. Soils with variable charge: Agronomy and fertility aspects. *In*: Theng, B.K.G. (Ed.), Soils With Variable Charge. New Zealand Society of Soil Science, New Zealand, pp. 195-220.
41. Alegre, J.C., Cassel, D.K., 1996. Dynamics of soil physical properties under alternative systems to slash and burn. Agric. Ecosyst. Environ. 58, 39-48.

Developments in Soil Science, Volume 28A
Editors: A. Violante, P.M. Huang, J.-M. Bollag and L. Gianfreda

ENERGY DISPERSIVE X-RAY MICROANALYSIS AND ITS APPLICATIONS IN BIOGEOCHEMICAL RESEARCH

T. A. Jackson and G. G. Leppard

National Water Research Institute, 867 Lakeshore Road, P.O. Box 5050
Burlington, Ontario L7R 4A6, Canada

The principles and modern techniques of energy dispersive X-ray microanalysis combined with transmission and scanning electron microscopy are briefly reviewed, and possible biogeochemical applications are discussed. This method and certain comparable techniques have the unique distinction of permitting measurement of the abundances of various metallic and nonmetallic elements in visually selected, individually analysed living and nonliving microscopic entities (for instance, bacterial cells and clay-sized particles) in heterogeneous natural materials, such as soils and sediments, in their native state. Indeed, specific subunits of these entities (for example, the cell wall and cytoplasm of a single bacterial cell) can be analysed separately. Hence, the technique has great potential usefulness in research on biogeochemical processes in natural environments and related experimental systems. Examples of published research results are presented, with particular reference to recent work on the accumulation and partitioning of heavy metals by bacteria and associated nonliving matter in polluted lake sediments.

1. INTRODUCTION

Soils, fine-grained sediments, and natural waters are extremely complex ecosystems characterised by enormous heterogeneity and variability (both physicochemical and biological) on a microscopic scale [1]. Consequently, the biogeochemical processes occurring in these ecosystems show marked short-range spatial variation (besides varying over time in response to changing conditions), and the larger-scale processes of each system represent the net effects of myriads of smaller-scale phenomena. Thus, a single bacterial cell may interact in particular ways with certain substances, such as heavy metals, in its immediate vicinity because of its specific physiological functions, surface properties, and biologically created chemical microenvironment, whereas only a few micrometres away bacteria belonging to different species or strains may interact very differently with these substances. For instance, one kind of bacterium may have a cell wall with ligands and mineral coatings which bind heavy metals strongly, thereby tending to prevent the metals from crossing the cell membrane into the cytoplasm, whilst another kind coexisting with it may specialise in taking the metals up and accumulating them in its cytoplasm. Similarly, some bacteria may solubilise metals by releasing water-soluble chelating agents, whilst others immobilise the metals by producing H_2S, causing them to be precipitated as highly insoluble sulfides, or coprecipitated with FeS. By the same token, the various kinds of colloidal organic and mineral particles mingled together in soils and sediments differ greatly in their properties, including the nature, abundance, and distribution of their ligands and other metal-binding sites, and hence in their ability to bind metals [2]. To take another example, it is possible that certain species of single-celled organisms in plankton communities preferentially accumulate certain metals which they have absorbed from the surrounding water, whereas other species coexisting with them have lower concentrations and, perhaps, different proportions of the metals. The complexity of situations such as these is compounded, moreover, by the fact that the various microbes, particles, and dissolved substances associated with each other in natural environments interact directly and indirectly in a multitude of ways. For instance, bacterial cells and particles commonly acquire organic or inorganic coatings (e.g. deposits of Fe oxyhydroxide precipitated *in situ*), or both, which drastically alter their surface properties, thereby profoundly affecting their ability to bind metals [2].

The multifarious small-scale and large-scale biogeochemical processes of soils, sediments, and water are mostly mediated by microorganisms. This is accomplished both directly through specific physiological and biochemical activities (e.g. biosynthesis of the readily bioaccumulated toxic compound methyl Hg from inorganic Hg(II)) and indirectly through the profound effects of microbial activities on environmental conditions (e.g. the creation of anoxic conditions by microbial consumption of O_2). Environmental factors, in turn, control the nature, activities, dynamics, and interspecific relations of the microbial community. Thus, environmental change may initiate ecological succession, with shifting results of interactions between mutualistic, competing, and antagonistic microbial species. A change in environmental conditions may be imposed from the outside (as in the stimulation of heterotrophs by labile organic matter) or created by the microbes themselves (as in the rise of anaerobes to positions of dominance following local depletion of O_2 by aerobes during decomposition of organic matter), or both [1]. Complex interactive phenomena such as these are of fundamental importance to all terrestrial and aquatic ecosystems, for they control the speciation, bioavailability, biological effects, and cycling of nutrients and toxic substances.

In our treatment of biogeochemical phenomena, particular emphasis will be placed on heavy metals, which have critically important biological effects, both beneficial and harmful [2]. In trace quantities, many heavy metals function as micronutrients, and some of them, such as Cu and Zn, are essential to all living things. At higher concentrations, however, they may be highly toxic, and the lowest concentrations at which harmful effects are observed are not necessarily very high; for instance, aqueous Cu^{2+} concentrations as low as 0.01 μg/mL may be lethal to certain species of green algae. Some of the more poisonous metals, such Hg and Pb, are not known to perform any essential biochemical functions.

Among the most important biogeochemical processes from an ecological standpoint are the speciation, binding, accumulation, and release of metallic and nonmetallic elements by microorganisms and by colloidal particles produced as a result of microbial activities (notably Fe and Mn oxyhydroxides, humic substances, extracellular biopolymers, and FeS), and effects of dissolved substances, such as organic acids, including chelating agents, secreted by microorganisms. Noteworthy examples of these processes include microbial decomposition of the remains of dead organisms, whereby their constituent elements, including heavy metals, are recycled, and the production, decomposition, mobilisation, and immobilisation of bioavailable heavy metal species by direct and indirect effects of microbial activities, as in (a) the formation and breakdown of organic complexes of metals and organometallic compounds (e.g. methyl Hg), (b) the scavenging of metal ions and metal complexes by Fe and Mn oxyhydroxides in oxygenated waters, and (c) the creation of anoxic environments (including microenvironments) during microbial decomposition of organic matter, leading to the solubilisation of oxyhydroxides (with release of their sorbed and coprecipitated heavy metals) and, under highly reducing conditions, the production of sulfides and thiols (which bind heavy metals strongly).

When heavy metals are introduced into natural waters and soils (as, for instance, in the pollution of rivers and lakes by mine wastes or smelter fallout), they undergo gross partitioning among the solid, liquid, and (in the case of Hg) gaseous compartments of the environment, besides being subject to a multitude of biogeochemical transformations, and are preferentially accumulated by organisms and fine particulate matter. However, smaller-scale partitioning also occurs owing to preferential accumulation by certain organisms and nonliving binding agents. Even within a single microbial cell or on the surface of a single microscopic particle partitioning of metals occurs owing to variations in the nature of the metal-binding sites and metal-concentrating mechanisms. Thus, bacteria in sediments may accumulate heavy metals mainly as a result of sorption and coprecipitation by oxyhydroxide coatings on their cell walls, together with complexing by ligands of the cell wall polymers themselves, or mainly through uptake of metals across the cell membrane followed by accumulation within the cytoplasm [3-14]; and the binding sites and binding mechanisms on the edge faces of clay crystals differ profoundly from those of the basal planes [2].

In general, then, the gross features of biogeochemistry are the net effects of myriads of small-scale phenomena occurring simultaneously or in succession, acting independently or interdependently, and interacting in various direct and indirect ways, some tending to reinforce, whilst others tend to cancel, each other's effects. Consequently, in-depth knowledge of biogeochemistry on a global scale requires understanding of countless complex processes occurring on a microscopic scale. Unfortunately, however, the in-depth investigation of small-scale biogeochemical phenomena and related aspects of microbial ecology under natural conditions has been severely impeded by technical limitations. Much of the published information about the chemistry of sediments, soils, and water was generated by relatively crude methods such as analysis of bulk samples and sample fractions, and the study of free-living microbial communities in their native state is still in its infancy. Thus, our understanding of biogeochemistry and microbial ecology is largely confined to gross effects, gross properties of soils, sediments, and water, and characteristics of certain isolated components of these ecosystems. Our knowledge of the nature and *in situ* activities of microorganisms, and of interactions between different kinds of microorganisms, dissolved substances, and particulate matter in natural environments is all too limited, considering the crucial geochemical and ecological functions of these phenomena [15]. Chemists and microbiologists have been striving to make good this deficiency [2, 15-18], but there is little direct, detailed knowledge of the complex relations between the various elements, such as heavy metals, and the countless microorganisms, dissolved substances, and particles of different kinds intermingled in soils, sediments, and natural waters (let alone subunits of microbes and particles - for instance, the cell walls and cytoplasmic constituents of bacteria) under natural conditions.

The refined and powerful technique of energy-dispersive X-ray microanalysis (EDXM) combined with transmission electron microscopy (TEM) (or, for some purposes the generally cruder method of scanning electron microscopy (SEM)) can, however, provide important information about the *in situ* distribution of many different metallic and nonmetallic elements amongst the various microbes and fine particles in soils, sediments, and natural waters; and from this information reasonable inferences may be drawn about biogeochemical processes occurring on a microscopic scale, especially if the microanalysis data are supplemented by data obtained by other methods. The underlying principle of the technique is simple (although in practice the method is demanding and challenging). The microscope's electron beam is focused on a chosen minute area of the specimen, and the abundances (though usually not the absolute concentrations) of all detectable elements in the irradiated region are determined rapidly by measuring the intensities of characteristic X-rays emitted by their atoms [19]. A major advantage of EDXM is that it allows microscopic components of heterogeneous natural materials, and even small subunits of these entities (e.g. a single inclusion in the cytoplasm of a bacterial cell) to be examined microscopically, selected visually for analysis, and then analysed individually for a wide range of metallic and nonmetallic elements [3, 13, 19-24]. There are other highly sophisticated methods, including electron spin resonance spectroscopy, X-ray photoelectron spectroscopy, and Mössbauer spectroscopy, for investigating extremely small-scale phenomena, such as the bonding of metal ions by different kinds of sorption sites on particle surfaces [2]; but only EDXM and comparable techniques called wavelength dispersive X-ray microanalysis (WDXM) [19] and electron energy loss spectroscopy (EELS) [25-27], in conjunction with electron microscopy, permit individual microscopic objects to be both seen and analysed at the same time. WDXM, however, is rendered impractical for analysing most environmental samples, because (a) it requires that the spectral peaks of individual elements be measured one at a time in a series of consecutive irradiations, and (b) successive irradiations cause progressive decomposition of the sample material. Consequently, this technique is treated only in a cursory manner in this review even though it has several advantageous features.

The purpose of this treatise is to review the principles and modern techniques of EDXM and to demonstrate the potential value of EDXM in biogeochemical research. Selected research results taken from recent literature are used for purposes of illustration.

2. EDXM: THEORY AND FUNDAMENTALS

2.1. The nature of EDXM

When the electron beam of an electron microscope strikes a solid object in the specimen plane, many important interactions occur (Figure 1). One kind of interaction leads to the production of "characteristic" X-rays which carry information about the atoms in the irradiated region (Figure 2). More specifically, atoms bombarded by electrons from an external source emit X-rays which are characteristic of those atoms and can be used to identify and quantify the elements present [19]. Thus, an electron microscope with EDXM capability permits correlation of ultrastructural information (from a well-differentiated image) (Figure 3) with a spectrum showing the presence and relative concentrations of the more abundant elements in small selected regions of the imaged object. One can either (a) analyse one spot at a time or (b) produce an element distribution map; spot analysis, however, is much more

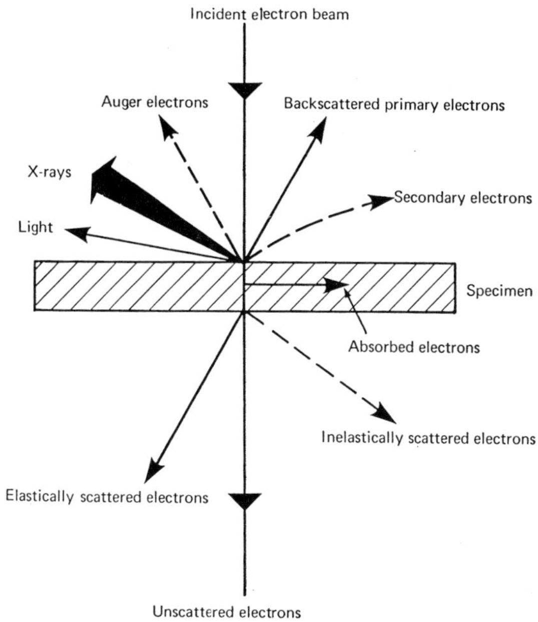

Figure 1. Diagram showing the various interactions between the incident electron beam and a specimen in an electron microscope. Elastically scattered electrons lose no energy, whereas inelastically scattered electrons lose some energy, and secondary electrons are of much lower energy than the primary electrons of the incident beam. (From [19]. Out of print; copyright has reverted to author, but attempt to contact him regarding permission to reprint illustration was unsuccessful).

sensitive than element mapping. Another important consideration is that when a thick specimen (hundreds of nanometres thick or greater, depending on the accelerating voltage and specimen composition) is irradiated, one's ability to localise the electron beam's interactions with a selected region of a specimen can become severely compromised as a result of beam spreading within the specimen and resultant lateral effects within the specimen (Figure 4). The spatial resolution for an analysis [19] can be greatly improved through the use of thin specimens (Figure 4). For environmental analyses, a very promising means of obtaining specimens of optimal thinness is to embed them in a resin and then cut ultrathin sections (< 100 nm thick) with an ultramicrotome [28]. In the discussion which follows we will focus primarily on EDXM in conjunction with TEM employing ultrathin sections.

2.2. Inherent limitations of the technique

An excellent general treatment of the physical and engineering constraints on EDXM apparatus and on electron microscope imaging has been presented by J.A. Chandler [19]. A number of inherent limitations are of specific concern to analysts of environmental samples.

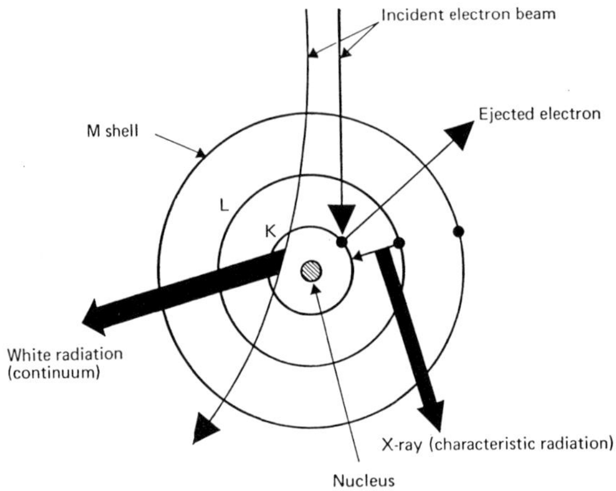

Figure 2. Diagram showing production of X-rays during bombardment of an atom by the incident electron beam of an electron microscope. When one of the atom's electrons falls to a lower quantum level to compensate for expulsion of an electron from that level by the beam, a discrete quantum of "characteristic radiation" is emitted. However, when incident electrons do not collide with electrons of the atom but are decelerated on passing the nucleus, they lose energy in the form of "white radiation." (From [19]. Out of print; copyright has reverted to author, but attempt to contact him regarding permission to reprint illustration was unsuccessful) .

To begin with, the spatial resolution of images from TEM is limited by the grain structure of the sample's embedding medium; currently, the best practical resolution obtainable is 1 nm using Nanoplast formulations [29]. With conventional SEM used for imaging whole mounts and bulk specimens, the practical resolution tends to be more than ten times worse; improved SEM instruments are available, but little advantage is taken of them owing to cost considerations and the non-specialist's fear of complication. Then there is the question of element detectability. In EDXM of a selected spot on a section the detection limit for any given element in the irradiated volume of the specimen is generally on the order of one part per thousand. In element mapping, in which the distribution of the element in the entire field of view is revealed, the detection limit is much higher. Hence, element mapping is feasible only for relatively high concentrations of the element. Exact detection limits are element specific and are discussed by Chandler [19] in relation to quantification standards. Also note that the problem of detectability is compounded by the fact that the determination of the less abundant elements requires longer irradiation times, heightening the risk that sections which have simply been laid directly on the grids will disintegrate prematurely. Therefore, mastery of techniques for stabilising ultrathin sections is of paramount importance [19].

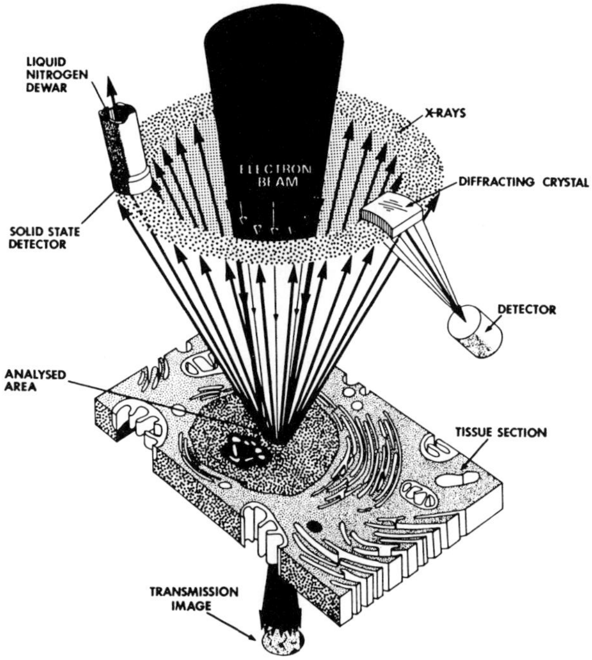

Figure 3. Diagrammatic picture of energy dispersive X-ray microanalysis (STEM-EDXM) of an ultrathin section of a specimen. (From [19]. Out of print; copyright has reverted to author, but attempt to contact him regarding permission to reprint illustration was unsuccessful) .

Furthermore, with EDXM involving TEM (or the TEM component of scanning-transmission electron microscopy (STEM), which is becoming increasingly popular amongst environmental scientists), one is generally limited (by the design of the X-ray detector) to measuring elements of atomic number > 10. Dedicated STEM-EDXM apparatus is available for the detection of lighter elements, but even with such equipment the determination of carbon is confounded by the high concentration of carbon in embedding resins.

Another limitation of EDXM is the fact that the electron beam has a finite, though indeterminate, thickness; even before it strikes the sample, the thickness of the beam cannot be determined precisely. On penetrating the sample, the electron beam spreads out, its width increasing sharply with depth of penetration, thus increasing the volume of sample material represented by the EDXM data. This "beam spreading effect" is minimised when ultrathin sections of the samples are used, but even under ideal conditions the thickness of the beam places a lower limit on the size of a spot pinpointed for analysis. This means that if two spots selected for analysis are sufficiently close together (which usually means on the order of tens to hundreds of nanometres apart, depending on the instrument), they will, to some extent, affect each other's EDXM results. The minimum acceptable distance separating, for the sake of argument, two nearby particles that are to be analysed individually depends on variables

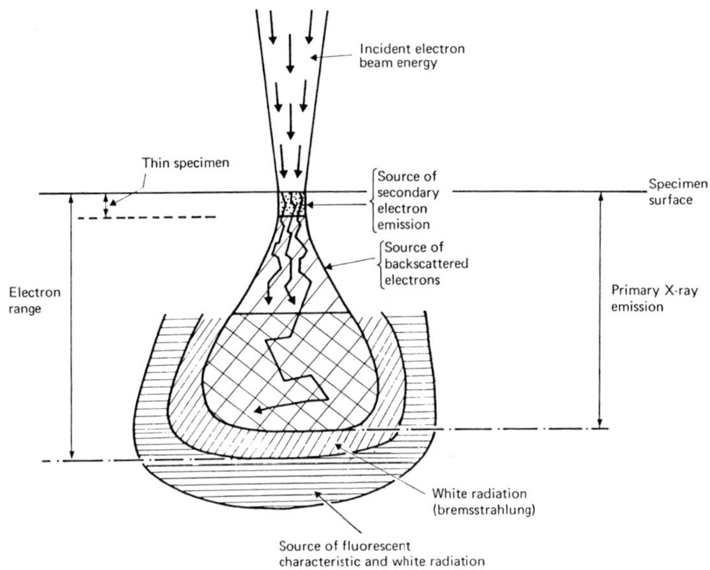

Figure 4. Diagram illustrating the lateral spreading of the beam of incident electrons on penetrating a relatively thick section in an electron microscope. As shown in the diagram, beam spreading would have been minimal if the specimen had been much thinner (e.g. ultrathin). (From [19]. Out of print; copyright has reverted to author, but attempt to contact him regarding permission to reprint illustration was unsuccessful).

such as the sizes and composition of the particles. For instance, the EDXM spectrum of a small Fe-poor particle next to a large Fe-rich particle is likely to yield Fe peaks that are too high owing to the effect of the other particle, whereas the Fe peaks of the Fe-rich particle will not be significantly affected by the small amount of Fe in the Fe-poor particle.

An additional limiting factor is that in an EDXM spectrum the principal characteristic peaks of two or more elements may overlap or coincide. For instance, the principal peaks of Hg and S are very close together and correspondingly hard to resolve.

It should also be mentioned that TEM-EDXM and STEM-EDXM are technically demanding and highly specialised. Mastery of the art requires years of training and experience. Besides, the methods are extremely expensive in terms of both capital and operating costs.

2.3. Constraints imposed by sample processing

To maximise the amount of information obtainable from an environmental sample, it is necessary to analyse the sample in a condition as close to its native state as possible. Hence, it is imperative that field samples be stabilised by special methods of preservation *immediately* after they have been collected. The requirements for achieving minimal perturbation of samples rich in submicrometre particles (colloids) have recently been discussed in detail by G.G. Leppard and J. Buffle [30]. To prevent creation of artifacts, sample storage prior to stabilisation must be avoided, and a high level of special expertise is necessary in sample processing with chemical solutions; furthermore, a multi-method approach permits improved interpretations of the data. Basic protocols for environmental TEM- and STEM-EDXM analyses, which guide the analyst from initial sampling through sample stabilisation to embedding have been developed [23, 31, 32]. The most common and most serious artifacts of sample processing are (a) dehydration-induced shrinkage of highly hydrated solids and (b) selective extraction of sample components by fluids used for processing the samples. These artifacts can be systematically identified, assessed, and minimised by multi-method approaches amenable to fine-tuning [23, 28, 32, 33].

2.4. Recent improvements relevant to environmental applications

Recent advances (see discussion of case studies below) have been made as a direct result of improved sample preparation protocols and the use of well-stabilized ultrathin sections in the 50-80 nm range; this range usually provides a combination of good image detail and sufficient thickness for informative spectra. The comparative study of alternating 50 nm and 100 nm sections of the same sample can yield similar results when element abundances are so close to the detection limit that utilisation of ~50 nm thick sections alone is precluded. Current use of the melamine resin formulation, Nanoplast FB 101 [23, 29, 31], permits the three-dimensional analysis of long range (multi-micrometre) associations between bacteria, their extracellular biopolymers, other microbes, organic debris, and colloidal minerals. Recent improvements in basic instrumentation (the development of STEM instruments with a field emission source of electrons [19] and user-friendly adaptations to conventional STEM-EDXM apparatus) along with computerized image analysis are facilitating advances in environmental applications. A new generation of STEM apparatus promises substantial instrumental improvements for quantitative analysis.

2.5. Quantification of results

Quantification of EDXM data is difficult [19] even with relatively simple, well understood "ideal" samples for which suitable certified standards are available. For complex, heterogeneous, imperfectly understood environmental samples the choice of standards is arbitrary and may not be appropriate. Therefore, it is usually not feasible to measure absolute quantities of elements in sample materials of this nature. Nevertheless, satisfactory answers to many important questions about biogeochemical processes on a microscopic scale can be obtained from semi-quantitative data [13]. Thus, knowledge of relative amounts of different elements in different kinds of microscopic structures can be highly informative, as shown by a series of case studies discussed below. In semi-quantitative TEM-EDXM spot analyses the abundance of each detectable element is represented by the area under the principal characteristic peak for the element in the EDXM spectrum or by the corresponding number of counts per second. Peak height has also been used, but it is less satisfactory, as it gives only an approximation of element abundance.

For the purposes of interpreting and quantifying biogeochemical processes, a particularly useful approach is to perform semi-quantitative spot analyses of statistically meaningful numbers of microscopic entities in each sample (and, if possible, statistically meaningful numbers of replicate analyses of individual entities), using the data to elucidate the associations of different elements with specific microscopic structures and with each other. The expression of EDXM data in cluster diagrams [27, 34] is becoming a useful method for revealing natural *in situ* associations between elements; ZAF standardless analysis is recommended [34]. In experimental situations where systematic sample dilution is permissible, ultrathin sections can be replaced by whole mounts of samples, whose coverage of a TEM grid with colloidal particles can be optimised for viewing [35]. Optimisation of coverage by dilution following centrifugation to preclude interference by large obscuring particles permits proper viewing, description, and quantification of individual colloid species and small aggregates of them in their native state prior to further TEM- or STEM-EDXM analysis [35]. With whole mounts, such quantification can be rapid, whereas it can be extremely time-consuming with ultrathin sections.

3. OPTIMAL USE OF EDXM FOR ANALYSIS OF ENVIRONMENTAL SAMPLES

3.1. The field site, and sampling and sample stabilisation at the site

As several case studies have shown, interpretation of TEM and STEM-EDXM information is facilitated by correlative use of chemical and physical data obtained at the field site [13, 16, 27, 34, 36, 37]. Artifacts of sampling, storage and sample handling are described succinctly elsewhere; these include artifacts associated with fractionation, microbial activities, and misguided attempts to minimise handling-induced aggregation [38]. Every means available should be used to avoid sample storage [30, 38].

3.2. Selection of protocols for fixation and embedding

In the preparation of ultrathin sections, several TEM protocols should be used for purposes of comparison, as has been emphasised recently, with [a] at least one protocol for production of Nanoplast-embedded samples and [b] at least one protocol for production of epoxy-embedded samples stabilised initially in a glutaraldehyde-based fixative [23, 32]. The rationale for such a correlative approach is widely accepted, and specific detailed instructions for preparing flocs and sediments [23, 32] have come into general use. A possible alternative to embedding which is worth considering is the use of cryogenic preparations, whereby ultrarapid freezing converts the water in the sample into the equivalent of a solid embedding medium.

3.3. Instrumentation

A revolution in instrumentation is occurring, especially with regard to improved versatility and quantification. It may seem as though one should invest one's resources to a maximum extent in the latest development in microscopes, but, given the inherent variability and patchiness that are typical of environmental samples, much of the potential of the improved instrument may go to waste. For many environmental research projects, equipment which is less sophisticated but more "user-friendly" may be a more valuable investment.

3.4. Recognition skill and search strategy

Because examination of microscopic structures necessarily involves visual recognition, and because environmental samples tend to have a rich diversity of visualisable components, the electron microscope technician must be highly skilled in the identification of specific structures, including artifacts. Failure in the past, on the part of principal investigators, to address this need for real skill has caused a lot of disappointment and wasted effort. Sound search strategies are also important to ensure that data collections properly represent the sample.

3.5. Correlative use of EDXM with other beam techniques

Electron energy-loss spectroscopy (EELS) is a technique involving analysis of the energy distribution of initially monoenergetic electrons after they have interacted with a specimen [25]. A beam of transmitted electrons is directed into a high-resolution electron spectrometer that separates the electrons according to their kinetic energy and produces an electron energy-loss spectrum. The spectrum shows the scattered electron intensity as a function of the extent to which the initial kinetic energy of the electrons (the energy of the "fast electrons") decreases when the electrons strike the sample. EELS is more demanding than EDXM but yields more information when technical constraints permit its use. It can be used not only to perform element analyses but also to obtain information about electronic structure; for instance, it is capable of distinguishing between Fe^{2+} and Fe^{3+} [39]. EELS and a variant of it called energy filtered transmission electron microscopy (EF-TEM) are now being adapted for the analysis of environmental samples [26, 27, 40, 41]. In addition, electron diffraction analysis of selected minute areas (previously termed "selected-area electron diffraction" analysis) can be used effectively in tandem with EDXM [3, 16], and the complementary use of EDXM with X-ray absorption spectroscopy (XAS) has begun [34]. XAS and synchrotron-based microscopy methods in general are rapidly being developed for application in biogeochemical research. As improvements in resolution approach the 25 nm level, the imaging of ultrastructural detail in heterogeneous polydisperse samples (microbial aggregates, colloidal mineral aggregates, complex flocs, and biofilms) should make it possible to obtain information on oxidation state and coordination for some elements of great importance in the smallest recognisable particles. Electron diffraction has been used to obtain crystallographic information about microscopic mineral deposits, such as mineral coatings on bacterial cells. Whereas EDXM yields measurements of element abundances without reference to mineral structure, electron diffraction analysis distinguishes between "amorphous" and crystalline minerals and permits identification of more or less crystalline ones on the basis of their crystal structure. Thus, EDXM combined with electron diffraction analysis is a powerful method for investigating inorganic materials on a microscopic scale. For biogeochemical research in which the various associations of abundant elements in particles analysed individually are already well documented, and where improved sensitivity and quantification for comparing just two (or a few) elements would be desirable, one might consider using WDXM along with EDXM. The advantages and disadvantages of this approach have been outlined by Chandler [19].

4. RESEARCH APPLICATIONS: EXAMPLES IN THE LITERATURE

4.1. The use of TEM and STEM-EDXM in research on the accumulation of heavy metals by bacteria and associated colloidal material

Free-living bacteria and their extracellular macromolecular products (e.g. fibrils) in natural environments can readily accumulate heavy metals and other elements and may have mineral coatings with bound metals on their surfaces [3-9, 11, 14, 42-48]. Moreover, the binding mechanisms may be highly selective [6, 9, 43]. Metal accumulation may occur by (a) passive and active uptake by bacterial cells, with immobilisation in the cytoplasm (except when the metal is immediately expelled from the cell after uptake) [4, 10, 14, 42, 48, 49]; (b) direct sorption, or surface complexation, by the cell wall and extracellular macromolecular products such as acid polysaccharides in colloidal fibrils [4-7, 9, 11, 14, 22, 42-48, 50, 51]; (c) precipitation of slightly soluble metal compounds, such as NiS, on the cell surface [3]; and (d) sorption and coprecipitation by mineral coatings, such as iron and manganese oxyhydroxides (FeOOH and MnOOH), ferrous sulfide (FeS), and clay minerals, formed *in situ* by passive or biologically mediated precipitation or adsorbed by the cell wall [3, 13, 37, 48, 52, 53]. The general importance of metal scavenging and immobilisation by FeOOH and MnOOH under oxidising conditions, and by inorganic and organic sulfides (H_2S, FeS, thiols, etc.) in reducing environments (including anoxic microenvironments within O_2-rich environments), is firmly established [2, 54-58]. In well oxygenated aquatic environments the formation of FeOOH and MnOOH coatings on bacterial cells and fibrils and on various other exposed surfaces (e.g. the surfaces of clay crystals, silt particles, sand grains, and planktonic organisms) is probably commonplace [13, 24, 48, 52, 53, 57], and minute particles of oxidation-resistant complex heavy metal sulfides have been found as well [59]. The binding of metals by humic matter and other complexing agents (nonhumic chelators, phosphate ions, etc.) sorbed to the cell, to mineral coatings on the cell, and to extracellular organic colloids could also be involved [2]. Although our knowledge of the subject is limited, the microbial accumulation of metals by these processes and the nucleation of mineral precipitates by bacterial cells are probably of widespread and frequent occurrence in nature. (Of course, the formation of FeOOH and MnOOH incrustations by bacterial species that specialise in the oxidation and precipitation of reduced aqueous Fe and Mn species under certain environmental conditions has been known for a long time [52, 53]; but more recently electron microscopy and EDXM have revealed that spontaneous or biochemically mediated deposition of mineral coatings on bacterial cells and fibrils may be a far more general and widespread phenomenon involving many, if not all, other kinds of free-living microbes.) Such processes are thought to be of great biogeochemical and ecological importance, both as a means of concentrating nutrient trace elements from dilute solution and as detoxifying mechanisms for the protection of microorganisms from harmful effects of heavy metals present in excessively high concentrations in the environment [10, 12-14, 49, 51]. Biotechnological applications, as in the prevention and amelioration of metal pollution in natural waters, the recovery of metals from ores, and the synthesis of useful inorganic substances such as catalysts, are possible as well [13, 42, 51, 52, 60].

EDXM in conjunction with TEM or STEM has been used by a number of workers in research on the accumulation of heavy metals by bacteria and their extracellular polymers, and the deposition of mineral coatings on their exposed surfaces, both in natural environments and in experimental systems involving pure cultures [3, 13, 20-22, 37, 46, 49, 61-64]. Important relevant information has also been obtained by various other means, such as the following: (a) techniques of microbial physiology and cytology as applied to the uptake, accumulation, and excretion of metals by microbes; (b) various chemical, biochemical, and biophysical methods of analysis used to determine the nature and properties of microbial cell walls, fibrils, and biofilms, and interactions of bacteria and their extracellular products with dissolved metals and other inorganic substances [4-9, 11, 43, 45, 46, 48, 49, 51-53, 65]; (c) electron diffraction analysis of selected small areas of mineral deposits [3, 62, 63, 66]; (d) Mössbauer spectroscopy for characterising Fe-bearing material (e.g. for differentiating between Fe^{2+} and

Fe^{3+} in Fe-rich mineral deposits) [36, 63]; and (e) various conventional methods of analysis for defining the field environment (measurements of the pH, dissolved O_2 content, hardness, alkalinity, etc. of water and the pH, Eh, concentrations of sulfide, FeOOH, MnOOH, and organic matter, etc. of sediments) [13, 34]. The biochemical and biophysical methods have revealed, among other things, a net negative charge on bacterial cell surfaces and an abundance of ligands, notably carboxyl ($-COO^-$) and phosphoryl ($-PO_4^{2-}$) groups (which account for the negative charge), on cell wall polymers of bacteria, both of which properties are conducive to the sorption and complexing of metal cations [4-7, 9, 11, 45, 48].

Obviously there is a great advantage in combining TEM and STEM-EDXM with other techniques that provide complementary information which is not obtainable by TEM and STEM-EDXM alone [2, 67, 68]. For example, TEM and STEM-EDXM can provide morphological information and measurements of total element abundances in a mineral coating on a microbe, but it cannot reveal the identity of the mineral (unless the mineral deposit happens to have a distinctive, diagnostic outward form which, together with its elemental composition, permits identification with a high degree of confidence). Let us consider, as an example, an Fe-bearing mineral whose morphology and elemental composition strongly suggest a particular mineral species. Electron diffraction analysis would yield crystallographic information from which the specific Fe-bearing mineral could be identified, and Mössbauer spectroscopy or EELS could be used to determine whether the Fe was Fe^{2+} or Fe^{3+} or a combination of both. By using some such combination of STEM-EDXM with other methods, we can greatly extend the range of attainable information and end up with a far more refined and detailed description of the sample components than is possible with only one of the methods alone. Furthermore, data acquired by different methods may all point independently to the same conclusion, thereby helping to distinguish genuine results from artifacts and confirming the validity of the conclusion [67]. It is useful, moreover, to compare empirical data for field samples in their native state with the results of complementary controlled experiments.

A series of important pioneer research projects employing a combination of "TEM/STEM" and EDXM techniques to investigate the sorption of metals and formation of metal-bearing mineral coatings by bacterial cells, both in selected natural environments and in experimentally manipulated cultures, has been carried out by T.J. Beveridge, F.G. Ferris, W.S. Fyfe, and their associates. These workers observed that free-living bacteria and their extracellular polymeric products in a variety of natural environments, including hot spring, lake, and stream sediments, and both freshwater and terrestrial biofilms, are commonly coated with mineral deposits [3, 4, 21, 63, 64, 66]. Such coatings are recognisable by inspection, because, in contrast to unmineralised cell structures, they are opaque to electrons. They vary widely in texture, ranging from extremely fine-grained, uniform incrustations, as seen, for instance, in cultures of the Mn-oxidising, Mn oxide-depositing bacterium *Leptothrix discophora* (Figure 5a), to armour-like layers of relatively coarse platy or lath-shaped adsorbed clay-sized mineral particles, a good example of which is a thick, evenly distributed mineral coating on an unidentified bacterium observed *in situ* in a sediment sample from Irvine Creek, Ontario (Canada) (Figure 5b). Moreover, EDXM showed that they vary greatly in elemental composition, and, as inferred from the EDXM data or determined directly by electron diffraction analysis, they comprise different kinds of minerals, including FeOOH, MnOOH, metal sulfides, and silicates.

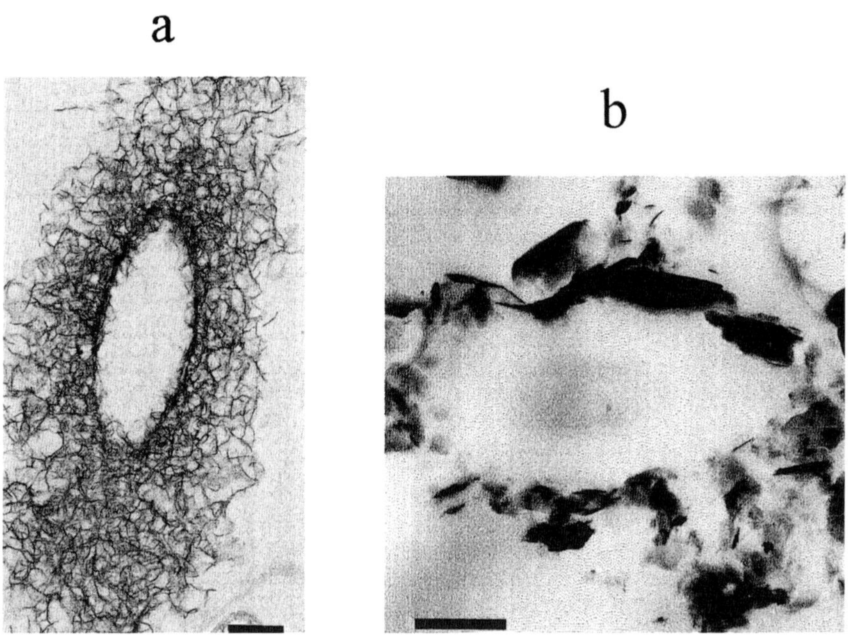

Figure 5. Transmission electron micrographs of thin sections showing (a) the sheath of a *Leptothrix discophora* cell encrusted with a fine deposit of MnO_2, and (b) an unidentified bacterium encased in a coating of relatively coarse clay-sized mineral particles observed *in situ* in sediment from Irvine Creek. In each picture, length of scale bar = 500 nm. (From [4]. Reprinted with the permission of John Wiley & Sons and T.J. Beveridge, who kindly provided the pictures).

Using TEM and STEM-EDXM to examine and analyse surficial sediment ("clay and siliceous mud") from a highly acidic, metal-rich stream draining a group of hot springs in Yellowstone National Park (U.S.A.), Ferris, Beveridge, and Fyfe found a number of bacterial cells coated with mineral deposits [21]. EDXM revealed that Si was the most abundant of the detectable elements in the mineral incrustations, suggesting nucleation of silica precipitation on the cell walls. Fe was plentiful as well, but it appeared to be bound directly to the cell wall biopolymers (quite possibly by surface complexation of cations) and was not part of the siliceous deposit. The authors surmised that under oxidising conditions "microdomains" of Fe oxides, carbonates, and silicate could, in time, develop from sorbed "hydroxide" precursors. They also directed attention to the possible relevance of their findings to the occurrence of well preserved silicified fossil microbes in Precambrian sedimentary rocks (notably chert). It should be mentioned, as well, that microorganisms (especially cyanobacteria) are thought to have been responsible for the deposition of the famous Precambrian banded iron formations and stromatolites [69, 70]; they may well have nucleated mineral precipitation partly by processes similar to the ones described by Ferris et al. and partly by the creation of distinctive microenvironments through physiological processes such as photosynthesis [69, 70].

In related research on a microbial mat in one of the hot springs in Yellowstone National Park, Ferris et al. found bacterial cells encrusted with Mn oxide precipitates [66]. Electron diffraction enabled them to identify the Mn oxide mineral as todorokite. Figure 6a is a striking electron micrograph of one of these cells, which is completely surrounded by a fine-grained, reticulated Mn oxide deposit extending far beyond the surface of the cell. Figure 6b is an EDXM-generated "X-ray dot map" (element map) displaying at a single glance the distribution of Mn in the picture shown in Figure 6a. The map reveals that practically all the detectable Mn in the field of view is in the extracellular deposit and is coextensive with it, proving that the deposit consists of Mn-rich material. This is a most compelling example of the sort of results that can be obtained by element mapping if the element of interest is sufficiently abundant. Unfortunately, the technique is of limited usefulness; it is so insensitive compared to spot analysis that it can only be applied to elements present in very high concentrations, as with Mn in massive incrustations of Mn oxide (see section 2.2).

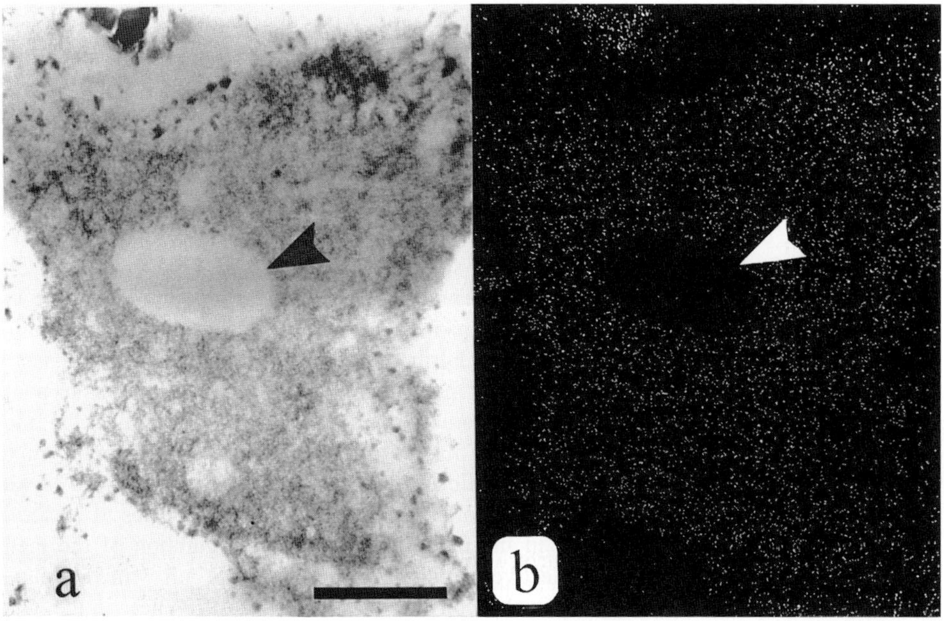

Figure 6. (a) Scanning transmission electron micrograph of an unidentified Mn oxide-encrusted bacterium in a specimen of microbial mat taken from a hot spring in Yellowstone National Park and (b) an X-ray dot map showing the distribution of Mn around the same microbe (the position of the cell being indicated in each picture by an arrow). Length of scale bar = 500 nm. (From [66]. Reprinted with the permission of Taylor & Francis, Inc. and F.G. Ferris, who kindly provided the pictures).

Following up on their hot spring research, Ferris et al. used TEM, STEM-EDXM, and electron diffraction to investigate the deposition of mineral coatings on bacteria in highly reducing, H_2S-generating sediment ("black mud") in Lower Moose Lake near Sudbury, Ontario (Canada) [3]. The lake is polluted with heavy metals and sulfate introduced by drainage from an adjacent lake where mine tailings have been dumped. The water draining the tailings-polluted lake is extremely acidic owing to oxidation of primary sulfide minerals to H_2SO_4, but measurement of the pH of Lower Moose Lake sediment yielded an average value of 7.3, because the polluted acidic water flowing into the lake had been neutralised by treatment with crushed limestone.

The sediment was found to contain many bacterial cells coated with mineral deposits. Interestingly, moreover, individual bacteria differed considerably from one another in the mineral composition of their coatings, suggesting species-related selectivity. This implies extremely short-range spatial variations in environmental conditions arising from differences in physiological functions, biochemical activities, and surface properties amongst different kinds of bacteria. The commonest minerals were described as Fe-Al silicates varying in their proportions of Fe, Al, and Si and characterised by corresponding variations in degree of crystallinity. The authors tentatively interpreted these deposits as clay-like aluminosilicates comparable to chamosite and concluded that they began as amorphous gels which subsequently underwent gradual crystallisation. As might be expected in the presence of free sulfide, the sediment samples also had a number of bacteria coated with metal sulfides, including ones which, on the basis of electron diffraction, were identified as "metastable mackinawite" (FeS_{1-x}) and millerite (NiS) (Figure 7). Some of the coatings had detectable trace quantities of Cu and Zn, probably owing to coprecipitation and sorption of these metals by the more abundant Fe and Ni sulfides. The mackinawite coatings were largely or wholly amorphous, but the millerite coatings were commonly crystalline. This is consistent with the possibility that certain substances, such as organic compounds released by the bacteria, selectively interfered with crystallisation of Fe sulfide (as postulated for CuS, but not for ZnS, in a similar environment elsewhere (71)) and predetermined the form of the Fe sulfide in a manner analogous to the role of biogenic organic acids in the preferential precipitation of ferrihydrite under oxidising conditions [72]. Furthermore, the degree of crystallinity has an important bearing on the scavenging of heavy metals by the sulfides, as amorphous or poorly crystallised colloidal mineral deposits are much more efficient sorbents than the more crystalline ones [2]. The different kinds of metal sulfide minerals were generally mixed together, but separate deposits of pure Fe sulfide and pure Ni sulfide were observed as well. This intriguing observation suggests, among other things, that different species of bacterium may differ greatly from one another in the nature and spatial configuration of the metal-binding cell wall sites which, collectively, comprise the template on which epitaxial nucleation of minerals takes place. It is also consistent with marked differences in the nature of the chemical microenvironments that different bacterial species create in the immediate vicinity of their cells through their widely differing physiological and biochemical activities. Presumably, the result of such interspecific variations is that bacteria in a naturally occurring mixed community may mediate highly selective and widely varying mineral-generating processes, depending on the characteristics of the individual species or strain. Environmental conditions imposed on the bacteria from the outside have crucial effects as well. Thus, the authors reasoned that Fe-Al silicates predominated wherever there was not enough free sulfide in the bacterium's immediate vicinity to bring about copious precipitation of metal sulfides. Possibly the microbes with sulfide coatings were sulfate-reducing (H_2S-generating) bacteria together with other kinds of bacteria that happened to be situated close enough to them to be exposed to high concentrations of H_2S.

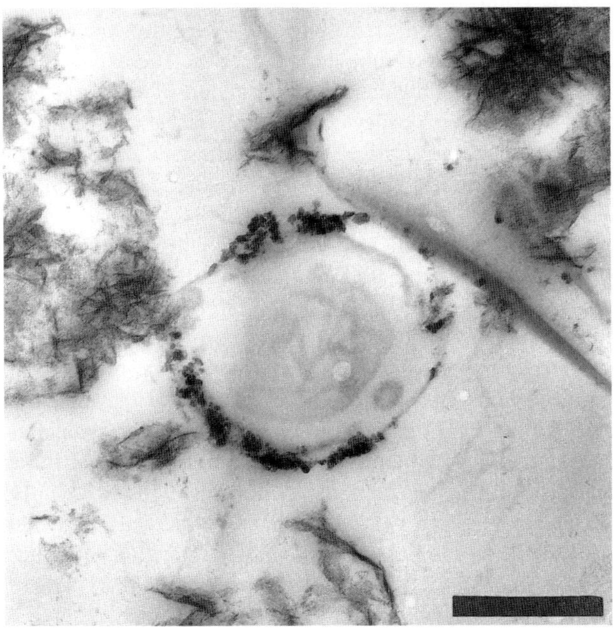

Figure 7. Transmission electron micrograph of an NiS (millerite) deposit on a bacterial cell in a thin section of sediment from Lower Moose Lake. Length of scale bar = 500 nm. (From [66]. Reprinted with the permission of Elsevier Science and F.G. Ferris, who kindly provided the picture).

In spite of the striking small-scale spatial variations in the products of mineral genesis and the postulated corresponding variations in the factors controlling precipitation of minerals, the authors believe that the fundamental mechanisms of biologically induced precipitation on the surfaces of bacterial cells are the same for all the minerals - silicates and sulfides alike – under all circumstances. According to their hypothesis, which is plausible and consistent with all the known facts, nucleation of the minerals on the outer surfaces of the cell walls invariably begins with the complexing of metal cations by ligands on cell wall polymers. This, in turn, promotes reaction of the cations with dissolved silica and sulfide species in the surrounding water. An hypothetical alternative explanation which they mentioned is the formation of cationic mineral colloids in the sediment pore water followed by sorption of these colloids by the negatively charged bacteria. Be that as it may, the findings of Ferris *et al.* constitute an elegant demonstration of the unique ability of TEM and STEM-EDXM, along with electron diffraction, to provide detailed biogeochemical information that is far beyond the scope of bulk analysis and other relatively crude conventional techniques.

In further research on biogeochemical phenomena in metal-polluted lakes near Sudbury, Ferris and coworkers performed a field experiment to investigate interactions between heavy metals and microbial biofilms in the water column at widely differing ambient pH values [63]. Growth of biofilms was induced at field sites in three lakes (Lower Moose Lake, Cranberry Lake, and Fraser Mine Pond) by suspending strips of epoxy-impregnated paper in the water 1.5 m below anchored buoys (i.e. near the surface, where the water was presumably rich in dissolved O_2) for different lengths of time up to 5 weeks. The experimental paper strips were accompanied by control strips sandwiched between nylon filters to prevent microbial growth. Biofilm samples and controls were collected both from lakes acidified by drainage from mine tailings (pH 3.1) and from lakes neutralised by treatment with limestone (pH 6.5-6.9), all lakes being situated in the same drainage basin. STEM-EDXM and electron diffraction were used to assess the results.

Biofilms consisting of bacterial cells together with reticulated masses of fibrillar extracellular polymers grew in both acidified and neutralised water, although the neutralised water was a more favourable habitat for the bacteria. In both types of environment the extracellular material was commonly coated with poorly crystallised ferrihydrite. Although all biofilms accumulated Mn, Fe, Ni, and Cu, metal uptake was orders of magnitude greater in neutralised than in acidified water, as would be expected if the metals were sorbed by the exposed surfaces of the biofilms (instead of being absorbed into bacterial cells). A relative dearth of dissociated (anionic) carboxyl and phosphoryl groups on the bacterial cell walls [63], as well as competition between H^+ ions and metal cations, at low pH probably accounts for this effect. Moreover, the ferrihydrite-encrusted extracellular polymers contained minor amounts of Si, Al, and Cl, but no detectable S, in neutralised water, whereas they contained minor amounts of S but no Si, Al, or Cl under acidic conditions. This difference in minor element content was correlated with a morphological difference, as the crystal habit of the ferrihydrite was granular in the circumneutral water but acicular in the acidic water. Thus, with the combination of techniques that they employed, Ferris et al. were able to (i) detect mineral-encrusted extracellular biopolymers, (ii) identify the mineral, (iii) and discover apparently pH-dependent variations in both the gross morphology and the minor element composition of the mineral deposit.

The work of Ferris et al. is a striking and elegant example of what can be accomplished by combining electron microscopic observation, EDXM, electron diffraction, and conventional methods of analysis (e.g. measurement of the pH of water). In addition, it demonstrates the value of comparing different field environments and integrating experimental methods with observation of phenomena actually occurring in nature. Investigations such as this can provide a wealth of otherwise inaccessible information, greatly extending the scope of our insight into basic biogeochemical processes.

In another investigation along the same general lines but involving an altogether different set of environmental conditions, TEM and STEM-EDXM combined with Mössbauer spectroscopy and conventional X-ray diffraction analysis were brought to bear on a subsurface terrestrial biofilm that developed on the damp wall of an underground chamber which had been excavated in an Archean granite batholith in the Canadian Shield [64]. The biofilm was found to be encrusted with ferrihydrite or hematite when in contact with air, but the oxidised layer was underlain by an anoxic microenvironment in which the oxide had evidently been reduced to siderite [64].

Thus, through the use of various complementary techniques, including TEM and STEM-EDXM, it has been shown that ferrihydrite and other minerals are commonly precipitated on biofilms, probably through bacterial action, in a wide range of environments, terrestrial as well as aquatic.

Up to this point we have been concerned solely with microbial communities in natural habitats. We now turn to a series of papers in which Beveridge and associates reported results of laboratory experiments on interactions of metals with cell walls of selected bacterial species grown in pure cultures on artificial media. Most of the experiments were performed on *Bacillus subtilis*, but a few other species, including *Escherichia coli* were used as well. This work involved TEM or STEM, either with EDXM [22, 46, 50, 61, 62, 75, 76] or without it [43, 73, 74].

Obviously, we have limited ability to draw conclusions, or formulate generalisations, about the interactions of metals with microbial communities in complex natural ecosystems by extrapolation from the results of experiments performed on cultures of one or a few isolated, arbitrarily selected microbial species in artificial and drastically simplified model systems. Nevertheless, controlled experiments of this kind serve a useful purpose, as they provide important complementary data which would be difficult or impossible to obtain by empirical analysis of field specimens alone. They permit direct, in-depth investigation of specific biochemical mechanisms of metal uptake and the effects of particular variables on these processes; thus, they expand and deepen our understanding of metal-microbe interactions in general and help to establish a sound, though limited, theoretical basis for tentative interpretation of data yielded by analysis of field samples in terms of possible or likely mechanisms. For instance, in work performed by Beveridge in collaboration with R.G.E. Murray, experiments on metal deposition by *B. subtilis* before and after chemical treatments for modification of cell wall ligands or removal of teichoic acid from the cell wall revealed that carboxyl groups were the principal metal-binding sites on the cell wall, whereas amino groups were ineffective [74]. Moreover, experiments on the sorption of many different metallic elements by cell walls of *B. subtilis* before and after partial lysozyme digestion demonstrated that some metals are bound preferentially with respect to others and that certain metal cations are subject to selective uptake by particular binding sites on the walls [43]. Empirical data for field samples in their native state have the salient advantage of representing phenomena actually occurring in nature; but by themselves they are apt to be more or less ambiguous, because what we observe is the net effect of so many different physicochemical and biological variables acting simultaneously or in succession, both directly and indirectly. We may draw reasonable inferences about the biogeochemical processes and cause-and-effect relations that produced the observed result (especially if specimens from different spatially or temporally varying field environments are compared), but experimental manipulation, whether in the field or in the laboratory, is required to obtain more definite evidence about theoretically possible mechanisms of metal-microbe interaction.

In their laboratory experiments involving TEM combined with STEM-EDXM, Beveridge et al. exposed whole cells as well as isolated cell walls and cell "envelopes" (cell wall + cell membrane) to solutions of a wide variety of heavy metals for purposes of comparison. The results showed that the cell exteriors scavenge dissolved heavy metals more or less effectively, tending to remove them from solution [22, 46, 50, 61, 62, 75, 76]. Under suitable conditions, moreover, cell exteriors nucleated the precipitation of mineral deposits, which the authors regarded (quite reasonably) as FeOOH, silicates, phosphates, and sulfides [22, 61, 76];

in one case involving a species of cyanobacterium, inorganic coatings formed on the cell were identified as carbonate minerals by electron diffraction [62]. Deposition of the presumptive FeOOH enhanced the uptake of other metals, suggesting sorption and coprecipitation of these metals by the oxyhydroxide [22]. Clearly, then, in more ways than one the experimental observations were consistent with what has been seen in nature [3, 4, 13, 21, 57, 63, 64, 66]. (Incidentally, the paper reporting uptake of heavy metals by colloidal mineral coatings on bacterial cells as opposed to bare cell walls [22] is one of the few in the literature that deal with this particular subject. Most of the publications on metal-microbe interactions as revealed by TEM and STEM-EDXM are concerned *either* with the deposition of mineral coatings such as oxyhydroxides and sulfides on bacteria and their extracellular products *or* with the sorption of heavy metals directly by bare bacterial surfaces, but not with the scavenging of metals by the coatings and the relative metal-binding capabilities of coated and uncoated cells, although the general importance of oxyhydroxides and sulfides in the binding of heavy metals has long been common knowledge. Another paper that expressly deals with this question (13) will be treated in full further on.)

In the experimental work of Beveridge et al. some interesting selective effects were seen. Thus, experiments with *E. coli* showed that certain metals were sorbed preferentially with respect to others; for example, the cell envelopes had higher sorption capacities for Zn, Pb and Co than for Cu, Hg, and Ni [75]. In addition, comparison of four different bacterial species revealed that they differed significantly from one another in their sorption capacity for metals and their preferences for particular metals. Under a given set of experimental conditions, for instance, *E. coli* was most effective in scavenging Cd^{2+}, whereas *B. subtilis* had the greatest affinity for Cu^{2+} [46]. Another point of interest is that living cells of *B. subtilis* were found to be less effective metal scavengers than dead cells, reflecting the fact that biochemically produced H^+ ions continually compete with metal ions for binding sites on actively metabolising cells [50].

These various findings illustrate basic principles which may very well be of general importance in natural microbial ecosystems (and could have practical applications, as in the removal of metals from wastewater). Although further work on natural communities is needed to test this hypothesis, the experimentally demonstrated selective effects imply the possible coexistence of different bacterial species that discriminate between different dissolved metals in their environment and differ amongst themselves in their affinities for particular metals, as well as in their sorption capacities for metals in general, within a single microbial community under natural conditions. Thus, there could be considerable spatial variation in metal partitioning between microbial cells and their external environment over microscopic distances. The experimental results also suggest that changes in the nature and activities of the microflora owing to processes such as ecological succession, changes in the proportion of dead cells to living ones, and variations in the intensity of metabolic activity in response to environmental changes imposed from the outside (changes in ambient temperature, pH, dissolved O_2 level, nutrient supply, etc.) could alter the metal-binding capabilities of the microbial populations. The evidence on which these hypotheses are based might never have come to light without the application of TEM and STEM-EDXM to experimentally manipulated model systems. Now we must face the technical challenge of finding ways to extend this promising work to microbial communities in natural environments. A useful approach would be to employ TEM and STEM-EDXM combined with a wide range of complementary methods (including biochemical techniques for characterising natural, *in situ* microbial communities) in a series of field experiments, or in laboratory experiments performed on bulk samples of natural sediment or soil in their native state, along with in-depth comparative study of field samples representing different environmental conditions (for instance, oxidised and reduced bottom sediment collected from a lake site where the oxidation-reduction potential alternates seasonally from positive to negative values).

We have seen that the literature dealing with experiments performed on microbial cultures includes a paper (by I.T. Mayers and T.J. Beveridge) reporting data that strongly suggest the scavenging of dissolved metals by FeOOH precipitated as coatings on cultured bacteria in artificial model systems [22]. One of the few other papers to present evidence for the accumulation of metals by oxyhydroxide coatings on individually analysed bacterial cells in natural environments was published by C.-P. Lienemann et al. [37]. (Another one already cited [13] will be discussed in full below.) Using conventional methods of chemical analysis, Lienemann et al. examined the vertical profiles of several dissolved substances, including Co, Mn, Fe, O_2, H_2S, and SO_4^{2-} in the water column of a stratified meromictic lake (Paul Lake, which lies on the border between Wisconsin and the upper peninsula of Michigan (U.S.A.)); this lake is characterised by a well oxygenated epilimnion and an anoxic hypolimnion containing free sulfide. The results showed a striking positive correlation between Co and Mn throughout the water column, with a higher Co/Mn ratio in particulate matter than in solution, implying that Co was scavenged by Mn oxide in the particulate matter. TEM and high-resolution STEM-EDXM were used to characterise the particulate matter, leading to the discovery of a considerable number of bacteria encrusted with Mn-rich mineral coatings presumed to consist of Mn oxide. These coated bacteria were concentrated in the transition zone between oxygenated and anoxic water, where dissolved O_2 from the epilimnion reacted with dissolved reduced Mn diffusing upward from the hypolimnion, oxidising it and causing precipitation of Mn oxide on the bacteria. Whether Mn oxide precipitation was biologically mediated remains to be determined, but the Mn oxide-encrusted microbes coexisted with uncoated microbes, suggesting involvement of processes mediated only by certain bacterial species specialising in the oxidation of Mn. The authors concluded that Co was accumulated by the Mn oxide-encrusted bacteria, quite possibly through scavenging of dissolved Co by the Mn oxide coatings. This paper is an excellent example of what can be accomplished by a multi-method approach integrating TEM and STEM-EDXM techniques with other methods, including ordinary methods of chemical analysis.

TEM and STEM-EDXM in conjunction with various other methods of analysis have been used to characterise colloidal Fe-rich mineral particles and incrustations (presumably oxyhydroxides precipitated by reaction of dissolved Fe(II) with dissolved O_2) in the transition zone between oxygenated and anoxic waters in small stratified lakes [36, 68, 77]. J. Buffle *et al.* demonstrated the power of the multi-method approach in a paper describing a complex precipitate in water samples from Lake Bret, Switzerland [36]. Elemental analysis was performed by both EDXM and conventional bulk analysis, both of which methods revealed abundant Fe, Ca, and P. Mössbauer spectroscopy showed that the Fe was a mixture of Fe(II) and Fe(III), and a laser microprobe mass analyser identified the P as phosphate. The authors concluded, reasonably enough, that the mineral colloid consisted of a negatively charged phosphate-bearing Fe oxyhydroxide with Ca^{2+} counter-ions. Another interesting multi-method study was reported by Taillefert et al. [68]. They used high-resolution TEM and STEM-EDXM to show that Fe-rich colloidal mineral particles presumed to be made up of

FeOOH form aggregates with natural organic matter in the waters of Paul Lake (see previous paragraph). On the basis of qualitative EDXM data they found an apparent association of Fe and Pb suggesting (though not proving) that Pb had been scavenged by the presumptive FeOOH. They observed that the Fe-bearing mineral had been precipitated as coatings on organic fibrils, which were associated with bacteria and were probably extracellular products of the bacteria. These fibrils were identified specifically as polysaccharides by TEM following treatment with an Ag-labeled stain, which was accumulated by the fibrils. This paper provides a striking example of the identification of organic colloids by treating them with certain electron-dense stains and then employing TEM to determine their affinity for the stains. D. Perret et al. [77] reported another example of the use of TEM, STEM-EDXM, and Ag-labeled stains (as well as conventional methods of chemical analysis) in a paper presenting evidence for the deposition of FeOOH coatings on high molecular weight organic templates in aquatic environments, specifically (i) extracellular polysaccharides in the waters of Paul Lake and Lake Lugano, Switzerland and (ii) humic substances in water associated with peat land crossed by the Bied River, Switzerland.

In all the publications discussed up to this point, the EDXM results are essentially qualitative. Selected EDXM spectra and corresponding transmission electron micrographs are displayed, and the peaks for the elements of interest are interpreted in terms of their relations to specimen morphology, and to each another, solely on the basis of inspection. Quantitative treatment (specifically, the measurement of element abundances at multiple selected points on the sample sections, and the collection and statistical treatment of sufficiently large numbers of such measurements to be statistically meaningful) has been conspicuous by its absence in the study of both natural and experimental systems. Besides, the STEM-EDXM method and other beam techniques for analysing small selected regions of specimens have been employed in the investigation of no more than a very few natural assemblages of bacteria and particles; and the most advanced techniques have been used only in a small number of relatively recent studies. Thus, our knowledge of interactions between metals, microbes, and colloidal particles on a microscopic scale in natural environments remains very limited. Though some extremely interesting and significant results have been obtained by TEM and STEM-EDXM (especially when combined with other methods), most of the research that has been done in this field is still at a preliminary stage.

The research of T.A. Jackson, M.M. West, and G.G. Leppard marks a new departure - a shift to quantitative STEM-EDXM in the field of environmental biogeochemistry. Thus, a recent publication by Jackson et al. [13] represents what appears to be the first truly quantitative work employing modern STEM-EDXM techniques to investigate interactions of heavy metals and other elements with bacteria and associated colloidal material in a natural assemblage. Surface sediment was collected from the bottom of Larder Lake, a small, well oxygenated circumneutral boreal forest lake in Northern Ontario (Canada) which had been polluted with tailings from a gold mine. The specimens were immersed in a special preservative in the field immediately after sample collection to keep them in their native state; they were subsequently impregnated with epoxy resin in a stepwise process, whereupon ultrathin (~0.08-0.11 μm) sections were cut. EDXM data for selected points on statistically meaningful numbers of bacterial cells (seventy-one altogether) and associated nonliving material were then collected; the exact locations of the spots selected for analysis were marked on electron micrographs of the specimens so that associations between particular elements and recognisable morphological features could be assessed. Moreover, cell exteriors (cell walls and fibrils, together with whatever mineral coatings they possessed) were subjected to quantitative comparison with the interior parts of the cells (the cytoplasm and any electron-dense inclusions within it). Relations of different metallic and nonmetallic elements with each other and with specific microstructures were then quantified by means of plots and statistical analysis. From these data and observations it was possible to draw reasonable inferences about the processes responsible for the observed partitioning and affinities of the elements. Note that the sediments and water of the lake had already been analysed in detail using a variety of conventional methods, and the results of these analyses assisted the interpretation of the EDXM data by providing information about the prevailing environmental conditions.

Examination of sediment sections by TEM revealed many scattered bacterial cells mingled with colloid-sized mineral grains and organic particles (Figures 8a-e). Some cells contained intact cytoplasm (Figures 8a,b,d,e), occasionally with electron-dense inclusions in it (Figure 8e), whereas others were empty owing to death and partial decomposition of the cells (Figure 8c), and certain cells were surrounded by fibrils (Figure 8c). Over 80% of the cells analysed had detectable electron-dense mineral coatings; these coatings varied greatly in gross morphology, suggesting a wide variety of biologically created microenvironments surrounding the cells. The different classes of coating ranged from extremely fine, evenly distributed deposits or impregnations (Figure 8c) to patchy or evenly distributed armour-like layers of relatively coarse platy or lath-like particles sorbed to the cell wall or penetrating it (Figures 8a,b). The coatings generally had abundant Fe, with or without Mn and with or without Si and Al, and were mostly enriched in Fe and (when Mn was present) in Mn with respect to the cytoplasm. Examples of EDXM spectra illustrating the variability of the cell coatings from the standpoint of chemical composition, along with a background spectrum (a baseline with no peaks superimposed on it), are shown in Figures 9a-d. The minerals of the coatings evidently range from aluminosilicates with subordinate Fe to Fe- and Mn-rich material devoid of Al and Si. Note that the baseline is very low, as would be expected for spectra obtained by STEM-EDXM employing ultrathin sections. The authors inferred that the cell coatings consisted of FeOOH and MnOOH deposits precipitated *in situ* and accompanied in many cases by detrital clay minerals sorbed by the cells or clay minerals formed *in situ*. The authors' interpretation of the Fe and Mn data is reasonable, as the bottom water in contact with the sediment contains abundant dissolved O_2, even in the summer, when the lake is thermally stratified; in such an environment precipitation of oxyhydroxides is to be expected.

Measurable amounts of Cu were detected in the exterior and interior parts of many bacterial cells. Amongst cells with mineral coatings one would expect Cu in the cell exterior to correlate positively with the associated Fe and Mn owing to the strong tendency of FeOOH and MnOOH to bind heavy metals. This, indeed, was observed, but only when a distinction was made between cell exteriors with and without detectable quantities of one or more of the metals K, Ca, and Mg (Figures 10a,b). Thus, Cu concentration gave highly significant positive correlations with Fe and Mn concentrations in cell exteriors containing at least one of the elements K, Ca, and Mg; in contrast, Cu in cell exteriors devoid of detectable alkali and alkaline earth metals showed no significant relationship with Fe and Mn. If the authors had not differentiated between the two chemically distinct subgroups of cell exteriors on the basis

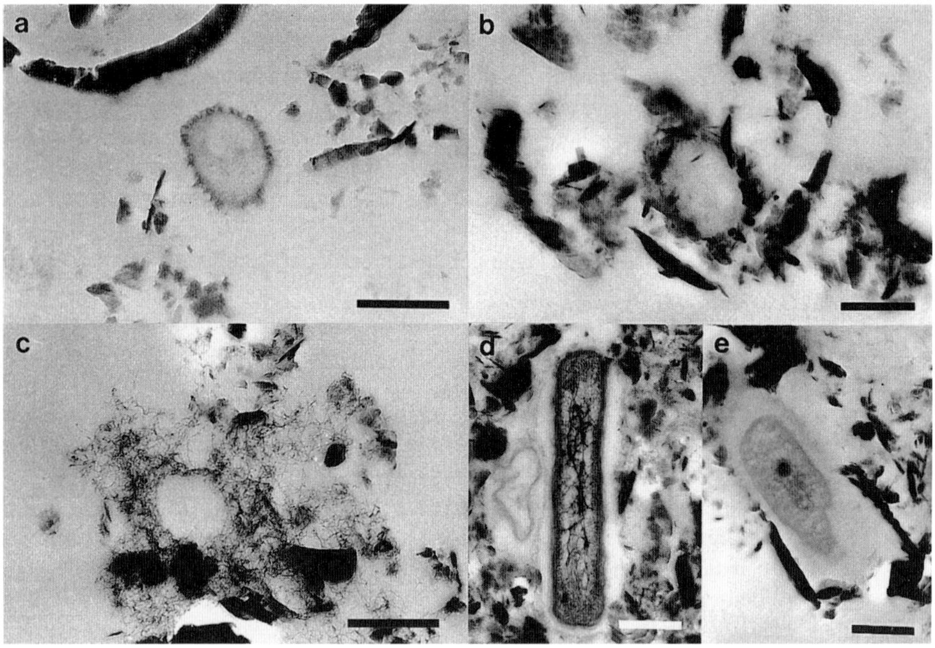

Figure 8. Transmission electron micrographs of bacterial cells in sediment from Larder Lake: (a) cell with coarse Fe-rich coating (probably FeOOH) devoid of detectable Mn, Si, or Al; (b) cell with coarse Si-rich, Al- and Fe-bearing, Mn-free coating (possibly a clay-FeOOH aggregate); (c) cell surrounded by fibrils and a fine Fe-rich coating (probably FeOOH) with some Si and Al (possibly clay) but no Mn; (d) uncoated cell; and (e) uncoated cell containing a cytoplasmic inclusion (probably a polyphosphate granule) enriched in P and Fe with respect to the cytoplasm. Length of each scale bar = 500 nm. (From [13]. Reprinted with the permission of the American Chemical Society).

of the presence or absence of elements other than the ones of immediate interest (Cu, Fe, and Mn), the strong association of Cu with Fe and Mn would not have emerged from the apparent chaos of the scatter diagrams. Further research (including controlled experiments) would be required to explain the unexpected difference between the two different populations of cell exteriors, but the authors advanced a tentative interpretation based on general knowledge of biogeochemical processes. According to their working hypothesis, Cu(II) is strongly bound to FeOOH and MnOOH deposits on the bacterial cell walls and fibrils through surface complexation and coprecipitation, with the formation of partly covalent bonds, but is also sorbed directly, though more weakly (perhaps to a greater extent electrostatically), by cation exchange sites (presumed to be ligands, such as $-COO^-$ and $-PO_4^{2-}$) on the bare biopolymers of the cell walls and fibrils. Cu(II) cations (Cu^{2+}, $CuOH^+$, etc.) are readily displaced from the cell wall exchange sites by the more numerous K^+, Ca^{2+}, and Mg^{2+} ions owing to mass action but are not displaced from the FeOOH and MnOOH. Consequently, the

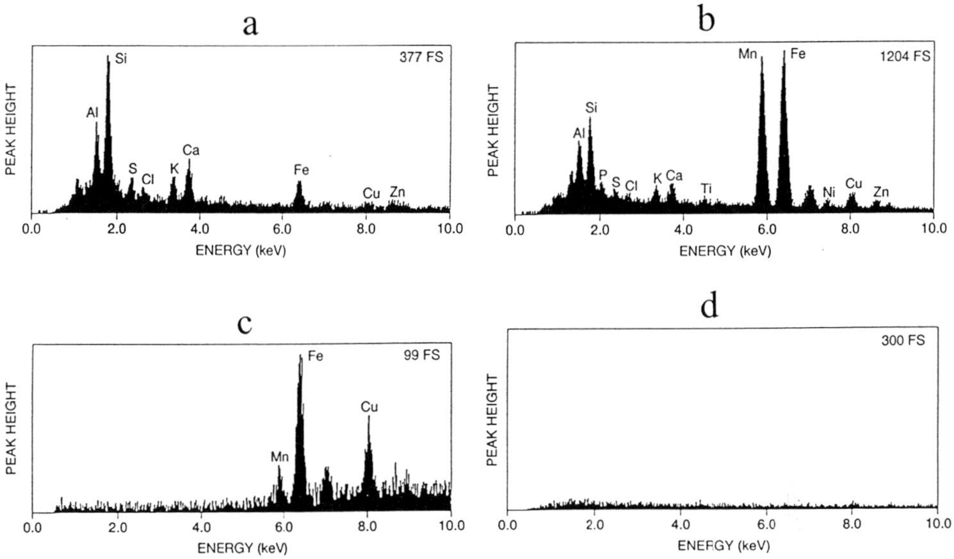

Figure 9. Selected EDXM spectra of mineral coatings on different individually analysed bacterial cells in Larder Lake sediment (a-c) and a sample of a background spectrum generated by analysing the embedding substance (Spurr's epoxy resin) (d). Note the wide variation in the elemental composition of the coatings. (From [13]. Jackson, T.A., West, M.M., Leppard, G.G., previously unpublished spectra).

highly significant correlation of Cu with Fe and Mn is revealed when the cell walls are swept clean of Cu by the alkali and alkaline earth cations, but in the absence of appreciable amounts of these cations the sorption of Cu by large numbers of weak binding sites on the cell wall masks the association of Cu with Fe and Mn. According to a possible alternative interpretation which the authors are contemplating in the light of additional data, the K, Ca, and Mg associated with the cell walls were mostly in the form of clay minerals (e.g. illite and chlorite) rather than free cations. If this hypothesis is correct, clay sorbed to the cell walls, or nucleated there, lowers the binding capacity of bacterial cell walls for Cu(II) ions because it (i) blocks many of the cell-wall ligands that have not already been covered by FeOOH and MnOOH precipitates and (ii) does not sorb Cu strongly except on the edge faces and has a relatively low sorption capacity for Cu(II) [2]. The FeOOH and MnOOH block these ligands too, but, unlike the clay, they have a high sorption capacity for Cu(II), and all their exposed sorption sites form strong, stable bonds with Cu(II). Consequently, when both clay and oxyhydroxides form the cell coatings, the Cu is preferentially accumulated by the oxyhydroxides and therefore correlates positively with Fe and Mn. In the absence of clay, however, sorption of Cu directly by exposed cell wall ligands tends to obscure the Cu-accumulating effect of the oxyhydroxide deposits. Whatever the correct explanation, these results illustrate the importance of gathering a wide range of data, taking ancillary information into account, having minds open to the possible existence of unexpected effects, and striving to achieve a coherent synthesis of the various interrelated pieces of information.

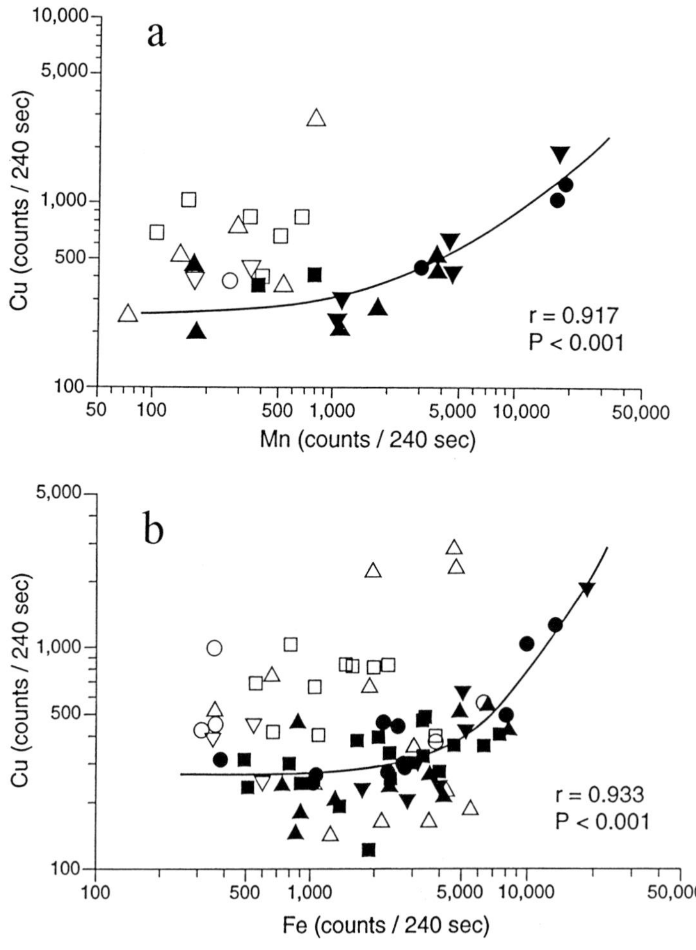

Figure 10. Variation of Cu content with respect to Mn (a) and Fe (b) content in coatings on bacterial cells in Larder Lake sediment. Solid symbols: at least one of the elements Ca, Mg, and K is present; open symbols: detectable Ca, Mg, and K are absent. Shape of symbol signifies type of coating: coarse, evenly distributed (●, O) ; coarse, patchy (■, □); fine, fibrillar (▲, △); or fine, non-fibrillar (▼, ▽). (From [13]. Reprinted with the permission of the American Chemical Society).

Radically different results were obtained for the cell walls defined as "uncoated" (perhaps more accurately described as cell walls with incipient coatings, whose existence was revealed by the presence of detectable Fe but not by visual examination). Among uncoated cell walls, Cu varied *inversely* with respect to Fe, Al, and Si, the significance of the correlation decreasing in the order Al > Si >> Fe (Figures 11a-c). The authors inferred tentatively that Al^{3+} ions complexed by ligands (notably $-COO^-$ and $-PO_4^-$ groups) on the cell wall biopolymers nucleate the precipitation of aluminosilicates (clay minerals) containing Fe (with accompanying deposition of FeOOH as well), and that these deposits interfere with the sorption of Cu(II) species by blocking the binding sites. They also reasoned that this effect is offset to some extent by the tendency of the mineral deposits themselves to sorb Cu(II), and that the seemingly profound difference between the results for coated and uncoated cells is a difference in the *net* effects of the two opposing phenomena rather than a fundamental disparity between the two classes of cells.

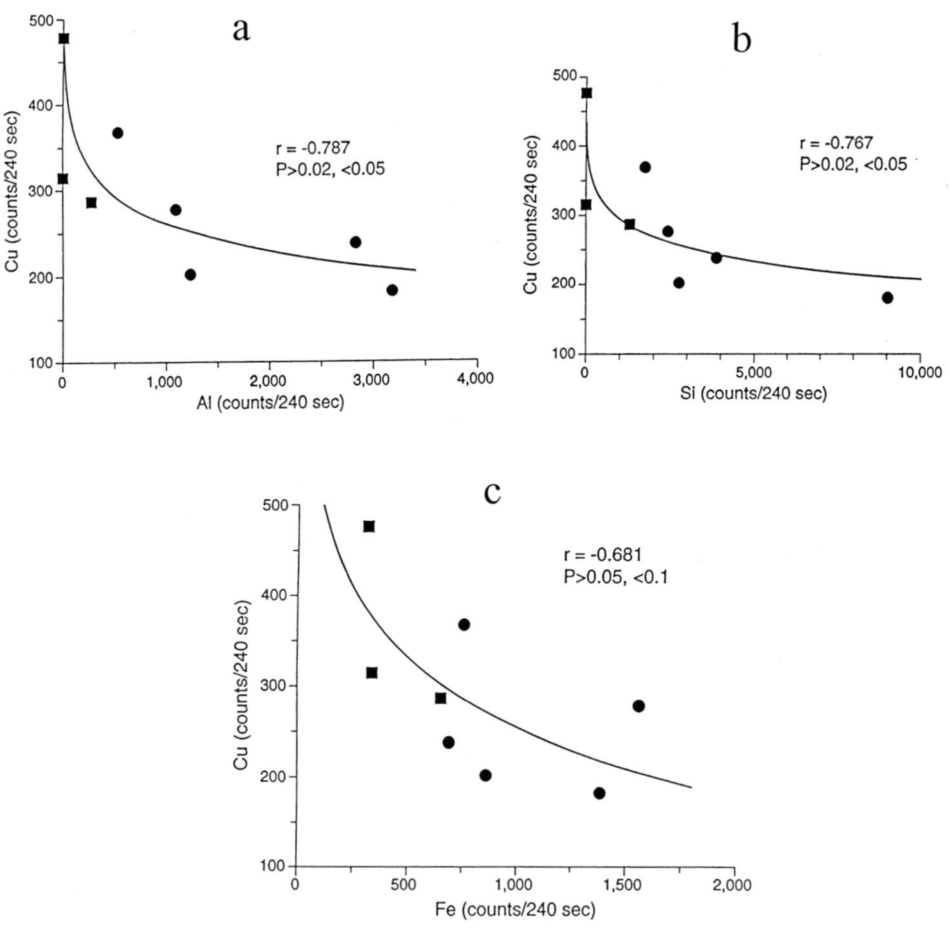

Figure 11. Variation of Cu content with respect to Al (a), Si (b), and Fe (c) in cell walls of "uncoated" bacteria in Larder Lake sediment. Explanation of symbols: ● means at least one of the elements Ca, Mg, and K is present; ■ means detectable Ca, Mg, and K are absent. (From [13]. Reprinted with the permission of the American Chemical Society)

Finally, a comparison of EDXM data for the exterior and interior parts of each cell in a series of coated bacterial cells containing intact cytoplasm yielded information on the partitioning of Cu in the cell. Some cells, as expected, were found to have higher concentrations of both Cu and Fe in the exterior of the cell than in the cytoplasm (Figure 12a), implying that the scavenging of Cu(II) by FeOOH coatings to some extent impedes the movement of Cu into the interior of the cell. In other cells, however, which contained cytoplasmic inclusions the Cu was mainly concentrated in the inclusions.

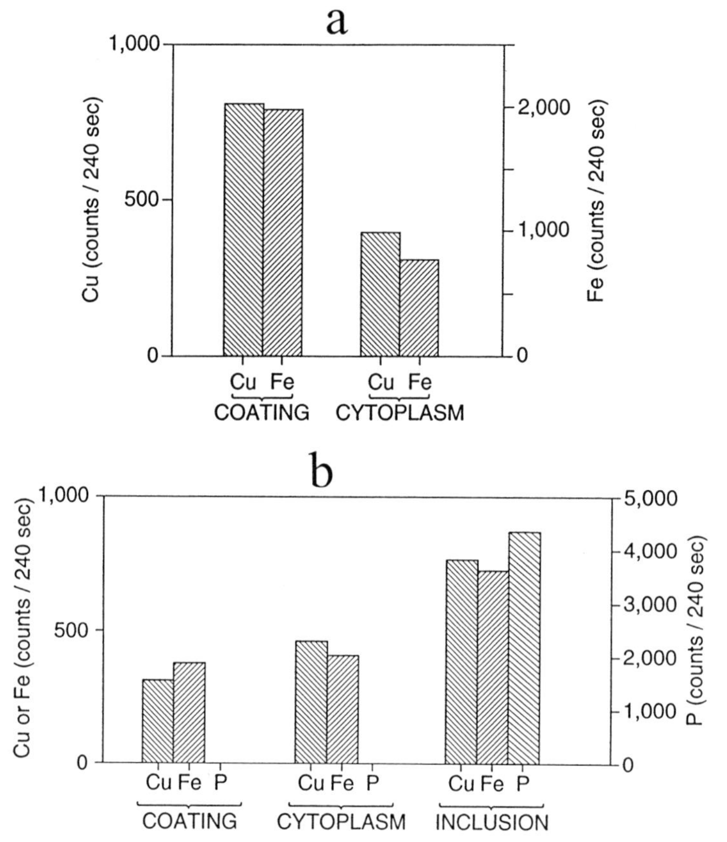

Figure 12. The abundances of Cu, Fe, and P in different components of individual bacterial cells in Larder Lake sediment: (a) cell with coarse, patchy coating (probably FeOOH) enriched in Cu and Fe with respect to the cytoplasm and (b) cell with coarse, evenly distributed coating whose cytoplasm contains an inclusion (probably polyphosphate) enriched in Cu, Fe, and, above all, P with respect to the cytoplasm and coating (neither of which contains detectable P). Note that in Figure 12b the order of decreasing Cu and Fe is inclusion > cytoplasm > coating. (From [13]. Reprinted with the permission of the American Chemical Society).

In a particularly striking example of this second type of cell the concentrations of Cu and Fe increased in the order coating < cytoplasm < inclusion, and P was abundant in the inclusion but undetectable in the coating and cytoplasm (Figure 12b). Evidently this bacterium had a strong tendency to take up Cu and Fe and concentrate them in polyphosphate inclusions in the cytoplasm. Both of these two entirely different mechanisms of bioaccumulation may represent strategies for detoxifying heavy metals like Cu in metal-polluted environments (or, in unpolluted environments, for concentrating nutrient trace elements, including Cu, from dilute solution).

A more recent, though less detailed, quantitative treatment of STEM-EDXM data was published by S.M. Webb, G.G. Leppard, and J.-F. Gaillard [34]. The authors used STEM-EDXM to investigate Zn concentrations and the relations of Zn with other elements in a variety of individually analysed, morphologically defined colloidal particles and microorganisms in samples of suspended matter and sediment from Lake DePue, Illinois (U.S.A.) and a stream that flows into it. The lake-stream system is contaminated with industrial wastes which had been discharged into the stream from several sources, including a zinc-smelting plant. Among other things, the authors observed definite trends in element associations in the sediments going from more polluted to less polluted localities. Thus, they found that in the vicinity of the sources of pollution (in the stream) Zn was commonly associated with Fe and P in colloidal particles which were, in large part, concentrated on the surfaces of bacteria and single-celled algae; from this they inferred that the surfaces of the microbial cells nucleate precipitation of the Fe- and P-rich deposits accompanied by coprecipitation of Zn. In contrast, Zn in particles collected relatively far from the sources of pollution (at the mouth of the stream and in the lake proper) was mainly associated with S, suggesting postdepositional formation of ZnS. This spatial variation in the element associations of Zn appears to be due to a shift in the sedimentary environment from strongly oxidising in the stream near the sources of pollution to anoxic or reducing at the mouth of the stream and in the open lake, not to the systematic decline in the severity of contamination going from the stream to the lake.

An important advantage of the design of the research carried out by Webb et al. is that it provided for collection of samples along an environmental gradient. There is a need for quantitative *comparative* studies such as this in environmental research employing TEM and STEM-EDXM.

4.2. Some biogeochemical applications of SEM-EDXM

Although conventional SEM-EDXM is a much cruder method than STEM-EDXM (being, in effect, bulk analysis on a microscopic scale owing to extensive beam spreading), applications of SEM-EDXM to natural materials have yielded some interesting and useful quantitative data.

Jackson and Keller [72, 78, 79] reported the results of a thorough, multi-method investigation of the role of the lichen *Stereocaulon vulcani* in the chemical weathering of basaltic lava flows on the Island of Hawaii (recording, in the process, Jackson's discovery, description, detailed physicochemical and mineralogical analysis, and biogeochemical interpretation of the previously unknown biogenic hydrous ferric oxide polymorph which, on being rediscovered elsewhere independently by other workers, was officially recognised as a new mineral named "ferrihydrite"). Weathering crusts formed in the presence and absence of the lichen, along with specimens of unweathered rock and lichen tissue, were characterised and compared using a wide range of techniques, including SEM-EDXM ("electron microprobe"), X-ray diffraction analysis (before and after heating at 1,000^{O}C), differential thermal analysis, Mössbauer spectroscopy, electron diffraction analysis, measurement of magnetic susceptibility, conventional bulk chemical analysis, microscopic examination of polished sections of weathered rock embedded in resin, and measurement of weathering crust thickness [78]. Only the weathering crusts were analysed by SEM-EDXM; the elemental analyses of the rock and lichen samples were done by conventional methods.

The research revealed the occurrence of conspicuous reddish weathering crusts associated exclusively with the lichens, and unquestionably produced by the lichens, in regions of relatively high rainfall; these "biogenic" weathering crusts were composed mainly of ferrihydrite and contained no detectable clay minerals. In contrast, the "abiotic" weathering crusts in adjacent lichen-free areas contained hematite but no detectable ferrihydrite, and clay minerals were present, implying less intense leaching of Si. Moreover, the biogenic crusts were orders of magnitude thicker than the abiotic crusts, which, at the most, consisted of extremely thin, faint reddish patinas on the rock surfaces. EDXM data from replicate spot analyses of weathering crusts in lichen-covered and lichen-free areas of two separate lava flows erupted in 1907 and 1926 showed that the ferrihydrite-rich biogenic crusts were highly enriched in Fe and correspondingly depleted in Si with respect to the abiotic crusts (although the two kinds of crust did not differ significantly in Al content) [78]. Unlike the biogenic crusts, the abiotic crusts were similar in composition to the unweathered rock. The striking difference in elemental composition between the biogenic weathering crusts and the abiotic weathering crusts and fresh rock, together with the similarities in composition between the abiotic crusts and the fresh rock, is vividly displayed in a triangular plot showing the proportions of Fe, Si, and Al in the weathering crusts, rocks, and lichen thalli [78].

From their various observations the authors concluded that the lichen enormously intensifies and accelerates chemical weathering and, through its biochemical processes, creates a distinctive microenvironment that predetermines the mineralogy and chemical composition of the weathering products. Thus, ferrihydrite is formed only in the immediate vicinity of the lichen, and biologically mediated chemical weathering, accompanied by the leaching of Si, is so intense that no clay minerals are formed in the presence of the lichen; thus, the biogenic weathering crust is composed exclusively of ferrihydrite and lesser amounts of other residual oxides. This study demonstrates the value of quantitative EDXM and the importance of using EDXM in conjunction with various complementary techniques.

Quantitative EDXM in conjunction with SEM was also used by R.L. Todd, K. Cromack, and J.C. Stormer, Jr. in a pilot study to compare the *in situ* concentrations of major inorganic nutrients (Ca, Mg, and K) in different microorganisms and plant and animal remains in microscopic domains within a sample of forest soil in its native state [80]. The sample was collected from the field site of the Coweeta Hydrologic Laboratory (U.S. Forest Service), North Carolina (U.S.A.). Mean values of replicate EDXM data obtained by spot analyses of bacteria, fungal hyphae, insect wing scales, and decomposing fallen leaves of trees showed marked differences as well as similarities in composition. For instance, the abundance of Ca was twice as high in a fungal hypha growing on an oak leaf as in the leaf tissue, and a pine needle, fungal hypha, and insect wing scale closely associated with each other differed widely in Ca content. This paper is preliminary and makes only brief, cursory mention of relations between different elements (a crucially important subject deserving fuller treatment, since the concentrations of a single element in altogether different kinds of microscopic entities cannot be reliably compared owing to differences in properties such as specimen thickness and density, whereas this limitation does not apply to element ratios and correlations between element abundances). Besides, a figure in the paper, the figure caption, and an accompanying table of EDXM data are marred by confusing errors, as revealed by discrepancies between letters used to designate some of the points on the sample preparation where analyses were done. Nevertheless, the paper serves the authors' primary purpose, which was to demonstrate the unique potential value of EDXM in research on the distribution of biologically important elements in soil on a microscopic scale under natural conditions - a subject of great interest in soil ecology which, as the authors rightly emphasised, is beyond the capability of conventional bulk analysis.

Finally, quantitative SEM-EDXM techniques, together with various other methods of analysis were employed by T.A. Jackson and T. Bistricki [57] to investigate the scavenging of heavy metals by plankton in the waters of three small Canadian Shield lakes (Schist Lake, Hamell Lake, and West Nesootao Lake) which have been polluted with metals (notably Cu and Zn) from a base-metal mining and smelting complex at Flin Flon, Manitoba (Canada). The biological entities analysed were mostly the remains of crustacean exoskeletons (Figure 13a) but included loricae of the euglenophytes *Trachelomonas* (Figure 13b) and *Strombomonas* (Figure 13c) and loricae of the rotifer *Keratella* (Figure 13d). Statistically meaningful numbers of individual specimens were analysed by EDXM, and relations between different metallic and nonmetallic elements were determined by regression analysis. Examples of EDXM spectra representing sample material (a *Strombomonas* lorica) and the background are shown in Figures 14a and 14b, respectively. Note that the baseline, as would be expected in SEM-EDXM spectra, is much higher than the typical baseline of a STEM-EDXM spectrum obtained from an ultrathin section (Figure 9).

The EDXM data for remains of crustacean exoskeletons revealed a strong positive correlation between Cu and Fe but a much weaker correlation between Cu and Mn, whereas, in contrast, Zn had a strong affinity for Mn but gave a relatively weak (and more complex) relationship with Fe (Figures 15a-d). The data for all three lakes, moreover, conformed to the same trends. The authors' interpretation of these results was that Cu and Zn ions in the water were scavenged by FeOOH and MnOOH coatings on the exoskeletons through sorption and coprecipitation, Cu being preferentially bound by the FeOOH whilst Zn was preferentially bound by the MnOOH. The data for *Keratella* and *Strombomonas* also yielded evidence for the binding of Cu and Zn by FeOOH and MnOOH (not shown). Owing to discrimination between Cu and Zn by the presumptive FeOOH and MnOOH phases of the mineral coatings, the Zn/Cu ratios of the crustacean exoskeletons varied correspondingly as functions of Mn and Fe (Figures 16a, b). The peculiar complex relationship between Zn and Fe (Figure 15b), and hence between Zn/Cu and Fe (Figure 16b), was viewed as a secondary manifestation reflecting the variation of Mn with respect to Fe (Figure 17), which, in turn, was found to be controlled by the oxidation-reduction potential of the sediment, and hence by factors such as free sulfide production. These findings were regarded as implying that oxyhydroxide coatings on exposed hard parts of plankton not only scavenge Cu and Zn (removing them from the solution phase of the water column and conveying them to bottom on settling out) but also affect the proportions of Cu and Zn in the water and sediment owing to their apparent ability to discriminate between the two metals.

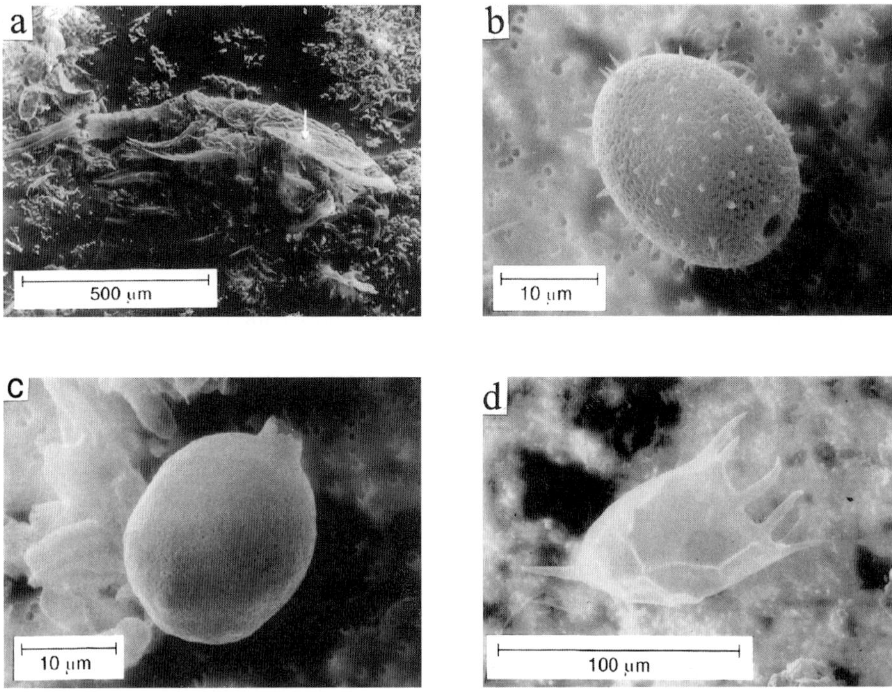

Figure 13. Scanning electron micrographs of an unidentified crustacean (a) and loricae of the euglenophytes *Trachelomonas* (b) and *Strombomonas* (c) and the rotifer *Keratella* (d) in plankton from Schist Lake (a, d) and Hamell Lake (b, c). (From [57]. Reprinted with the permission of Elsevier Science).

Another interesting result was that the lake whose sediments were poorest in free sulfide was the only one in which the Zn bound to crustacean exoskeletons was associated to a significant degree with S (57). The relative affinities between Zn and S, and between Cu and S, in each of the three lakes were expressed as correlation coefficients computed by regression analysis; the correlation coefficients, in turn, were found to correlate *inversely* with the free sulfide content of the bottom sediment. These unexpected observations were tentatively interpreted as meaning that organisms in sulfide-poor environments protect themselves from toxic effects of heavy metals in the water by releasing thiol compounds that complex the metals, making them less bioavailable, whereas in environments where there is abundant ambient sulfide from sources in the sediments there is less adaptive pressure to develop this detoxifying mechanism, because the sulfide suppresses biological uptake of the metals by immobilising them.

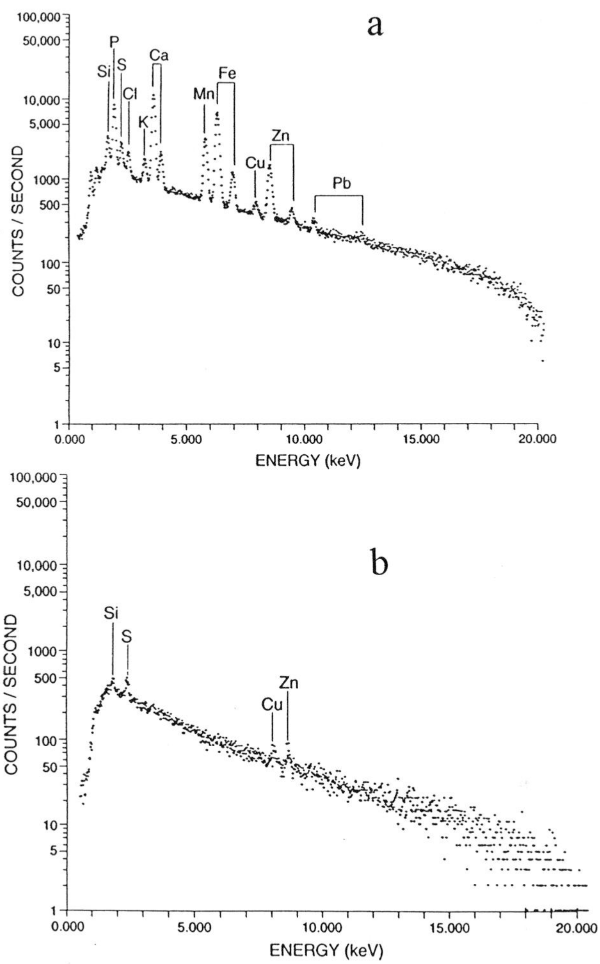

Figure 14. SEM-EDXM spectra of (a) a *Strombomonas* lorica in a plankton sample from Hamell Lake and (b) associated background material. (From [57]. Reprinted with the permission of Elsevier Science).

This paper demonstrates the advantages of quantitative EDXM applied to relatively large numbers of visually identifiable microscopic entities of different kinds followed by statistical treatment of the results (in particular, regression analysis to measure the degree of association of different elements with one another). It also illustrates the usefulness of comparing samples from multiple field sites differing from each other in environmental conditions, comparing different heavy metals, and integrating EDXM data with relevant data generated by various other methods. Of course, the paper has certain limitations too. Ideally, TEM would have been preferable to SEM, and the results would have been all the more informative if it had been feasible to apply additional advanced techniques to complement the EDXM work (for instance, electron diffraction and Mössbauer spectroscopy, which might have made it possible to verify the presence of oxyhydroxides and to characterise and, perhaps, identify them); it would also have been advantageous to analyse, if possible, a much wider range of planktonic organisms (along with associated nonliving particles), to compare materials from a larger number of different aquatic environments (and from the same sites at different seasons), to compare a larger number of heavy metals so that differences in their behaviour in the environment could be linked by regression analysis to metal properties such as electronegativity, ionisation potentials, etc., and to acquire much more biological information (such as biochemical data to test the hypothesis that certain organisms secrete thiol compounds for the purpose of detoxifying heavy metals in sulfide-poor environments). However, with the data that they were able to amass under the circumstances, the authors were able to draw reasonable inferences about the biogeochemical processes controlling the interactions of heavy metals with individual components of the plankton, and to address the research topic more comprehensively and in far greater depth than would have been possible without EDXM and with a more limited range of data acquired by other means.

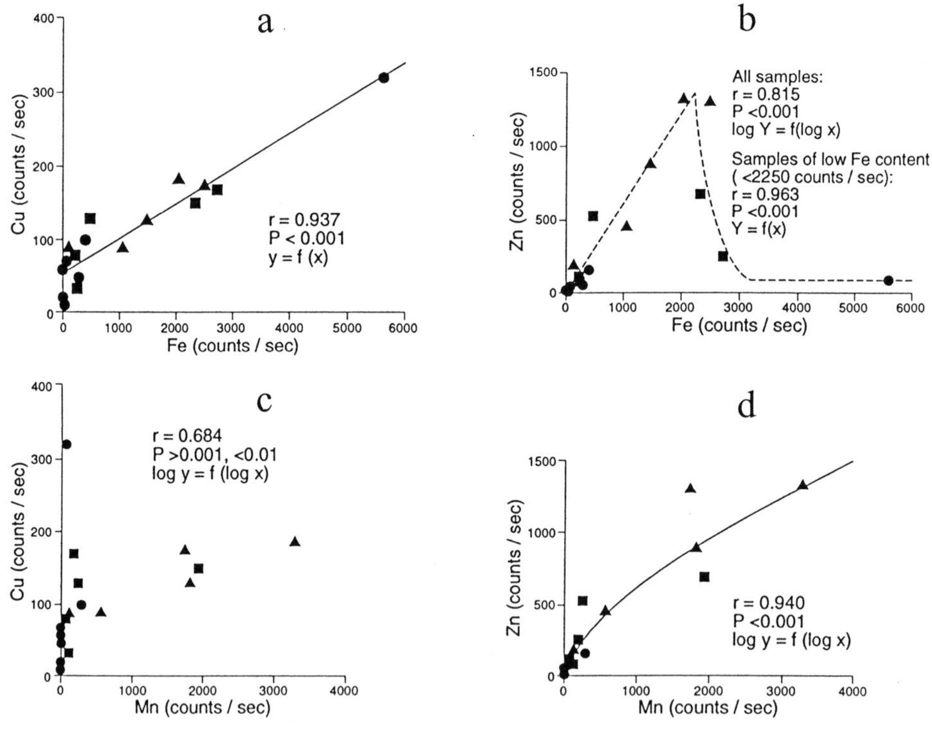

Figure 15. Plots of SEM-EDXM data showing relationships between Cu and Fe (a), Zn and Fe (b), Cu and Mn (c), and Zn and Mn (d) in remains of crustacean exoskeletons in plankton samples from Schist Lake (●), Hamell Lake (■), and West Nesootao Lake (▲). (From [57]. Reprinted with the permission of Elsevier Science).

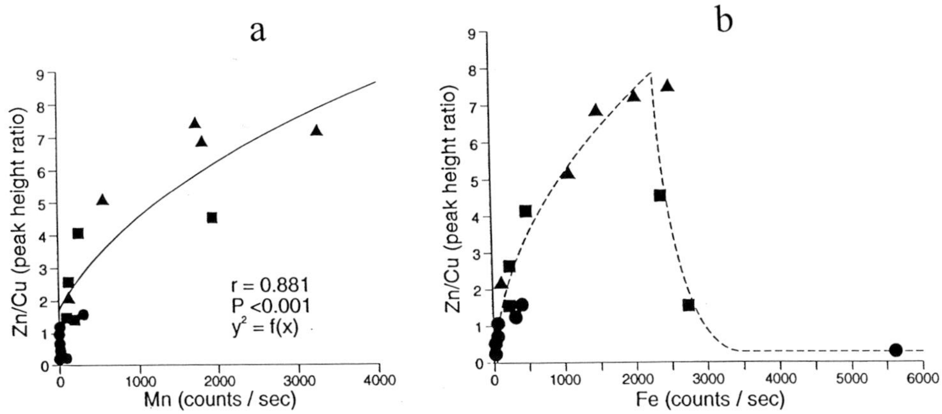

Figure 16. Plots of SEM-EDXM data showing variations of the Zn/Cu ratio with respect to Mn (a) and Fe (b) in crustacean exoskeletons in plankton samples from Schist Lake (●), Hamell Lake (■), and West Nesootao Lake (▲). (From [57]. Reprinted with the permission of Elsevier Science).

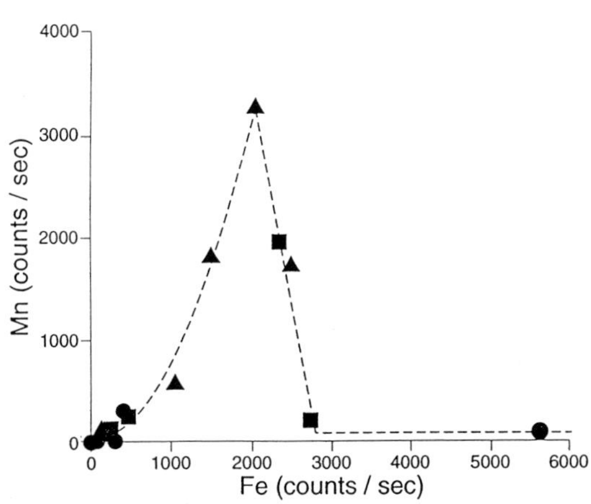

Figure 17. Plot of SEM-EDXM data showing the relationship between Mn and Fe in crustacean exoskeletons in plankton samples from Schist Lake (●), Hamell Lake (■), and West Nesootao Lake (▲). (From [57]. Reprinted with the permission of Elsevier Science).

5. CONCLUSIONS

To summarise briefly, STEM-EDXM and SEM-EDXM (and the closely related techniques WDXM and EELS) offer the unique advantage of permitting specific microscopic components of heterogeneous natural materials (such as soils, sediments, and plankton) preserved *in situ* in their native state to be selected visually and then analysed individually (and almost instantaneously) for a wide range of metallic and nonmetallic elements. The best results are obtained by spot analyses employing the most refined and advanced STEM-EDXM techniques applied to ultrathin sections, as these methods can be used to pinpoint and analyse individual domains much less than a micrometre in diameter within the sample material. With this technique specific subunits of individual microscopic entities, such as small regions of the cell walls and cell contents of single bacterial cells, can be analysed separately. Improved modern SEM-EDXM techniques are also satisfactory, and some useful results have been obtained even with the relatively crude earlier techniques of SEM-EDXM (which could be described as bulk analysis on a microscopic scale). In the STEM-EDXM technique, a single electron microscope has the capability of providing both TEM images and an SEM probe; it is ideal for the analysis of ultrathin sections. In any case, techniques such as STEM-EDXM and SEM-EDXM have the potential to revolutionise biogeochemistry, microbial ecology and other fields, as they are able to provide a wealth of important information on phenomena occurring on a microscopic scale - information that would be impossible to obtain with conventional methods of bulk analysis or, indeed, with any other technique, however powerful and sophisticated.

It must be kept clearly in mind, however, that EDXM, even at its best, has certain limitations. One of the principal limitations is that in order to be detectable, elements must be present in fairly high concentrations within the region of the sample selected for analysis (the detection limit being on the order of 1 part per 1,000 in the irradiated volume of the specimen). Another limiting factor is that it is generally (though not invariably) impossible to determine elements of atomic weight < 11 (the atomic weight of Na) by EDXM. A third consideration is that in EDXM spectra peaks of two or more elements may, in some cases, overlap or coincide (as with the principal peaks of Hg and S). A fourth inherent limitation of EDXM is that only total element abundance can be determined by this method; the forms of the elements are not revealed. Other limiting factors include the thickness of the electron beam, beam spreading (which is minimised but not eliminated when ultrathin sections are used), and misleading results (artifacts, distortions, and selective losses of sample constituents) arising from deficiencies of certain sample preparation techniques. It could also be mentioned that from the perspective of most environmental scientists TEM and STEM-EDXM are very expensive in terms of both capital and operating costs. Furthermore, they are technically very demanding and specialised; thus, mastery of the art requires years of training and experience.

Another point that must be stressed is the advisability of combining EDXM with an assortment of other techniques (ranging from other specialised, sophisticated instrumental methods to classical methods of bulk analysis) which can provide important complementary information. An interdisciplinary multi-method approach, with a synthesis of all the different but related lines of evidence obtained thereby, is bound to make possible a much deeper and more comprehensive understanding of the subject matter than reliance on a single technique such as EDXM, however powerful it may be. For instance, EDXM may reveal that a mineral coating on a certain bacterial cell is rich in Fe, and, on the basis of electron diffraction performed on the same deposit, the material, if sufficiently crystalline, may be identified as a specific polymorph of Fe oxide or oxyhydroxide. Moreover, if electron diffraction revealed that different kinds of coexisting bacteria were coated with different polymorphs of Fe oxide or oxyhydroxide (e.g. ferrihydrite in some cases and goethite in others), this would constitute strong evidence for the existence of specific biologically created templates or microenvironments that predetermine the nature of the Fe-bearing mineral precipitated on the cell. To take another example, EDXM data showing a close association between Zn and S in electron-dense deposits on a bacterial cell in lake sediment may reasonably be interpreted as implying that the deposit consists of zinc sulfide (ZnS) if other kinds of data, such as the Eh and sulfide content of the sediment, have established that the sedimentary environment is highly reducing and produces copious amounts of free sulfide; identification of the sulfide mineral as sphalerite (one of the polymorphs of ZnS) by electron diffraction "fingerprinting" would clinch the argument. Thus, from EDXM data alone one can at least draw reasonable inferences about the biogeochemical processes responsible for the observed distribution of elements, but supplementary data obtained by other methods are required to enable one to progress from plausible working hypotheses to more or less firmly established conclusions.

As we have seen, important pioneer work has been done using essentially qualitative EDXM, but to realise the full potential of the EDXM technique a quantitative approach is needed. In quantitative EDXM the abundance (i.e. *relative* concentration) of each detectable element is determined by measuring the size of the appropriate peak in the EDXM spectrum (preferably the area under the peak or the number of counts per second), and statistically meaningful numbers of spot analyses are done. More specifically, analyses of large numbers of microscopic entities or replicate analyses of individual entities, or both, are performed. By this means, associations of elements with each other, as well as the partitioning of the elements amongst different visually recognisable microstructures, can be properly assessed and treated statistically (using procedures such as regression analysis).

Finally, our ability to use EDXM and complementary analytical techniques to maximum advantage in biogeochemical research depends not only on employment of superior, up-to-date methods and instruments, together with the required technical expertise, but also on the design of our research projects, the breadth and depth of our knowledge and insight, and the scope of our originality and imagination. In future research along these lines the greatest amount of useful information on the partitioning of elements on a microscopic scale, on element associations, and on the biogeochemical processes responsible for these phenomena will probably be obtained by employing quantitative EDXM and other methods to (i) compare many different kinds of coexisting microscopic entities in a given sample, (ii) compare samples from a number of different environments (e.g. lakes with contrasting physicochemical conditions, or different points along a pollution gradient in a river), (iii) compare samples taken from a single field site at different times (e.g. a lake sampled in different seasons or a polluted lake sampled before and after implementation of a pollution abatement scheme), and [iv] compare samples in their native state collected from natural environments with sample material that has been altered experimentally, either in the field or in the laboratory. In general, we strongly recommend in-depth interdisciplinary research employing many different methods, including EDXM, with interpretations based on synthesis of widely differing but interrelated lines of evidence.

ACKNOWLEDGMENTS

We wish to express our gratitude to T.J. Beveridge (University of Guelph) and F.G. Ferris (University of Toronto) for their kindness in providing previously published electron micrographs for inclusion in this paper. We also thank the publishing companies that gave us permission to use these and other illustrations taken from the literature. M.M. West (McMaster University) assisted in the preparation of the illustrations, and our work was carried out with the support of the Department of the Environment, Government of Canada.

REFERENCES

1. Stolp, H., 1988. Microbial Ecology. Cambridge University Press, Cambridge.
2. Jackson, T.A., 1998. The biogeochemical and ecological significance of interactions between colloidal minerals and trace elements. *In:* Parker, A., Rae, J.E. (Eds.), Environmental Interactions of Clays. Springer, Berlin, pp. 93-205.
3. Ferris, F.S., Fyfe, W.S., Beveridge, T.J., 1987. Bacteria as nucleation sites for authigenic minerals in a metal-contaminated lake sediment. Chem. Geol. 63, 225-232.
4. Beveridge, T.J., 1989. Metal ions and bacteria. In: Beveridge, T.J., Doyle, R.J. (Eds.), Metal Ions and Bacteria. John Wiley & Sons, New York, pp. 1-29.
5. Beveridge, T.J., 1989. Role of cellular design in bacterial metal accumulation and mineralization. Ann. Rev. Microbiol. 43, 147-171.
6. Doyle, R.J., 1989. How cell walls of Gram-positive bacteria interact with metal ions. *In:* Beveridge, T.J., Doyle, R.J. (Eds.), Metal Ions and Bacteria. John Wiley & Sons, New York, pp. 275-293.
7. Ferris, F.G., 1989. Metallic ion interactions with the outer membrane of Gram-negative bacteria. *In:* Beveridge, T.J., Doyle, R.J. (Eds.), Metal Ions and Bacteria. John Wiley & Sons, New York, pp. 295-323.
8. Ferris, F.G., Shotyk, W., Fyfe, W.S., 1989. Mineral formation and decomposition by microorganisms. *In:* Beveridge, T.J., Doyle, R.J. (Eds.), Metal Ions and Bacteria. John Wiley & Sons, New York, pp. 413-441.
9. Geesey, G.G., Jang, L., 1989. Interactions between metal ions and capsular polymers. *In:* Beveridge, T.J., Doyle, R.J. (Eds.), Metal Ions and Bacteria. John Wiley & Sons, New York, pp. 325-357.
10. Kushner, D.J., 1993. Effects of speciation of toxic metals on their biological activity. Water Pollut. Res. J. Canada 28, 111-128.
11. Schultze-Lam, S., Thompson, J.B., Beveridge, T.J., 1993. Metal ion immobilization by bacterial surfaces in freshwater environments. Water Pollut. Res. J. Canada 28, 51-81.
12. Silver, S., Misra, T.K., Laddaga, R.A., 1989. Bacterial resistance to toxic heavy metals. *In:* Beveridge, T.J., Doyle, R.J. (Eds.), Metal Ions and Bacteria. John Wiley & Sons, New York, pp. 121-139.
13. Jackson, T.A., West, M.M., Leppard, G.G., 1999. Accumulation of heavy metals by individually analyzed bacterial cells and associated nonliving material in polluted lake sediments. Environ. Sci. Technol. 33, 3795-3801.
14. Roane, T.M., 1999. Lead resistance in two bacterial isolates from heavy metal-contaminated soils. Microb. Ecol. 37, 218-224.

15. Madsen, E.L., 1998. Epistemology of environmental microbiology. Environ. Sci. Technol. 32, 429-439.
16. Tessier, A., Fortin, D., Belzile, N., De Vitre, R.R., Leppard, G.G., 1996. Metal sorption to diagenetic iron and manganese oxyhydroxides and associated organic matter: narrowing the gap between field and laboratory measurements. Geochim. Cosmochim. Acta 60, 387-404.
17. Warren, L.A., Ferris, F.G., 1998. Continuum between sorption and precipitation of Fe(III) on microbial surfaces. Environ. Sci. Technol. 32, 2331-2337.
18. Wen, X., Du, Q, Tang, H., 1998. Surface complexation model for the heavy metal adsorption on natural sediment. Environ. Sci. Technol. 32, 870-875.
19. Chandler, J.A., 1977. X-ray Microanalysis in the Electron Microscope. *In:* Glauert, A.M. (Ed.), Practical Methods in Electron Microscopy Series. North-Holland, Amsterdam.
20. Heldal, M., Norland, S., Tumyr, O., 1985. X-ray microanalytic method for measurement of dry matter and elemental content of individual bacteria. Appl. Environ. Microbiol. 50, 1251-1257.
21. Ferris, F.G., Beveridge, T.J., Fyfe, W.S., 1986. Iron-silica crystallite nucleation by bacteria in a geothermal sediment. Nature 320, 609-611.
22. Mayers, I.T., Beveridge, T.J., 1989. The sorption of metals to *Bacillus subtilis* walls from dilute solutions and simulated Hamilton Harbour (Lake Ontario) water. Can. J. Microbiol. 35, 764-770.
23. Leppard, G.G., Heissenberger, A., Herndl, G.J., 1996. Ultrastructure of marine snow. I. Transmission electron microscopy methodology. Mar. Ecol. Prog. Ser. 135, 289-298.
24. Fortin, D., Leppard, G.G., Tessier, A., 1993. Characteristics of lacustrine diagenetic iron oxyhydroxides. Geochim. Cosmochim. Acta 57, 4391-4404.
25. Egerton, R.F., 1996. Electron Energy-Loss Spectroscopy in the Electron Microscope. 2nd Ed. Plenum, New York.
26. Mavrocordatos, D., Lienemann, C.-P., Perret, D., 1994. Energy filtered transmission electron microscopy for the physico-chemical characterization of aquatic submicron colloids. Mikrochim. Acta 117, 39-47.
27. Mavrocordatos, D., Mondi-Couture, C., Atteia, O., Leppard, G.G., Perret, D.J., 2000. Formation of a distinct class of Fe-Ca (-C_{org})-rich particles in a complex peat-karst system. J. Hydrol. 237, 234-247.
28. Leppard, G.G., 1992. Evaluation of electron microscope techniques for the description of aquatic colloids. *In:* Buffle, J., van Leeuwen, H.P. (Eds.), Environmental Particles. Vol. 1. Lewis, Boca Raton, pp. 231-289.
29. Frosch, D., Westphal, C., 1989. Melamine resins and their application in electron microscopy. Electron Microsc. Rev. 2, 231-255.
30. Leppard, G.G., Buffle, J., 1998. Aquatic colloids and macromolecules: effects on analysis. *In:* Meyers, R.A. (Ed.), Encyclopedia of Environmental Analysis and Remediation. Vol. 1. John Wiley & Sons, New York, pp. 349-377.
31. Perret, D., Leppard, G.G., Muller, M., Belzile, N., DeVitre, R., Buffle, J., 1991. Electron microscopy of aquatic colloids: non-perturbing preparation of specimens in the field. Water Res. 25, 1333-1343.
32. Liss, S.N., Droppo, I.G., Flannigan, D.T., Leppard, G.G., 1996. Floc architecture in wastewater and natural riverine systems. Environ. Sci. Technol. 30, 680-686.

33. Leppard, G.G., 1992. Size, morphology and composition of particulates in aquatic ecosystems: solving speciation problems by correlative electron microscopy. Analyst 117, 595-603.
34. Webb, S.M., Leppard, G.G., Gaillard, J.-F., 2000. Zinc speciation in a contaminated aquatic environment: characterization of environmental particles by analytical electron microscopy. Environ. Sci. Technol. 34, 1926-1933.
35. Lienemann, C.-P., Heissenberger, A., Leppard, G.G., Perret, D., 1998. Optimal preparation of water samples for the examination of colloidal material by transmission electron microscopy. Aquat. Microb. Ecol. 14, 205-213.
36. Buffle, J., DeVitre, R.R., Perret, D., Leppard, G.G., 1989. Physico-chemical characteristics of a colloidal iron phosphate species formed at the oxic-anoxic interface of a eutrophic lake. Geochim. Cosmochim. Acta 53, 399-408.
37. Lienemann, C.-P., Taillefert, M., Perret, D., Gaillard, J.-F., 1997. Association of cobalt and manganese in aquatic systems: chemical and microscopic evidence. Geochim. Cosmochim. Acta 61, 1437-1446.
38. Buffle, J., Leppard, G.G., 1995. Characterization of aquatic colloids and macromolecules. 2. Key role of physical structures on analytical results. Environ. Sci. Technol. 29, 2176-2184.
39. Garvie, L.A.J., Buseck, P.R., 1998. Ratios of ferrous to ferric iron from nanometre-sized areas in minerals. Nature 396, 667-670.
40. Mavrocordatos, D., 1997. Développement de la Microscopie Electronique Analytique pour la Caractérisation Physico-Chimique de Colloïdes Aquatiques. Ph. D. Dissertation, University of Lausanne, Switzerland.
41. Couture, C., Lienemann, C.-P., Mavrocordatos, D., Perret, D., 1996. New directions towards the understanding of physico-chemical processes in aquatic systems. Chimia 50, 625-629.
42. Brierley, C.L., Brierley, J.A., Davidson, M.S., 1989. Applied microbial processes for metals recovery and removal from wastewater. *In:* Beveridge, T.J., Doyle R.J. (Eds.), Metal Ions and Bacteria. John Wiley & Sons, New York., pp. 359-382.
43. Beveridge, T.J., Murray, R.G.E., 1976. Uptake and retention of metals by cell walls of *Bacillus subtilis*. J. Bacteriol. 127, 1502-1518.
44. Mittelman, M.W., Geesey, G.G., 1985. Copper-binding characteristics of exopolymers from a freshwater-sediment bacterium. Appl. Environ. Microbiol. 49, 846-851.
45. Ferris, F.G., Beveridge, T.J., 1986. Site specificity of metallic ion binding in *Escherichia coli* K-12 lipopolysaccharide. Can. J. Microbiol. 32, 52-55.
46. Mullen, M.D., Wolf, D.C., Ferris, F.G., Beveridge, T.J., Flemming, C.A., Bailey, G.W., 1989. Bacterial sorption of heavy metals. Appl. Environ. Microbiol. 55, 3143-3149.
47. Schlekat, C.E., Decho, A.W., Chandler, G.T., 1998. Sorption of cadmium to bacterial extracellular polymeric sediment coatings under estuarine conditions. Environ. Toxicol. Chem. 17, 1867-1874.
48. Johnson, D.B., 1998. Microorganisms and the biogeochemical cycling of metals in aquatic environments. *In:* Langston, W.J., Bebianno, M.J. (Eds.), Metal Metabolism in Aquatic Environments. Chapman & Hall, London, pp. 31-57.
49. Hashemi, F., Leppard, G.G., Kushner, D.J., 1994. Copper resistance in *Anabaena variabilis*: effects of phosphate nutrition and polyphosphate bodies. Microb. Ecol. 27, 159-176.

50. Urrutia Mera, M., Kemper, M., Doyle, R., Beveridge, T.J., 1992. The membrane-induced proton motive force influences the metal binding ability of *Bacillus subtilis* cell walls. Appl. Environ. Microbiol. 58, 3837-3844.
51. Leppard, G.G., 1997. Colloidal organic fibrils of acid polysaccharides in surface waters: electron-optical characteristics, activities and chemical estimates of abundance. Colloids Surfaces, A: Physicochem. Eng. Aspects 120, 1-15.
52. Ghiorse, W.C., 1984. Biology of iron- and manganese-depositing bacteria. Ann. Rev. Microbiol. 38, 515-550.
53. Nealson, K.H., Rosson, R.A., Myers, C.R., 1989. Mechanisms of oxidation and reduction of manganese. *In:* Beveridge, T.J., Doyle, R.J. (Eds.), Metal Ions and Bacteria. John Wiley & Sons, New York, pp. 383-411.
54. Jenne, E.A., 1968. Controls on Mn, Fe, Co, Ni, Cu, and Zn concentrations in soils and water: the significant role of hydrous Mn and Fe oxides. *In:* American Chemical Society (Eds.), Trace Inorganics in Water. Advances in Chemistry series no. 73. American Chemical Society, Washington, pp. 337-387.
55. Lee, G.F., 1975. Role of hydrous metal oxides in the transport of heavy metals in the environment. *In:* Krenkel, P.A. (Ed.), Heavy Metals in the Aquatic Environment. Pergamon Press, Oxford, pp. 137-147.
56. Tessier, A., Fortin, D., Belzile, N., DeVitre, R.R., Leppard, G.G., 1996. Metal sorption to diagenetic iron and manganese oxyhydroxides and associated organic matter: narrowing the gap between field and laboratory measurements. Geochim. Cosmochim. Acta 60, 387-404.
57. Jackson, T.A., Bistricki, T., 1995. Selective scavenging of copper, zinc, lead, and arsenic by iron and manganese oxyhydroxide coatings on plankton in lakes polluted with mine and smelter wastes: results of energy dispersive X-ray micro-analysis. J. Geochem. Explor. 52, 97-125.
58. Jackson, T.A., 1998. Mercury in aquatic ecosystems. *In:* Langston, W.J., Bebianno, M.J. (Eds.), Metal Metabolism in Aquatic Environments. Chapman & Hall, London, pp. 77-158.
59. Rozan, T.F., Lassman, M.E., Ridge, D.P., Luther, G.W. III, 2000. Evidence for iron, copper and zinc complexation as multinuclear sulphide clusters in oxic rivers. Nature 406, 879-882.
60. Davis, S.A., Burkett, S.L., Mendelson, N.H., Mann, S., 1997. Bacterial templating of ordered macrostructures in silica and silica-surfactant mesophases. Nature 385, 420-423.
61. Urrutia, M.M., Beveridge, T.J., 1993. Remobilization of heavy metals retained as oxyhydroxides or silicates by *Bacillus subtilis* cells. Appl. Environ. Microbiol. 59, 4323-4329.
62. Schultze-Lam, S., Beveridge, T.J., 1994. Nucleation of celestite and strontianite on a cyanobacterial S-layer. Appl. Environ. Microbiol. 60, 447-453.
63. Ferris, F.G., Schultze, S., Witten, T.C., Fyfe, W.S., Beveridge, T.J., 1989. Metal interactions with microbial biofilms in acidic and neutral pH environments. Appl. Environ. Microbiol. 55, 1249-1257.
64. Brown, D.A., Kamineni, D.C., Sawicki, J.A., Beveridge, T.J., 1994. Minerals associated with biofilms occurring on exposed rock in a granitic underground research laboratory. Appl. Environ. Microbiol. 60, 3182-3191.

65. Beveridge, T.J., Graham, L., 1991. Surface layers of bacteria. Microbiol. Rev. 55, 684-705.
66. Ferris, F.G., Fyfe, W.S., Beveridge, T.J., 1987. Manganese oxide deposition in a hot spring microbial mat. Geomicrobiol. J. 5, 33-42.
67. Leppard, G.G., 1993. Organic flocs in surface waters: their native state and aggregation behavior in relation to contaminant dispersion. *In:* Rao, S.S. (Ed.), Particulate Matter and Aquatic Contaminants. Lewis Publishers, Boca Raton, pp. 169-195.
68. Taillefert, M., Lienemann, C.-P., Gaillard, J.-F., Perret, D., 2000. Speciation, reactivity, and cycling of Fe and Pb in a meromictic lake. Geochim. Cosmochim. Acta 64, 169-183.
69. Cloud, P., 1976. Beginnings of biospheric evolution and their biogeochemical consequences. Paleobiol. 2, 351-387.
70. Reid, R.P., Visscher, P.T., Decho, A.W., Stolz, J.F., Bebout, B.M., Dupraz, C., Macintyre, I.G., Paerl, H.W., Pinckney, J.L., Prufert-Bebout, L., Steppe, T.F., DesMarais, D.J., 2000. The role of microbes in accretion, lamination and early lithification of modern marine stromatolites. Nature 406, 989-992.
71. Jackson, T.A., 1978. The biogeochemistry of heavy metals in polluted lakes and streams at Flin Flon, Canada, and a proposed method for limiting heavy-metal pollution of natural waters. Environ. Geol. 2, 173-189.
72. Jackson, T.A., 1993. Comment on "Weathering, plants, and the long-term carbon cycle" by Robert A. Berner. Geochim. Cosmochim. Acta 57, 2141-2144.
73. Beveridge, T.J., 1978. The response of cell walls of *Bacillus subtilis* to metals and to electron-microscopic stains. Can. J. Microbiol. 24, 89-104.
74. Beveridge, T.J., Murray, R.G.E., 1980. Sites of metal deposition in the cell wall of *Bacillus subtilis*. J. Bacteriol. 141, 876-887.
75. Beveridge, T.J., Koval, S.F., 1981. Binding of metals to cell envelopes of *Escherichia coli* K-12. Appl. Environ. Microbiol. 42, 325-335.
76. Beveridge, T.J., Meloche, J.D., Fyfe, W.S., Murray, R.G.E., 1983. Diagenesis of metals chemically complexed to bacteria: laboratory formation of metal phosphates, sulfides, and organic condensates in artificial sediments. Appl. Environ. Microbiol. 45, 1094-1108.
77. Perret, D., Gaillard, J.-F., Dominik, J., Atteia, O., 2000. The diversity of natural hydrous iron oxides. Environ. Sci. Technol. 34, 3540-3546.
78. Jackson, T.A., Keller, W.D., 1970. A comparative study of the role of lichens and "inorganic" processes in the chemical weathering of recent Hawaiian lava flows. Amer. J. Sci. 269, 446-466.
79. Jackson, T.A., Keller, W.D., 1970. Evidence for biogenic synthesis of an unusual ferric oxide mineral during alteration of basalt by a tropical lichen. Nature 227, 522-523.
80. Todd, R.L., Cromack, K., Stormer, J.C., Jr., 1973. Chemical exploration of the microhabitat by electron probe microanalysis of decomposer organisms. Nature 243, 544-546.

Developments in Soil Science, Volume 28A
Editors: A. Violante, P.M. Huang, J.-M. Bollag and L. Gianfreda

INFLUENCE OF pH AND OF SEVERAL ORGANIC ACIDS ON THE INTERACTION BETWEEN ESCULETINE AND IRON(III)

S. Deiana, B. Manunza, M.G. Molinu, A. Palma, A. Premoli and V. Solinas

DISAABA, Università di Sassari, V.le Italia 39, 07100 Sassari, Italy
Phone +39 79 229210, Fax no. +39 79 229276, E-mail: s_deiana@antas.agraria.uniss.it

The availability of iron to plants is influenced by organic molecules of low molecular weight with complexing or reducing capacity such as organic acids and phenolic acids and their derivates. To provide information about the interactions between Fe(III) and esculetine (ESC) and the influence of organic acids on these interactions, the redox activity of ESC at different pH values in aqueous solution was investigated in the presence and in the absence of citric, malic, oxalic and pyruvic acid on systems with different Fe(III)/ESC molar ratios. At pH < 4.0, all the systems showed the highest redox activity. At pH ≥ 4, the redox activity decreased and was negligible at pH 6.0. This trend was attributed to the formation of hydrolyzed Fe(III) species, which interact with ESC by forming soluble complexes that subsequently transform into insoluble Fe(III)-ESC polymers. The malic, citric and oxalic acid compete with ESC in the Fe(III) co-ordination and form soluble complexes with the metal ion thus impeding its precipitation. The organic acids were found to compete with ESC in the iron co-ordination according to the following affinity order: citric > oxalic > malic >> pyruvic acid.

1. INTRODUCTION

One of the main problems concerning the mineral nutrition of plants deals with the low availability of iron, which is mainly due to the reactions that iron undergoes in the soil [1-2]. They are generally hydrolysis reactions or reactions that lead to the formation of stable and sparingly soluble organo-mineral complexes. In the rhizosphere Fe(III) interacts with the pectic component by forming stable complexes that hinder its diffusion towards the root epidermal cells [3]. However, iron can be mobilized and thus made available to plants by the root exudates through complexation, reduction or ligand exchange reactions [4-7]. Complexing agents, such as mugineic acid, a compound released by several plants, as well as desferrioxamine-B (DFOB), a compound synthesized by soil micro-organisms, can mobilize iron by forming soluble complexes that are stable in a wide pH range [8-11].

The absorption of iron by plants is dependent upon its reduction, which can occur outside the plasmalemma by cytoplasmatic reductants or in the apparent free space by organic molecules such as phenolic compounds [12]. Olsen et al. [13] gave evidence for the release of caffeic acid by several plants as a response to stress conditions due to iron deficiency and for a high redox activity of the biomolecule towards iron. Recently, the Fe(III) complexation and reduction mechanisms by caffeic acid and the effect of pH on such mechanisms were hypothesized [14-15]. One molecule of caffeic acid was found to reduce 9 ions of Fe(III); this compound can interact with different Fe(III) species and cause its reduction by formation of intermediates with reducing activity and of insoluble Fe(III)-caffeic acid polymers. Among the oxidation products of caffeic acid, 6,7-dihydroxycoumarin (Figure 1), also known as esculetine (ESC), can be active in the transport of iron as well as of other nutrients in the rhizosphere [16].

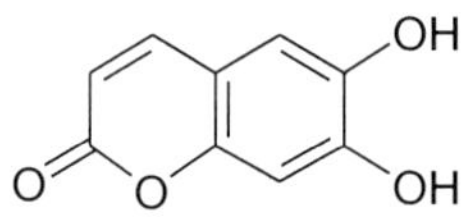

Figure 1. Structure of esculetine (6,7-dihydroxycoumarin).

A previous study [17] about the interaction between ESC and Fe(III) found that in the 2-3 pH range one molecule of ESC is able to reduce seven atoms of Fe(III) through two different mechanisms depending on the Fe(III)/ESC molar ratio. At Fe(III)/ESC ratios lower than 3, semiquinonic radicals form, which successively transform into dimers through condensation reactions. The other mechanism, which is active at Fe(III)/ESC ratios greater than 3, involves the oxidation of ESC to quinone which then oxidizes to organic acids.

With the aim of improving understanding of the interactions between the phenolic substances and the iron(III) immobilized in the rhizosphere, the Fe(III)-ESC reaction was investigated as a function of pH in the presence and in the absence of organic acids commonly found in the plant root exudates.

2. MATERIALS AND METHODS

All the reagents, if not otherwise specified, were obtained from Fluka. The solutions of ESC and iron(III) perchlorate monohydrate (Aldrich) were prepared by using deionized water just before the beginning of each experiment. Sodium perchlorate monohydrate was used as the supporting electrolyte at 0.01 M. The kinetic measures were carried out in the 2.5-6.0 pH range, at 20±1°C under continuous stirring, on systems containing 100 μM ESC and different amounts of Fe(III) (the metal to ligand molar ratio varied from 1.2 to 10), by monitoring the Fe(II) and ESC content. The systems were prepared by mixing the ESC and Fe(III) solutions separately brought to the working pH by addition of $HClO_4$ or NaOH.

The Fe(II) content was determined in the form of 1,10-phenanthroline (Phen) (J.T. Baker) complex [18] by using a small volume of solution buffered by acetate at pH 4.5 with added EDTA as an Fe(III) sequestering agent to avoid its reduction by the ESC present in the reaction medium. In the presence of Phen the reduction of Fe(III) by ESC occurs also in the presence of the organic acids considered in this work, and Fe(II) is quantitatively reduced even at about the neutrality pH value [19]. This happens because when Phen is added, it forms a complex with Fe(III) (characterized by a redox potential equal to + 1.20V [20] which is quantitatively reduced to Fe(II) by ESC. Thus, if a sample contains Fe(III) and reducing substances the amount of Fe(II) measured by the Phen method will be higher than the true value. We found that the reduction of Fe(III) in the presence of Phen and reducing substances can be avoided by the presence of EDTA at a Phen/EDTA molar ratio equal to 15 [21]. Samples were kept in the dark to avoid the photochemical reduction of Fe(III) [22]. The absorbance of the 1,10-phenanthroline-Fe(II) complex was measured at 510 nm, which is the absorption maximum in our experimental conditions.

The ESC concentration and that of its oxidation products was determined by an HPLC Dionex DX-300 system, equipped with a UV-VIS Merck Hitachi Diode Array detector, and an Alltech Alltima C18 5U column. A H_2O-acetonitrile-acetic acid (77.5% - 17.5% - 5.0%) mixture brought to pH 3.2 was employed as an eluent at a flow rate of 0.4 mL/min and at room temperature. Samples of 20 μL were applied to the column. The FT-IR analysis of the precipitates, centrifuged at 19,000 g (5°C), dehydrated and stored under vacuum, was performed on KBr disks (2 mg of sample with 100 mg of KBr). The spectra were recorded with a Nicolet 210 spectrophotometer.

The reaction between Fe(III) and ESC, in the presence of malic, oxalic, citric and pyruvic acid, was studied in the 4.5-7.0 pH range. The ternary systems had constant Fe(III) and organic acid concentrations (0.1 mM) and a variable ESC concentration in order to obtain organic acid: Fe(III):ESC molar ratios equal to 1:1:0.5, 1:1:1 and 1:1:2.

The reduction kinetics were determined by monitoring the Fe(II) and ESC content, as reported above, whereas the organic acids were determined by HPLC analysis using a Dionex DX-300 system equipped with a UV-VIS detector operating at 210 nm and a Biorad Aminex Ion Exclusion XPX-87H column. Sulfuric acid (0.4 mM) was used as an eluent.

The redox potentials of ESC, in the presence and in the absence of Fe(III) and malic acid, were determined under nitrogen by using a platinum electrode (Orion) and a Ag/AgCl electrode as a reference, as reported by Nicoli et al. [23].

The pH was monitored by an Orion pH-meter mod. 420A.

3. RESULTS AND DISCUSSION

3.1. Influence of pH on the redox activity of the Fe(III)-ESC system

The reduction kinetics of the systems with Fe(III)/ESC molar ratios between 1.2 and 10.0 in the 2.5-5.0 pH range (Figure 2) indicate that in all the systems the Fe(II) formation mostly occurs in the first hours of reaction and that equilibrium is reached in about 15 hours.

The yield of Fe(II) as a function of the Fe(III)/ESC molar ratio in the same pH range is reported in Figure 3. The highest redox activity of esculetine towards Fe(III) is recorded at pH 2.5 and 3.0. In particular, at pH 2.5, the yield of Fe(II) increases linearly with increasing Fe(III) concentration, tending to a constant value at Fe(III)/ESC molar ratios greater than 7. This is consistent with a previous study of the stoichiometry of the Fe(III)-ESC reaction in which one molecule of ESC was found to release 7 electrons [17].

The redox reaction leads to the formation of two main products, which are oligomers of esculetine whose concentration depends on the Fe(III)/ESC molar ratio. In particular, their concentration is the highest at Fe(III)/ESC molar ratios < 2.5 and negligible at Fe(III)/ESC molar ratios > 3. At pH ≥ 4.0, the redox activity dramatically decreases and becomes negligible at pH values near neutrality. Furthermore the formation of ESC oxidation products

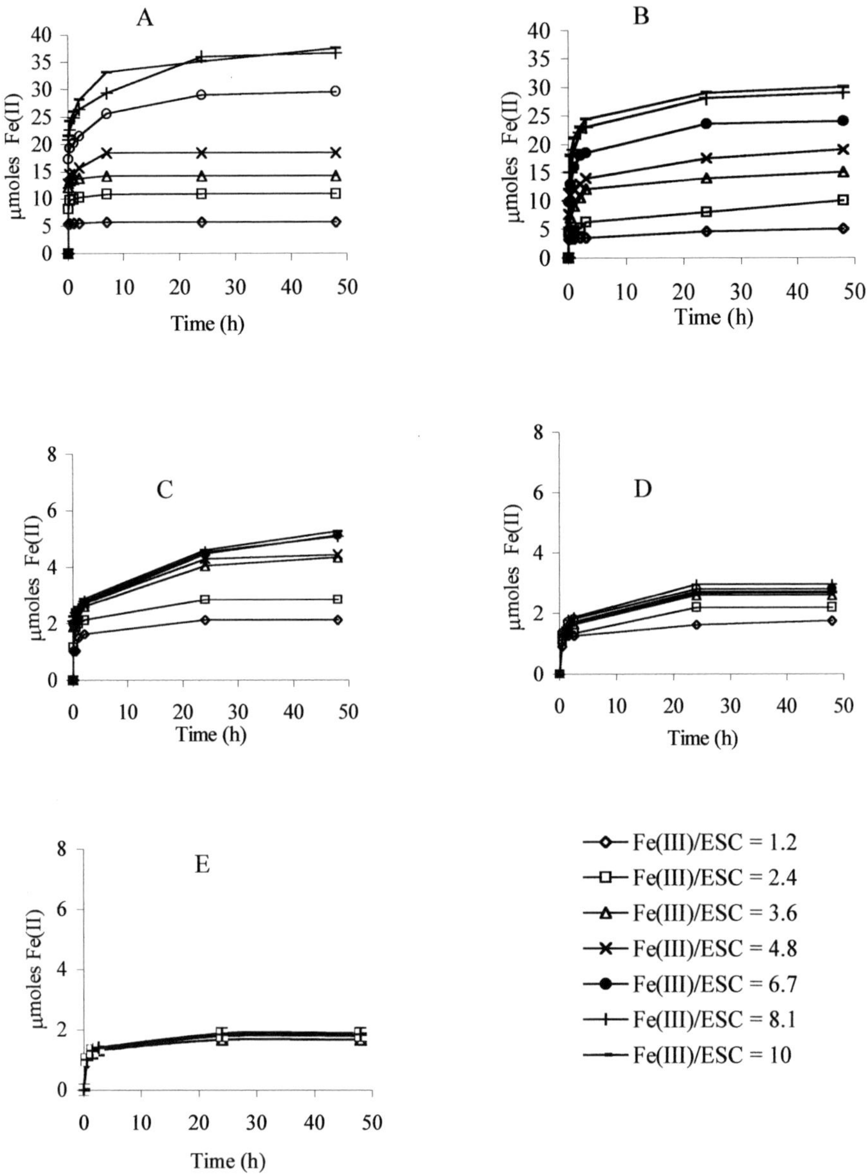

Figure 2. Yield of Fe(II) in the Fe(III)-ESC systems as a function of time. Starting conditions: 5 µmoles ESC; 0.01 M $NaClO_4$; reaction volume 50 mL; (A) pH = 2.5; (B) pH = 3.0; (C) pH = 3.5; (D) pH = 4.0; (E) pH = 5.0.

does not occur. Such a trend, as already observed for caffeic acid [15], is mainly attributable to the Fe(III) hydrolysis reactions as well as to the formation of $Fe(OH)_3$ precipitates and Fe(III)-ESC complexes.

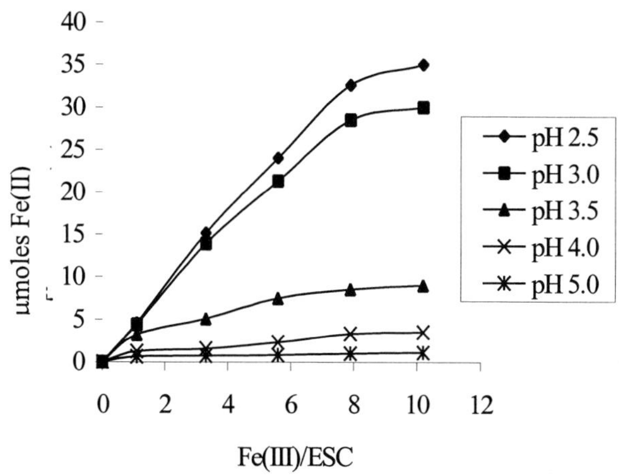

Figure 3. Yield of Fe(II) at equilibrium relative to the Fe(III)-ESC systems at different pH values as a function of the Fe(III)/ESC molar ratio.

Taking account of the concentration of the Fe(III) species in solution, calculated by the Haltafall program [24] by employing the formation constants of the $Fe(OH)^{2+}$, $Fe(OH)_2^+$ and $Fe(OH)_3$ species [25] (Table 1), and for the concentration of Fe(II) produced at different pH values (Table 2), we can hold that the species Fe^{3+} and $FeOH^{2+}$ are the only ones active in the oxidation of the organic molecule. In fact, at all pH tested, the yield of Fe(II) nearly equals the sum of the concentration of the Fe^{3+} and $FeOH^{2+}$ species.

Table 1
Percentage of all Fe(III) species calculated with the Haltafall program

	Fe(III) species			
pH	Fe^{3+}	$FeOH^{2+}$	$Fe(OH)_2^+$	$Fe(OH)_{3\,(S)}$
2.5	25.44	65.88	8.68	0.00
3.0	10.44	67.52	22.04	0.00
3.5	1.68	43.48	54.84	0.00
4.0	0.36	23.84	75.24	0.56
5.0	0.00	0.68	22.44	76.88

Table 2
Percentage of the Fe(II) formed at equilibrium in the Fe(III)-ESC systems

	Fe(III)/ESC molar ratio				
pH	1.2	2.4	4.8	6.7	8.1
2.5	85.5	91.2	90.5	80.7	70.1
3.0	84.9	89.7 ‡	82.1 ‡	75.5 ‡	68.2 ‡
3.5	56.9	32.4 ‡	26.8 ‡	22.2 ‡	25.7 ‡
4.0	23.1	10.1 ‡	7.2 ‡	5.8 ‡	5.0 ‡
5.0	12.4	4.6 ‡	3.2 ‡	2.8 ‡	2.0 ‡

‡ Formation of a black precipitate. The amount of Fe(II) produced was almost constant during the precipitation process indicating that this species is not involved in the formation of the precipitate.

The decrease in the yield of Fe(II) that occurs with increasing pH (Figure 3) could appear in contrast with the values of the ESC redox potential, equal to 0.324, 0.287, 0.205 and 0.147 V at pH 3.0, 4.0, 5.0 and 6.0, respectively, which would suggest a higher redox activity of ESC with increasing pH. This does not really occur, since such an increase is balanced by the iron hydrolysis reactions, as well as by the formation of Fe(III)-ESC complexes, which lead to a decrease in the redox potential of the Fe^{3+}/Fe^{2+} semicouple (+0.77 V) [26].

Esculetine, as shown by the UV-VIS spectra (Figure 4), interacts with Fe(III) by forming complexes whose concentration increases with increasing pH.

The lack of isosbestic points in the UV-VIS spectra indicates that one Fe(III)-ESC complex prevails. The band at 680 nm of the system at pH 3.3, attributed to the d-d transitions of the Fe(III)-ESC complexes [27-28], shifts towards lower wavelengths with increasing pH until it reaches 580 nm in the system at pH 6.0 due to a change in the co-ordination between ESC and Fe(III). In fact, the shift of the band from 680 to 580 nm is attributable to the deprotonation of the OH phenolic groups involved in the co-ordination of Fe(III) at low pH values and therefore a co-ordination of phenolate type can be established [29]. The comparison between the UV-VIS spectra of the Fe(III)-ESC system and those of other systems containing Fe(III) and catechol, or cinnamic and acetic acid, supports the involvement of the phenolic OH groups in the co-ordination sphere of the metal center. Indeed, the systems containing the cinnamic and acetic acid do not show any absorption band in the 800-400 nm range [15], whereas the UV-VIS spectra of the Fe(III)-catechol spectra, are similar to those of the Fe(III)-ESC systems.

To determine the co-ordination stoichiometry between Fe(III) and ESC, we carried out a spectrophotometric survey on systems with different Fe(III)/ESC molar ratio at pH 5.0.

The UV-VIS spectra show that the intensity of the absorption band at 586 nm increases with increasing ESC concentration. The Job's plot [30] (Figure 5), which reports the maximum of the absorption of the band at 586 nm as a function of the ESC/Fe(III) molar ratio, indicates that Fe(III) is able to co-ordinate three ESC molecules. A similar stoichiometry was found for the catechol-Fe(III) system, where catechol interacts with the metal ion through a co-ordination of phenolate type [29].

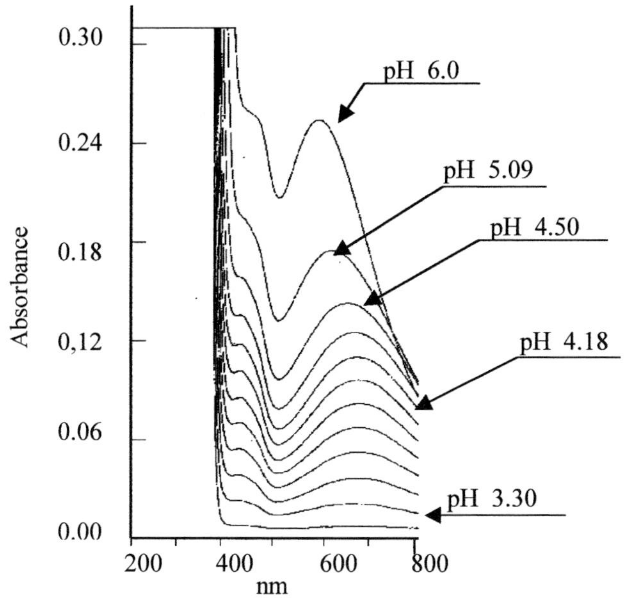

Figure 4. UV-VIS spectra of the reaction solution of the Fe(III)-ESC system. The curves refer to pH values varying from 3.0 to 6.0. Starting conditions: Fe(III)/ESC molar ratio = 0.25; ESC = 100 μM.

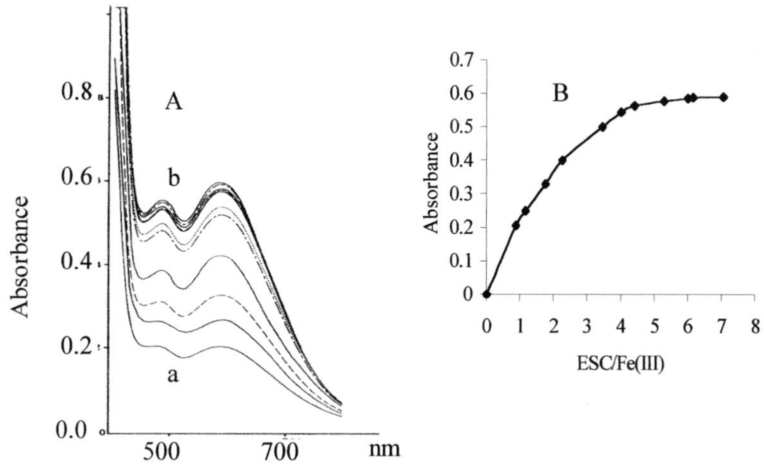

Figure 5. Spectra UV-VIS (A) and Job's plot (B) of the Fe(III)-ESC system at pH 5.0. The Job's plot was obtained by reporting the absorbance of the band at 586 nm against the ESC/Fe(III) molar ratio. Fe(III) = 0.1 mM; ESC concentration varies from 0.02 (a) to 0.7 mM (b).

The UV-VIS spectra of the systems with Fe(III)/ESC molar ratios > 1.0, in the 4.0-6.0 pH range, show a trend similar to that reported above. However, in these systems, during the first 100 minutes of reaction a strong decrease in the absorption bands at 350 and in the 635-580 nm range is recorded. The decrease in the absorption band at 350 nm, due to ESC, as shown by the HPLC analysis, cannot be attributed to the oxidation of the biomolecule by Fe(III) (the amount of Fe(II) produced is about 8%), but to the adsorption of ESC by the iron hydroxides surfaces that form following the hydrolysis of the metal ion. This observation is supported by preliminary tests as well as by several studies that indicate the ability of the iron hydroxides to adsorb phenolic compounds and to promote the formation of polymers [31-32]. The decrease in the band attributed to the Fe(III)-ESC complexes is due to their precipitation. As an example Figure 6 reports the UV-VIS spectra of the system at pH 5.0 with the Fe(III)/ESC molar ratio = 5.5 at different reaction times.

The amount of precipitates increases with increasing Fe(III)/ESC molar ratio. Solubility tests in acidic medium showed that precipitates are stable even at pH values as low as 1.0, values at which Fe(III)-hydroxides are soluble. Therefore, these precipitates can be held as polymers constituted by ESC units bound to each other though iron bridges. The existence of the interaction between Fe(III) and ESC is confirmed by the FT-IR spectra reported below. Studies about the nature of these precipitates are in progress.

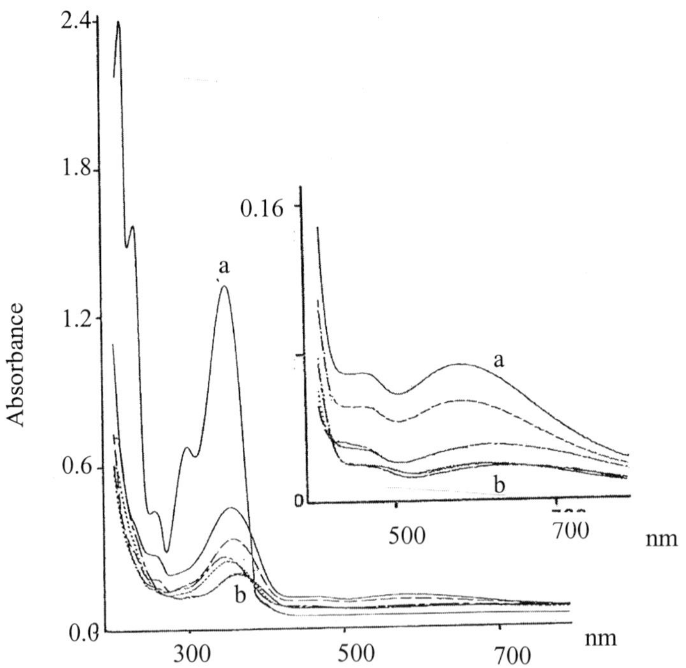

Figure 6. UV-VIS spectra of the system at pH 5.0 with Fe(III)/ESC molar ratio = 5.5 at different reaction times. Reaction time = 0 h (a); reaction time = 48 h (b).

The FT-IR spectra of the free ESC exhibit the stretching vibrations of the ν(OH) group in the 3200-3400 cm^{-1} range and at 1400 cm^{-1} the bending vibrations of the δ(OH) group, vibrations which are missing in the spectra of the Fe(III)-ESC precipitates. These data, supported by Griffith and Mostafa [33], indicate that the phenolic groups of ESC are probably involved in the iron co-ordination sphere. The persistence at 1280 cm^{-1} of the stretching vibration of the ν(C=O) carbonylic group of both the free ESC and Fe(III)-ESC precipitates excludes the involvement of this group in the co-ordination of the metal ion.

The FT-IR spectra of ESC and of the precipitate that forms at pH 5.0 in the Fe(III)-ESC system with a molar ratio equal to 5.5 are reported as an example in Figure 7.

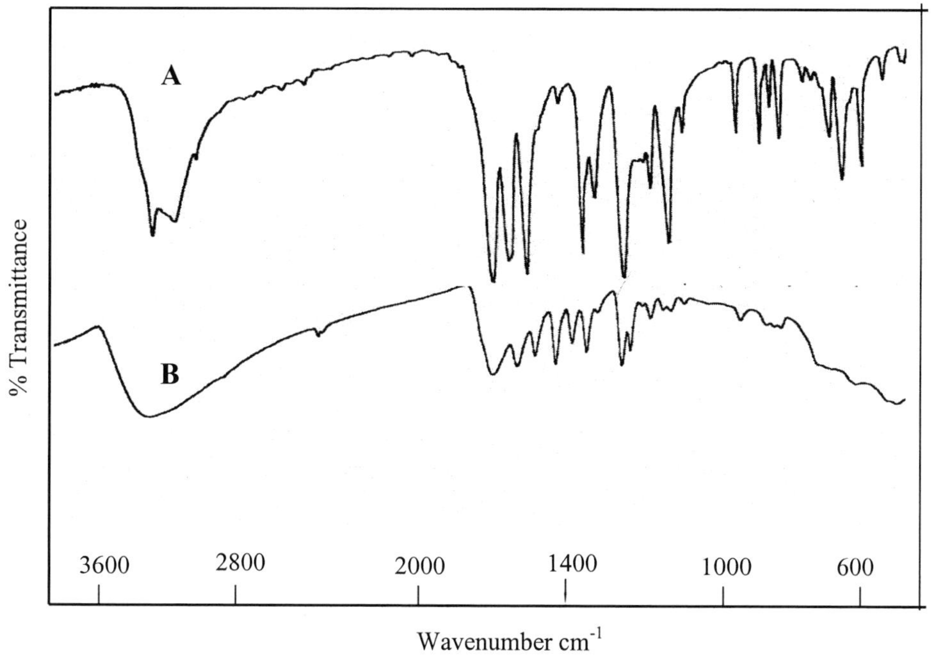

Figure 7. FT-IR spectra of ESC (A) and of the precipitate (B) that forms at pH 5.0 in the system with an Fe(III)/ESC molar ratio equal to 5.5.

3.2. Influence of malic, pyruvic, citric and oxalic acid on the redox activity of the Fe(III)-ESC system

The reduction of Fe(III) by ESC in the presence of malic, pyruvic, citric and oxalic acid was studied at pH 4.5, 5.0, 5.5, 6.0 and 7.0, values at which the Fe(III)-ESC binary systems showed a scarce reducing activity and the formation of precipitates occurred.

The kinetic data show that the organic acids considered do not affect significantly the yield of Fe(II) compared to that found in the Fe(III)-ESC binary systems. This is probably because the formation of the Fe(III)-organic acid complexes does not affect the redox potential of the metal ion. Furthermore, in contrast to the Fe(III)-ESC binary systems, in the presence of these organic acids the formation of precipitates does not occur. This could be explained by considering that these organic acids form soluble complexes with Fe(III) [30] and that a partial or total competition between the organic acids and ESC for the Fe(III) co-ordination can occur, which prevents the precipitation of Fe(III)-ESC complexes. The distribution diagrams of the most significant soluble Fe(III)-organic acid complexes are reported as a function of pH in Figures 8.

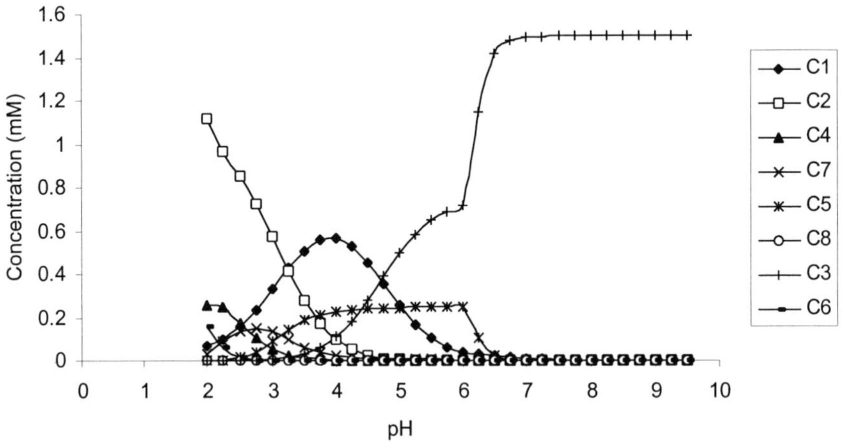

Figure 8a. Distribution diagrams of the most significant soluble species as a function of pH in the Fe(III)-malic acid system calculated by the Haltafall program [24]. C1 = malic acid (H_2L); C2 = monodeprotonated malic acid (HL); C3 = malate (L); C4 = ML; C5= $M_2(H_{-1}L)_2L$; C6 = Fe^{3+}; C7 = $M_2(H_{-1}L)_2$; C8 = $Fe(OH)_2^+$. The distribution of the species was calculated by using initial malic acid and Fe(III) concentrations equal to 0.1 mM and the formation constants of the Fe(III)-malic acid complexes and of the Fe(III) hydrolysis reported by Martell and Smith [34]. $H_{-1}L$ = malate with deprotonated alcohol group; M = metal ion.

To evaluate such a competition, the complexation reaction between ESC, organic acids, and Fe(III) at pH 5.0 was studied. The elaboration of the potentiometric data by employing programs, such as the Superquad program, which allow calculation of the formation constants of the complexes, and as a consequence, the determination of the stoichiometry of these complexes, could not be reliable here since a little reduction (8%) of Fe(III) to Fe(II) occurs. Therefore, the Job's plot [30] was chosen to have a better graphical representation of the UV-VIS data. Iron (II) does not show absorption bands that lay on those of ESC so that it does not interfere in the Fe(III)-ESC absorption bands. Furthermore, the UV-VIS spectra of Fe(III) in the presence of the organic acids do not show absorption bands in the region where those of Fe(III)-ESC appear.

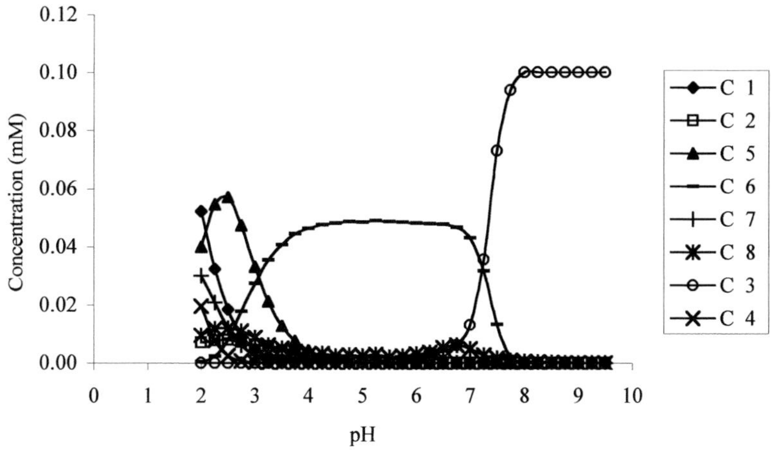

Figure 8b. Distribution diagrams of the most significant soluble species as a function of pH in the Fe(III)-citric acid system calculated by the program [24]. C1 = citric acid (H_3L); C2 = H_2L; C3 = L; C4 = $Fe(OH)^{2+}$; C5 = ML; C6 = ML_2; C7 = Fe^{3+}; C8 = $M_2(H_{-1}L)$. The distribution of the species was calculated by using initial citric acid and Fe(III) concentrations equal to 0.1 mM and the formation constants of the Fe(III)-citric acid complexes and of the Fe(III) hydrolysis reported by Martell and Smith [34]. $H_{-1}L$ = citrate with deprotonated alcohol group; L = citrate; M = metal ion.

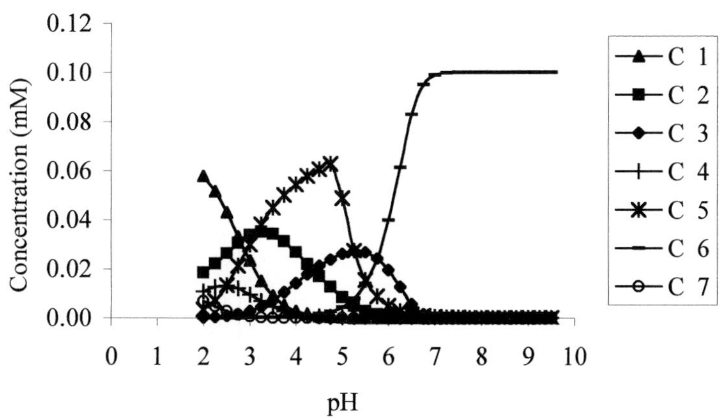

Figure 8c. Distribution diagram of the species in the Fe(III)-oxalic acid system as a function of pH calculated by the Haltafall program (24). C1 = $Fe(OH)^{2+}$; C2 = $Fe(OH)_2^+$; C3 = ML; C4 = ML_2; C5 = MHL; C6 = oxalate (L); C7 = Fe^{3+}. The distribution of the species was calculated by using initial oxalic acid and Fe(III) concentrations equal to 0.1 mM and the formation constants of the Fe(III)-malic acid complexes and of the Fe(III) hydrolysis reported by Martell and Smith [34]. M = metal ion.

The comparison between the trend of the absorbance of the Fe(III)-ESC systems in the absence (Figure 5) and in the presence of malic, oxalic, pyruvic and citric acid (Figure 9) shows that the hypothesized competition exists. In fact, the absorbance of the complex Fe(III)-ESC is lower in the presence of organic acids, indicating that some competition occurs between the organic acid tested and ESC in the Fe(III) co-ordination. The pyruvic acid does not compete significantly with ESC. This aspect is well represented by the Job's plot obtained by reporting the absorbance of the band at 586 nm against the ESC/Fe(III) molar ratio (Figure 10).

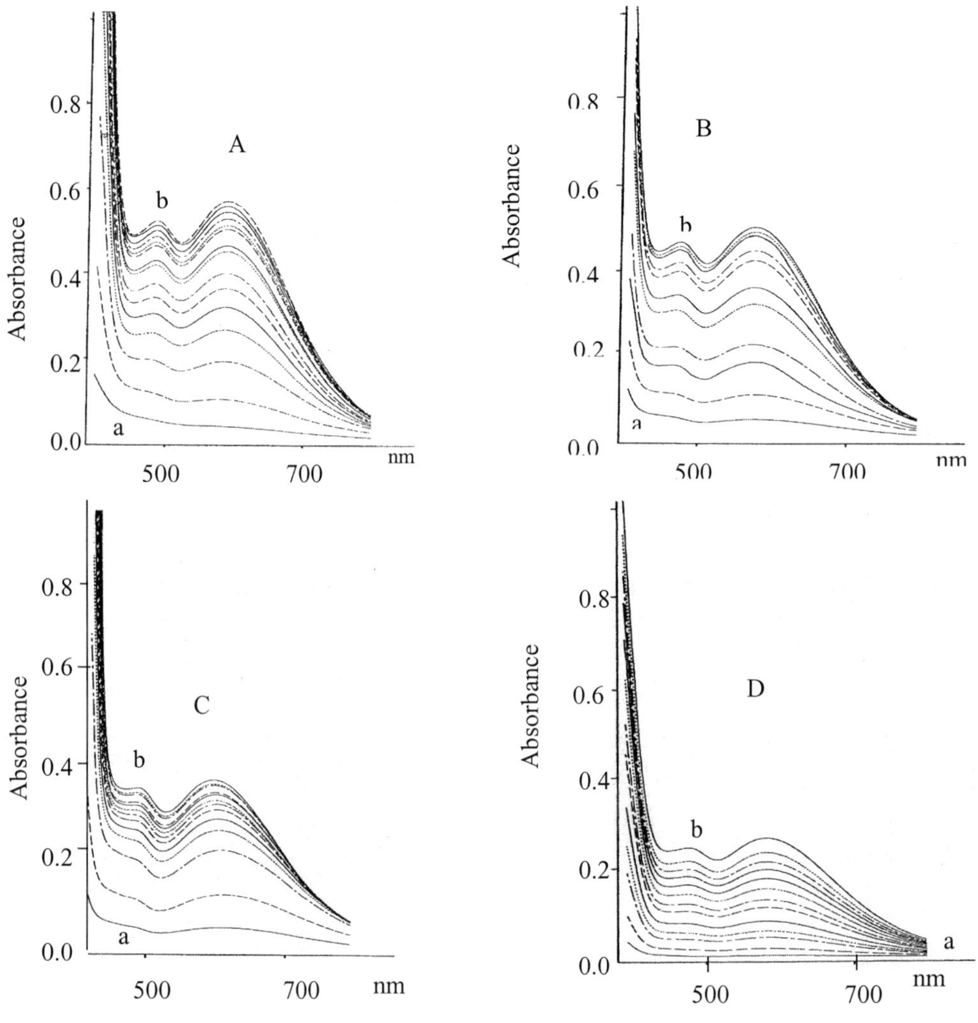

Figure 9. Spectra UV-VIS of the Fe(III)-ESC-organic acid systems at pH 5.0. Fe(III)-ESC-pyruvic acid (A), Fe(III)-ESC-malic acid (B), Fe(III)-ESC-oxalic acid (C) and Fe(III)-ESC-citric acid (D) systems. The initial concentration was: Fe(III) = 0.1 mM; organic acid = 0.1mM. The ESC concentration varied from 0.02 (a) to 0.7 mM (b).

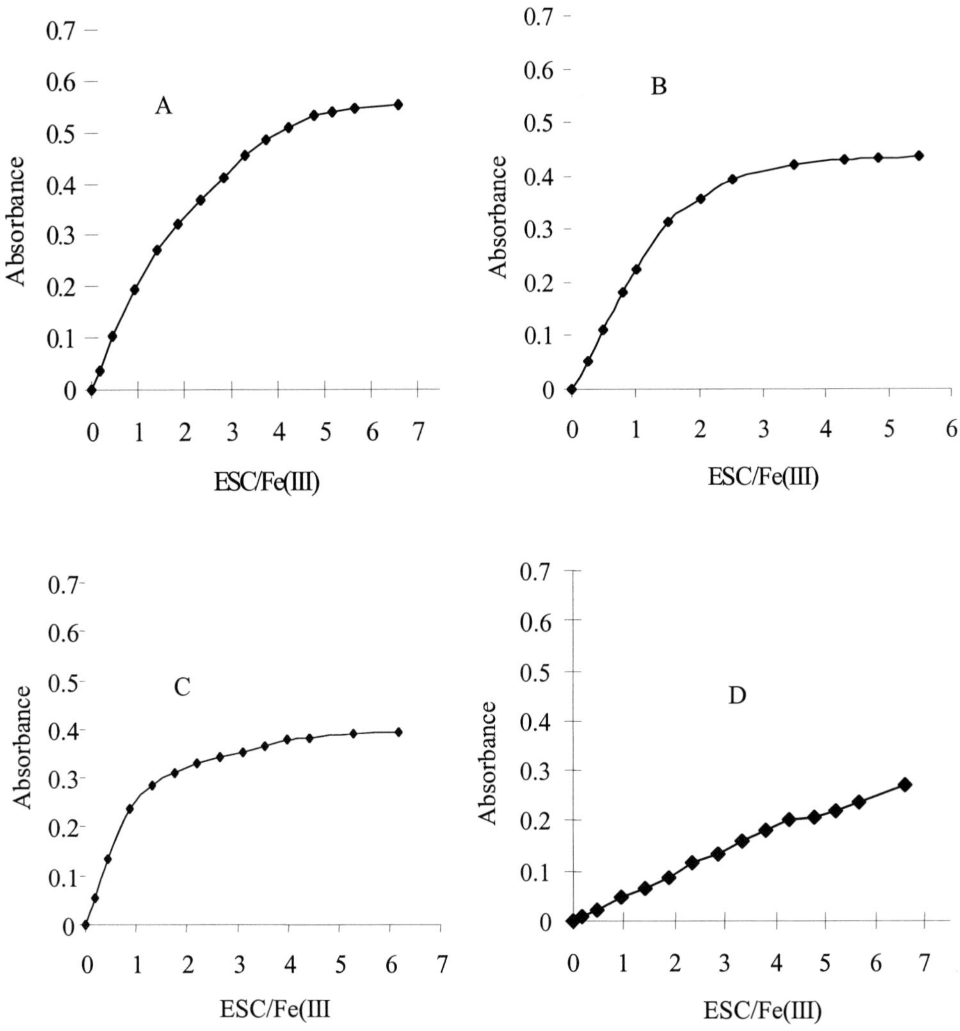

Figure 10. Job's plot of the Fe(III)-ESC-organic acid systems at pH 5.0. Fe(III)-ESC-pyruvic acid (A), Fe(III)-ESC-malic acid (B), Fe(III)-ESC-oxalic acid (C) and Fe(III)-ESC-citric acid (D) systems. The Job's plot was obtained by reporting the absorbance of the band at 586 nm against the ESC/Fe(III) molar ratio. The initial concentration was: Fe(III) = 0.1 mM; organic acid = 0.1mM. The ESC concentration varied from 0.02 to 0.7 mM.

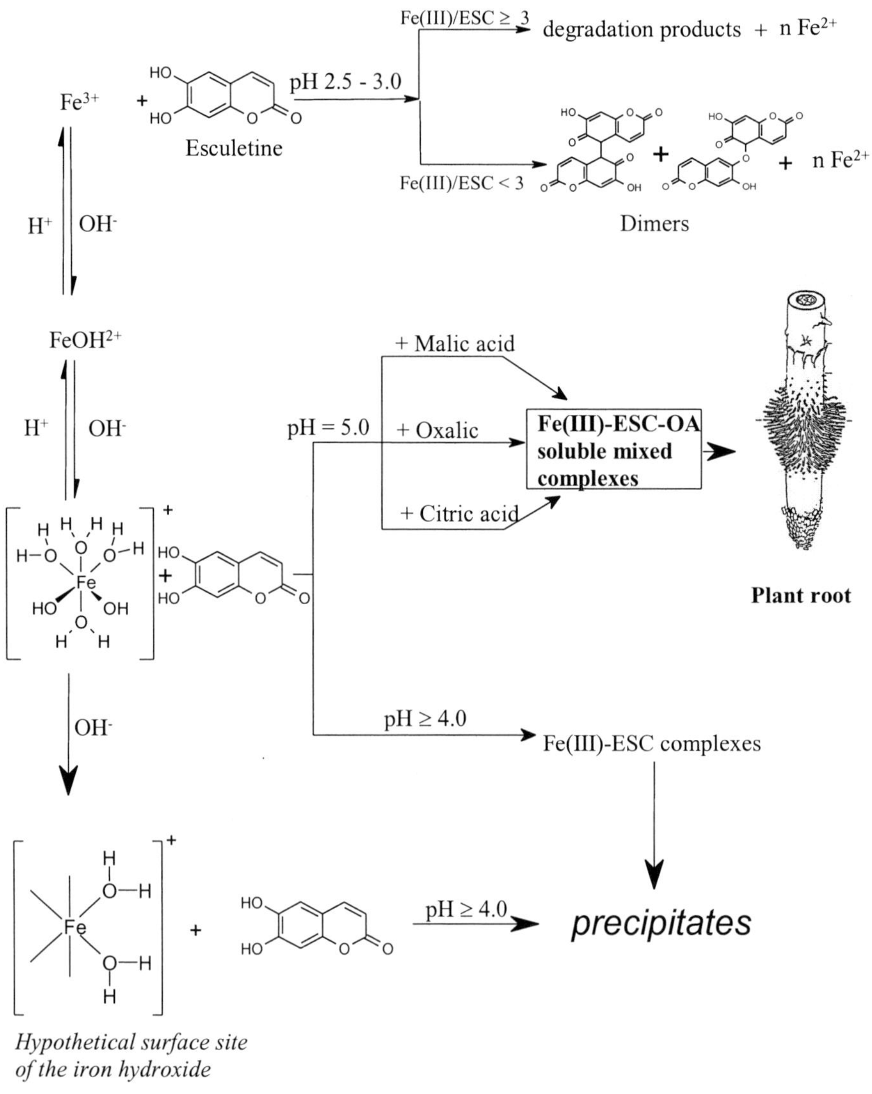

Figure 11. Hypothetical scheme of the reduction and complexation processes in the Fe(III)-ESC and Fe(III)-ESC-organic acid systems. OA = organic acid.

4. CONCLUSIONS

The results indicate that ESC interacts with different Fe(III) species, causing their reduction mainly in the 2.5-3.5 pH range. At pH ≥ 4.0, the redox activity of ESC may be depressed due to the formation of Fe(III)-ESC insoluble complexes or polymeric products, as well as by its adsorption on the surfaces of Fe(III) hydroxides.

The formation of an Fe(III)-ESC insoluble compound is of great importance in the mineral plant nutrition as it makes iron unavailable to plants. Furthermore, the organic acids examined, which are present in both the soil and the rhizosphere, competing with ESC in the co-ordination of iron, and forming Fe(III) soluble complexes, can make this ion available to the redox reactions that take place at the plasmalemma [35].

These findings are indicative of the great chemical flexibility of ESC which in the soil-plant system can activate redox, complexation, and polymerization reactions depending on the environmental conditions. The results can also contribute to the comprehension of the mechanisms that regulate the formation of polymers in the soil [36-39] and in the rhizosphere [35] following the interaction between the inorganic components, such as Fe or Mn-Fe compounds, and the phenolic substances, as well as to the evaluation of the influence of organic acids on such processes [40].

ACKNOWLEDGMENTS

Financial support was provided by Ministero dell'Università e della Ricerca Scientifica (MURST 40%).

REFERENCES

1. Chen, Y., Hadar, Y., 1991. Iron Nutrition and Interactions in Plants. Kluwer Academic Publishers, London.
2. Lindsay, W.L., Schwab, A.P., 1982. The chemistry of iron in soils and its availability to plants. J. Plant. Nutr. 5, 821-840.
3. Gessa, C., Deiana, S., Premoli, A., Ciurli, A., 1997. Redox activity of caffeic acid towards iron (III) complexed in a polygalacturonate network. Plant Soil 190, 289-299.
4. Brown, J.C., 1969. Agricultural use of synthetic metal chelates. Soil Sci. Soc. Am. Proc. 33, 59-61.
5. Mench, M., Morel, J.L., Guckert, A., Gillet, B., 1988. Metal binding with root exudates of low molecular weight. J. Soil Sci. 39, 521-527.
6. Deiana, S., Gessa, C., Piu, P., Seeber, R., 1991. Iron(III) reduction by D-Galacturonic acid, part 3. Influence of the presence of additional metal ions and of 2-amino-2-deoxy-D-Gluconic acid. J. Chem. Soc. Dalton Transl. 1237-1241.
7. Takagi, S., Nomoto, K., Takamoto, T., 1984. Physiological aspect of mugineic acid, a possible phytosiderophore of graminaceus plants. J. Plant Nutr. 7, 469-477.
8. Ma, J.F., Kusano, G., Kimura, S., Nomoto, K., 1993. Specific recognition of mugineic acid ferric complex by barley roots. Phytochemistry 34, 599-603.
9. Loehr, T.M., 1989. Iron Carriers and Iron Proteins. VCH, New York.
10. Shenker, M., Hardar, Y., Chen, Y., 1999. Kinetics of iron complexing and metal exchange in solutions by rhizoferrin, a fungal siderophore. Soil Sci. Soc. Am. J. 63, 1681-1687.
11. Solinas, V., Deiana, S., Gessa, C., Pistidda, C., Rausa, R., 1996. Reduction of the iron(III)-desferrioxamine-B complexes by caffeic acid: a reduction mechanism of biochemical significance. Soil Biol. Biochem. 28, 649-654.
12. Abadia, J., 1995. Iron Nutrition in Soils and Plants. Kluwer Academic Publishers, Netherlands.

13. Olsen, R.A., Brown, J.C., Bennet, J.H., Blume, D., 1982. Reduction of the Fe(III) as it relates to Fe chlorosis. J. Plant Nutr. 5, 433-445.
14. Deiana, S., Gessa, C., Manunza, B., Marchetti, M., Usai, M., 1992. Mechanism and stoichiometry of the redox reaction between iron(III) and caffeic acid. Plant Soil 145, 287-294.
15. Deiana, S., Gessa, C., Marchetti, M., Usai, M., 1995. Phenolic acid redox properties: pH influence on iron(III) reduction by caffeic acid. Soil Sci. Soc. Am. J. 59, 1301-1307.
16. Deiana, S., Gessa, C., Pilo, M.I., Premoli, A., Solinas, V., 1995. Role of caffeic acid oxidation products on the iron mobilization at the soil-root interface. Giorn. Bot. It. 29, 941-942.
17. Deiana, S., Gessa, C., Manunza B., Palma, A., Premoli, A., Solinas, V., 1997. Stoichiometry and reduction mechanism of iron(III) by esculetine. J. Biol. Research, Suppl. 9.10, 30-31.
18. Willard, H.H., Merritt, L.L., Dean, J.A., 1974. Instrumental Methods of Analysis. D. Van Nostrand Company, New York.
19. Stucki, J.W., Anderson, W.L., 1981. The quantitative assay of minerals for Fe^{2+} and Fe^{3+} using 1,10-Phenanthroline: I. source and variability, Soil Sci. Soc. Am. J. 45, 633-637.
20. Bell, C., 1977. Principles and Applications of Metal Chelation. Oxford Chemistry, Series 25. Clarendon Press, Oxford.
21. Palma, A., Deiana, S., Gessa C., Manunza B., Solinas V., 1998. Influence of organic acids and N-heteroaromatic compounds on the iron(III)-esculetine interaction. Polyph. Comm. 98, 2, 577-578.
22. Novak, J., Arend, H., 1964. The photosensitivity of the complex of iron(III) with 1,10-phenanthroline. Talanta 11, 898-899.
23. Nicoli, M.C., Manzocco, L., Anese, M., Lerici, C.R., 1997. Cambiamenti delle proprietà antiossidanti di alimenti sottoposti a trattamenti di trasformazione e conservazione. Ric. Inn. Ind. Alim. 3, 436-442.
24. Ingri, N., Kakolowicz, W., Sillen, L.G., Warnqvist, B. 1967. High-speed computers as supplement to graphical methods. V. Haltafall, a general program for calculating the composition of equilibrium mixtures. Talanta 14, 1261-1286.
25. Lindsay, W.L., 1979. Chemical Equilibria in Soils. John Wiley & Sons, New York.
26. Bienfait, H.F., 1988. Mechanisms in Fe-efficiency reactions of higher plants. J. Plant Nutr., 11, 605-629.
27. Avdeef, A., Sofen, S.R., Bregante, T.L., Raymond, K.N., 1978. Coordination chemistry of microbial Fe transport compounds. 9. Stability constants for catechol models of enterobactin. J. Am. Chem. Soc. 100, 5362-5370.
28. Harris, W.R., Carrano, C.J., Cooper, S.R., Sofen, S.R., Avdeef, A.E., 1979. Coordination chemistry of microbial iron transport compounds. 19. Stability constants and electrochemical behaviour of ferric enterobactin and model complexes. J. Am. Chem. Soc. 101, 6097-6104.
29. Hider, R.C., 1984. Siderophore mediated adsorption of iron. Struct. Bond. 58, 38-42.
30. Hill, Z.D., MacCarthy, P., 1986. Novel approach to Job's method. J. Chem. Educ. 63, 162-167.
31. Shindo, H., Huang, P.M., 1984. Catalytic effect of manganese(IV), iron(III), aluminum and silicon oxides on the formation of phenolic polymers. Soil Sci. Soc. Am. J. 48, 927-934.

32. Huang, P.M., Schnitzer, M., 1986. Interactions of Soil Minerals with Natural Organics and Microbes. SSSA Spec. Publ. 17. SSSA, Madison, WI.
33. Griffith, W.P., Mostafa, S.I., 1992. Complexes of esculetin with second and third row transition elements. Polyedron 11, 871-877.
34. Martell, A.E., Smith, R.M., 1982. Critical Stability Constants. Plenum Press, London.
35. Deiana, S., Manunza, B., Palma, A., Premoli, A., Gessa, C., 2000. Interaction and mobilization of metal ions at the soil-root interface. *In:* Gobran, G.R., Wenzel, W.W., Lombi, E. (Eds.), Trace Elements in the Rhizosphere. CRC Press, London, pp. 127-148.
36. McBride, M.B., 1987. Adsorption and oxidation of phenolic compounds by iron and manganese oxides. Soil Sci. Soc. Am. J. 51, 1466-1472.
37. McBride, M.B., 1989. Oxidation of dihydroxybenzenes in aerated aqueous suspensions of birnessite. Clays Clay Min. 37, 341-347.
38. Ukrainczyk, I., McBride, M.B., 1992. Oxidation of phenol in acidic aqueous suspensions of manganese oxides. Clays Clay Min. 40, 157-166.
39. Naidja, A., Huang, P.M., and Bollag, J.M., 1998. Comparison of reaction products from the transformation of catechol catalyzed by birnessite or tyrosinase. Soil Sci. Soc. Am. J. 62, 188-195.
40. Gobran, G.R., Clegg, S., Courchesne, F., 1999. The rhizosphere and trace element acquisition. *In:* Selim, H.M., Iskander A. (Eds.), Fate and Transport of Heavy Metals in the Vadose Zone. CRC Press, London, pp. 225-250.

Developments in Soil Science, Volume 28A
Editors: A. Violante, P.M. Huang, J.-M. Bollag and L. Gianfreda

ADSORPTION OF PHOSPHATE ON VARIABLE CHARGE MINERALS AND SOILS AS AFFECTED BY ORGANIC AND INORGANIC LIGANDS

A. Violante, M. Pigna, M. Ricciardella and L. Gianfreda

Dipartimento di Scienze Chimico-Agrarie, Università di Napoli Federico II,
via Università 100, 80055 Portici (Napoli), Italy

Metal oxides, noncrystalline or short-range ordered iron and aluminum hydroxides, poorly crystalline aluminosilicates, which are found within a wide range of soil orders, as well as organo-mineral complexes, are responsible for phosphate retention in soil environments. Strongly chelating organic acids produced by microorganisms or by plants (i.e., root exudates), as well as humic and fulvic acids, may strongly influence the adsorption of phosphate and its availability for plants. Maximum reduction in phosphate adsorption occurs when organic ligands are previously adsorbed on variable charge minerals or soils. The competitive adsorption of phosphate and organic ligands (e.g., oxalate, tartrate, malate, citrate) is influenced by pH, the nature of the ligands, and the nature of the surfaces of clay minerals and soils. Organic ligands may coprecipitate with OH-Al or OH-Fe species, forming organo-mineral complexes, which differ in chemical composition, surface properties, and reactivity toward phosphate. Nutrients and pollutants also compete with phosphate for the sorption sites of soil constituents. Sulfate inhibits phosphate adsorption or is not completely removed from the surfaces on which it is previously adsorbed only at low pH values. However, sulfate present in hydroxy-Al sulfate complexes is only partially removed even by large amounts of phosphate. Arsenate strongly competes with phosphate, but its efficiency in inhibiting phosphate adsorption is influenced by pH, concentration, order of anion addition, and nature of the surface of clay minerals and soils.

1. INTRODUCTION

The term "variable charge" is used to describe soil constituents whose charge varies with the pH of the soil solution [1-5]. Important inorganic variable charge constituents of soils are the hydroxides and oxyhydroxides of aluminum, iron, titanium, and manganese, as well as some short-range ordered aluminosilicates such as allophane and imogolite. Phyllosilicates, coated by OH-Al or OH-Fe species (i.e., monomers, polymers, noncrystalline or poorly crystalline Al or Fe precipitation products), are also considered variable-charge soil constituents [6-7]. Because of their large surface area, disordered network, and high charge density, these soil constituents react readily with anions, cations, and organic molecules, which include biomolecules (e.g., enzymes, toxins, DNA, RNA, polysaccharides) and xenobiotics [3, 6-9].

Inorganic ligands, usually found in soil solutions, are differently adsorbed on variable charge minerals and soils. The anions NO_3^-, Cl^-, Br^-, I^-, and ClO_4^- and the organic monodentate ligands (i.e., acetate, formate, benzoate) are adsorbed as outer-sphere complexes and loosely adsorbed on surfaces that exhibit a positive charge. Adsorption is then sensitive to ionic strength. These anions do not compete with phosphate for adsorption on clay minerals [1, 3, 5, 6].

When the chemical interaction between the mineral surface and anion cannot be described by electrostatic forces, the process is known as specific adsorption and can be viewed as a chemical surface reaction involving ligand exchange and formation of covalent bonds between the metal cations of the oxides and the anion. Phosphate is specifically adsorbed by replacing the coordinated $-OH_2$ and $-OH$ groups of variable charge minerals [3, 5-7, 9, 10]. Humic and, fulvic acids, low molecular-mass organic ligands (e.g., oxalic, citric, tartaric, salicylic, and malic acids, and a series of aminoacids), and inorganic ligands (e.g., sulfate, selenate, arsenate, arsenite, molybdate, chromate), which are also specifically adsorbed, form inner-sphere complexes and strongly compete with phosphate for adsorption sites [1, 3, 5-10]. Sulfate can be adsorbed as an outer-sphere or inner-sphere complex [1, 11-15], but only a few studies have provided direct in situ spectroscopic evidence to suggest inner-sphere complexation [15, 16].

In the last decade, we have carried out extensive studies on the factors that influence the adsorption of phosphate on soil components [7-9, 17-23]. In this paper, we describe some of our findings on the competitive adsorption of phosphate with inorganic (i.e., sulfate, arsenate) and organic (i.e., oxalate, tartrate, aspartate, citrate, tannate) ligands on variable charge minerals and soils and synthetic organo-mineral complexes as affected by pH, nature and concentration of foreign ligands, initial phosphate/ligand molar ratio, and order of anion addition. We also present the effect of selected ligands on phosphate removal from variable charge minerals and soils on which phosphate was previously adsorbed or coprecipitated.

2. MATERIALS AND METHODS

The variable charge minerals and soils used were goethite [13], gibbsite, noncrystalline Al precipitation products [17], poorly crystalline boehmites [19], Al(OH)x-montmorillonite complexes, containing 3.3 or 6.0 mol Al kg^{-1} [21, 22], an hydroxy-sulfate-aluminum precipitation product, containing 1.35 mol SO_4 and 11.5 mol Al kg^{-1} [23], hydroxy-aluminum-oxalate precipitates containing 0.96, 1.58 and 2.67 mol oxalate kg^{-1} [18] and a subsurface volcanic soil (Andisol), containing a high amount of allophanic materials (42 %).

The natural and synthetic samples were characterized by X-ray diffraction (XRD). The specific surface of most the samples was determined by the gravimetric method based on the retention of the ethylene glycol monoethyl ether [18]. The point of zero salt charge (PZSC) of some samples was determined according to the method of Sakurai et al. [cited by 18, ?].

The sorption of phosphate in the absence or presence of selected organic and inorganic ligands on the above cited samples was carried out in the pH range of 3.0-9.0, as described in previous works [13, 17,18, 21-23]. Information on some analytical procedures are given, when necessary, in the text.

The amounts of phosphate, tartrate, oxalate, aspartate, malate, sulfate, and arsenate that remained in the supernatant solution after reaction were determined by ion chromatography (Dionex model 2000i/SP, Dionex Co, Sunnyvale, CA).

3. RESULTS AND DISCUSSION

3.1. Competitive adsorption of phosphate and inorganic ligands

3.1.1. Competition in adsorption between sulfate and phosphate

Most authors suggested that the mechanisms of sulfate and phosphate adsorption are similar, and that both ions compete for the same sorption sites, although adsorbed sulfate does not compete strongly with phosphate[1, 6]. However, until today, the exact mechanism of sulfate adsorption is not known. Some researchers showed evidence that sulfate can be adsorbed as an inner-sphere complex [13, 15-16, 24], whereas others have concluded that sulfate can be adsorbed only as outer-sphere complexes [11,12]. Probably pH, sulfate concentration, and the nature of the clay mineral influence the complexation of sulfate with the surfaces of the sorbents [24].

Recently, we have carried out studies on the competitive adsorption of phosphate and sulfate by an Andisol. The amounts of phosphate adsorbed by this variable charge soil, when added alone (470 mmol kg^{-1}), slightly decreased with increasing pH, from 460 mmol kg^{-1} at pH 3.5 to 340 mmol kg^{-1} at pH 8.0. The amounts of sulfate adsorbed by the soil (190 mmol kg^{-1} at pH 3.5) were much lower than those of phosphate and strongly decreased with increasing pH. At pH > 7.0, negligible amounts of sulfate were adsorbed (data not shown). When phosphate was added to the soil as a mixture with sulfate (470 mmol kg^{-1}) at a phosphate/sulfate molar ratio ranging from 0.02 to 2.0, the quantities of adsorbed sulfate decreased significantly (Figure 1A), whereas phosphate was totally adsorbed.

The efficiency of phosphate in depressing sulfate adsorption was pH-dependent and increased with increasing pH. Figure 1B shows the percentage of sulfate adsorbed on the Andisol as referred to the relative amounts of sulfate adsorbed in the absence of phosphate at different pH values (i.e., 2.5, 3.5, and 4.5) versus the initial phosphate/sulfate molar ratios. It appears evident that phosphate depressed sulfate adsorption more easily at pH 4.5 than at lower pH values (i.e., 3.5 and 2.5). In fact, at a phosphate/sulfate molar ratio of 2, the adsorption of sulfate was completely inhibited at pH 4.5, whereas at pH 3.5 and 2.5, ≅ 25 and 50%, respectively, of the sulfate adsorbed in the absence of phosphate was still present.

These findings corroborate the conclusions of Elzinga and Sparks [24] that sulfate is mainly adsorbed as an outer-sphere complex, but at very low pH values, it may be strongly adsorbed on the positively charged surfaces of variable charge minerals, forming inner-sphere complexes. As a consequence, inorganic sulfate ions adsorbed on the surfaces of soil minerals may be easily desorbed by phosphate only at pH > 5.0. Some scientists observed that phosphate could replace very large percentages or all of the sulfate previously sorbed in their soils or clay minerals only at neutral or alkaline pH values [14, 25, 26].

In addition to sulfate adsorption on clay minerals and Al- or Fe-oxides, the precipitation of Al hydroxy sulfate minerals is suggested to be a relevant pathway of sulfate retention in acid sulfate soils and temperate acid forest soils subjected to heavy atmospheric sulfate deposition [27]. The formation of aluminum hydroxysulfate precipitates in acid soils containing sulfate or receiving sulfate has been widely reported [28-31]. Courchesne and Hendershot [32] demonstrated that the formation of basic aluminum sulfate minerals in the mineral horizons of two Spodosols is involved in the retention of sulfate. The precipitation of hydroxy Al sulfates can be promoted by applying gypsum as an ameliorant for acid subsoils [31].

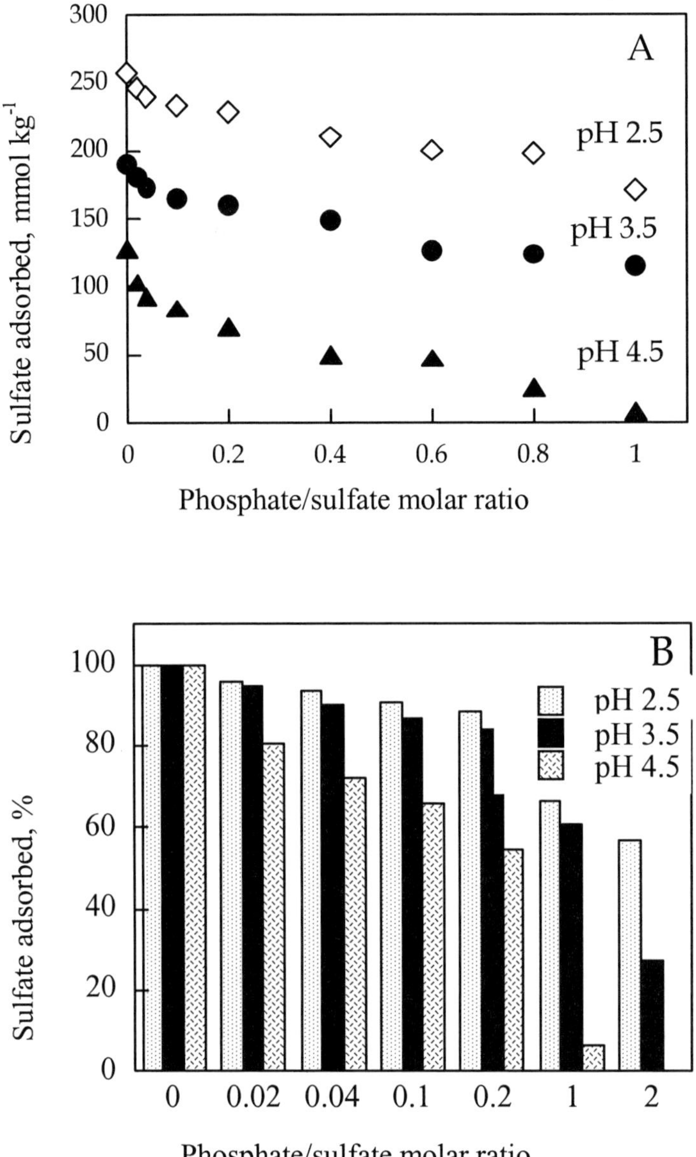

Figure 1. A) Sulfate adsorbed (470 mmol sulfate added kg^{-1}) by an Andisol at pH 2.5, 3.5 and 4.5 in the presence of increasing phosphate concentrations. B) Percentage of sulfate adsorbed at pH 2.5, 3.5 and 4.5 in the presence of increasing phosphate concentrations (initial phosphate/sulfate molar ratio ranging from 0 to 2). The percentages of sulfate adsorbed (470 mmol sulfate added kg^{-1}) are referred to the relative amounts of sulfate adsorbed in the absence of phosphate.

When sulfate, as well as other inorganic or organic ligands, are co-precipitated with aluminum (or iron), they are either adsorbed on the external surfaces or present in the network of the initially formed short-range ordered aluminum (or iron) precipitation products [23, 31, 32]. However, it is not possible to determine how much of the ligand is surface-adsorbed and how much is part of the structure of the precipitate [8].

The removal of sulfate by phosphate (or other ligands) is substantially different when sulfate ligands are present in aluminum (or iron) hydroxysulfate precipitation products. Violante et al. [23] demonstrated that large amounts of phosphate or even repeated extractions with phosphate only partially removed the sulfate ions present in a synthetic aluminum hydroxysulfate precipitate. Some researchers also found that extractions with KH_2PO_4 did not remove all the sulfate present in a latosol [26], and brown forest soils [33]. Haque and Walmesley [33] observed that sulfate adsorbed on surface groups of hydrous oxides is desorbed, whereas sulfate that penetrates into some amorphous region of the crystal surface is retained.

According to Prietzel and Hirch [27], inorganic sulfate is probably underestimated in soils containing aluminum hydroxysulfates.

3.1.2. Competition in adsorption between arsenate and phosphate

Many studies have shown that the phosphate and arsenate adsorption behaviors on oxides are similar, either considering the amounts of anions adsorbed or relating the adsorption to pH [34-37].

Recently, Liu et al. [38] have studied the competition in adsorption between arsenate and phosphate on a goethite. Table 1 shows the amounts of arsenate (As) and phosphate (P) adsorbed at pH 5.0 on a goethite (surface area 82 m^2 g^{-1}) when the ligands were added alone and when arsenate was added as a mixture with phosphate (*As + P*), before phosphate (*As before P*), or after phosphate (*P before As*).

Table 1
Amounts (mmol kg^{-1}) of phosphate (P) and arsenate (As) adsorbed on a goethite at pH 5 when added alone or together (240 mmol kg^{-1} P or As added)

	P adsorbed	As adsorbed	Total	Rf**
		mmol kg^{-1}		
P added alone	172	-	172	-
As added alone	-	161	161	-
P + As *	90	101	181	1.12
As before P	46	147	183	3.19
P before As	128	74	202	0.58

*The phosphate was added together (P + As), after (As before P) or before (P before As) arsenate.

** Rf is As adsorbed/P adsorbed molar ratio.

In these experiments, equal amounts (240 mmol kg^{-1}) of each anion were added to goethite. The amounts of anions adsorbed decreased compared with those of only arsenate and phosphate (161 and 172 mmol kg^{-1}, respectively). In fact, in *As + P* systems, the amounts of arsenate and phosphate were substantially reduced but similar (101 and 90 mmol kg^{-1}). By adding arsenate 24 h before phosphate (*As before P*), the amount of arsenate was reduced by 19%, whereas that of phosphate was reduced by 73% with respect to the quantities of arsenate and phosphate adsorbed when added alone. In contrast, when phosphate was added to goethite 24 h before arsenate (*P before As*), the amount of phosphate was reduced by 25%, whereas that of arsenate was reduced by 54%, with respect to the quantities of phosphate and arsenate adsorbed when added alone. These authors also found that the decrease in adsorption of phosphate in the presence of increasing concentration of arsenate was greater than that of arsenate in the presence of increasing concentrations of phosphate.

These findings appear surprising because both arsenate and phosphate form the same surface complexes on goethite: monodentate, bidentate-binuclear, and/or bidentate-mononuclear complexes in different proportion, depending on surface coverages [13, 34, 36-41]. However, according to Lumsdon et al. [42], arsenate and phosphate occupy equivalent sites on the goethite surfaces, but for its larger size, arsenate interacts more strongly than phosphate with some of the –OH groups on the surface of the goethite. Furthermore, recent studies by extended X-ray adsorption fine structure (EXAFS) and transmission-Fourier transform infrared (T-FTIR)-attenuated total reflectance-FTIR (ATR-FTIR) [36, 37] showed that the strong retention of arsenate on goethite (and ferrihydrite) is most likely caused by the formation of binuclear (mainly) and trinuclear complexes with the iron oxides. It is possible that mainly in acidic environments arsenate forms stronger complexes on Fe-oxides than phosphate does.

Our more recent studies (Violante and Pigna, unpublished data) have shown that in spite of the fact that phosphate and arsenate adsorption behavior on soil components is usually similar when added alone, the competition in adsorption between these ligands is substantially different on different variable charge adsorbents (e.g., gibbsite, goethite, allophane, mixed Al-Fe gels, birnessite, pyrolusite, montmorillonite, and kaolinite) and soils. These studies demonstrated that for goethite, pyrolusite, and birnessite, the affinity of these ligands tends to be arsenate > phosphate; these studies confirmed the findings of Liu et al. [38] on the competition in adsorption onto goethite. However, for gibbsite, kaolinite, montmorillonite, and Andisols, the affinity of these ligands tends to be phosphate > arsenate. Table 2 shows the amounts of arsenate and phosphate adsorbed at pH 3.5, 5.0, and 7.0 on goethite, gibbsite, and a subsurface volcanic soil. The quantities of arsenate and phosphate adsorbed on each adsorbent (mainly for goethite and the Andisol) were substantially similar when added alone.

However, when added in equimolar amounts as a mixture, arsenate and phosphate competed for the sorption sites of the minerals, but more phosphate than arsenate was adsorbed on gibbsite and on the Andisol, whereas more arsenate than phosphate was adsorbed on goethite. At a given pH, the adsorbed arsenate/adsorbed phosphate molar ratio (Rf) decreased in the order goethite > Andisol > gibbsite.

The Rf values decreased with increasing pH, indicating that arsenate competes with phosphate more in acidic than in neutral (and alkaline; data not shown) environments. Our recent studies have demonstrated that arsenate replaces phosphate, previously adsorbed on different sorbents more easily from Fe- or Mn-oxides than from Al-oxides, allophanes, kaolinite, or Andisols. The competitive adsorption of these ligands on soil components deserves closer attention.

Table 2
Amounts (mmol kg^{-1}) of phosphate (P) and arsenate (As) adsorbed on a goethite, a gibbsite and a subsurface Andisol at pH 3.5 , 5.0, and 7.0 when added alone or as a mixture (*P* + *As* system)

	Anion added alone		Anion added as a mixture		Rf*
	P adsorbed	As adsorbed	P adsorbed	As adsorbed	
	mmoli kg^{-1}				
Goethite	(200 mmol P or As added kg^{-1})				
pH 3.5	177	168	79	92	1.16
pH 5.0	158	175	78	87	1.11
pH 7.0	136	152	73	72	0.99
Andisol	(470 mmol P or As added kg^{-1})				
pH 3.5	457	458	453	359	0.79
pH 5.0	449	457	436	267	0.61
pH 7.0	400	370	390	158	0.41
Gibbsite	(400 mmol P or As added kg^{-1})				
pH 3.5	350	259	277	130	0.47
pH 5.0	280	205	198	82	0.41
pH 7.0	225	n.d	161	60	0.37

* Rf is As adsorbed/P adsorbed molar ratio.

3.2. Competitive adsorption of phosphate and organic ligands

The competitive adsorption of phosphate and chelating organic ligands is particularly important in soil environments because the supply of nutrients to plants should be strongly influenced by the presence of these ligands [7-9,17-23, 43-47].

According to Bar-Yosef [48], phosphate solubilization by organic acids in soils may be due to one or more of the following mechanisms: (1) competition of phosphate on common adsorption sites, (2) modification of the soil surface characteristics, and (3) complexation of cations (Al or Fe) with which phosphate coprecipitates. The three mechanisms overlap, but the first mechanism seems to be the most important.

The studies of Swenson et al. [49] and Johnston [50] were among the first works on the effect of organic ligands on phosphate adsorption on variable charge minerals and soils. These authors found that aliphatic and aromatic hydroxy-acids are capable of forming stable complexes with cations responsible for phosphate fixation (Al, Fe, Ca) and thereby are effective in preventing phosphate adsorption. Nagarajah et al. [43, 44] found that the competitive ability of carboxylic acids in preventing phosphate adsorption on kaolinite, gibbsite and goethite is in the order citrate > oxalate > malonate > tartrate > acetate; later, Lopez-Hernandez et al. [46] demonstrated that malate and oxalate strongly reduce phosphate adsorption by tropical soils. More recently, many studies showed evidence that the competitive adsorption of phosphate and organic ligands on variable charge minerals and soils is influenced by pH, nature of the ligand, order of anion addition, and initial ligand/phosphate molar ratio [17-22, 51-53]. Figure 2 shows the amounts of phosphate (P) adsorbed, at pH 5.0 in the presence of oxalate (OX) on a chlorite-like mineral (an $Al(OH)_x$-montmorillonite complex containing 3.3 mol Al kg^{-1} clay) when 500 mmol kg^{-1} of phosphate and increasing quantities of oxalate were added. Phosphate ions were added together (*P + OX* systems), after (*OX before P* systems), or before oxalate (*P before OX* systems). Oxalate reduced phosphate adsorption more when added first than when added as a mixture with phosphate or after phosphate. Ligands characterized by low affinity for Al and Fe (e.g., acetate, formate, phthalate, aspartate, malonate, gluconate, succinate, and benzoate) compete poorly with phosphate and only at ligand/phosphate molar ratio ≤ 2 [51-53].

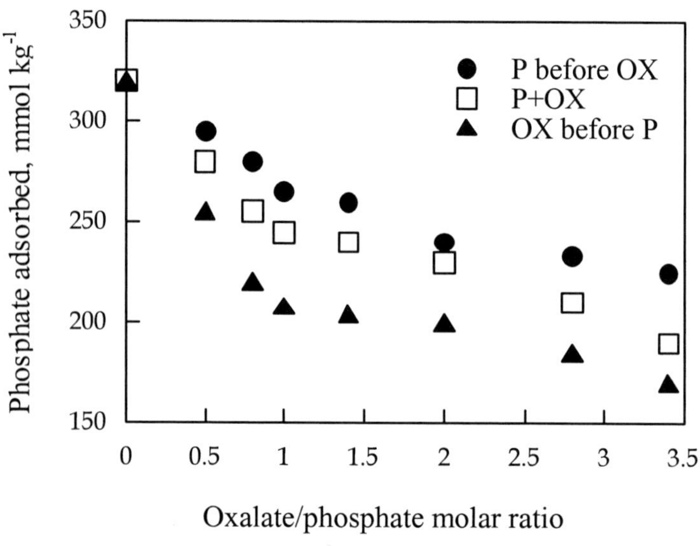

Figure 2. Amounts of phosphate (P) adsorbed (500 mmol P added kg^{-1}) on a chlorite-like complex (3.3 mol Al kg^{-1} montmorillonite) at pH 5.0 in the presence of increasing concentrations of oxalate (OX). The phosphate was added together (P+OX), after (OX before P) and before (P before OX) oxalate.

The surface loading also plays an important role in ligand competition. Figure 3 shows the amounts of phosphate adsorbed at pH 5.5 on an $Al(OH)_x$-montmorillonite complex, in the presence of tartrate (r_i, initial tartrate/phosphate molar ratio). Fifty, 150, or 400 mmol kg^{-1} of phosphate were added to the mineral. Phosphate adsorption generally decreased with increasing tartrate concentration. However, no evidence of competitive adsorption was observed when low quantities of phosphate (50 mmol kg^{-1}) were added to the complex in the presence of tartrate.

This may result because there were sufficient sites available for both of the anions. By adding 150-400 mmol phosphate kg^{-1} clay, more phosphate than tartrate was adsorbed on the complexes even when the initial r_i was 3.0.

Mixtures of tartrate and oxalate added in equimolar quantities were much more effective in inhibiting phosphate fixation than tartrate or oxalate alone under the same ligand concentrations. Clearly, many sites on the complexes were specific only for phosphate, whereas many others, common to phosphate, tartrate, and oxalate, had a greater affinity for phosphate than for oxalate or tartrate. However, some sites showed a very high affinity for tartrate and some others for oxalate. Evidently, when a mixture of ligands was added to the complexes, their effectiveness in preventing phosphate adsorption was greater because more sites at high affinity for both the organic ligands were occupied by tartrate and oxalate than by oxalate or tartrate alone.

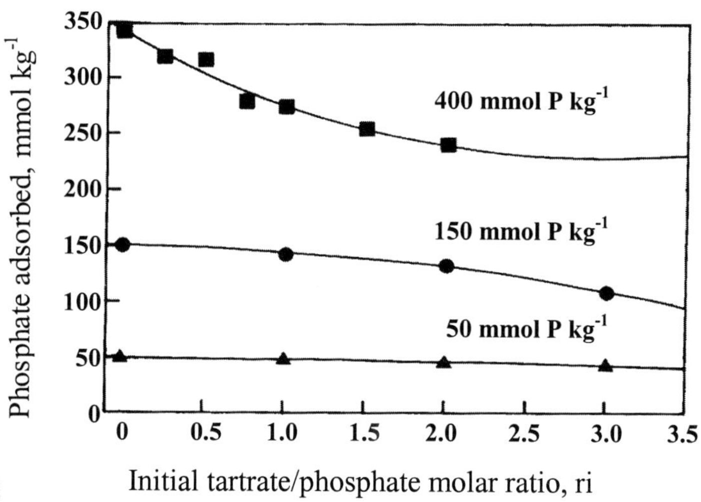

Figure 3. Added tartrate/added phosphate molar ratios versus adsorbed phosphate (mmol P kg^{-1}) at pH 5.5 on an Al(OX)x-montmorillonite complex, containing 1.6 mol Al kg^{-1} montmorillonite. Fifty, 150 or 400 mmol p kg^{-1} were added. Tartrate and phosphate were added as a mixture (modified from [22]).

Long-term laboratory studies have demonstrated that the efficiency of organic ligands in reducing phosphate adsorption is different on adsorbents characterized by different surface and physico-chemical properties (e.g., noncrystalline Al-hydroxides, goethite, ferrihydrite, clay fractions of variable charge soils, chlorite-like complexes, montmorillonite-AlOOH complexes) [8-9, 51]. The efficiency of organic ligands in reducing phosphate adsorption on variable charge minerals and soils increases with increasing initial ligand/phosphate molar ratios and decreases with increasing pH, especially at pH > 6.0.

Similar results were reported by Hue [52], who studied phosphate sorption on two Andisols, an Oxisol, an Ultisol, and a Vertisol from Hawaii to evaluate the effects of soil mineralogy and organic acids (acetic, malic, and protocatechuic acid) on phosphate retention. Findings in the laboratory were verified by bioassay in the greenhouse, using lettuce as a test crop. Soil minerals controlled phosphate sorption. Andisols (short-range ordered materials) sorbed most phosphate; the Vertisol (smectites), least; the Oxisol and Ultisol (kaolinite, Al and Fe oxides), intermediate. In reducing phosphate adsorption, malic acid was the most effective, protocatechuic acid had an intermediate effect, and acetic acid was the least effective. However, in the greenhouse, protocatechuic acid was more effective in increasing phosphate phytoavailability than malic acid because protocatechuic acid is more resistant to microbial decomposition. According to Hue [52], competition of the acids with phosphate for sorption sites played a major role in releasing phosphate in solution. This means that the efficiency of phosphate fertilizers increases significantly if they are applied along with acid-producing materials such as green manures and animal wastes.

Fox and Comerford [45] and Fox et al. [53] identified oxalate and formate as the most abundant low molecular weight organic ligands in the soil solutions in Spodosols under forest vegetation in northern Florida and investigated the effect of the organic anions on the kinetics and inorganic- or organic-P release from A, Bh, and Bt horizons of a representative forested Spodosol. Results showed that oxalate had little influence on phosphate release in the A horizon, owing to the lack of Al oxides to hold phosphate, but, on the contrary, it greatly increased solution phosphate in the Bh and Bt horizons. In the subsoil horizons, the release of phosphate was rapid and followed the disappearance of oxalate from solution. These facts suggested that phosphate was released via a ligand-exchange reaction. Conversely, the presence of formate did not increase phosphate release.

3.3. Adsorption of phosphate on organo-mineral complexes

The physico-chemical properties, chemical composition, and reactivity of Al and Fe precipitation products are influenced by pH, nature, and concentration of biomolecules present in soil environments [7-9,19,54]. Multi-functional organic ligands associated with Al and Fe may act as stabilizing agents in the formation of aggregates. During the formation of short-range-ordered Al or Fe precipitation products, organic ligands are strongly adsorbed on the external surfaces and incorporated into the network of these minerals. As a consequence, the organic molecules are not easily oxidized or are not completely replaced by other ligands [20]. The amounts of organic ligands present in organo-mineral complexes depend on the initial ligand/Al (or Fe) molar ratio (R), pH, and nature of the ligand [7, 19, 20, 55].

The role of organic molecules present in Al-organic matter associations in the adsorption of anions and cations has received scant attention. However, it seems possible that depending on the mechanisms of their interfering reactions, humic, organic acids, and phenols have a dual role of both hindering and promoting phosphate adsorption by the Al precipitation products. First, organic ligands complex Al in aqueous solution and subsequently hamper the crystallization of Al and Fe oxides. The effectiveness of an organic ligand in perturbing hydrolytic reactions of Al is related to its chemical composition, molecular structure, size, functional groups, and subsequent affinity for Al and Fe [7, 19, 20]. Maintenance of short-range structure of the precipitates with large surface structures by the presence of critical concentrations of some biomolecules helps to promote a high phosphate retention capacity of organo-mineral complexes.

Recently, De Cristofaro et al. [18] found that hydroxy-Al-oxalate precipitates, containing different amounts of oxalate, showed different chemical and physicochemical properties and reactivity toward phosphate (Tables 3 and 4). The quantities of phosphate sorbed on the hydroxy-Al-oxalate precipitates were related either to the amount of oxalate coprecipitated with Al or to the specific surface. In fact, it was found that the greater the oxalate content in the precipitates, the higher were the specific surface and solubility of the solids, in accordance with the findings of other authors [7-9, 19]. Phosphate sorption on the hydroxy-Al-oxalate precipitates that contained greater amounts of oxalate remained nearly constant in the range of pH 5.0-9.0, probably due to the release of oxalate from the surfaces of the solids with increasing pH and formation of new sorption sites (Table 4). The above cited results seem to strengthen the observation of Bloom [56]. This author showed that in a pH range 4.7-6.1, phosphate sorption on an Al-peat complex and on a P-fertilized soil was not affected by pH. On the contrary, Haynes and Swift [57] found that the phosphate sorption capacity increased with the pH of Al-organic matter-soil associations. Liming has been reported to increase, decrease, or not affect the phosphate sorption by highly weathered acid soils [58 and references therein reported].

Table 3
Chemical composition, surface area (SA), and point of zero salt charge (PZSC) of the aluminum precipitation products formed at pH 7.0 in the presence of different initial oxalic acid/Al molar ratios (Ri)* (modified from [18])

Property	$Al(OH)_3$	Al(OH)OX1	Al(OH)OX2	Al(OH)OX5
Oxalate mol kg^{-1}	-	0.960	1.580	2.670
Al, mol kg^{-1}	12.74	10.51	10.27	8.67
OX/Al**	-	0.09	0.15	0.31
PZSC	7.70	6.80	6.80	5.14
SA, m^2 g^{-1}	192	200	465	698

* Al(OH)OX1, Ri = 0.1; Al(OH)OX2, Ri = 0.2; Al(OH)OX5, Ri = 0.5.
** OX/Al = oxalate/Al molar ratios in the precipitates.

De Cristofaro et al. [18] also showed that relatively high amounts of oxalate were released from the hydroxy-Al-oxalate precipitates by phosphate, and much more from the complexes containing higher amounts of oxalate (Table 4). At alkaline pHs, the much greater percentage of oxalate, relative to the total content of oxalate coprecipitated with Al in the solids, was released from the solids containing greater amounts of the organic anion. These results may be explained by considering that the higher the amounts of oxalate initially coprecipitated with Al, the greater the percentage of oxalate present in the external surfaces of the samples [55]. In other words, the $Al(OH)_x$-oxalates with a lower organic molecule content had greater amounts of oxalate present within the network of the organo-mineral complexes, where the organic ligands were much more protected and not easily replaced by OH^- ions.

Table 4.
Amounts of phosphate (P) sorbed (1000 mmol P added kg^{-1}) on hydroxy-Al-oxalate precipitates* and oxalate (OX) released from the solids at different pH values (modified from [18]). Values in parenthesis are the percentages of oxalate released by the Al-OH-oxalate precipitates relative to the total content of oxalate coprecipitated

pH	OX released in the absence of P	P sorbed	OX released in the presence of P	R_p
		mmol kg^{-1}		
Al(OX)OX1*				
4.0	45 (<5)	105	45 (<5)	-
5.0	15 (<5)	125	40 (<5)	0.20
6.0	65 (7)	105	75 (8)	0.10
7.0	70 (7)	100	85 (9)	0.15
8.0	90 (7)	85	110 (11)	0.23
Al(OX)OX2*				
4.0	175 (11)	716	163 (10)	-
5.0	75 (<5)	555	137 (9)	0.11
6.0	60 (<5)	590	210 (13)	0.25
7.0	130 (8)	580	375 (24)	0.42
8.0	435 (27)	490	520 (33)	0.17
Al(OX)OX5*				
4.0	1230 (46)	950	1185 (44)	-
5.0	450 (17)	950	1120 (42)	0.7
6.0	490 (18)	945	1180 (44)	0.73
7.0	775 (29)	950	1410 (53)	0.66
8.0	1300 (48)	955	1920 (72)	0.65

* See Table 3

The counteracting effects of exposed organic ligands on the surfaces of the precipitation products have also been demonstrated [20,55]. Violante and Huang [20] synthesized crystalline boehmites (AlOOH) in the presence of selected organic ligands (aspartate, R = 0.2; citrate, R = 0.01; tartrate, R = 0.01 and tannate R = 0.01 and 0.02). These oxyhydroxides showed comparable specific surfaces ranging from 460 (citrate) to 560 m^2 g^{-1} (tannate; R = 0.01). From 138 to 148 mmol kg^{-1} of ligands were found in the samples formed in the presence of citrate, tartrate, or tannate; a much greater amount of aspartate ligand (437 mmol kg^{-1}) was found in the samples formed in the presence of aspartate. Figure 4 shows the adsorption of phosphate by these short-range ordered Al precipitation products.

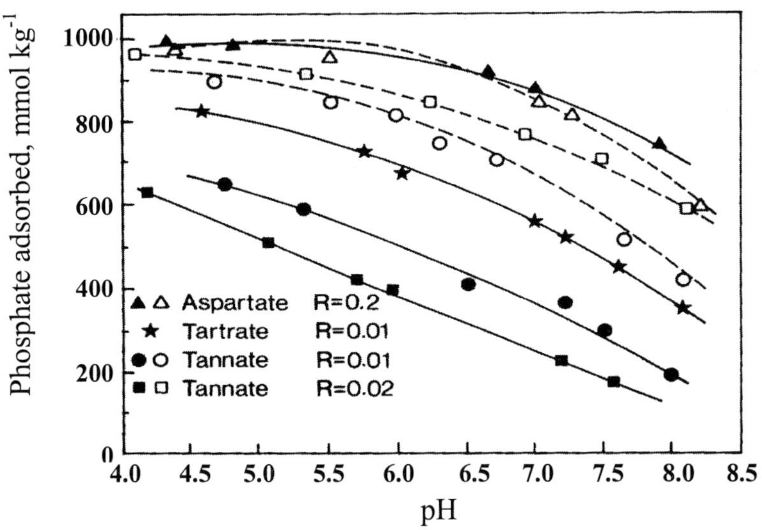

Figure 4. Adsorption of phosphate by poorly crystalline boehmites (ALOOH), formed in the presence of aspartate, tartrate and tannate, before and after (dashed lines) NaOCL treatment. R indicates the initial ligand/Al molar ratio (modified from [20]).

The sample formed in the presence of aspartate retained more phosphate than the samples obtained in the presence of tartrate, citrate, or tannate, even though a much larger amount of organic anion was present in the former than in the latter. Clearly, the nature and the amount of perturbing ligands coprecipitated with Al and the surface area and crystallinity of the samples played a significant role in phosphate sorption. However, aspartate, which shows a moderate affinity for Al, would poorly counteract phosphate sorption and would be easily replaced by phosphate [20,23]. On the other hand, citrate, tartrate, and tannate, which have a high affinity for Al, would not easily be removed and inhibit phosphate sorption. Amounts of phosphate retained by precipitation products formed in the presence of tannate were relatively small, in spite of their large surface area. Tannate anions, which have a high molecular weight (MW = 1701), should cover a large area of the surfaces of Al precipitation products, masking many phosphate adsorption sites. After oxidation by NaOCl, adsorption of phosphate by samples formed in the presence of citrate, tartrate, and tannate was substantially increased (Figure 4; dashed lines). The amount of phosphate adsorbed on the precipitation product formed in the presence of tannate (R = 0.02) was greater than that adsorbed on the materials formed in the presence of citrate or tartrate. The amount of phosphate adsorbed by the precipitate formed in the presence of aspartate after NaOCl treatment was still greatest among the samples studied. These findings are attributed to oxidation of organic ligands, which counteracted phosphate adsorption as well as disintegration of aggregates promoted by organic ligands. The data thus reveal that the influence of oxidation pretreatment on phosphate adsorption ability of organo-mineral complexes varied with the nature and amount of biomolecules present in the samples.

4. CONCLUSIONS

Adsorption of phosphate on variable charge minerals and soils and organo-mineral complexes is affected by the pH, and the presence, concentration, and nature of inorganic and organic ligands. We have demonstrated that:

- Sulfate poorly competes with phosphate for adsorption sites of minerals and soils at pH > 5.0, but at very low pH values, it is not very easily replaced by phosphate, probably forming inner-sphere complexes. Sulfate coprecipitated with Al or Fe that is part of the structural network is only partly removed by large amounts of phosphate.
- Phosphate and arsenate adsorption behavior on different sorbents is usually similar. However, the relative affinity of these anions for goethite, pyrolusite, and birnessite is arsenate > phosphate, whereas for gibbsite, kaolinite, montmorillonite, and Andisols, it is phosphate > arsenate. The desorption of phosphate by arsenate is also affected by pH, anion concentration, and nature of the variable charge minerals.
- Chelating organic acids influence the adsorption of phosphate. Maximum reduction in phosphate adsorption occurs in acidic systems and when organic ligands are previously adsorbed on clay minerals and soils. Strongly organic ligands present on the surfaces of metal oxides counteract phosphate adsorption more than chelating ligands with moderate or poor affinity for Al or Fe.
- Hydroxy-Al-oxalate precipitates, containing different amounts of organic ligands, show different reactivity toward phosphate. The quantities of phosphate sorbed on organo-mineral complexes are related either to the amount of organic ligands coprecipitated with Al or to the specific surface.

Our studies clearly show that many different organic ligands, usually released in the rhizosphere by roots of plants and microorganisms, as well as nutrients and xenobiotics, play an important role in the adsorption/desorption of phosphate on variable charge soil minerals.

ACKNOWLEDGMENTS

This work was supported in part by Ministero dell' Università e della Ricerca Scientifica e Tecnologica (MURST), Programmi di Ricerca Scientifica di interesse nazionale (PRIN). Contribution N. 203 from Dipartimento di Scienze Chimico-Agrarie (DISCA).

REFERENCES

1. Sparks, D.L., 1995. Environmental Soil Chemistry. Academic Press, San Diego.
2. Sumner, M.E., 2000. Handbook of Soil Science. CRC Press, Boca Raton. FL.

3. Hsu, P.H., 1989. Aluminum hydroxides and oxyhydroxides. *In*: Dixon, J.B., Weed, S.B. (Eds.), Minerals in Soil Environments. 2nd Ed., Book Ser. no.1, Soil Sci. Soc. Am., Madison, WI., pp. 331-378.
4. Barnhisel, R.I., Bertsch, P.M., 1989. Chlorite and hydroxy-interlayered vermiculite and smectite. *In*: Dixon, J.B., Weed, S.B. (Eds.), Minerals in Soil Environments. 2nd Ed., Book Ser. no.1, Soil Sci. Soc. Am., Madison, WI., pp. 729-788.
5. Wada, K., 1989. Allophane and imogolite. *In*: Dixon, J.B., Weed, S.B. (Eds.), Minerals in Soil Environments, 2nd Ed., Book Ser. no.1, Soil Sci. Soc. Am., Madison, WI, pp.1051-1087.
6. Sposito, G., 1996. The Environmental Chemistry of Aluminum. 2nd Ed., CRC Press, Lewis Pub., London.
7. Violante, A., Krishnamurti, G.S.R., Huang, P.M., 2002. Impact of organic substances on the formation and transformation of metal oxides in soil environments. *In*: Huang P.M., Bollag, J.-M., Senesi. N., (Eds.), Interactions Between Soil Particles and Microorganisms and the Impact on Terrestrial Environment, John Wiley & Sons Ltd. (in press).
8. Huang, P.M., Violante, A., 1986. Influence of organic acids on crystallization and surface properties of precipitation products of aluminum. *In*: Huang, P.M., Schnitzer, M. (Eds.), Interactions of Soil Minerals with Natural Organics and Microbes, Spec. Publ. No. 17, Soil Sci. Soc. Am., Madison WI., pp. 159-221.
9. Violante, A., Gianfreda, L., 2000. Role of biomolecules in the formation of variable-charge minerals and organo-mineral complexes and their reactivity with plant nutrients and organics in soil. *In*: Bollag, J.-M., Stotzky, G. (Eds.), Soil Biochemistry. Vol. 10, Marcel Dekker, New York, pp. 207-270.
10. Goldberg, S., Davis, J.A., Hem, J.D., 1996. The surface chemistry of aluminum oxides and hydroxides. *In*: Sposito, G. (Ed.) The Environmental Chemistry of Aluminum. 2nd Ed., Lewis Publishers, Boca Raton, FL, pp. 271-331.
11. Curtin, D., Syers, J.K., 1990. Mechanisms of sulphate adsorption by two tropical soils. J. Soil Sci. 41, 295-304.
12. Zhang, P.C., Sparks, D.L., 1990. Kinetics and mechanisms of sulfate adsorption/ desorption on goethite using pressure-jump relaxation. Soil Sci. Soc. Am. J. 54, 1266-1273.
13. Liu, F., He, J., Colombo, C., Violante, A., 1999. Competitive adsorption of sulfate and oxalate on goethite in the absence or presence of phosphate. Soil Sci., 164, 180-189.
14. Pasricha, N.S., Fox, R.L., 1993. Plant nutrient sulfur in tropics and subtropics. Adv. Agron. 50, 209-269.
15. Peak J.D., Sparks, D.L., Ford, R.G., 1999. An *in situ* ATR-FTIR investigation of sulfate bonding mechanisms on goethite: J. Coll. Interf. Sci. 218, 289-299.
16. Turner, L.J., Kramer, J.R., 1991. Sulfate ion binding on goethite and hematite. Soil Sci. 152, 226-230.
17. Violante, A., Colombo, C., Buondonno, A., 1991. Competitive adsorption of phosphate and oxalate by aluminum oxides. Soil Sci. Soc. Am. J. 55, 65-70.
18. De Cristofaro, A., He, J.Z., Zhou, D.H., Violante, A., 2000. Adsorption of phosphate and tartrate on hydroxy-aluminum-oxalate precipitates. Soil Sci. Soc. Am. J. 64, 1347-1355.
19. Violante, A., Huang, P.M., 1984. Nature and properties of pseudoboehmites formed in the presence of organic and inorganic ligands. Soil Sci. Soc. Am. J. 48, 1193-1201.

20. Violante, A., Huang, P.M., 1989. Influence of oxidation treatments on surface properties and reactivity of pseudoboehmites formed in the presence of organic ligands. Soil Sci. Soc. Am. J., 53, 1402-1407.
21. Violante, A., Gianfreda, L., 1993 Competition in adsorption between phosphate and oxalate on an aluminum hydroxide montmorillonite complex. Soil Sci. Soc. Am. J., 57, 1235-1241.
22. He, J.Z., De Cristofaro, A., Violante A., 1999. Comparison of adsorption of phosphate, tartrate, and oxalate on hydroxy aluminum montmorillonite complexes. Clays Clay Miner. 47, 226-233.
23. Violante, A., Rao, M.A., De Chiara, A., Gianfreda, L., 1996. Sorption of phosphate and oxalate by a synthetic aluminium hydroxysulphate complex. Eur. J. Soil Sci. 47, 241-247.
24. Elzinga E. J., Sparks, D., 1999. Sulfate sorption on goethite in the presence of Pb. 1999 ASA Annual Meeting, Salt Lake City. pp. 218.
25. Chao, T.T., Harwad, M.E., Fang, S.C., 1962. Anionic effects on sulfate adsorption by soils. Soil Sci. Soc. Am. Proc. 28, 581-583.
26. Bornemisza, E., Llanos, R., 1967. Sulphate movement, adsorption, and desorption in three Costa Rica soils. Soil Sci. Soc. Am. Proc. 31, 356-360.
27. Prietzel, J., Hirch, C., 1998. Extractability and dissolution kinetics of pure and soil-added synthesized aluminium hydroxy sulphate minerals. Eur. J. Soil Sci. 49, 669-681.
28. Adams, F., Rawajfih, Z., 1977. Basaluminate and alunite: a possible cause of sulphate retention by acid soils. Soil Sci. Soc. Am. Proc. 41, 686-692.
29. Evans, A. Jr., 1991. The interactions of aliphatic acids with basic aluminum sulphates in a forested Ultisol. Soil Sci. 152, 53-60.
30. Wolt, J.D., Hue, N.V., Fox, R.L., 1992. Solution sulphate chemistry in three sulfur-retentive Hydrandepts. Soil Sci. Soc. Am. J. 43, 118-121.
31. Sumner, M.E., 1993. Gypsum and acid soils: the world scene. Adv. Agron. 51, 1-32.
32. Courchesne, F., Hendershot, W.H., 1990. The role of basic aluminum sulphate minerals in controlling sulphate retention in the mineral horizons of two Spodosols. Soil Sci. 150, 571-578.
33. Haque, I., Walmstey, D., 1973. Adsorption and desorption of sulphate in some soils of the West Indies. Geoderma. 9, 269-278.
34. Manning, B.A., Goldberg, S., 1996. Modeling arsenate competitive adsorption with phosphate and molybdate on oxide minerals. Soil Sci. Soc. Am. J. 60, 121-131.
35. Smith, E. R., Naidu R., Alston, A. M., 1998 Arsenic in the soil environment: A review. Adv. Agron. 64, 149-195.
36. Sun, X., Doner, H.E., 1996. An investigation of arsenate and arsenite bonding structures on goethite by FTIR. Soil Sci. 161, 865-872.
37. Waychunas, G.A., Rea, B.A., Fuller, C.C., Davis, J.A., 1993. Surface chemistry of ferrihydrite: Part 1. EXAFS studies on the geometry of coprecipitated and adsorbed arsenate. Geochim. Cosmochim. Acta 57, 2251-2269.
38. Liu F., De Cristofaro A., Violante A., 2001. Effect of pH, phosphate and oxalate on the adsorption/desorption of arsenate on/from goethite. Soil Sci. 166, 197-208.
39. Fuller, C.C., Davis, J.A., Waychunas, G., 1993. Surface chemistry of ferrihydrite: Part 2. Kinetic of arsenate adsorption and coprecipitation. Geochim. Cosmochim. Acta. 57, 2271-2282.
40. Fendorf, S.M., Eich, M.J., Grossl, P., Sparks, D.L., 1997. Arsenate and chromate retention mechanisms on goethite. 1. Surface structure. Environ. Sci. Technol. 31, 315-320.

41. Hsia, T.H., Lo, S.L., Lin, C.F., Lee, D.Y., 1994. Characterization of arsenate adsorption on hydrous iron oxides using chemical and physical methods. A: Physicochemical Eng. Aspects. Coll. Surf. 85, 1-7.
42. Lumsdon, D.G., Fraser, A.R., Russell, J.D., Livesey N.T., 1984. New infrared band assignments for the arsenate ion adsorbed on synthetic goethite (α-FeOOH). J. Soil Sci. 35, 381-386.
43. Nagarajah, S., Posner, A.M., Quirk, J.P., 1968. Desorption of phosphate from kaolinite by citrate and bicarbonate. Soil Sci. Soc. Am. Proc. 32, 507-510.
44. Nagarajah, S., Posner, A.M., Quirk, J.P., 1970. Competitive adsorption of phosphate with polygaracturonate and other organic anions on kaolinite and oxide surfaces. Nature (London). 228, 83-84.
45. Fox, T.R., Comerford, N.B., 1990. Low-molecular-weight organic acids in selected forest soils of southeastern USA. Soil Sci. Soc. Am. J. 54, 1139-1144.
46. Lopez-Hernandez, D., Siegert, G., Rodriguez, J.V., 1986. Competitive adsorption of phosphate with malate and oxalate by tropical soils. Soil Sci. Soc. Am. J. 50, 1460-1462.
47. Marschner, H., (1995). Mineral Nutrition of Higher Plants. 2nd Ed., Academic Press, London.
48. Bar-Yosef, B., 1991. Root excretions and their environmental effects. Influence on availability of phosphorus. *In*: Waisal, Y., Eshel, A., Kafkafi, U. (Eds.), The Plant Root: The Hidden Half. Marcel Dekker Inc, New York, pp. 529-557.
49. Swenson, R.M., Cole, C.V., Sieling, D.H. 1949. Fixation of phosphate by iron and aluminum and replacement by organic and inorganic ions. Soil Sci. 67, 3-22.
50. Johnston, H.W., 1959. The solubilization of insoluble phosphate 5. The action of some organic acids on iron and aluminium phosphates. New Zealand J. Sci. 2, 215-218.
51. Violante, A., Gianfreda, L., 1995. Adsorption of phosphate on variable charge minerals: Competitive effects of organic ligands. *In*: Huang, P.M., Berthelin, J., Bollag, J.-M., McGill, W.B., Page, A.L. (Eds.), Environmental Impact of Soil Component Interactions. CRC Lewis Publisher, Boca Raton, FL, pp. 27-36.
52. Hue, N.V., 1991. Effects of organic acids/anions on P sorption and phytoavailability in soils with different mineralogies. Soil Sci. 152, 463-471.
53. Fox, T.R., Comerford, N.B., Mc Fee, W.W., 1990. Kinetics of phosphorus release from Spodosols: effects of oxalate and formate. Soil Sci. Soc. Am. J. 54, 141-147.
54. Violante, A., Violante, P., 1980. Influence of pH, concentration and chelating power of organic anions on the synthesis of aluminum hydroxides and oxyhydroxides. Clay Clay Miner. 30, 431-437.
55. Violante, A., Huang, P.M., 1992. Effect of tartaric acid and pH on the nature and physicochemical properties of short-range ordered aluminum precipitation products. Clays Clay Miner. 40, 462-469.
56. Bloom, P.R., 1981. Phosphorus adsorption by an aluminum-peat complex. Soil Sci. Soc. Am. J. 45, 267-272.
57. Haynes, R.J., Swift, R.S.,1989. The effects of pH and drying on adsorption of phosphate by aluminum-organic matter associations. J. Soil Sci. 40, 773-781.
58. Haynes, R.J., 1982. Effects of liming on phosphate availability in acid soil. Plant Soil. 68, 289-308.

Developments in Soil Science, Volume 28A
Editors: A. Violante, P.M. Huang, J.-M. Bollag and L. Gianfreda

RELATIONSHIPS BETWEEN ORGANIC AND INORGANIC P FRACTIONS WITH SOIL Fe AND Al FORMS IN FOREST SOILS OF SIERRA DE GATA MOUNTAINS (WESTERN SPAIN)

M. B. Turrión[a], J. F. Gallardo[b] and M. I. González[c]

[a] Universidad de Valladolid. Área de Edafología y Química Agrícola. 34004 Palencia, Spain

[b] C. S. I. C., Apdo, 257. 37071 Salamanca (Spain); E-mail: jgallard@gugu.usal.es

[c] Universidad de Salamanca. Área de Edafología y Química Agrícola, 37080 Salamanca, Spain

The aim of this work was to study the interactions of Al, Fe and P forms in four forest soils in a Mediterranean environment. A sequential procedure was used to distinguish soil P pools, and different extractions were used to determine Fe and Al forms. The acid nature of the bedrock, in addition to the high rainfall regime, favours high contents of free forms of Al and Fe. The soils studied (Humic Cambisols) had a high total P concentration (500–1200 mg kg^{-1}), but low contents of P extractable with anionic exchange membranes, (2–34 mg kg^{-1}). The labile P (P extracted with $NaHCO_3$) ranged from 74–189 mg kg^{-1} and accounted for 15–25% of total extractable P. Soils developed on granite showed higher concentrations of labile P than did those developed on slate. The adsorbed forms of P (P extracted with NaOH) ranged from 183–420 mg kg^{-1} and accounted for more than 50% of total extractable P. A high percentage of soil P was occluded in soil Al and Fe forms. In the soils over slate, the most stable fractions (P extracted with NaOH after applying ultrasound) accounted for more than 80% of the total extractable P, while in the soils over granite, the stable fractions were less than 60% of the total extractable P. The ratio between the most available forms of P and total extractable P was a negative function of the total amount of Fe and Al oxides. These compounds, either by occluding or by adsorbing phosphate on high-affinity sites, are crucial for the dynamics of P in these soils.

1. INTRODUCTION

Phosphorus is present in soil in a variety of inorganic (P_i) and organic (P_o) compounds. The pool of readily available P_i is small, and therefore plant productivity relies on replenishment of P_i in the soil solution at a rate sufficient for plant growth [1]. Dissolved inorganic P is known to be absorbed by plants directly, while labile organic P is utilized after mineralization. The dissolved inorganic P comes from the release of sorbed P_i from soil colloids, solubilization of precipitated P_i, and biological mineralization of soil P_o. Although the P_o content in soil may account for 20–90% of total P (P_t), only a small portion of the P_o is labile [2].

To evaluate the availability of soil P_i and P_o, various extraction methods have been developed. The fractionation procedure of Hedley et al. [3] relies on sequential extraction of soil P_i and P_o using an anion exchange resin for the most biologically available P_i, $NaHCO_3$ extraction for labile P_i and P_o, and successively stronger reagents based on NaOH, HCl, and finally, H_2SO_4/H_2O_2 for residual P. Sequential P fractionation enables us to determine P availability in soils and to assess the biochemical or geochemical cycling of P [4]. The advantage of the Hedley fractionation scheme is that information can be obtained about short- and long-term P availability by quantifying organic and inorganic P fractions [5, 6]. Ecological interpretations of these chemically defined fractions is controversial [4, 7, 8, 9]. The terminology used was based on the conceptual model (Figure 1) of soil P transformations [6, 8].

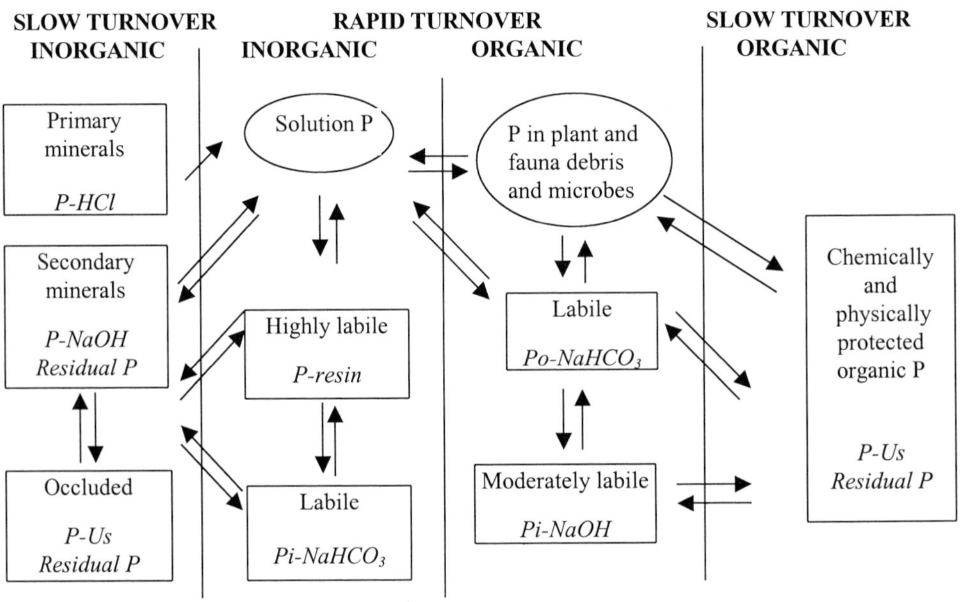

Figure 1. Conceptual model of soil phosphorus transformations [8] with its measurable components (modified from [6]). P-resin is extracted with resin membranes; P-$NaHCO_3$ is extracted by 0.5 M $NaHCO_3$; P-NaOH is extracted by 0.1 M NaOH; P-Us is occluded P, extracted by NaOH after ultrasonification; P-HCl is primary P, removed by 0.1 M HCl.

The low concentration of available phosphate in acid soils is a consequence of the formation of insoluble forms (Fe and Al phosphates and P forms sorbed by sesquioxides); thus, mineralization of organic P forms could be important in the cycle of this bioelement in this type of soil [8]. In acid soils, P sorption is generally attributed to hydrous oxides of Fe and Al and to (1:1) clays [10]. Knowledge of the forms and quantity of Fe and Al in the soils facilitates the interpretation of the physicochemical processes in which P is involved.

The aim of this work was to determine the relationships between organic and inorganic P forms with Fe and Al in forest soils (under *Quercus pyrenaica* and *Castanea sativa*) of the Sierra de Gata Mountains, western Spain.

2. MATERIALS AND METHODS

2.1. Site description

The soils under study are forest soils located in the Sierra de Gata Mountains (40° 2' N; 3° 0' W, Salamanca province, estern Spain). Four experimental plots, close to one another, were selected along a rainfall gradient. Three of these forests were *Quercus pyrenaica* oak coppices, and the fourth was a *Castanea sativa* coppice (Table 1).

The climate of the area is characterized by rainy winters and hot, dry summers and is classified as temperate to warm Mediterranean, with an average rainfall of 1570 mm yr^{-1} and a temperature of 11.3°C at Navasfrías, and 720 mm yr^{-1} and 13.3°C at Fuenteguinaldo. The four soils are Humic Cambisols [11] over acid bedrock (granite or slate). Hereafter, the following symbols will be used: SM for San Martín; NF for Navasfrías; VR for Villasrubias; and FG for Fuenteguinaldo.

Table 1
Characteristics of the forest plots studied

	Fuenteguinaldo (FG)	Villasrubias (VR)	Navasfrías (NF)	San Martín (SM)
Vegetation	----------- *Quercus pyrenaica*		---------	*Castanea*
Soil	------------------------ Humic Cambisol			-----------------------
Altitude [m.a.s.l].	870	900	1000	940
Bedrock	Granite	Slate	Slate	Granite
Mean precip. [$Lm^{-2}yr^{-1}$]	720	872	1570	1150
Mean T [°C]	13.3	N.d.	11.3	14.2
Density [trees ha^{-1}]	738	1043	820	3970
DBH [cm]	15.2	25.4	16.5	10.0
Mean tree height [m]	12.0	8.5	13.0	13.0
Basal area [m^2 ha^{-1}]	21.2	13.5	15.6	30.0
LAI [m^2 m^{-2}]	2.6	2.0	1.8	3.7
Aboveground production [Mg ha^{-1} yr^{-1}]	4.09	2.83	2.60	5.25

Note: N.d., no data available; m.a.s.l.; meters above sea level; Mean precip. mean annual rainfall; Mean T, mean annual temperature; DBH, mean trunk diameter; LAI, leaf area index.

2.2. Soil sampling

Six replicate samples were taken from each plot at depths of 0–10 cm, 10–20 cm, and 20–40 cm. Several points in every plot (8–10) were sampled and later pooled in a common sample for each soil depth. Soil samples were ground to < 2 mm mesh size. The results were calculated on a 105°C dry basis.

2.3. General analyses

The pH was measured in H_2O at a soil:solution ratio of 1:2.5 using a glass electrode. Organic carbon (C_{org}) was determined by dry combustion in a Carmhograph 12 Wösthoff analyzer; total nitrogen (N_t) was measured by Dumas oxidative digestion on a Macro N Heraeus apparatus. The particle-size distribution was determined by the pipette method [12]. The most important physical and chemical properties of the soils are shown in Table 2.

2.4. Fe and Al analyses

Total aluminum (Al_t) and iron (Fe_t) were determined by calcination at 500°C for 4 h and subsequent digestion in acid medium (HCl, HF, H_3BO_4) according to Hartstein et al. [13]. Free Al (Al_d) and Fe (Fe_d) contents were extracted with a mixed complexing and reducing buffer solution of Na citrate and Na ditionite as described by Holmgren [14]. The amorphous forms of Al (Al_o) and Fe (Fe_o) were determined by measuring Al and Fe dissolved in oxalic acid at pH 3.0 [15]. Aluminum and iron bonded with organic matter (Al_p, Fe_p) were extracted with 0.1 M $Na_4P_2O_7$ [16]; in this extract, the content of C (C_p) was determined using a Beckman 915A analyzer. The exchangeable Al (Al_{ex}) was extracted with 1 M KCl [17]. Fe and Al concentrations in the solutions were determined by atomic absorption spectrophotometry.

2.5. Sequential extractions and P analyses

Different soil P forms were determined using a modified fractionation method of Hedley et al. [3]. The fractionation delineates the following forms:

- Plant-assimilable P (P-resin) was extracted with anion exchange membranes according to Turrión et al. [18].
- Labile P (part of the P adsorbed onto surfaces of crystalline sesquioxides) was extracted by 0.5 M $NaHCO_3$ (P-$NaHCO_3$).
- P held strongly by chemisorption on Fe and Al soil components was removed by 0.1 M NaOH extraction (P-NaOH).
- P on the internal surfaces of soil aggregates (P-Us) was removed by ultrasonification then extraction of the soil residue in 0.1 M NaOH. Ca-bound P (P-HCl) was mainly removed with an acid extractant (1M HCl).
- Finally, the most chemically stable P forms were extracted by oxidation and acid digestion (H_2O_2+ H_2SO_4).

Aliquots of $NaHCO_3$, NaOH, and sonic NaOH extracts were acidified to precipitate extracted organic matter, and P_i was determined in the supernatant. Total P of these extracts was determined after digestion with persulfate [19]. P_o was calculated as the difference between total and inorganic P.

A separate soil sample was analyzed for total P concentrations (P_t) according to Saunders and Williams [20].

The P concentration in the solutions was analyzed colorimetrically with the molybdate-ascorbic acid procedure [21].

To compare soils, the percentages of each P form with respect to the total extracted P were calculated [% P-form = (mg P-form kg^{-1}) x 100 / (mg total extracted P kg^{-1})].

Table 2
Chemical and physical properties of the forest soils

Plots	Depth [cm]	pH (H_2O)	C_{org} [g kg^{-1}]	C_p [g kg^{-1}]	N_t [g kg^{-1}]	C/N	Sand [%]	Silt [%]	Clay [%]
FG	0-10	5.4	41.5	16.0	3.22	13	50	26	14
	10-20	5.1	25.0	11.0	2.08	12	53	29	14
	20-40	5.3	12.3	6.0	1.35	9	62	26	12
VR	0-10	4.6	67.3	40.0	3.99	17	20	62	14
	10-20	5.1	12.0	N.d.	1.25	10	10	69	17
	20-40	5.2	6.1	N.d.	0.91	7	9	74	16
NF	0-10	4.9	105.0	51.0	4.98	21	12	48	17
	10-20	4.8	58.0	41.0	3.37	17	21	53	14
	20-40	5.0	5.0	N.d.	0.47	11	12	77	7
SM	0-10	5.1	45.0	26.0	2.53	18	58	19	13
	10-20	4.7	34.0	23.0	1.85	21	59	21	15
	20-40	4.9	18.0	N.d.	1.08	17	62	26	10

Note: FG: Fuenteguinaldo; VR: Villasrubias; NF: Navasfrías; SM: San Martín; N.d., no data available; C_{org}, total organic carbon; C_p, pyrophosphate-extractable carbon; N_t, total nitrogen.

2.6. Statistical analyses

The means for replicates are presented with standard deviations. Differences in chemical measurements among the four sampling sites were tested by analysis of variance (ANOVA) when variables showed a normal distribution and homogeneity of variances. Otherwise, a Kruskall-Wallis procedure (nonparametric method) was applied. Subsequent multiple comparisons were made by a Scheffé test; significance testing was performed at the $p < 0.05$ level. Relationships between the P, Fe and Al forms, and other physical and chemical properties of the soils were investigated by correlation and regression analysis.

3. RESULTS AND DISCUSSION

3.1. Soil Fe and Al

Total Fe (Fe_t), free Fe (Fe_d) and amorphous Fe oxides (Fe_o) concentrations were significantly higher in the soils developed over slate than in the soils over granite (Table 3); thus, there were higher contents in Fe_t and amorphous oxides (Table 3) in the plots with higher amounts of soil organic matter (SOM).

The Fe_p are very similar to the Fe_o values (Table 3). At first, these results may seem to indicate that practically all the extracted amorphous Fe (Fe_o) is linked to SOM (Fe_p). However, this happens even for the deepest horizons, where the C_{org} content is low and therefore not much organically complexed Fe can be expected. Fe_p may also include the Fe that is not organically complexed, mainly ferrihydrite remaining in suspension. Fine suspended particles in the pyrophosphate extracts might also explain why the Al_p values are similar to or higher than the Al_o values (Table 3).

Table 3
Concentrations [g kg^{-1}] (except Al_{ex} in [mg kg^{-1}]) and standard deviations (n = 6) of the Fe and Al forms in the forest soils

Plot	Fe_t	Fe_d	Fe_o	Fe_p	Al_t	Al_d	Al_o	Al_p	Al_{ex}
FG									
0-10	19.7±1.6	7.3±0.6	2.4±0.1	1.9±0.2	78.6±3.4	6.5±1.0	2.4±0.6	3.4±0.4	1.2±0.1
10-20	18.6±0.7	5.6±1.2	2.5±0.1	1.9±0.1	84.1±1.9	8.1±0.6	2.7±0.5	3.7±0.7	1.3±0.8
20-40	18.2±0.9	5.5±1.8	2.3±0.2	1.6±0.2	92.0±2.2	7.2±0.7	2.5±0.3	3.5±0.3	2.0±0.2
VR									
0-10	36.6±1.8	23.8±1.6	5.7±0.3	5.5±0.2	85.7±1.8	25.2±0.4	4.0±0.3	8.9±0.5	272±27
10-20	43.0±0.8	21.3±0.5	5.6±0.2	5.3±0.1	104.1±2.3	26.0±0.9	4.2±0.4	9.0±0.1	239±20
20-40	43.5±3.8	19.4±0.9	5.2±0.5	5.1±0.2	106.4±6.9	26.9±2.1	4.5±0.8	10.0±0.4	207±22
NF									
0-10	35.3±1.8	14.4±2.0	5.8±0.4	5.1±0.3	82.7±3.6	33.5±0.6	8.1±0.2	10.2±0.2	282±28
10-20	43.8±1.6	15.6±1.7	6.1±0.6	5.0±0.5	96.1±1.9	39.4±1.5	9.4±0.3	11.6±0.2	244±45
20-40	45.6±2.5	13.7±2.7	5.8±0.4	4.8±0.4	115.8±3.3	36.0±0.4	10±0.6	11.9±0.4	157±11
SM									
0-10	26.1±1.0	7.8±0.5	4.4±0.1	3.8±0.1	83.8±3.4	13.5±0.3	5.1±0.1	6.0±0.2	220±34
10-20	25.3±1.0	8.5±1.0	4.8±0.2	4.1±0.1	91.6±2.8	18.5±1.4	6.4±0.2	7.3±0.3	222±25
20-40	23.5±1.6	9.6±2.4	5.2±0.5	4.2±0.3	97.4±2.7	19.1±0.9	6.9±0.3	7.4±0.5	200±24

Note: FG: Fuenteguinaldo; VR: Villasrubias; NF: Navasfrías; SM: San Martín; Fe_t, total iron; Fe_d, free iron; Fe_o, amorphous iron; Fe_p, SOM-bound iron ; Al_t, total aluminum; Al_d, free aluminum; Al_o, amorphous aluminum; Al_p, SOM-bound aluminum; Al_{ex}, exchangeable aluminum.

All the studied soils showed similar Al_t concentrations. However, the amounts of some forms of Al showed differences between plots. The soil of NF showed the highest content of Al_d, Al_o, and Al_{ex}, mainly due to the different degree of weathering of the bedrock, affected by the highest annual rainfall (Table 1) and also the highest content in C_{org}. A high Al_{ex}

concentration was found in all the soils (whose pH was lower than 5), except in FG (Table 3). The high Al_{ex} contents for NF, VR, and SM suggest Al toxicity [22]; however, the high content of C_{org} can maintain this Al in a non-active form [23] and no toxicity is observed in any of the plots.

Significant positive correlations were observed between Fe forms, Al forms (except Al_t) and C_{org} and C_p; these significant correlations could be an indirect reflection of the rainfall. Significant negative correlations ($p < 0.001$) were observed between Fe forms and Al forms (except Al_t) with pH (Table 4), reflecting the intensity of leaching. Correlations of Fe forms with silt (positive) and with sand (negative) are higher than with clay; the low correlation obtained with clay is due to the fact that this fraction is scarce in these soils, the silty fraction being more abundant. However, cause-effect relationships from the correlation analyses cannot be derived.

Table 4
Correlation matrix between Fe and Al forms and physical and chemical properties of the forest soils

	C_{org}	C_p	pH	Sand	Silt	Clay
Fe_t	0.67***	0.78***	-0.76***	0.84***	0.85***	0.54**
Fe_d	0.68***	0.71***	-0.82***	-0.89***	0.91***	0.51**
Fe_o	0.79***	0.89***	-0.83***	-0.56***	0.59***	N.s.
Fe_p	0.78***	0.84***	-0.91***	-0.73***	0.76***	N.s.
Al_t	N.s.	N.s.	-0.40*	-0.41*	0.41*	0.35*
Al_d	0.81***	0.94***	-0.83***	-0.69***	0.67***	0.64***
Al_o	0.59***	0.71***	-0.57***	N.s.	N.s.	0.45**
Al_p	0.76***	0.90***	-0.81***	-0.65***	0.64***	0.59***
Al_{ex}	0.76***	0.78***	-0.78***	N.s.	N.s.	N.s.

Note: FG: Fuenteguinaldo; VR: Villasrubias; NF: Navasfrías; SM: San Martín; N.s. not significant; * significant at $p < 0.05$; **, significant at $p < 0.01$; ***, significant at $p < 0.001$; C_t, total carbon; C_p, pyrophosphate-extractable carbon.

Johnson and Todd [24] also showed significant correlations between Al forms and SOM for forest soils. A condition for the formation of the metal-organic complexes is a high release of these metals, in particular of Al, from the parent material. Soils with such properties are, therefore, usually formed from volcanic ashes, but also from acid Al-rich substrates, such as granites and slates, under high rainfall and good drainage. Leaching of bases and silicon leads then to a relative accumulation of Al, which subsequently reacts with functional groups of organic molecules [23]. In the forest soils studied, the coincidence of annual precipitations (720–1520 L m^{-2} yr^{-1}) and an acid Fe- and Al-rich bedrock as geological substrate favors the formation of metal-organic complexes. Many authors [23, 25] have suggested that the metal complexation of humic matter protects the organic molecules from microbial attack, thus contributing to the humus accumulation in soil.

3.2. P fractionation

The forest soils under study showed high extractable P (P_{ex}; sum of extractable P forms) and P_t concentrations (Table 5).

3.2.1. P-resin

In all the plots, the highest P-resin contents were found in the surface horizons (Table 5). In two plots (VR and NF), there was a drastic decrease of P-resin with depth, where it accounted for only 1% of the total P_i at 20–40 cm, perhaps due to adsorption of P by the soil sesquioxides and metal-organic complexes (Table 3). The results of in situ studies [26] also showed that the contents of available P in the soils of VR and NF were lower than in those of FG and SM, thus indicating a lower availability of this element for plants and a low microbial activity [26].

The percentage of P-resin (% P-resin) was significantly positively correlated with the percentage of P-$NaHCO_3$ (% P-$NaHCO_3$, $p < 0.001$) and negatively correlated with the percentage extracted with NaOH and after ultrasonification ($p < 0.001$, and $p < 0.01$, respectively). These correlations reveal that in these soils the directly available P pool (P-resin) is mainly controlled by: a) the moderately labile fraction (P-$NaHCO_3$), which acts as a source of P that, after mineralization, becomes available; and b) by the P-NaOH and P-Us fractions, which act as P sinks for labile inorganic P [27].

3.2.2. P-$NaHCO_3$

The P extractable with $NaHCO_3$ ranged from 74–189 mg kg^{-1} (Table 5) and accounted for 15–25% of the P_{ex}. Tiessen et al. [6] suggested that this fraction is quickly recyclable or available for plants.

The inorganic P extractable with $NaHCO_3$ ranged from 29–136 mg kg^{-1} (Table 5). The highest values were observed in the FG plot, which has the highest soil pH. This inorganic $NaHCO_3$-extractable fraction is considered to be a labile form [6], and it accounted for 13–30% of the total extractable P_i.

The P_o extractable with $NaHCO_3$ ranged from 8–113 mg kg^{-1} (Table 5) and accounted for 4–37% of soil P_o. Concentrations of P_o extractable with $NaHCO_3$ decreased with depth, a trend pointed out by Frossard et al. [28] for forest soils in Canada and by Trasar-Cepeda et al. [27] for Humic Leptosols and Cambisols in Galicia (northwestern Spain) due to the decrease of SOM with soil depth.

The relatively high percentage of P-$NaHCO_3$ was found to be negatively correlated with different forms of Al and Fe (Fe_t, Fe_d, Fe_p, Al_t, Al_d, Al_p; Table 6); thus, when the content of these Al and Fe fractions in soil was high, the proportion of labile P fraction (P-$NaHCO_3$) in the soil was low.

Table 5

P concentrations [mg kg^{-1}] and standard deviations (n = 6) for the forest soils

Plot	Depth [cm]	P-resin	P-NaHCO$_3$ Total	P-NaHCO$_3$ Inorganic	P-NaHCO$_3$ Organic	P-NaOH Total	P-NaOH Inorganic	P-NaOH Organic	P-Us Total	P-Us Inorganic	P-Us Organic	P-HCl	P$_{ex}$	P$_t$
FG	0-10	23±5	189±15	136±11	53±4	258±41	93±13	165±20	94±12	53±2	41±4	154±22	718	767±169
	10-20	15±3	136±17	90±15	46±3	211±45	77±15	134±9	108±13	75±13	33±5	109±20	579	645±185
	20-40	7±2	92±17	62±9	30±10	183±37	79±18	104±23	112±16	72±5	40±5	110±9	504	499±152
VR	0-10	6±2	94±10	51±3	43±15	405±63	196±45	209±30	78±16	34±6	44±6	15±3	598	965±150
	10-20	4±2	95±11	74±6	21±9	392±60	190±11	202±67	82±12	49±8	33±8	9±3	582	869±135
	20-40	3±1	93±11	76±7	17±5	382±54	185±15	197±51	85±14	49±9	36±7	9±3	572	794±139
NF	0-10	8±2	74±11	29±9	45±12	279±40	92±9	187±20	97±18	51±15	46±6	23±3	481	720±143
	10-20	3±1	80±12	48±7	32±15	288±31	125±5	163±15	95±10	71±10	24±6	15±3	481	677±149
	20-40	2±1	76±9	68±7	8±4	272±37	123±11	149±17	89±14	68±14	21±3	14±2	453	648±93
SM	0-10	34±6	186±33	73±12	113±9	420±46	240±47	180±24	110±15	31±2	79±13	67±17	817	1206±77
	10-20	16±3	178±25	67±14	111±18	378±26	273±72	105±81	123±16	31±2	92±5	69±25	764	1005±118
	20-40	7±3	151±25	52±16	99±21	364±29	261±80	103±66	107±17	30±3	76±7	64±20	693	903±77

Note: FG: Fuenteguinaldo; VR: Villasrubias; NF: Navasfrías; SM: San Martín; P-resin is available P, extracted with resin membranes; P-NaHCO$_3$ is labile P, extracted by 0.5 M NaHCO$_3$; P-NaOH is extracted by 0.1 M NaOH; P-Us is occluded P, extracted after ultrasonification; P-HCl is primary P, removed by 0.1 M HCl ; P$_{ex}$ is the sum of extractable P forms determined by Hedley et al. [3]; P$_t$ is total P determined by Saunders and Williams [20].

3.2.3. P-NaOH

The fraction of P extracted with NaOH showed the highest values of the P fractions, ranging from 183–420 mg kg^{-1} (Table 5). P-NaOH accounted for a greater proportion of P_{ex} (more than 50%) in NF, SM, and VR. The soils of NF and VR showed greater amounts of Fe and Al forms (Table 3) and also higher silt and clay contents than FG and SM soils (Table 2), causing greater P sorption and thus lesser lability of P in the first two soils. The high percentage of P-NaOH with respect to P_{ex} means that this fraction can be considered as the quantity factor of the potential soil P [28]. Frossard et al. [28] found that this fraction represented 34% of P_{ex} in forest soils (pH 6.2) and Zubillaga et al. [29] obtained percentages of P-NaOH (in relation to the P_{ex}) between 45–78% in soils of Argentina.

It is thought that NaOH-extractable P and P-Us contain secondary forms of P and their predominance is typical in developed soils [6, 30]. The predominance of the secondary forms observed (P_t-NaOH and P_t-Us), as opposed to primary ones (P-HCl), could be due to the fact that during rock weathering, abundant amorphous forms of Fe and Al are produced, and there is also an increase in the soil acidity due to the small base content; in these conditions, the soil P becomes stabilized under secondary forms [31]. The metal-organic complexes not only play an important role in the accumulation of SOM but also affect the P dynamic in the soil because they are strong adsorbents for P_i [32]. Soils rich in humic substances, with a very similar distribution of the P fractions and a predominance of P-NaOH, were also reported in evolved soils in northwestern Spain [27], where the climate is humid. A shift from Ca-P to Al- and Fe-P forms with the increasing degree of soil weathering was also found by Sharpley et al. [30]. The role of P_o sources in the reconstitution of the pool of available P has been evoked in numerous publications and will be particularly important at relatively advanced stages of soil development [30].

A highly significant positive correlation was obtained between the percentage of P-NaOH (% P-NaOH) and C_{org} contents of the soils (Table 6), whereas a highly significant negative correlation was obtained with soil pH (Table 6, $p < 0.001$).Tthis corroborates the greater sorption activity shown by Fe and Al at low pH values [10].

The role of SOM in the stabilization of P in the soil has been attributed to the association between SOM, P and the soil sesquioxides [10, 32]. Harter [33] suggested that P could be directly joined to the SOM, replacing OH groups. Other authors believe that there is a competitive relationship between the P and the SOM for sorption sites [34].

3.2.4. P-Us

The amounts of P extracted with NaOH following previous ultrasonic treatment (P-Us) ranged from 78–123 mg kg^{-1} (Table 5), accounting for 12–20% of the P_{ex}. The largest percentages of this fraction were found in the NF soil, which also had the greatest content in Fe and Al forms (Table 3). The P-Us fraction corresponds to aggregate protected P [6] and is therefore not directly available to the plants. Trasar Cepeda et al. [27] found that this fraction ranged from 2–33% of the total P extracted in soils of Galicia (northwestern Spain).

The sum of P contained in the less labile fractions (P-NaOH + P-Us) accounts for more than 80% of the P_{ex} in NF and VR and less than 60% in FG and SM. The first two sites also showed the lowest contents in labile P forms (P-resin and P-$NaHCO_3$) and the highest content in Fe and Al oxides. Thus, a chemical and physical stabilization of soil P under secondary forms occurs in NF and VR, and P is less available to plants.

Table 6
Correlation coefficients and significance levels between P forms and physicochemical and chemical properties of the forest soils

	C_p	N_t	pH	% Sand	% Silt	%Clay	Fe_t	Fe_d	Fe_p	Al_t	Al_d	Al_p
P_{ex}	-0.52**	N.s.	0.46**	0.70***	-0.66***	-0.72***	-0.60***	-0.42**	-0.36*	-0.45**	-0.64***	-0.56***
% P-resin	-0.37*	N.s.	0.68***	N.s.	N.s.	N.s.	-0.68***	-0.59***	-0.59***	-0.71***	-0.70***	-0.69***
% P-$NaHCO_3$	-0.60***	-0.46**	0.70***	0.82***	-0.81***	-0.63***	-0.70***	-0.67***	-0.61***	-0.47**	-0.69***	-0.64***
% P-NaOH	0.79***	0.71***	-0.86***	0.81***	0.84***	0.36*	0.85***	0.90***	0.95***	0.48**	0.77***	0.80***
% P-Us	N.s.	N.s.	N.s.	N.s.	N.s.	N.s.	N.s.	N.s.	N.s.	N.s.	N.s.	N.s.
% P-HCl	-0.86***	0.68***	0.87***	0.75***	-0.78***	-0.35*	-0.88***	-0.83***	-0.97***	-0.51**	-0.86***	-0.89***

Note: FG: Fuenteguinaldo; VR: Villasrubias; NF: Navasfrías; SM: San Martín; P-resin is available P, extracted by resin membranes; P-$NaHCO_3$ is labile P, extracted by 0.5 M $NaHCO_3$; P-NaOH is extracted by 0.1M NaOH; P-Us is occluded P, extracted after ultrasonification; P-HCl is primary P, removed by 0.1M HCl ; P_{ex} is the sum of extractable P forms determined by Hedley et al. [3]; C_p, pyrophosphate-extractable carbon; N_t, total nitrogen; Fe_t, total iron; Fe_d, free iron; Fe_p, iron bonded with organic matter; Al_t, total aluminum; Al_d, free aluminum; Al_p, aluminum bonded with organic matter; N.s. not significant; * significant at $p < 0.05$; **, significant at $p < 0.01$; ***, significant at $p < 0.001$.

3.2.5. P-HCl

The P-HCl ranged from 9–154 mg kg^{-1} (Table 5), which accounts for 2–21% of the P_{ex}. The highest values for this fraction were those obtained for granitic soils (FG and SM), which show the greatest amounts of apatite in the parent rock [6]. Sharpley et al. [30] found P-HCl concentrations between 6–259 mg kg^{-1} in poorly developed soils and between 1–70 mg kg^{-1} in highly weathered soils.

4. CONCLUSIONS

The studied soils showed high total P content. However, the concentration of available P represents a very small proportion of total P. The ratio between the most available forms of P and total extractable P is a negative function of the total amount of Fe and Al oxides. These compounds, either by occluding or by adsorbing phosphate on high-affinity sites, are crucial for the dynamics of P in these soils. In addition to a high rainfall regime, the acid nature of the bedrock of the studied soils favours the abundance of free forms of Al and Fe that promotes the stabilization of P.

ACKNOWLEDGMENTS

The authors wish to thank the Junta de Castilla y León for allowing them to use the forest plots and the European Union (MEDCOP/AIR and CAST/ENVIRONMENT Projects) and the Spanish C.I.C.Y.T. for financial support. We are indebted to Dr. M. Rico for her help with the statistical analysis and to Prof. N. Senesi for reviewing an earlier version of the manuscript. English revision was done by D.I. Garvey and B.C. Knowels.

REFERENCES

1. Harrison, A.F., 1987. Soil Organic Phosphorus. A Review of World Literature. C.A.B. International. Wallingford, UK, pp. 1-257.
2. Guggenberger, G., Christensen, B.T., Rubaek, G., Zech, W., 1996. Land-use and fertilization effects on P forms in two European soils: resin extraction and ^{31}P-NMR. Eur. J. Soil Sci. 47, 605–614.
3. Hedley, M.J., Stewart, J.W.B., Chauhan, B.S., 1982. Changes in inorganic and organic soil phosphorus fractions induced by cultivation practices and laboratory incubations. Soil Sci. Soc. Am. J. 46, 970–976.
4. Cross, A.F., Schlesinger, W.H., 1995. A literature review and evaluation of the Hedley fractionation: Applications to the biogeochemical cycle of the soil phosphorus in natural ecosystems. Geoderma 64, 197–214.
5. Schmidt, J.P., Buol, S.W., Kamprath, E.J., 1997. Soil P dynamics during 17 years of continuous cultivation: A method to estimate long-term P availability. Geoderma 78, 59–70.
6. Tiessen, H., Stewart, J.W.B., Cole, C.V., 1984. Pathways of P transformations in soils of differing pedogenesis Soil Sci. Soc. Am. J. 48, 853–856.
7. Tiessen, H., Moir, J.O., 1993.Characterization of available P by sequencial extraction. *In*:

Carter, M.R. (Ed.), Soil Sampling and Methods of Analysis, Part 2, Lewis, Boca Raton, FL. pp. 75-86.

8. Turrión, M.B., Gallardo, J.F., González, M.I., 2000. Distribution of P forms in natural and fertilized forest soils of the Central Western Spain: Plant response to superphosphate fertilization. Arid Soil Res. Rehabil. 14, 159–173.
9. Turrión, M.B., Glaser, B., Solomon, D., Ni, A., Zech, W., 2000. Effect of deforestation on P pools in mountain soils of the Alay Range, Khyrgyzia. Biol. Fert. Soil 31, 134–142.
10. Sanyal, S.K., De Datta, S.K., 1991. Chemistry of phosphorus transformations in soil. *In*: Stewart, B.A. (Ed.), Adv. Soil Sci. 16, 2–119.
11. Food and Agriculture Organization (F.A.O.) 1989. Mapa mundial de suelos. Leyenda Revisada. F.A.O., Roma.
12. Soil Conservation Service, 1972. Soil Survey Laboratory Methods and Procedures for Collecting Soil Samples. U.S. Department of Agriculture, Washington, D.C.
13. Harstein, A.M., Freedmon, R.W., Platter, D.W., 1973. Novel wet digestion procedure for trace-metal analysis of coal by atomic absorption. Anal. Chem. 45, 611–620.
14. Holmgren, C.G.S., 1967. A rapid citrate-ditionite extractable iron procedure. Soil Sci. Soc. Am. Proc. 31, 210–211.
15. Blakemore, L.C., Searle, P.L., Daly, B.K., 1987. Methods for Chemical Analysis of Soils. N.Z. Soil Bureau Sci. Rep. 80. Soil Bureau, Lower Hutt, New Zealand.
16. van Reeuwijk, L.P., 1993. Procedures for Soil Analysis. 4th Ed. ISRIC, International Soil Reference and Information Centre. Wageningen.
17. Barnhisel, R., Berstch, P.M., 1982. Aluminium. *In*: Page, A.L., Miller, R.H., Keeney, D.R., (Eds.), Methods of Soil Analysis, Part 2: Chemical and Microbiological Properties. Madison, WI, pp. 275-300.
18. Turrión, M.B., Gallardo, J.F., González, M.I., 1999. Extraction of availability forms of phosphate, nitrate and sulfate using anion exchange and determination by ionic chromatography. Comm. Soil Sci. Plant Anal. 30, 1137–1152.
19. Greenberg, A.E., 1980. Standard Methods for the Examination of Water and Wastewater. APHA/AWNA/WPCF, Washington.
20. Saunders, W.M.H., Williams, E.G., 1955. Observations on the determination of total organic phosphorus in soil. J. Soil Sci. 6, 254–267.
21. Murphy, J., Riley, J.P., 1962. A modified single solution method for the determination of phosphate in natural waters. Anal. Chem. Acta 27, 31–36.
22. Coppenet, M., Juste, C., 1987. Oligoelementos indispensables para la vida de las plantas. Fenómenos de toxicidad. *In*: Bonneau, M., Souchier, B. (Eds.), Edafología. 2. Constituyentes y Propiedades del Suelo. Masson, Barcelona, pp. 410-418.
23. Blaser, P., Klemmedson, J.O., 1987. Die Bedeutung von hohen Aluminiumgehalten für die Humusanreicherung in sauren Waldböden. Z. Pflanzenernähr. Bodenk. 150, 334–341.
24. Johnson, D.W., Todd, D.E., 1983. Relationships among iron aluminum, carbon and sulphate in a variety of forest soils. Soil Sci. Soc. Am. J. 47, 792–800.
25. Carballas, M., Cabaneiro, A., Gutián-Ribera, F., Carballas, T., 1980. Organo-metallic complexes in Atlantic humiferous soils. Anal. Edaf. Agrobiol. 39, 1033–1043.
26. Turrión, M.B., Gallardo, J.F., González, M.I., 1997. Nutrient availability in forest soils as measured with anion exchange membranes. Geomicrobiol. J. 14, 51–64.
27. Trasar Cepeda, M.C., Gil-Sotres F., Gutian Ojea, F., 1989. Relación entre algunas propiedades físico-químicas y las fracciones de fósforo en suelos naturales de Galicia. Anal.

Edaf. Agrobiol. 48, 665–679.
28. Frossard, E., Stewart, J.W.B., Arnaud, R.J., 1989. Distribution and mobility of phosphorus in grassland and forest soils of Saskatchewan. Can. J. Soil Sci. 69, 401–416.
29. Zubillaga, M.S., Giufré, L., 1996. Phosphorus fractions in Argentine soils of different pedogenesis. Commun. Soil Sci. Plant Anal. 27, 2137–2145.
30. Sharpley, A.N., Tiessen, H., Cole, C.V., 1987. Soil phosphorus forms extracted by soil tests as a function of pedogenesis. Soil Sci. Soc. Am. J. 51, 362–365.
31. Robles, C., 1991. Transformaciones y traslocaciones del fósforo como indicadores del desarrollo del suelo. Suelo y Planta 1, 793–800.
32. Blaser, P., Klemmedson, J.O., 1987. Die Bedeutung von hohen Aluminiumgehalten für die Humusanreicherung in sauren Waldböden. Z. Pflanzenernähr. Bodenk. 150, 334–341.
33. Harter, R.D., 1969. P adsorption sites in soils. Soil Sci. Soc. Am. Proc. 33, 630–632.
34. Ryden, J.C., McLaughlin, J.R., Syers, J.K., 1977. Mechanisms of phosphate sorption on soils and hydrous ferric oxide gel. J. Soil Sci. 28, 72–92.

Developments in Soil Science, Volume 28A
Editors: A. Violante, P.M. Huang, J.-M. Bollag and L. Gianfreda

EFFECTS OF ORGANIC LIGANDS ON ADSORPTION OF PHOSPHATE ON A NONCRYSTALLINE AL HYDROXIDE

H.Q. Hu, J. Z. He and X.Y. Li

Department of Resources, Environment and Agrochemistry, Huazhong Agricultural University, Wuhan 430070, P.R. China

The effects of oxalate (OX), tartrate (TR) and citrate (CT) on the adsorption of phosphate (P) on a synthetic noncrystalline Al-hydroxide [Al(OH)x] were studied at different initial phosphate/ligand molar ratios and different reaction conditions. The quantities of P adsorbed by the Al(OH)x varied with the ligand concentrations and the reaction conditions. When OX was introduced to the reaction system before P (OX before P), the effect of OX in decreasing P adsorption was greater than that when OX was added as a mixture with P (OX + P) or when P was introduced before OX (P before OX). When P was added after OX adsorption and removal of residual OX solution (OX(-) before P), the effect of OX in preventing P adsorption was lower than that in the system of OX before P. When P was added to the Al(OH)x as a mixture with two or three organic ligands, the reduction of P adsorption was related to the nature and concentration of the organic ligands. The combined effect of organic acids on P adsorption was not equal to the sum of the individual effect of each organic ligand. P adsorption increased with the equilibrium time from 1 to 7 days more in the presence of OX than CT. These findings have important implications for P availability in acidic soils.

1. INTRODUCTION

Low molecular mass organic acids present in rhizospheric soil [1-3] may have an important role in P availability for plants in acid soils, which contain large amounts of Fe and Al oxides. Many studies have been carried out on the influence of organic acids in reducing P adsorption on variable charge minerals and soils [4-10]. The influence of the nature and concentration of organic acids, pH and sequence of addition of organic ligands and phosphate to the sorbents has been considered by many researchers [7-10]. However, the effect of mixtures of several organic acids on P adsorption by the sorbents has received scant attention.

This work aims to investigate the adsorption of P by a synthetic noncrystalline Al(OH)x as affected by the presence of low molecular mass organic acids (oxalic, citric or tartaric acid) and their mixtures to provide useful information on P availability to plants in rhizospheric soil.

2. MATERIALS AND METHODS

2.1 Preparation of Al-hydroxide

A stock solution of 0.2 mol· L^{-1} $AlCl_3$ was titrated to pH 5.50 with 0.5 mol· L^{-1} NaOH at the rate of 5 ml· min^{-1} (OH/Al molar ratio of 2.7). The final suspension was aged for 48 h at room temperature and then centrifuged at 10,000 rpm for 30 min. The precipitate was dialyzed until Cl^- free, air-dried and passed through a 0.25 mm sieve. This oxide appeared noncrystalline as ascertained by x-ray diffraction analysis and showed a point of zero charge (PZC) of 6.78. The Al content in the precipitate was 281 g· kg^{-1}.

2.2 Effect of oxalate on the adsorption of P as affected by various addition orders

Twenty--five tubes, each containing 10mg Al(OH)x, were divided into 5 groups. To make the concentration ratio of the P (Cp) to the OX (Cox) reach 0.026, 0.13, 1.3 (these for the 1.29mmol· L^{-1} P), 10.3 and 103 (these for the 5.16 mmol· L^{-1} P), various oxalate concentrations were designed and five addition orders were carried out in each group. They were: [1] OX was added after P adsorption and removal of residual P solution [P(-) before OX]; [2] P was added after OX adsorption and removal of residual OX solution [OX(-) before P]; [3] P was added before OX addition [P before OX]; [4] OX was added before P addition [OX before P]; and [5] OX and P were added together [OX+P]. The volume of the solution was 5.0 ml and pH was controlled at 5.5 with dilute KOH and HCl. After shaking for 24 h at 25°C, the suspensions were centrifuged at 10000 rpm for 15 min. The concentrations of P in the supernatants were determined colorimetrically [11]. The amounts of adsorption were calculated according to the differences between initial and equilibrium concentrations.

2.3 Effect of organic ligands on the adsorption of P as affected by the mixture of two or three kinds of organic ligands

The Al(OH)x was mixed with a solution containing 1.29 mmol· L^{-1} P and two or three organic acids, such as OX, TR and CT. The ratio of solid:solution was 1:500. The pH was controlled at 5.5, and the reaction time was 24 h. The amounts of P adsorption were determined by the method described above.

2.4 P adsorption at different equilibrium times in the presence of OX or CT

20 mg Al(OH)x samples were equilibrated for 4 h with 5.0 ml 1 mmol· L^{-1} OX or CT. Then another 5.0 ml KH_2PO_4 solution was added to the system for initial P concentrations of 1.29 or 5.16 mmol· L^{-1}. A few drops of toluene were added to prevent microbial activity. The system was equilibrated for 1, 2, 3, 5 or 7 days with shaking for 2 h every day during the period. The suspensions were centrifuged at the end of each reaction period and P concentration in the supernatant was determined.

3. RESULTS AND DISCUSSION

3.1. Adsorption of P as influenced by OX at different addition orders and Cp/Cox concentration ratios

The amount of P adsorbed by the Al(OH)x decreased with decreasing Cp/Cox ratio for the same addition order (Table 1). Meanwhile, the amount of adsorption changed with the addition orders of P and OX. For example, at Cp/Cox ratio of 1.3, the amounts of P adsorption decreased as follows: P before OX > OX+P > OX(-) before P > OX before P > P(-) before OX. At Cp/Cox

= 103, the introduction of 0.05 mmol· L^{-1} OX had a negligible influence on P adsorption. At Cp/Cox = 0.026, the introduction of 50 mmol· L^{-1} OX had the greatest influence on P adsorption. Obviously, the influence of OX on P adsorption by Al(OH)x became stronger at higher OX concentration.

Table 1. P adsorption (mmol· g^{-1}) on Al(OH)x in the presence of OX at different addition orders and concentration ratios

Addition Conc. (mmo/L^{-1})		Addition order							
C_P	C_{OX}	C_P/C_{OX}	P alone	OX *bf** P	P *bf* OX	OX+P	P(-) *bf* OX	OX(-) *bf* P
5.16	0.05	103	0.765	0.753	0.789	0.775	0.750	0.758
5.16	0.5	10.3	0.765	0.634	0.737	0.720	0.656	0.683
1.29	1.0	1.3	0.505	0.381	0.506	0.472	0.329	0.436
1.29	10.0	0.13	0.505	0.247	0.495	0.273	0.310	0.371
1.29	50	0.026	0.505	0.168	0.478	0.207	0.270	0.242

bf = before.

It is of interest that in the system OX(-) before P, more P was adsorbed than in the system OX before P. Futhermore, in the system P(-) before OX, less P was adsorbed than that in the systems P before OX. Obviously, the concentrations of the organic ligand in equilibrium solutions had a vital influence on P adsorption. Our findings seem to demonstrate that in the systems containing both OX and P, the earlier introduced ion occupied first many common surface sites and could be only partially desorbed by the other ion that entered later. Clearly, OX was desorbed by P more easily than P by OX.

Previous researchers have found that low molecular mass organic acids can reduce the amount of P adsorption by soil and mineral [7,12-14], but the adsorption of an organic acid would greatly change its effectiveness to mobilize nutrients from the rhizosphere [15]. Whether the variation of concentration ratios between organic acid and P at definite addition orders has an effect on P adsorption or not is poorly understood.

This study revealed that the amount of P adsorption was related to the addition order of P and OX. OX addition first without removal of residual OX resulted in the least P adsorption, but P before OX addition produced the most P adsorption. When the ratio of Cp/Cox was 0.026, the sequence of P adsorption quantity was: P before OX > OX(-) before P > P(-) before OX > OX+P > OX before P. At the ratio of 103, the sequence of P adsorption quantity was: P before OX > OX+P > P(-) before OX OX before P OX(-) before P. It was obvious that at high ratio of Cp/Cox, P adsorption was affected less by OX addition than at low ratio, i.e., only if there was a large amount of organic acid in the solution (higher OX concentration than P) could P be adsorbed less.

3.2. Adsorption of P as affected by the presence of several organic ligands

The introduction of a single organic acid, such as CT, OX or TR decreased P adsorption, as previously reported by Hu et al. [16]. Figure 1 shows that P adsorption was affected by the co-existing organic acids. There were different P adsorption amounts at various combinations of kinds and concentrations of organic ligands. When the concentration of organic acids increased from 0.6 to 6 mmol L^{-1}, P adsorption decreased from 0.396 to 0.196 mmol g^{-1}. Thus, organic acids reduced P adsorption by 21.6-61.2%, compared with that of P adsorption at control, which was only 1.29 mmol P L^{-1}. When the combination consisted of CT, the amount of P adsorption was reduced about 60%, with the concentration of CT going from 0.1 mmol L^{-1} up to 2 mmol L^{-1}.

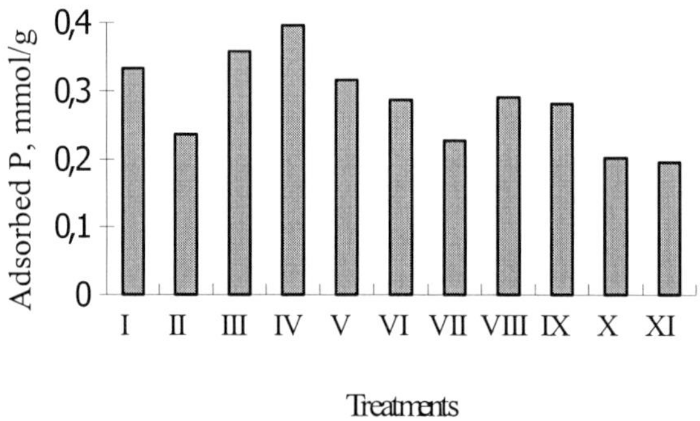

Figure 1. The amount of P adsorption on Al(OH)x surface in the presence of two/three kinds of organic acids. The treatments are: I = 0.5 (mmol.L^{-1}) CT + 0.5 (mmol.L^{-1}) TR; II = 2 CT + 2 TR; III = 0.5 OX + 0.5 TR; IV = 0.5 OX + 0.1 CT; V = 0.5 OX + 0.5 CT; VI = 0.5 OX + 0.5 CT + 0.5 TR; VII = 0.5 OX + 2.5 CT; VIII = 2 OX + 2 TR; IX = 2 OX + 0.4 CT; X = 2 OX + 2 CT; XI = 2 OX + 2 CT + 2 TR. The units for numbers in the treatments are concentrations in mmol L^{-1}. OX stands for oxalate, CT for citrate, and TR for tartrate.

Comparison of P adsorption amount in the presence of the same concentration of organic acids suggests that CT could reduce P adsorption most remarkably, followed by OX and TR. This was similar to the effect of a single organic acid in decreasing P adsorption by the sorbent.

The amount of P adsorption by Al(OH)x decreased with increasing concentrations of mixtures of organic ligands, indicating that the mixtures were more effective than using only the individual organic acid system. We could draw a conclusion that a combination of several organic acids could improve P efficiency more than a single organic acid. Of course, both the concentration and kind of organic acid must be considered.

In the presence of a single organic acid, linear equations were obtained on phosphate adsorption and organic acid concentration as follows [16]:

$$1/y_1=0.4469C_{ox}+2.0077, \quad (r=0.923, n=8)$$

$$1/y_2=0.5134C_{TR}+1.8833, \quad (r=0.969, n=6),$$

where y_1 and y_2 (mmol· g^{-1}) are P adsorption amounts in the presence of 0--2 mmol· L^{-1} OX and 0--2 mmol· L^{-1} TR, respectively.

If the effect of two organic acids on P adsorption was additive, based on the equations described above, the P adsorption amounts should be 0.408 mmol· g^{-1} for treatment of 0.5

mmol· L^{-1} OX+0.5 mmol· L^{-1} TR and 0.182 mmol· g^{-1} for treatment of 2 mmol· L^{-1} OX+2 mmol· L^{-1} TR. However, the actual P adsorption amounts for the two treatments were 0.358 and 0.290 mmol· L^{-1}, respectively. There was some difference between the calculated value from the single organic acid system and the actual value obtained from the experiment. At low organic acid concentration, the combined effect of two organic acids on reduction of P adsorption was larger than the sum of individual organic acids, but at high concentration, the result was opposite. This finding confirmed and extended the results obtained from hydroxy aluminum montmorillonite complexes [17].

In addition, in the treatment of 0.5 mmol· L^{-1} OX+0.5 mmol· L^{-1} CT, the P adsorption amount was 0.315 mmol g^{-1}, and in the treatment of 0.5 mmol· L^{-1} OX+0.1 mmol· L^{-1} CT, the P adsorption amount was 0.396 mmol g^{-1}. The difference of P adsorption between the above two treatments was not equal to the effect of 0.4 mmol· L^{-1} CT. The difference in P adsorption between treatment of 0.5 mmol· L^{-1} TR+0.5 mmol· L^{-1} CT and 0.5 mmol· L^{-1} OX+0.5 mmol· L^{-1} CT was not equal to the difference in P adsorption between 0.5 mmol· L^{-1} TR and 0.5 mmol· L^{-1} OX. Thus, the combined effect of several co-existing organic acids on P adsorption was not the sum of the individual organic acids.

3.3. Influence of equilibrium time on P adsorption in the presence of an organic acid

Figure 2 shows the effects of the equilibrium time on P adsorption in the presence of 1 mmol L^{-1} OX or CT. P adsorption at day 7 was 36.4--43.2% more than that at day 1. This implied that with the increase of equilibrium time, the influence of organic acid sorbed first on P adsorption became weaker, and the affinity between the surface of Al(OH)x and ligands was becoming more and more significant. However, it was possible that some organic ligands would be desorbed slowly by P addition.

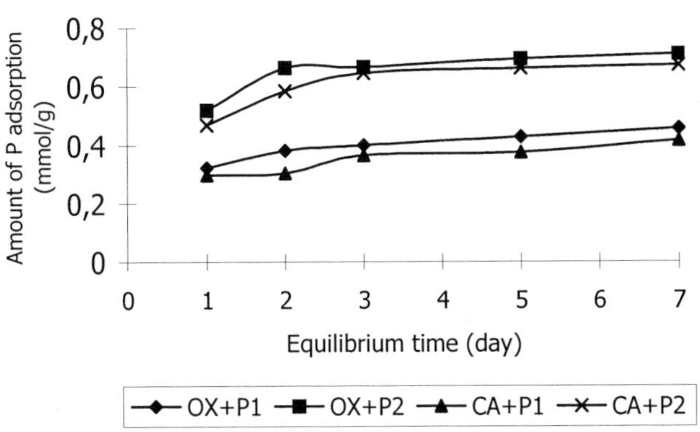

Figure 2. Phosphate adsorption at different equilibrium times in the presence of 1 mmol L^{-1} oxalate or citrate. P_1 represents 1.29 mmol L^{-1} P and P_2 represents 5.16 mmol L^{-1} P.

Although CT could reduce P adsorption more remarkably than OX at different reaction periods, the variation would be weaker with time. P adsorption that became stable in the presence of an organic acid varied with the P concentration, i.e., C_P/C_{OA}. For instance, comparing 7 days with 5 days at 1.29 mmol L^{-1}, the amount of P adsorption increased by 6.5%

(for OX) and 10.9% (for CT). However, comparing 7 days with 3 days at 5.16 mmol L^{-1}, the amount of P adsorption increased by 6.7% (for OX) and 4.0% (for CT). These results indicated that the greater the ratio of Cp to C_{OA}, the faster P adsorption reaches equilibrium.

The equilibrium time of P adsorption was affected by the nature of organic acid. CT had a stronger influence than OX at 3-day and 2-day equilibrium periods. Comparing P adsorption within 7 days and that within 2 days, it was clear that P adsorption became stable at 2 days for OX, but P adsorption was slower to reach stability for CT. This could be because the affinity of Al(OH)x for CT is stronger than that for OX.

ACKNOWLEDGEMENTS

This study was supported by the Natural Science Foundation of China (Grants No: 49971050 and 49871043) and the International Scientific Cooperation (Contract No:CII*-CT94-0048) of the European Community. The authors wish to express appreciation to Prof. A. Violante, Prof. P. M. Huang and two anonymous referees for their critical reviews of the manuscript.

REFERENCES

1. Stevenson, F.J., 1967. Ogranic acids in soil. *In*: McLarne, A.D., Peterson, J.H. (Eds.), Soil Biochemistry. Marcel Dekker, NY. pp119-146.
2. Jones, D.L., Darrah, P.R., 1994. Role of root derived organic acids in the mobilization of nutrients from the rhizosphere. Plant Soil 166, 247-257.
3. Jones, D.L., 1998. Organic acids in the rhizosphere - a critical review. Plant Soil 205, 25-44.
4. Bolan, N.S., Naidu, R., Mahimairaja, S., Baskaran, S., 1994. Influence of low-molecular-weight organic acids on the solubilization of phosphates. Biol. Fertil. Soils 18, 311-319.
5. Geelhoed, J.S., Hiemstra, T., Van Riemsdijk, W.H., 1998. Competitive interaction between phosphate and citrate on goethite. Environ. Sci.Tech. 32, 2119-2123.
6. Nagarajah, S., Posner, A.M., Quirk, J.P., 1970. Competitive adsorption of phosphate with polygalacturonate and other organic anions on kaolinite and oxide surfaces. Nature 228, 83-85.
7. Violante, A., Colombo, C., Buondonno, A., 1991. Competitive adsorption of phosphate and oxalate by aluminum oxides. Soil Sci. Soc. Am. J. 55, 65-70.
8. Earl, K.D., Syers, J., McLaughlin, R., 1979. Origin of the effect of citrate, tartrate and acetate on phosphate sorption by soils and synthetic gels. Soil Sci. Soc. Am. J. 43, 674-678.
9. Hu, H.Q., He, J. Z., Li, X.Y., Liu, F., 2001. Effect of several organic acids on phosphate adsorption by variable charge soils of central China. Environ. Int. 25, 351-356.
10. Violante, A., Gianfreda, L., 1995. Adsorption of phosphate on variable charge minerals: competitive effect of organic ligands. *In*: Huang, P.M., Berthelin, J., Bollag, J.-M., McGill, W.B., Page, A.L. (Eds.), Environmental Impact of Soil Component Interactions, Vol 2. CRC Press. Boca Raton, FL. pp.29-38.
11. Murphy, J., Riley, J.P., 1962. A modified single solution method for the determination of phosphate in natural waters. Anal. Chem. Acta 27, 31-36.

12. Hue, N.V., 1991. Effects of organic acids/anions on P sorption and phytoavailability in soils with different mineralogies. Soil Sci. 152, 463-471.
13. Lopez-Hernandez, D., Siegert, G., Rodriguez, J.V., 1986. Competitive adsorption of phosphate with malate and oxalate by tropical soils. Soil Sci. Soc.Am. J. 50, 1460-1462.
14. Yuan, T.L., 1980. Adsorption of phosphate and water-extractable soil organic material by synthetic aluminum silicates and soils. Soil Sci. Soc. Am. J. 44, 951-955.
15. Jones, D.L., Brassington, D.S, 1998. Sorption of organic acids in acid soils and its implications in the rhizosphere. European J. Soil Sci. 49, 447-455.
16. Hu, H.Q., Li, X.Y., He, J.Z., 2000. Effect of organic acids on phosphate adsorption by aluminum hydroxides. Plant Nutrition Fert. J. (in Chinese) 6, 35-41.
17. He, J.Z, A. De Cristofaro, Violante, A., 1999. Comparison of adsorption of phosphate, tartrate and oxalate on hydroxy aluminum montmorillonite complexes. Clays Clay Miner. 47, 226-233.

Developments in Soil Science, Volume 28A
Editors: A. Violante, P.M. Huang, J.-M. Bollag and L. Gianfreda

REACTIONS OF SOME SHORT-RANGE ORDERED ALUMINOSILICATES WITH SELECTED ORGANIC LIGANDS

E. Hanudin, N. Matsue and T. Henmi

Laboratory of Environmental Soil Science, Ehime University,
3-5-7 Tarumi, Matsuyama 790-8566, Japan.

Allophane, a short-range ordered aluminosilicate with reactive aluminol groups has an important role in reaction with organic compounds in volcanic ash soils. Adsorption of low molecular weight organic compounds (acetate, oxalate and citrate) on allophane showed that oxalate has the highest affinity among the three for allophane at lower pH (4-5). At higher pH (6-9), however, citrate adsorption was highest among the three organic compounds. For all organic compounds, the amount of adsorption on allophane increased with decreasing pH. Increase in solution pH with adsorption was observed in most cases, indicating dehydroxylation reaction via ligand exchange. On the contrary, a decrease in pH after acetate and citrate adsorption at lower pH was observed due to reaction between $Al\text{-}OH_2^+$ and COOH. Oxalate and citrate are possibly adsorbed in bidentate and/or binuclear form, but molecular orbital calculation indicated that the binuclear form is more stable for oxalate and the bidentate form is more stable for citrate. These results have an important implication in study on pedogenesis, soil fertility and environmental chemistry.

1. INTRODUCTION

Short-range ordered aluminosilicates and organic compounds, which are the main components of volcanic ash soils, have significant effects on the physical, chemical and biological properties of soils. Among the short-range ordered aluminosilicates, allophane has various chemical compositions with large specific surface and high chemical reactivity [1-2]. This aluminosilicate was believed not to have any definite morphology and structure. However, Henmi and Wada [3] found that allophane separated from volcanic ash soils and weathered pumice grains has definite hollow spherical morphology with diameter of about 5 nm. They named this type of allophane nano-ball allophane. A similar feature was also observed for a synthetic allophane [4]. Two end-members are recognized for nano-ball allophane on the basis of chemical composition, physicochemical properties and structure, although they have the same morphology [3, 5]. The first end-member of allophane has only monomeric SiO_4 tetrahedra in its structure. The wall of its spherules are built up of imogolite structural units, $(OH)_3Al_2O_3SiOH$, with some defects or pores of about 0.5 nm in size. It has a

Si/Al atomic ratio of 0.5, and is also referred to as proto-imogolite allophane [5-6]. The other end-member, with a Si/Al atomic ratio of 1.0, has dimeric or polymeric SiO_4 tetrahedra in the wall of the spherules [6].

Among organic compounds, low molecular weight organic acids, including acetic, oxalic and citric acid, are found in soils and aquatic environments [7]. The concentration of these organic acids is generally very low and depends upon the conditions in the soil [8-9]. These organic acids, which are naturally abundant in root sap, are excreted in large quantities into soil that is deficient in Fe and P [10-13]. Although the concentration of low molecular weight organic acids in soils is not as high as that of high molecular weight organic compounds, such as humic acid and fulvic acid, their role in the soil environment is of great importance. Among the low molecular weight organic acids implicated in rhizosphere processes, citric, malic and oxalic acids seem to be the most important. Low molecular weight aliphatic and aromatic acids may block reactive adsorption sites on soil materials and thus reduce P adsorption [14-17].

Studies on the interaction of organic compounds with clay minerals have focused on adsorption and reactions of humic and fulvic acid [10,18-21], or on the effects of low molecular weight organic acid on formation of short-range ordered aluminosilicates or precipitation products of aluminum [7,10,22-27]. However, the detailed adsorption mechanism of low molecular weight organic acids on allophane has not been elucidated with respect to the morphology and chemical structure of allophane. The objective of the present study was to understand the adsorption mechanism of acetate, oxalate and citrate on nano-ball allophane, using the detailed information on the morphology and chemical structure of the allophane.

2. MATERIALS AND METHODS

2.1. Sample preparations

Two pumice grain samples containing nano-ball allophane were collected from volcanic ash soils from two different locations in Japan: Kakino, Kumamoto prefecture near Mount Aso (KnP), and Kurayoshi, Tottori prefecture near Mount Daisen (KyP). To obtain nano-ball allophane samples (< 0.2 μm) without contaminants such as volcanic glasses, opaline silica, imogolite and organic matter, only the inner part of the pumice grains was used [3]. The separation was carried out by ultrasonification at 28 kHz and dispersion at pH 4 for KyP and at pH 10 for KnP. The KyP sample, which represents allophane with a low Si/Al ratio, contains more Al than Si and hence is easier to disperse under acidic conditions. However, the KnP represents allophane with a high Si/Al ratio and has a higher level of Si. It is, therefore, easier to disperse under alkaline conditions. The collected samples were flocculated by saturated NaCl solution, washed with water, and freeze-dried. The resulting allophane samples were examined by X-ray diffraction, infrared spectroscopy and thermal analysis, and were found to be free from the aforementioned contaminants. The Si/Al atomic ratio of KyP and KnP was 0.67 and 0.99, respectively. Schematic chemical structures for unit particles of nano-ball allophane with a Si/Al ratio of 0.5 and with a Si/Al ratio greater than 0.5 are presented in Figure 1.

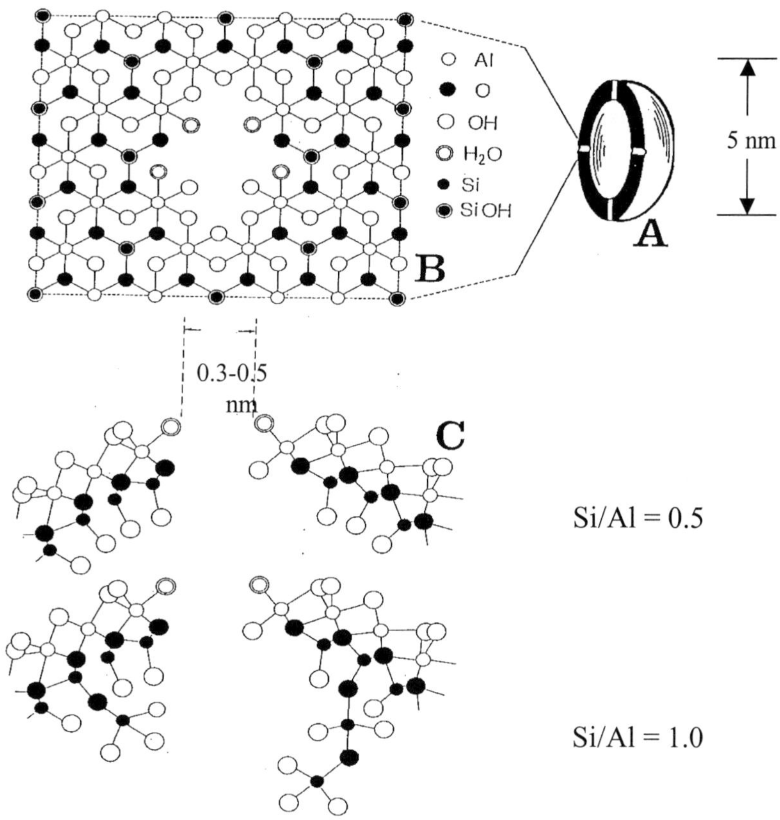

Figure 1. Morphology and chemical structure of nano-ball shaped allophane (A: molecular morphology in section; B: atomic arrangement near the pore of hollow particles; C: atomic arrangement in the cross section of allophane particles [6]).

2.2. Adsorption experiments and molecular orbital analysis

The adsorption experiments were carried out by mixing 50 mg of freeze-dried allophane sample with 100 mL of 0 to 2 mM aqueous solution of the selected organic compounds (Na-acetate, Na-oxalate, and Na-citrate) at initial pH between 4 and 10 under 10 mM NaCl background solution. The pH adjustment was done with NaOH or HCl. The suspensions were shaken for 24 h at room temperature and then centrifuged. The supernatant was analyzed for released Si and Al by atomic absorption spectrophotometry and the concentrations of the organic compounds by UV spectrophotometry at 190 nm. In this experiment, only one type of organic acid was present in solution at any particular time. The UV spectrophotometry was, therfore, applicable in the determination of organic acid concentrations in solution. The amounts of carboxylate compounds adsorbed by the allophane sample were calculated by subtracting the amounts in the equilibrium solution from that in the initial solution.

For molecular orbital analysis, the MOPAC program was used with the semi-empirical MNDO-PM3 basis set [28-29], which is incorporated in the CAChe system for Windows (CAChe Scientific Inc., Sony Tektronix Corporation). This basis set gives heat of formation closer to experimental values for actual molecules than any other semi-empirical basis set does [29]. Cluster models for allophane were built up with Al normal octahedra and Si tetrahedra by using bond distance of Al-O = 0.1912 nm, Si-O = 0.1618 nm and O-H = 0.0944 nm.

3. RESULTS AND DISCUSSION

3.1. Adsorption isotherms

The pH-dependent speciation of the three organic anions used is shown in Figure 2. The pKa of acetic acid is 4.76; pK_1 and pK_2 of oxalic acid are 1.25 and 4.27, respectively; pK_1, pK_2 and pK_3 of citric acid are 3.13, 4.76 and 6.40, respectively. Figure 2 shows that each organic acid possesses a unique speciation profile as a function of solution pH, which is characteristic of the number and distribution of carboxyl groups within each molecule [30].

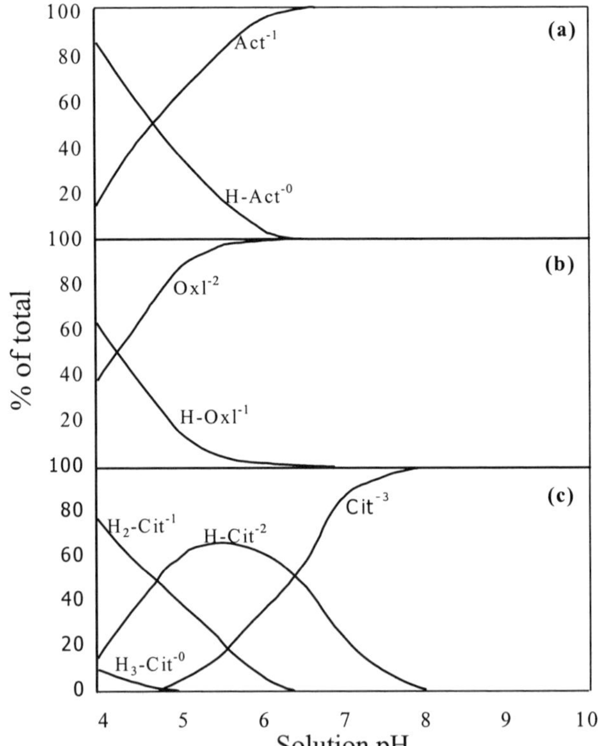

Figure 2. General properties of the organic acids under test. The main panels show the pH-dependent speciation of (a) acetate, (b) oxalate and (c) citrate [modified from 30].

Figure 3 shows the adsorption isotherm of acetate, oxalate and citrate on KyP and KnP at different initial pH values. At initial pH 4, the amount of oxalate adsorbed on allophane was the highest among the three organic compounds. A similar trend was also observed for adsorption of organic acids on Spodosols and ferrihydrite [30]. At initial pH 6-10, citrate adsorbed more than oxalate and acetate. The amount of each organic compound adsorbed by allophane increased with increasing equilibrium organic compound concentration. The increase was steep at lower equilibrium concentrations and reached a plateau at higher concentrations. The amount of the organic compounds adsorbed by allophane decreased with increasing pH of the solution in all cases.

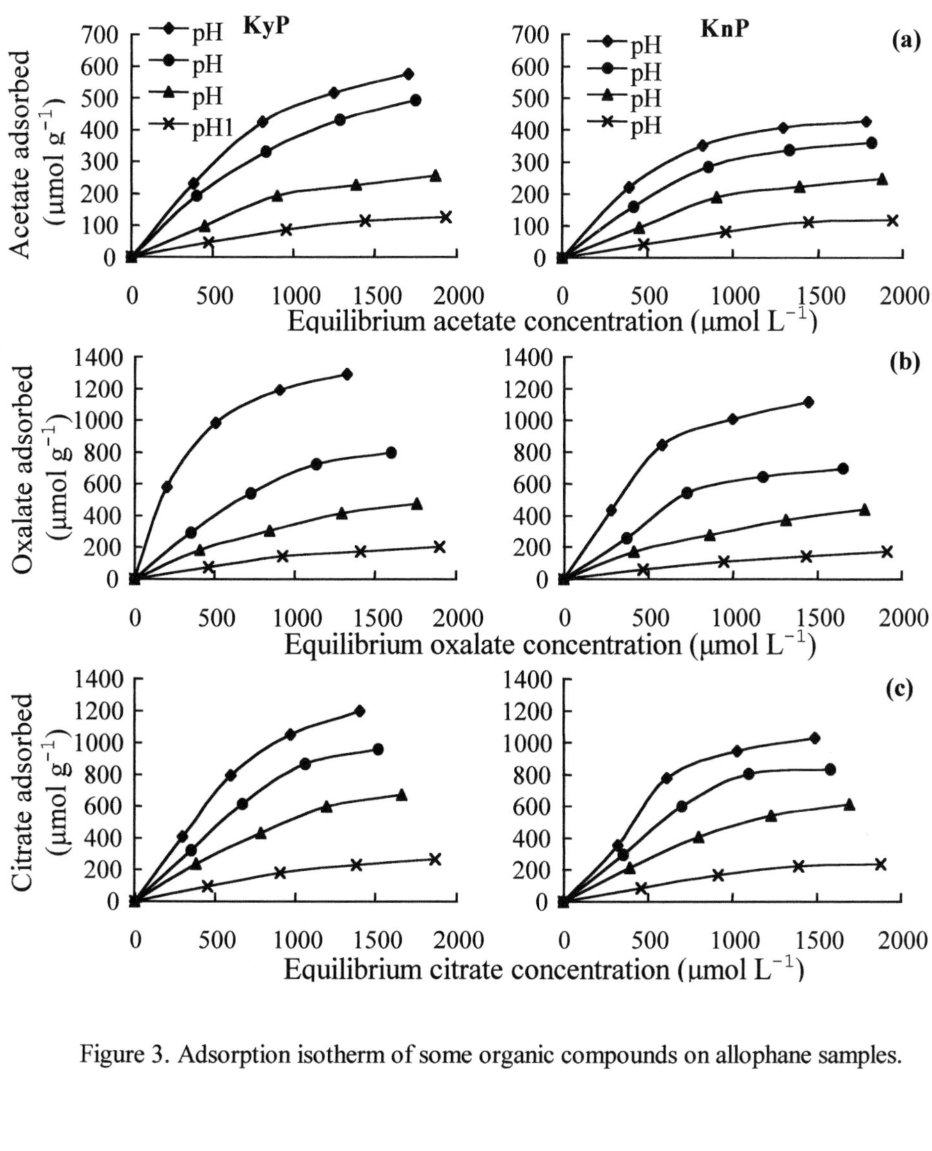

Figure 3. Adsorption isotherm of some organic compounds on allophane samples.

This phenomenon is related to the charge characteristics of the nano-ball allophanes, which at lower pH (4-5) have positive charges ($Al-OH_2^+$) at the pore site of the ball irrespective of the Si/Al ratio of the sample [31]. At higher pH (6-9), allophanes have no positive charge, but only negative charges ($Si–O^-$) at the inner surface of the ball; these negative charges increase with increasing solution pH [31].

The amount of the organic compounds adsorbed by KyP was higher than that of KnP, because KyP contains a greater number of aluminol groups per unit mass than does KnP. Aluminol groups, Al-OH and $Al-OH_2^+$, are responsible for phosphate and organic matter adsorption in soils and clay minerals [32]. In the structure of allophane, aluminol groups are located only at the pores of the wall of hollow spherules [24]. In addition to its lower aluminol content, KnP has accessory condensed Si tetrahedra attached to the main frame of allophane at the pore region. This prevents the organic acids from getting access to the aluminol groups at the pore and thereby decreases adsorption.

3.2. Release of Si and Al

Silicon and aluminum as main components of the allophane samples were always released into the equilibrium solution after adsorption of the organic compounds. The amount of Si and Al released to the solution depended upon the concentration and species of organic compounds, pH, and physicochemical properties of the allophane itself. Figures 4 and 5 show the amount of released Si and Al, respectively. The amounts tended to increase with increasing amounts of organic compounds adsorbed by the allophane, and decrease with increasing pH of the organic solutions. These relationships occur because, at lower pH, proton as well as organic ligands contributed to the dissolution of Si and Al from the allophane samples. The amount of Si and Al released from allophane after treatment with citrate and oxalate was higher than that with acetate. This indicates that the dissolution power of citrate and oxalate was stronger than that of acetate.

At initial pH 4, the three carboxylate compounds selectively dissolved Si more than Al from KnP, but more Al than Si from KyP. The content of Al in KyP was higher than Si, and aluminum as an amphoteric element is easily dissolved from clay minerals at lower pH. The KnP, as an allophane with a high Si/Al ratio, contains a lot of polymerized Si tetrahedra weakly bonded to the fundamental structure (Si/Al=0.5). The Si tetrahedra could be a kind of shelter for protecting the fundamental structure of allophane from the proton attack. In acidic conditions, the role of protons in dissolution of amorphous aluminosilicate was more dominant than that of the chelating system. There are two theories about the role of protons in dissolving the clay minerals: (a) the protons dissociated by the acids cause the surface acidity of the clay minerals to increase, and the H^+ saturated clay spontaneously decomposes; or (b) the protons penetrate into the octahedral sheet, replacing Al^{3+} ions, which are then preferentially adsorbed on the clay surface [33]. At low pH, the Si/Al molar ratio in solution tended to increase with increasing amount of the organic ligand adsorption, except for treatment with acetate. The ratio in the acetate solution tended to decrease across the pH range tested, for the both allophane samples. However, the Si/Al molar ratio in the solutions of the organic ligands after reaction with KnP was higher than that with KyP at the pH range tested.

In treatments at initial pH between 6 and 10, Si was selectively dissolved much more than Al. The amount of Si released from KnP was higher than that from KyP, because KnP contains a lot more accessory polymeric Si weakly attached to the Si tetrahedra in the structure. As shown in Figure 1, the fundamental chemical structure of a unit particle of allophane has a Si/Al ratio of 0.5. Most of the additional Si, which increases the value of the Si/Al ratio, is weakly bonded Si attached to the fundamental structure. Therefore, the higher release of Si than Al from the two-allophane samples indicated that the organic compounds studied can easily break down the accessory polymerized Si that is weakly attached to the fundamental allophane structure. The capability of organic compounds to break down the accessory Si depends upon the amount of functional groups (COO^-) contained in the compound. Organic compounds that have higher amounts of COO^- have greater capability to breakdown the bond. Increasing solution pH increased the amount of the dissociated forms of organic acid. However, that also resulted in an increase in the amount of negative charge possessed by allophane, culminating in decreased adsorption due to electrical repulsion.

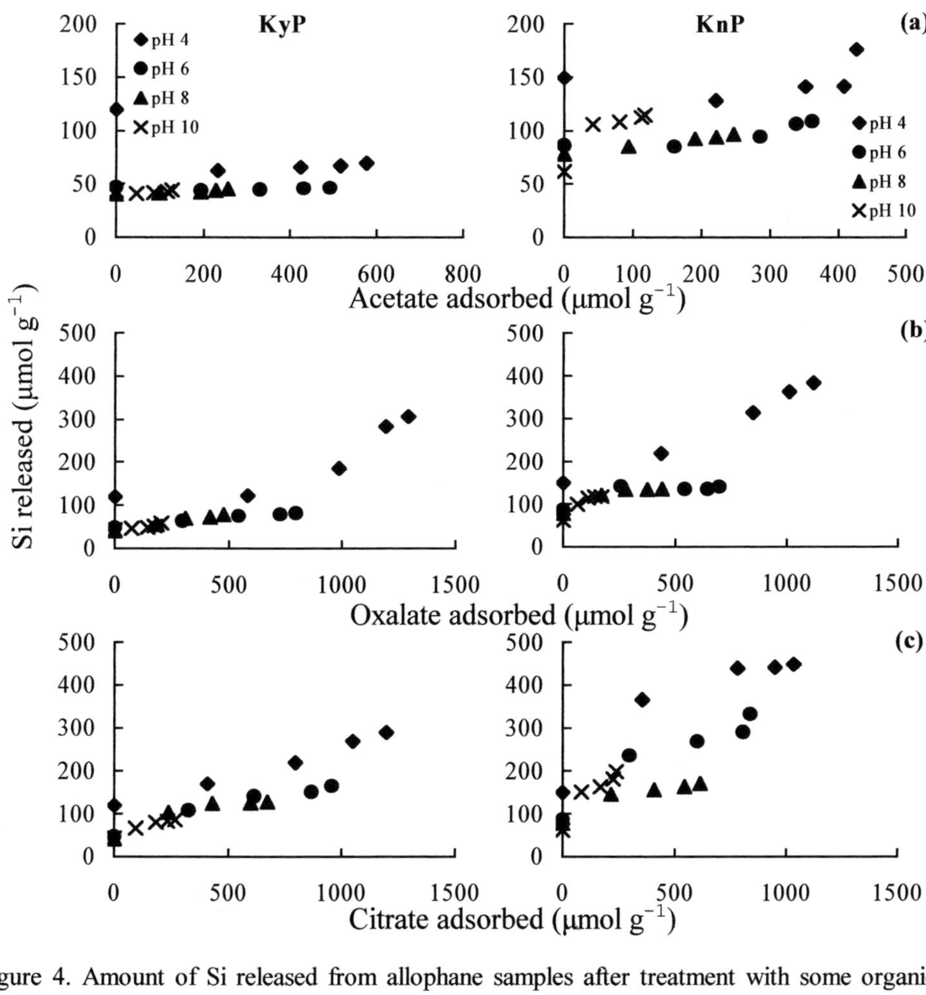

Figure 4. Amount of Si released from allophane samples after treatment with some organic compounds.

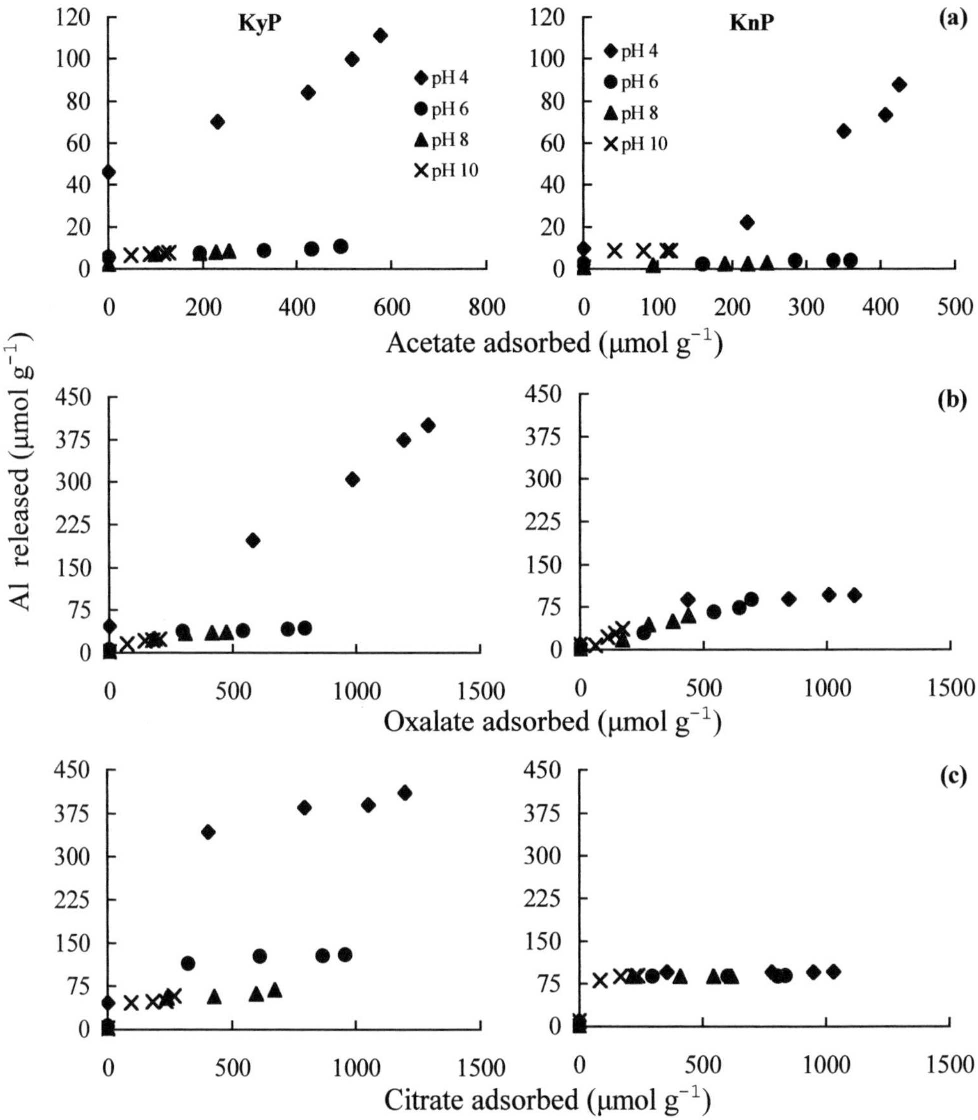

Figure 5. Amounts of Al released from allophane samples after treatment with some organic compounds.

3.3. Change in pH, molecular orbital analysis and proposed reaction mechanism

Adsorption of some organic compounds on the allophane samples caused the solution pH to change (Figure 6). In each case, the pH value of the blank run was considered to be the pH of the solution without organic compound adsorption on allophane. The allophane samples treated with acetate and citrate at pH 4 showed decreased equilibrium pH with respect to the blank run. This decrease in pH tended to deminish with increasing concentrations of acetate and citrate. This indicated the release of protons to the bulk solution. High H^+ activity at lower solution pH caused an increase in the amount of positive charges on the surface of allophane. On the other hand, under acidic conditions, solutions of acetic and citric acid are predominated by undissociated forms (pKa and pK_2 of both organic acids   4.76). Reactions took place between undissociated forms of both organic compounds and $Al-OH_2^+$. During the reaction, protons were released to the solution, and pH decreased.

Allophane treated with acetate and citrate at pH 8-10 showed that the equilibrium pH after adsorption was higher than that of the blank run and tended to increase with increasing concentrations of acetate and citrate. The same result was also obtained for treatment with oxalate across the pH range tested. At pH higher than pKa, the organic solutions were dominated by the negatively charged anion species. Thus, a reaction between aluminol groups and the negatively charged organic compound led to OH^- release.

It is evident from Figure 6 that the ΔpH values are higher in KyP than in KnP. This result was observed for the three organic compounds studied, and also occurred across the pH range tested. This higher ΔpH could be related to the difference in chemical structure of both types of allophane. The amount of aluminol groups in allophane irrespective of sample type is higher than Si. KyP, as a representative of allophane with a low Si/Al ratio, contains much more Al octahedra per unit mass than KnP (allophane with high Si/Al) does. Therefore, it seems that in reaction with the organic compounds, aluminol groups play a more important role than silanol groups. Thus with higher aluminol content, the ΔpH values will be higher.

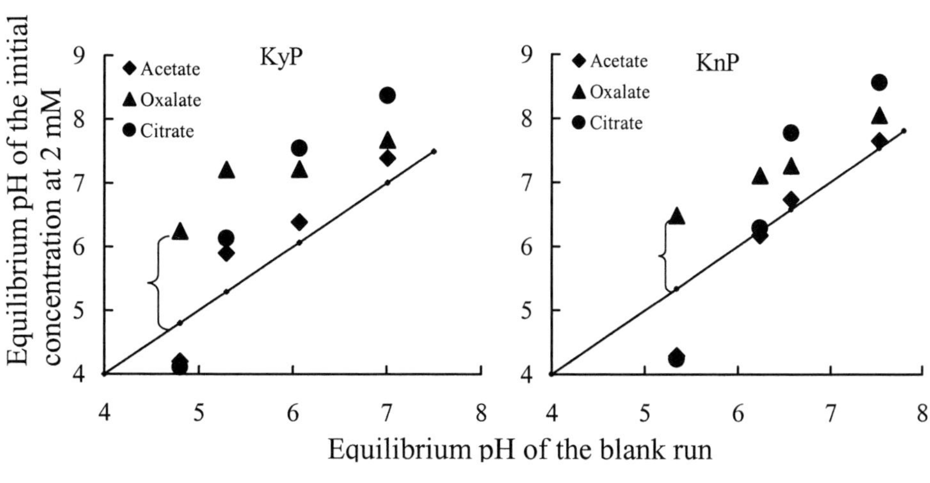

Figure 6. Changes in pH ($\Delta pH = pH_{treatment}-pH_{blank}$) of organic solutions after reaction with allophane.

Greater increase in pH after oxalate adsorption compared with acetate adsorption (Figure 6) indicates that both COO^- reacted with aluminol groups on allophane, releasing OH^-. At low pH and at concentrations $\leq$ 100 μmol g^{-1}, oxalic acid is strongly adsorbed on goethite as a binuclear complex, replacing two singly coordinated OH groups by ligand exchange. Higher concentrations, on the contrary, result in the formation of a monodentate complex [25]. In the case of gibbsite, however only the bidentate form exists [26]. Oxalate is also adsorbed on Spodosols in monodentate, bidentate and binuclear forms at low pH [1]. There is, therefore, a possibility of both bidentate and binuclear reactions between COO^- groups of oxalate and citrate on one hand and allophane on the other hand (Figure 7).

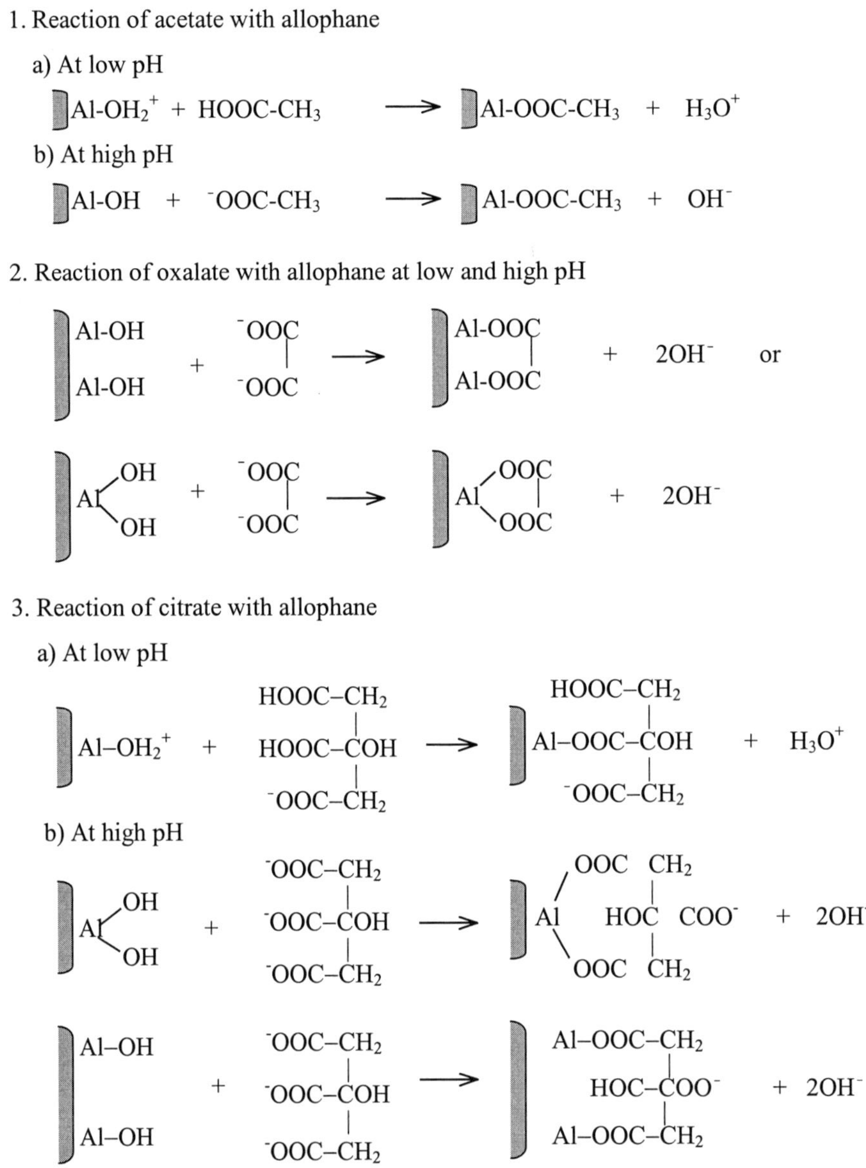

Figure 7. The proposed reaction mechanisms between Al-OH or Al-OH_2^+ with the organic compounds.

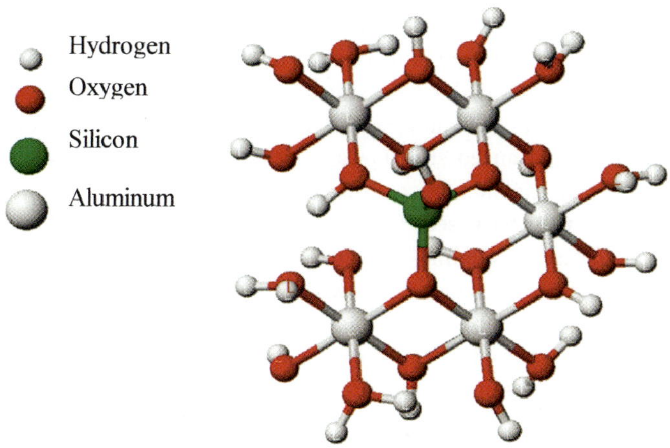

Figure 8. Model cluster of allophane used for molecular orbital analysis.

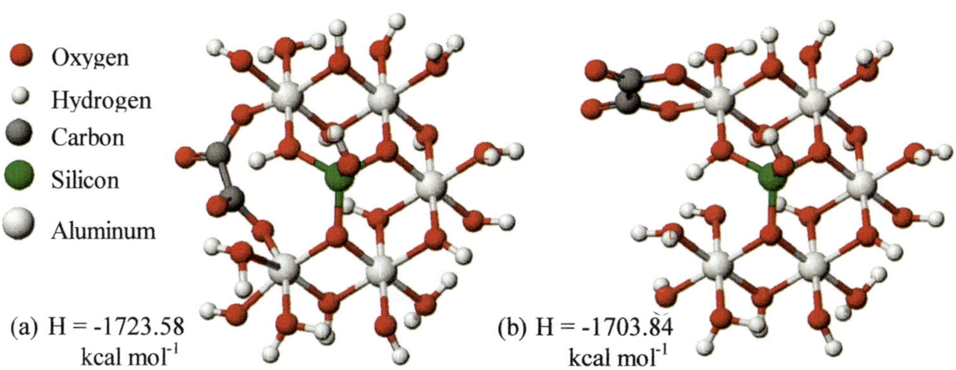

Figure 9. Oxalate adsorbed on allophane in binuclear (a) or bidentate form (b).

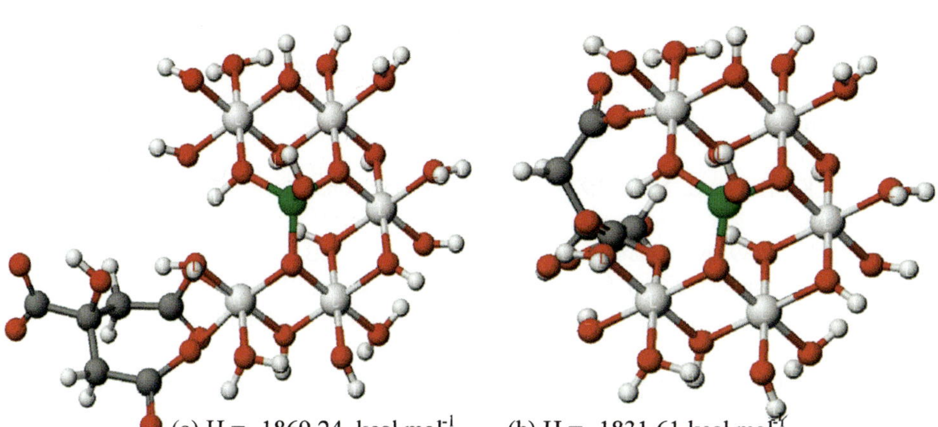

Figure 10. Citrate adsorbed on allophane in bidentate (a) or binuclear form (b).

It is however, very difficult to determine experimentally which of the two formation reactions is more likely. Therefore, a theoretical molecular orbital calculation that simulates the pore region of the nano-ball allophane structure was applied.

In the molecular orbital calculation, we used a model cluster of allophane shown in Figure 8. Figure 9 shows the optimized structure and heats of formation (H) for the bidentate and binuclear adsorption of oxalate. The optimized structure is the most stable geometry of the allophane model with PM3 basis set. The lower H value of the binuclear form than that of the bidentate form indicates that the binuclear complexation of oxalate is more likely. However, the calculation for citrate adsorption indicated the bidentate form as being more stable than the binuclear form (Figure 10). The difference in mode of reaction between oxalate and citrate may arise from the geometry of the organic molecules. The distance between oxygen atoms of the two COO^- groups in free oxalate (0.29 nm) is more suitable for binuclear complexation than that of free citrate (0.48 nm). This is because the O-O distance of aluminol groups at the allophane pore is about 0.3 nm. For citrate adsorption, a monodentate bond is also possible because of the longer distance between the COO^- groups in the acid. However, it must be emphasized that the possibility exists for a monodentate reaction between CH_3COOH and allophane.

4. CONCLUSIONS

Adsorption of oxalic acid on allophane was much greater than citric and acetic acids at lower pH, but at high pH, citrate had the highest affinity. Due to the higher Al-OH group content per unit mass, KyP, with a low Si/Al ratio showed higher capacity for adsorption of these three organic compounds than KnP, which has a higher Si/Al ratio. A deprotonation reaction occurred for acetate and citrate at low pH, as opposed to a dehydroxylation process at high pH. Through ligand exchange reaction, oxalate adsorption on allophane possibly occurs as binuclear form rather than as the bidentate form. Understanding reaction mechanism short-range ordered aluminosilicate with organic ligands has an important implication in study on pedogenesis, soil fertility and environmental chemistry.

REFERENCES

1. Wada, K., Harward, M.E., 1974. Amorphous clay constituents of soils. Adv. Agron. 26, 211-260.
2. Wada, K., 1989. Allophane and imogolite. *In*: Dixon, J.B., Weed, S.B. (Eds.). Minerals in Soil Environments. 2nd ed. Soil. Sci. Soc. Am., Madison, WI, pp. 1051-1087.
3. Henmi, T., Wada, K., 1976. Morphology and composition of allophane. Am. Mineral. 61, 379–390.
4. Linder, G.G., Nakazawa, H., and Hayashi, S., 1998. Hollow nanospheres, allophanes 'all-organic' synthesis and characterization. Microporous Mesoporous Mater. 21, 381-386.
5. Farmer, V.C., Russel, J.D., Berrow, M.L., 1980. Imogolite and proto-imogolite allophane

in spodic horizons: evidence for a mobile aluminum silicate complex in podzol formation. J. Soil Sci. 31, 673-684.

6. Parfitt, R.L., Henmi, T., 1980. Structure of some allophanes from New Zealand: Clays Clay Miner. 28, 285-294.
7. Kwong, K.F. Ng Kee, Huang, P.M., 1979. The relative influence of low-molecular-weight complexing organic acids on the hydrolysis and precipitation of aluminum. Soil Sci. 128, 337-342.
8. Stevenson, F.J., 1967. Organic acids in soil, *In*: McLaren, A.D., Peterson, G.H. (Eds.). Soil Biochemistry. Vol. 1. Marcel Dekker, New York, pp. 119-146.
9. Forstner, U., 1981. Metal transfer between solid and aqueous phase. *In*: Forstner, U., Wittman, G.T.W. (Eds.). Metal Pollution in the Aquatic Environment. Springer-Verlag, New York, pp. 197-270.
10. Inoue, K., Huang, P.M., 1984. Influence of citric acid on the natural formation of imogolite . Nature (London) 308, 58-60.
11. Marschner, H., 1986. Mineral Nutrition of Higher Plants. Academic Press, London.
12. Dinkelaker, B., Romheld, V., Marschner, H., 1989. Citric acid excretion and precipitation of calcium citrate in rhizosphere of white lupin (*Lupinus albus* L.). Plant Cell Environ. 12, 285-292.
13. 13.Gerke, J., Romer, W., Jungk, A., 1994. The excretion of citric and malic acid by proteoid root of *Lupinus albus* L.: Effects on soil solution concentrations of phosphate, iron, and aluminum in the proteoid rhizosphere in samples of an oxisol and a luvisol. Z. Pflanzenern. Bodenk. 157, 289-294.
14. Nagarajah, S., Posner., A.M., Quirk, J.P., 1968. Desorption of phosphate from kaolinite by citrate and bicarbonate. Soil Sci. Soc. Am. Proc. 32, 507-510.
15. Kafkafi, U., Bar-Yosef, B., Rosernberg, R., Sposito, G., 1988. Phosphorus adsorption by kaolinite and montmorillonite: II. Organic anion competition. Soil Sci. Soc. Am. J., 52, 1585-1589.
16. Fox, T.R., Comerford, N.B., Mc Fee, W.W., 1990. Phosphorus and aluminum release from a spodic horizon mediated by organic acids. Soil. Sci. Soc. Am. J. 54, 1763-1767.
17. Violante, A., Colombo, C., Buondonno, A., 1991. Competitive adsorption of phosphate and oxalate by aluminium oxides. Soil. Sci. Soc. Am. J., 55, 65-70.
18. Mortland, M.M., 1970. Clay-organic complexes and interactions. Adv. Agron. 22, 75-117.
19. Greenland, D.J., 1971. Interactions between humic and fulvic acids and clays. Soil Sci. 111, 34-41.
20. Huang, C.P., 1991. Ionic factors affecting the formation of short-range ordered aluminosilicates. Soil Sci. Soc. Am. J. 55, 1172-1180.
21. Schulthess, C.P., Huang, C.P., 1991. Humic and fulvic acid adsorption by silicon and aluminium oxide surfaces on clay minerals. Soil Sci. Soc. Am. J. 55, 34-42
22. Kwong, K.F. Ng Kee, Huang, P.M., 1977. Influence of citric acid on the hydrolitic reactions of aluminium. Soil Sci. Soc. Am. J. 41, 692-697.

23. Violante, A., Violante, P., 1980. Influence of pH, concentration, and chelating power of organic anions on the synthesis of aluminum hydroxides and oxyhydroxides. Clay Clay Miner. 28, 425-434.
24. Henmi, T., Huang, P.M., 1985. Removal of phosphorus by poorly ordered clays as influenced by heating and grinding. Appl. Clay Sci. 1, 133-144.
25. Inoue, K., Huang, P.M., 1985. Influence of citric acid on the formation of short-range ordered aluminosilicates. Clays Clay Miner. 33, 312-322.
26. Henmi, T., Huang, P.M., 1987. Effect of phosphate anion on the formation of imogolite. *In*: Schultz, L.G. (Ed.), Proc. Int. Clay Conf. 8th, Denver, CO. 27 July-2 Aug. 1986. Clay Miner. Soc., Bloomington, IN, pp. 231-236.
27. Inoue, K., Huang, P.M., 1987. Effect of humic and fulvic acids on the formation of allophane. *In*: Schultz, L.G. (Ed.). Proc. Int. Clay Conf. 8th, Denver, CO. 27 July-2 Aug. 1986. Clay Miner. Soc., Bloomington, IN. pp. 221-226.
28. Stewart, J.J.P., 1989a. MOPAC ver. 6. QCPE#455. F. J. Seiler Research Laboratory, United States Air Force Academy., CO.
29. Stewart, J.J.P., 1989b. Optimization of parameters for semi-empirical methods. I. Method. J. Comput. Chem. 10, 209-220.
30. Jones, D.L., Brassington, D.S., 1998. Sorption of organic acids in acid soils and its implications in the rhizosphere. Europ. J. Soil Sci. 49, 447-455.
31. Wada, S., 1987. Adsorption of Al (III) on allophane, imogolite, goethite, and noncrystalline silica and the extractability of the adsorbed Al (III) in 1 M KCl solution. Soil Sci. Plant Nutr. 33, 487-491.
32. Parfitt, R.L., 1978. Anion adsorption by soils and soil materials. Adv. Agron. 30, 1-50.
33. Bolt, G.H., Bruggenwert, M.G.M., Komphorst, A., 1978. Adsorption of cation by soil. *In*: Bolt, G.H., Bruggenwert, M.G.M. (Eds.). Soil Chemistry. A. Basic Elements. Elsevier Science, New York, pp.54-90.
34. Parfitt, R.L., Farmer, V.C., Russell, J.D., 1977. Adsorption on hydrous oxides: I. Oxalate and benzoate on goethite. J. Soil Sci. 28, 29-39.
35. Parfitt, R.L., Farmer, V.C. and Russell, J.D., 1977. Adsorption on hydrous oxides: II. Oxalate, benzoate and phosphate on gibbsite. J. Soil Sci. 28, 40-47.
36. Bhatti, J.S., Comerfold, N.B., Johnston, C.T., 1998. Influence of soil organic matter removal and pH on oxalate sorption onto a spodic horizon. Soil Sci. Soc. Am. J. 62, 152-158.

Developments in Soil Science, Volume 28A
Editors: A. Violante, P.M. Huang, J.-M. Bollag and L. Gianfreda

THE ROLE OF CLAYS IN THE RESTORATION OF PERTURBED ECOSYSTEMS

G. J. Churchman
CSIRO Land and Water, Private Mail Bag No. 2,
Glen Osmond, South Australia 5064, Australia

Examples are given of 3 types of perturbed ecosystems that can be restored with the help of clays. Some examples were obtained from work in our own laboratory, both published and unpublished, while others are from the published work of other laboratories. The essential part played in soil ecosystems by associations of clays with other entities is illustrated by comparisons of electron micrographs of virgin and cultivated counterparts of the same soil type. Soils that have been perturbed by some agricultural practices may suffer degradation of physical properties and loss of their component clays. Reference to the literature suggests that re-introduction of clays needs to be accompanied by the use of biological agents to restore the essential associations. Research and practice has shown that simple addition of clays to soils that have become hydrophobic (non-wetting) is sufficient to overcome this particular problem. Clays are capable of removing cations and also positively-charged species, such as many proteins, from contaminated water. Desorption of heavy metal cations can be inhibited by heating metal-exchanged smectite clays. Smaller cations become fixed on clays following heating at lower temperatures than are required for larger cations. Complexes of (smectite) clays with proteins cannot be completely dissociated by cation exchange. Clays can be modified to enable the uptake of non-ionic organic species from water by their exchange with organic cations, including both quaternary ammonium cations (QACs) and polymeric cations. The complexes between the (smectite) clays and highly-charged polymeric cations show a positive surface charge after the addition of only a little polymer to clay. These complexes can remove anionic pollutants from water. Both QAC-modified clays and clay-based wastes from industries that process biological materials show an affinity for petroleum oil. Both may be used to clear films of oil on water and as barriers to curtail the leakage of plumes of oil from storage and transmission facilities.

1. INTRODUCTION

There are many different kinds of ecosystems and a myriad of ways in which they may be perturbed. It would be a mammoth task to discuss all possible combinations. Instead, this paper focuses on three important general types of perturbation, viz.

1. the loss of integrity and of materials in soils,
2. the development of hydrophobic (non-wetting) soils, and
3. the contamination of water.

With even these being quite broad topics, the paper provides examples of particular instances where clays have been shown to have a role in the restoration of these types of perturbation.

Clays are generally defined by particle size, most commonly comprising <2 μm material. Minerals in this size range are generally known as clay minerals. Most of these are aluminosilicates, although there are also oxides, hydroxides, oxyhydroxides, among others e.g. phosphates. In this paper, "clays" and "clay minerals" are used synonymously unless otherwise stated.

Among the most useful properties of clays are their high specific surface areas. Those of allophanes have been measured at nearly 600 m^2 g^{-1} (1), while smectites can have specific surface areas of ~800 m^2 g^{-1} if intracrystalline surface are included as well as external surfaces, e.g. [2]. Their extensive surfaces encourage the adsorption of other materials.

Adsorption is also encouraged by the electrical charges on clays. These vary in both quantity and sign, although most aluminosilicate clays are negatively charged in solutions at neutral or near-neutral pH, as well as at high pH. All clays have some charge that varies with pH and many also have a permanent negative charge. Some clay minerals with a low permanent negative charge e.g. kaolinite can acquire sufficient positive charge from the adsorption of protons to display a net positive charge under acid conditions. The most highly charged commonly occurring clays (smectites) have an adsorption capacity of ~1 mmol. of positive charge per gram of clay.

This paper first compares the disposition of clays in both unperturbed ("natural") soils and soils that have been perturbed by agricultural use. The comparison is aimed to help establish at least one of the roles of clays in the operation of natural, undisturbed ecosystems. It then discusses a possible role for clays in the reversal of damage that can result from some agricultural practices. A description is then made of the use of clays to enhance the productivity of hydrophobic, hence unproductive soils. It finally demonstrates some of the properties of clays that may make them useful for the clean-up of polluted water. The paper present some results from the author's laboratory, both published and unpublished, and also refers to the published work of others.

2. CLAYS IN NATURAL AND PERTURBED SOILS AND STRATEGIES FOR THEIR RESTORATION

Figures. 1 and 2 respectively show transmission electron micrographs (TEMs) of ultrathin sections of aggregates within the upper 0.05 m of one type of soil (a Calcic Haploxeralf) from a virgin site (Figure 1), and a nearby site that has been farmed, including with cultivation for cropping, for ~ 130 years (Figure 2). The soil, sites, and methods of sampling and of preparation of the ultrathin sections are described by Churchman and Foster [3].

Figure 1 shows that aggregates in this Alfisol in its virgin state are composed of complex multilaminate microaggregates generally of the order of a micrometre, or less, in linear extent. Many of the microaggregates appear to have a rounded shape. Selective staining has shown that the cores of these microaggregates are variously composed of quartz, plant cell remains, extracellular polysaccharides and microbial colonies [3]. Virtually all of the fine material appears to be bound into the microaggregates in the virgin soil. The clay, shown as the darker material in the micrographs, typically surrounds other components of the soil.

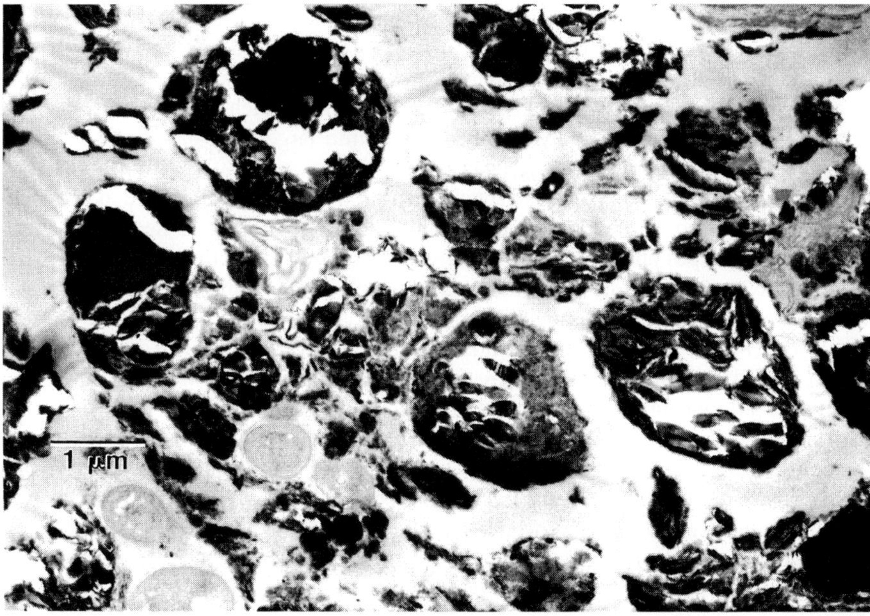

Figure 1. Transmission Electron Micrograph of an ultrathin section from 0-10 mm depth in a virgin Calcic Haploxeralf

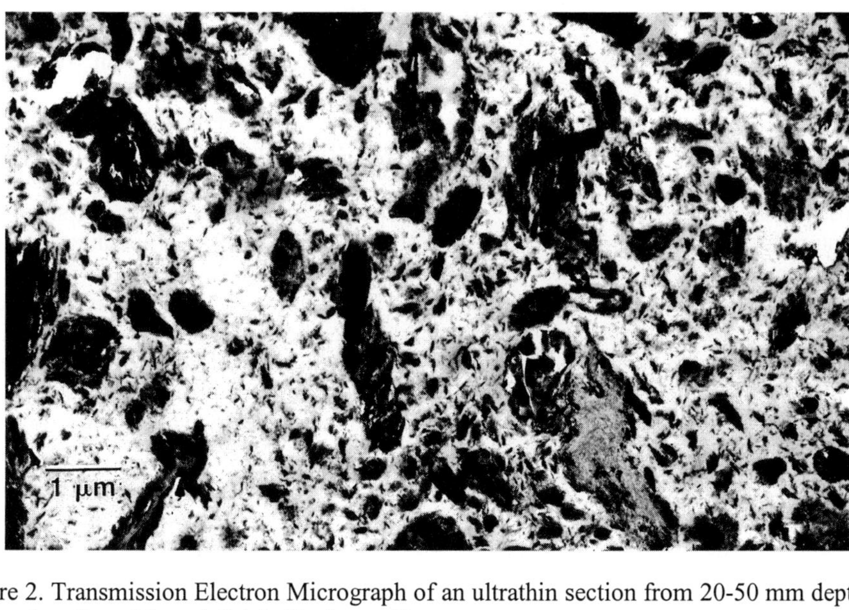

Figure 2. Transmission Electron Micrograph of an ultrathin section from 20-50 mm depth in a conventionally cultivated Calcic Haploxeralf.

In Figure 1 clay often surrounds quartz, which is indicated by shards resulting from sectioning. Elsewhere in the same sample, it is seen to surround bacteria and extracellular polysaccharides (Figure 1.13 in [4]) and also plant cells. In general, clay mineral layers, which are typically ~ 1 nm in thickness, are bound into much larger entities, either by association with other material or by self-association into larger clay particles e.g. in borders formed by clays around many microaggregates.

By contrast, Figure 2 shows that much dispersed clay occurs in the pores between microaggregates in the farmed (cultivated) soil. The microaggregates seem to occur in a greater variety of shapes and sizes than in the virgin counterpart of this soil (Figure 1). This suggest that they many may have disintegrated. Much of the dispersed clay occurs in very small particles, suggesting that associations between clay layers are fewer and are also often smaller. While Figure 1, for the virgin soil, and Figure 2, for the cultivated soil, are taken of samples at slightly different depths from each other, both are within the depths in which organic matter is concentrated. The cultivated soil sample (for Figure 2) is derived from within the depth zone over which ploughing occurs.

The comparisons between Figures 1 and 2 indicate that clays can be extensively bound to other materials. This is the case in natural (virgin) soils (Figure 1). To underline this effect, a TEM of a sample taken from much deeper (100-150mm) in the subsoil of the virgin soil also showed largely complete microaggregates and pores between the microggregates that are largely clear of fine particles (G.J. Churchman and R.C. Foster, unpublished). In the farmed soil, however. perturbation, apparently as a result of cultivation and/or the action of livestock, has led to a breakdown of associations of the layers of clay minerals with one another as well as with other entities in the soil (Figure 2).

The disintegration of associations that is seen in Figure 2 at micron and sub-micron levels may lead to such deleterious effects for the soil as pore blockage, hence poor aeration and transmission of water and nutrients to plant roots, and also erosion by water and/or wind. Farming on these and similar soils that are widespread in southern Australia began with the beginning of European settlement. A considerably earlier study from South Australia [5] found that erosion was widespread and that >75% of the depth of surface soil had been lost from some areas. Restoration of the type of perturbation seen in Figure 2 may benefit from addition of clay to the soils to replace erosion losses. However, any clay added needs to be re-integrated into the soil system, hence re-formed into aggregates that are themselves composed of microaggregates, following the hierarchical scheme for stable soil structure or architecture [6]. The experimental work of Dorioz et al. [7], points to a possible strategy for the restoration of this particular form of perturbation of soil ecosystems.

Electron micrographs presented by Dorioz et al. [7] show that the action of bacteria and yeasts, followed by drying then rehydration, has led to the aggregation of particles of clay minerals (kaolinite and montmorillonite). Roots were also shown to have a strong effect. They caused orientation of clay particles and their associations, with fissures separating different groups of associated particles (i.e. microaggregates). Observations close to hyphae showed that these had similar effects. The agents causing the orientation and association of particles include mucigels identified at root apices and also polysaccharide strands which clearly provided links between particles in fungal cultures.

Dorioz *et al.*'s (1993) work [7], like those in many earlier papers [8-13] shows clear visual evidence for the key role played by polysaccharides as glues cementing clays into associations. However, there are other agents besides polysaccharides that may play a role in stabilising associations of clays to form a firm foundation for the structure of soils. Among these are proteins, which undoubtedly interact strongly with clays [14,15], and these interactions are exemplified by results presented in Part 4 herein. Wright and Upadhayaya [16] suggested that a protein, glomalin, which is produced by arbuscular mycorrhizal fungi, could be hydrophobic and could thereby contribute to the stabilisation of aggregates. Hydrophobicity is commonly thought to help stabilise soil aggregates by slowing rates of wetting [17-20]. Proteins may play a role in the stabilisation of soil aggregates by their possible incorporation in soil humic substances [14], which can act as organic glues [21]. On the other hand, glues for aggregates in soils may not be wholly organic. Organic matter in soils is often linked to polyvalent cations such as Ca, Al and Fe, to oxides, hydroxides and oxyhydroxides of Al and Fe, especially, and also to aluminosilicates in soils. Many workers [22-28] have shown that these combinations confer stability to microaggregates in soils.

In summary, clays provide the skeleton for the fundamental associations forming the structure of soils. Many agents, both organic and inorganic, and their combinations, act to bring about and stabilise these fundamental associations. Their breakdown can have serious effects on the soil ecosystem, leading, in the extreme, to the loss of the clays forming the basic building blocks. Restoration of this ecosystem may require the replacement of clay and its re-integration into the soil ecosystem.

3. THE USE OF CLAYS FOR THE IMPROVEMENT AND RESTORATION OF HYDROPHOBIC (NON-WETTING) SOILS

While hydrophobicity on the small (micrometre) scale can confer stability to aggregates, it can present a serious problem for both the productivity and conservation of soils when it is encountered at the large (metre or kilometre) scale [29-34]. It is a widespread problem, having been reported in Florida [35], New Zealand [36], Australia [29], and California [31], for example. It occurs most commonly in sandy soils. It has been observed that the problem of water repellency is exacerbated by recent fires in above-ground vegetation [31], but it also occurs often where there is no evidence for recent fires.

In Australia, hydrophobicity in sandy soils has been attributed to the occurrence of particular, waxy types of organic components [34]. The problem is often more severe in summer and may even disappear over wet winters.

The solution to the problem, at least as it occurs in Australia, has proven to be a relatively simple one, involving the addition of clays to the generally sandy hydrophobic soils. Farmers themselves discovered the usefulness of the addition of clay for overcoming this problem and clay, generally locally sourced and often from subsoils, is now used extensively in southern and western Australia [37]. However, not all clays are similarly effective as additives for decreasing hydrophobicity. It has been found that kaolinites are more effective than montmorillonites for overcoming water-repellency [32, 33]. Other factors besides clay type appear to have a minor effect on the effectiveness of clays for this purpose. Ma'shum *et al.* [33] and Ward and Oades [34] both found that increased dispersibility of kaolinites improved their effectiveness, while Ward and Oades [34] found improvements in their effectiveness after wetting and drying cycles following applications of clay.

The restoration of hydrophobic soils by the addition of clays is a practical, widely accepted example of the application of clay as a relatively low-cost, yet long-term solution to a locally-important perturbation of the efficient functioning of the soil-plant ecosystem.

4. THE USE OF CLAYS AND MODIFIED CLAYS FOR THE RESTORATION OF CONTAMINATED WATER

4.1. Removal of positively-charged species by clays

There have been many studies of the sorption of cations by clay minerals, as summarised by Tiller [38]. By virtue of their large cation exchange capacities, smectites can remove large amounts of cations from solution by ion exchange, although the sorption of heavy metal cations may involve surface complexation reactions as well as simple ion exchange [39].

While many different clay minerals can adsorb proteins [14], smectites have a particularly high capacity as a result of their relatively high charge. Especially as wastes from food processing, proteins are significant sources of contamination. They contribute nitrogen, and hence increase biological oxygen demand in water. A concern with the high soluble protein load in wastewaters from abattoirs in Australia which is added to the environment, together with the energy costs of alternative treatments, led to an investigation of the effectiveness of different smectitic clays for removing the waste protein [40] that is summarised briefly here. The different clays, which were studied as they had been mined or marketed, showed cation exchange capacities (CECs) of between 39 and 86 $cmol_{+}kg^{-1}$ in); quartz contents of between 3 and 20 %; exchangeable Na as a percentage of exchangeable cations (ESP) of between 36 and 83; and percentage values for total charge that is tetrahedral of between 13 and 100%.
The uptake of protein from a wastewater sample with 0.33 g L^{-1} protein for the set of 8 samples is plotted against ESP in Figure 3. With the values for the raw smectitic samples shown as crosses, Figure 3 mainly shows that there was a scatter in uptake values in relation to ESP. However, when each of the samples was further exchanged with an excess of NaCl, to confer ESP values between 88 and 96, all gave values for protein uptake (shown by solid triangles in Figure 3) within only a narrow range (110-150 mg g^{-1}), except for the Arumpo bentonite. In the particular case of Arumpo bentonite, there was little increase in uptake as a result of sodium saturation. However, when a Na-saturated sample of Arumpo bentonite was ultrasonicated prior to reaction with the wastewater, its uptake of protein (shown by a square symbol in Figure 3) was increased considerably so that it was comparable to that of the other smectitic materials studied. Ultrasonic treatment led to a decrease in uptake of protein by one of the other samples (Saponite E) (G.J. Churchman, unpublished). A separate determination of the uptake of bovine serum albumin protein at its isoelectric point of pH 4.7, showed that ultrasonic treatment alone increased uptake of this protein by the Arumpo sample more than three-fold, that ultrasonic treatment marginally decreased protein uptake by the Na-saturated Saponite E sample and also that Na saturation led to an increased protein uptake by 6 of the 8 smectitic samples, with little effect on the other two (G.J. Churchman, unpublished).

The generally positive effect of a high ESP on protein uptake indicates the effect of interlayer Na in enhancing the swelling and dispersion of smectites, hence the separation of their layers, enabling easier displacement of the interlayer cation by bulkier protein molecules [14]. However, uptake is generally lower than many values reported for individual proteins [14]. Maximum uptake tends to exceed 1.5 g protein g^{-1} smectite e.g. for haemoglobin at a pH near 7 [14] . Low protein: clay ratios used in the current study have probably resulted in incomplete uptake. Nonetheless, there were distinct peaks at 24 Å, indicating intercalation of protein, in XRD patterns of products.

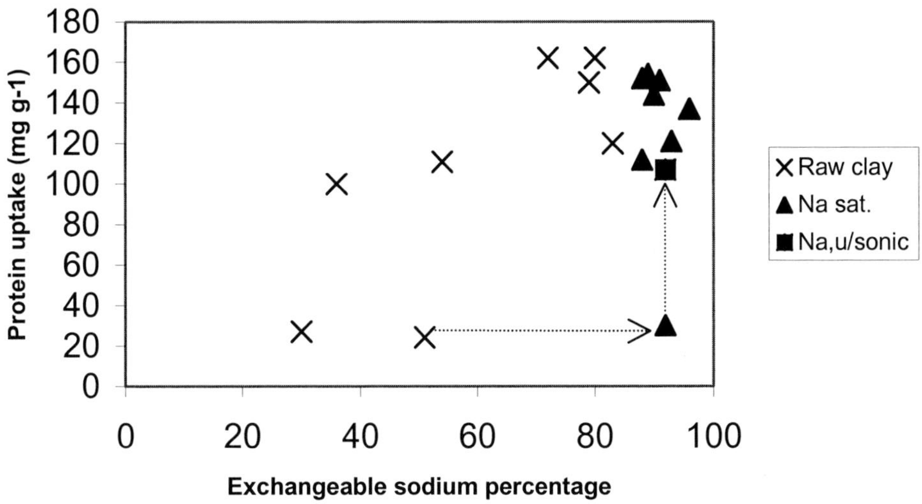

Figure 3. Uptake of protein from a wastewater with 0.33 g L^{-1} protein by the raw smectitic clay samples (crosses) and also the samples after Na saturation (triangles) and for one sample (Arumpo bentonite) after ultrasonic treatment of the Na-saturated material (square). The points for the 3 treatments of Arumpo bentonite are joined by arrowed lines.

The effect of ultrasonic treatment on increased protein uptake by the Arumpo sample is explained by other work on bulk Arumpo bentonite. In the first place, a suspension of the raw Arumpo sample showed little more viscosity than water alone [46-48]. Although Na-exchange increased the viscosity, ultrasonic treatment was particularly effective. The viscosity of the raw sample was increased by ~ 50 times while that of the Na-saturated sample was increased ~ 4 times as a result of ultrasonic treatment [46-48]. These effects can be explained by the observation of microaggregates of ~0.2-0.5 μm with sub-spherical shapes in transmission electron micrographs of a bulk Arumpo sample [46-48]. Viscosity changes suggest that these are broken down through the application of ultrasonic energy. Furthermore, both its chemical analysis and X-ray diffraction patterns [46-48] indicate that the smectite layers in the bulk Arumpo bentonite sample are interstratified with layers of illite, which appear to comprise ~20% of the total [48]. It appears likely that the remaining smectite layers are highly charged [48]. Together with mechanical constraints from microaggregation, the high charge on the layers could have inhibited the intracrystalline swelling of smectite layers and hence the extensive swelling of the bentonite [49].

The different smectitic clays showed a contrast of >5 times in uptake between the most adsorbent of 8 Australian smectitic samples. However, when Na-saturated, 7 of the 8 samples removed amounts of protein that were within 40% of each other in spite of the large differences (up to 220%) between their CECs. However, one exceptional sample (Arumpo bentonite) required ultrasonic treatment to break down microaggregates apparent in transmission electron micrographs before adsorption of protein could occur to a comparable extent to the other samples. Activation of clays was necessary to ease access to interlayers by bulky protein molecules. Generally, this could be achieved by sodium saturation but mechanical activation with ultrasound was necessary in one case.

4.2. Retention of adsorbed cationic species by clays

4.2.1. Heavy metals

The most suitable adsorbent to restore aqueous ecosystems that have become contaminated with soluble pollutants will be one that provides not only easy uptake of contaminants, but also their restraint against subsequent desorption back into solution. The adsorbed phase needs to be held against displacement by other components that may occur or become introduced into the solutions.

Churchman *et al.* [50] examined the ease of displacement of three different heavy metal cations (Cr^{3+}, Ni^{2+} and Pb^{2+}) that were adsorbed by a Wyoming bentonite. No more than 3% of the heavy metal cations were released by leaching with water in each case. A 1M Na salt solution tended to release more of each of these metals than was released by a 0.1M solution of the same salt. However, heating, at a suitable temperature for each cation, can help to prevent desorption of the cations (Figure 4).

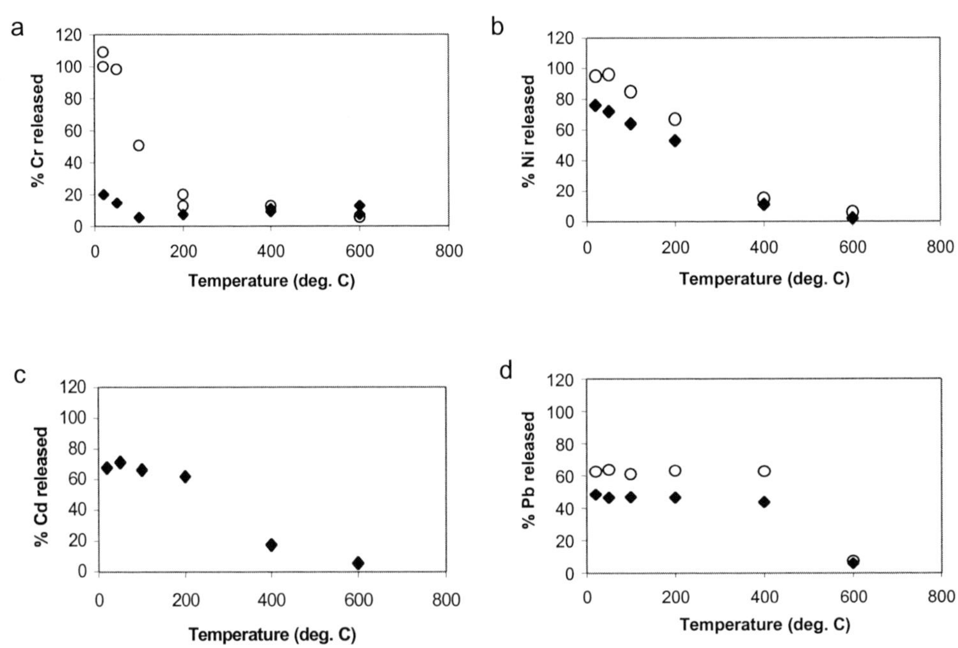

Figure 4. The percentages of each heavy metal originally held on Wyoming bentonite exchanged with each heavy metal (a. Cr, b. Ni, c. Cd, d. Pb) that was released by either 0.1M (filled symbols) or 1M Na salts (open circles) after the heavy metal-exchanged bentonites had been heated at different temperatures. Individual duplicate values are plotted for Cr-bentonite (a). No determinations were made with Cd-bentonite and 1M salt (c).

While Cr was hardly released by dilute salt regardless of the temperature of pre-heating, heating had a strong effect on the ease of release of Cr from Cr-bentonite by 1M Na salt solution (Figure 4a). Heating at 200°C brought about a strong decrease in the amount released compared with that from samples heated at lower temperatures, while very little Cr was released from a sample heated at 600°C. The percentage of Ni released from Ni-bentonite generally showed a steady decrease with heating (Figure 4b). The percentage of Cd released from Cd-bentonite (by 0.1M NaCl) decreased strongly after the bentonite was heated at 400°C relative to that released after heating at lower temperatures, which had been quite invariant with temperature (Figure 4c). There appears to have been a slight further decrease in Cd released following heating at 600°C. The proportion of Pb released from Pb-bentonite with 1M Na salt was almost constant following heating at temperatures up to and including 400°C, but decreased dramatically after heating at 600°C (Figure 4d).

Heating could bring about the dehydration of the adsorbed ion, its migration into the lattice and/or the dehydroxylation of the clay mineral and consequent formation of oxycations. The results of heating a clay mineral with different adsorbed cations are likely to be influenced by differences between the ions in their ease of dewatering, their size, and the ratio of their charge to their size.

The different effects of heating on the four different adsorbed heavy metal cations on bentonite apparently reflects their different (crystal) ionic sizes. Immobilisation of adsorbed Ni^{2+}, with an crystal ionic radius of 0.69Å, required heating to a higher temperature than for Cr^{3+}, with a smaller ionic radius of 0.63Å, but a lower temperature than for Pb^{2+}, which has a much larger crystal ionic radius (1.20Å). While the temperature at which the exchangeability of Cd, with a crystal ionic radius of 0.92Å was decreased most strongly by heating i.e. 400°C, was similar to that for Ni^{2+}, the decrease with temperature was gradual over the whole range for the smaller Ni ion but limited to higher temperatures for Cd.

Chorom and Rengasamy [51] noted that cations with crystal ionic radii <0.7Å (Li^{+}, Mg^{2+} and Al^{3+}) became strongly attached to mineral surfaces at considerably lower temperatures than larger cations. In view of the Hoffmann-Klemen effect whereby Li migrates to octahedral sites on heating, these authors considered that Mg and Al also migrated to octahedral sites with sufficient heating. For the same reasons, Cr^{3+} may have migrated into octahedral sites on heating. However, the larger crystal ionic radii of Pb^{2+} and Cd^{2+}, would have prevented their intracrystalline migration according to this reasoning. Na^{+} (crystal ionic radius = 0.97Å), K^{+} (1.33Å) and Ca^{2+} (0.99Å) are also too large to allow intracrystalline migration [51]. Chorom and Rengasamy [51] suggested that heating increases the covalent bonding of such large cations to mineral surfaces.

Although there may be a number of different mechanisms for the immobilisation of adsorbed metals on clay surfaces as a result of heating [50] the results suggest that less heating is required to immobilise small cations that are adsorbed by bentonites than is needed to immobilise larger adsorbed cations.

4.2.2.Proteins

Another recent study in our laboratory [52] has investigated the possible desorption of proteins from clay-protein complexes. Since bentonites are used world-wide for the removal of heat-sensitive proteins from wines during the wine-making process, a considerable proportion of the solid wastes from the industry comprises bentonite-protein complexes as components of the lees. This particular study was motivated by a perceived need to minimise the volume of solid waste from wineries and distilleries that is disposed in landfill when space available for landfill is decreasing. Desorption of the protein from complexes with bentonite, allowing the re-use of the bentonite in the wine industry or elsewhere would decrease the volume of solid waste to be disposed in landfill.

A hypothesised procedure for the desorption of protein in bentonite-protein complexes using alkalis and salts – thereby reversing the process for protein uptake by bentonites – proved to be ineffective for substantial removal of the protein (Figure 5).

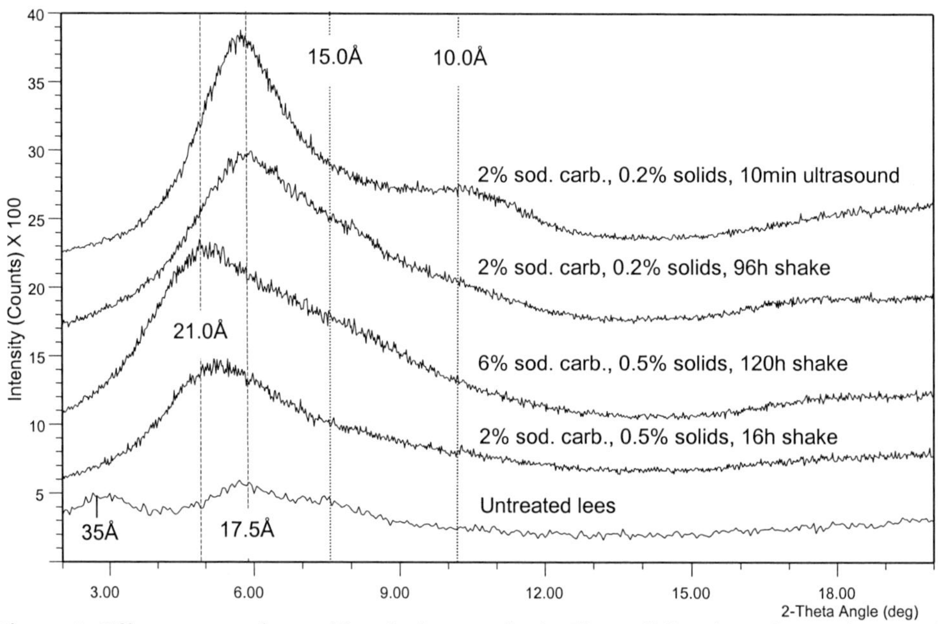

Figure 5. Effects on spacings of basal plane peaks in X-ray diffraction of smectite-protein complex in wine lees that were treated with sodium carbonate at different concentrations, different solid:solution ratios and following different agitation treatments. Products were examined as unoriented powders after oven-drying at 105°C. Spacings of treated complexes ranged from 21.0 to 17.5Å (vertical lines with large dashes), while expected spacings for an uncomplexed Na-smectite would range between ~15 and ~10 Å (spacings of 15.0 and 10.0Å are shown by vertical lines with short dashes). The complex in the untreated lees shows a spacing of 35Å.

Figure 5 shows that the basal spacings of any of the oven-dried products of sodium carbonate treatments remained at 17.5Å or above. This suggests that substantial amounts of material, either proteins or their break-down products, remain within the interlayers of the smectite regardless of any treatments with sodium carbonate. Furthermore, analysis of the protein released using [45] showed that treatments with other salts and alkalis viz. 10% KCl, 2M $CaCl_2$, 0.01M NaOH and 0.1M NaOH were less effective than the most effective of the treatments shown in Figure 5 i.e. 10 min ultrasonic treatment of a 0.2% suspension of solids with 2% Na_2CO_3. This latter treatment removed 66.5 mg protein g^{-1} lees. At the other extreme, a 16hr shaking treatment of a suspension having the same solids concentration with 2M $CaCl_2$ at pH 10.5 released no measurable protein.

It appears that, while charge interactions may play a major role in attracting proteins to bentonite clay, the bonds that form between the aluminosilicate clay and the protein are not simply electrostatic, as with many simple inorganic cations. Instead, they are likely to include van der Waals interactions and also involve favourable entropy changes [14]. Clearly, the associations between clays and proteins can be very strong and difficult to break.

4.3. Removal of non-ionic species by modified clays

Clays are naturally hydrophilic. However, because of their ready uptake of cations, layered aluminosilicate clays may become hydrophobic through their adsorption of organic cations containing non-polar groups. Quaternary ammonium cations (QACs) have been used most often to render smectites hydrophobic and many studies have shown their efficacy for this purpose [53].

Complexes of bentonites with some cationic polymers are also capable of removing reasonably high proportions of non-ionic pollutants from water [54-58]. Most usefully, the cationic polymers that have been used for this purpose include poly-diallyldimethylammonium chlorides (poly-DADMACs), which have been recognised as safe materials for application to waters for drinking [59]).

Isotherms shown by Churchman [58] indicate that the uptake of toluene from solution at a common equilibrium concentration of 100 mg L^{-1} was ~12 mg g^{-1} by a cationic polystyrene – Wyoming bentonite complex and ~5 mg g^{-1} by a poly-DADMAC complex with the same bentonite. Graphical interpolation of isotherms shown by Jaynes and Boyd [60]) indicate that a complex formed between Wyoming bentonite and the large hexadecyltrimethlammonium (HDTMA) quaternary ammonium cation showed an uptake of ~7 mg g^{-1} also at an equilibrium concentration of 100 mg L^{-1}. It appears that complexes with polycations can be similarly as effective as those with large organic cations for the removal of hydrocarbons from water.

4.4. Removal of negatively-charged species by modified clays

Bentonites complexed with several cationic polymers, including poly-DADMACs developed a positively-charged surface when sufficient polymer had been added to the clay [54, 55, 58, 61-64]. Figure 6 shows the zeta potential, indicating surface charge, of complexes of 4 different cationic polymers in different proportions with a Wyoming bentonite. It shows that only small additions (<0.2g polymer g^{-1} clay) of cationic polyelectrolytes with a high proportion of cationic charge (a polystyrene, a poly-DADMAC and a polyamide) needed to be added to the bentonite in order to produce a complex with a net positive charge. The positive charge was maintained, and often increased following further additions of the polymer to the clay. However, addition of a polymer with only a low cationic charge (a co-polymer of polyacrylamide) does not confer a net positive charge to the complex even when large amounts of the polymer is added.

It has been shown that positively-charged clay-polymer complexes removed the colour from solutions of anionic dyes [54, 58]. This type of complex has also been found to have a high affinity for phosphate in concentrations in water that are realistic for water that is contaminated from agricultural sources i.e. between 20 and 200 parts per billion P (T. W. Klenig and G.J. Churchman, unpublished data). Hence clays that have been modified with the appropriate cationic polymers could be used for the removal of both non-ionic and anionic pollutants from water.

It has been shown that positively-charged clay-polymer complexes removed the colour from solutions of anionic dyes [54, 58]. This type of complex has also been found to have a high affinity for phosphate in concentrations in water that are realistic for water that is contaminated from agricultural sources i.e. between 20 and 200 parts per billion P (T. W. Klenig and G.J. Churchman, unpublished data). Hence clays that have been modified with the appropriate cationic polymers could be used for the removal of both non-ionic and anionic pollutants from water.

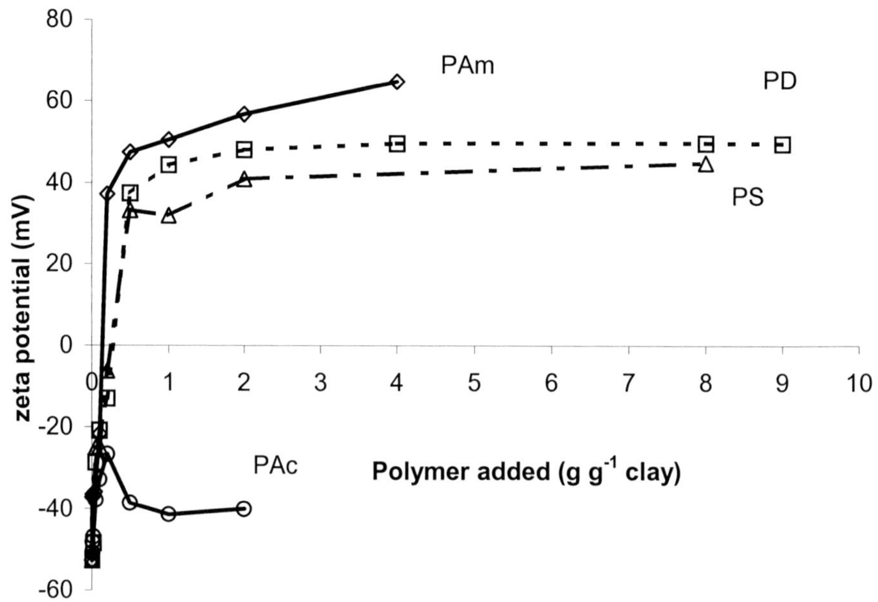

Figure 6. Changes in zeta potential of products with increasing addition of polymers to Wyoming bentonite. Cationic polyelectrolytes are: PAm: a polyamide; PD: a poly-DADMAC; PS: a polystyrene; PAc: a (co-polymer of) polyacrylamide

4.5 Removal of oils by modified and waste clays

Organo-clays formed by the modification of smectitic clays with quaternary ammonium cations have been shown to be oleophilic so that they attract oil on the surface of water, such as occurs in oil spills at sea. They break the films of oil and instead form clumps of the oil that float on the surface and can be removed from the surface of the water. These organo-clays make up the product "Petro-lock", which is marketed by the Lockheed Corporation for clearing oil spills. Some have been patented for the same purpose, including for oil spills occurring in especially cold climates [65], where the natural degradation of spilt oil occurs very slowly. However, it may be considered that organo-clays are too expensive for widespread use on oil spills, which may occur anywhere and at any time. Work in our laboratory have shown that mixtures of clays with organic compounds of biological origin can also show oleophilic properties and hence may be employed to help clear oil spills and to provide reactive barriers for the attenuation of plumes of hydrocarbons from storage and transmission facilities [66]. Several industries use clays in considerable amounts. The clays once used, now combined or mixed with appropriate (organic) industrial materials, may be available as wastes from these industries to be used as inexpensive environmental materials.

5. CONCLUSIONS

Clays are naturally occurring minerals that are widespread, hence generally inexpensive, and are non-toxic. They are integral components of soil ecosystems. As hydrophilic materials with large surface areas and charge, they contribute greatly to the chemical, physical and nutritional properties of soils. They are seen to occur in soils in associations with both organic matter and other inorganic species that provide the architecture and stability of soils at the fundamental microaggregate scale.

Their associations in soils may be disturbed by agricultural practices, leading to deleterious physical consequences as well possibly to the loss of the clays by erosion. Addition of further clay to soils perturbed in this way may enable their restoration but restoration of their associations that are essential components of soil ecosystems will require the introduction of biological agents.

The hydrophilicity of clays has meant that their simple addition to soils that are hydrophobic at the large scale, i.e. non-wetting soils, can restore these to enhance plant growth and soil conservation.

Especially when they have large cation exchange capacities, as in bentonites, clays are useful for the removal of positively-charged pollutants from water. These include heavy metals, ammonium ions and also proteins with a net positive charge at the pH of the water. The usefulness of clays for the restoration of contaminated water is enhanced when desorption of adsorbed pollutants is prevented, or at least minimised. This can be done by heating metal-exchanged bentonites, with smaller cations requiring heating at lower temperatures than larger cations to fix them on the smectites comprising the bentonites. Associations between smectites and proteins can be irreversible against displacement of the adsorbed proteins by simple inorganic cations, even at high pH.

While naturally hydrophilic and generally net negatively-charged, the extensive surfaces of clays can be exploited, by means of their adsorption of organic cations, so that they become hydrophobic and can adsorb non-ionic organic species such as hydrocarbons from water. (Smectite) clays can be modified for this purpose by their adsorption of quaternary ammonium cations and also cationic polyelectrolytes. Provided these latter have a high proportion of positive charge, their complexes with the clays can acquire a net positive charge after only small amounts of the polyelectrolyte is added. These complexes then have the ability to remove anionic pollutants from water.

Since most useful clays such as smectites (as bentonite) occur in quite high concentrations in surface or near-surface deposits, so are generally quite inexpensive to mine, the work shows that clays can provide environmental solutions that are relatively inexpensive. Together with their generally low toxicity, this provides one of the advantages of clays for environmental uses. The possible re-use of clay-based wastes for environmental purposes serves to enhance this particular advantage of the use of clays.

The different applications of clays to the restoration of perturbed environments shows that they may play a direct role in the process, as in the removal of heavy metals and proteins from water. However, their role is often indirect insofar as they serve as templates for active agents in the various restoration processes. Thus they provide the skeleton on which soil structure is restored by biological agents and their products. Their addition attracts water into otherwise hydrophobic soils. Nonetheless, it is the associated organic materials that attract non-ionic, partially water-soluble non-ionic organic compounds, oil and/or anions to modified clays and also clay-rich wastes from industries processing materials of biological origin. However, it is their association with clays that enables the easy transfer of the liquid contaminants into the solid phase so that they can be removed for the restoration of a clean aqueous phase.

ACKNOWLEDGEMENTS

This work received financial support from: Grains Research and Development Corporation; Meat Research Corporation; Grape and Wine Research and Development Corporation. I thank the following CSIRO staff: Ralph Foster, for electron microscopy and interpretation, Nina deLacy and Jenny Anderson, for laboratory work, and Richard Merry for advice on field work and for helpful comments on a version of the manuscript. I thank the following companies for samples: Integrated Mineral Technology Ltd., Arumpo Bentonite Pty Ltd., Commercial Minerals Ltd, Volclay Pty Ltd. and Watheroo Minerals Pty Ltd, all for clay samples, and Ciba Speciality Chemicals for the Magnafloc samples and for information and advice

REFERENCES

1. Vandickelen, R., Roy, G. de, Vansant, E.F., 1980. New Zealand allophanes: a structural study. J. Chem. Soc. Faraday I 76, 2542-2551..
2. Gast, R.G. 1977. Surface and colloid chemistry. *In:* Dixon, J.B., Weed, S.B. (Eds.), Minerals in Soil Environments. Soil Science Society of America, Madison, WI, pp. 27-73.
3. Churchman, G.J., Foster, R.C. 1994. The role of clay minerals in the maintenance of soil structure, Trans, 15th World Congress of Soil Science, Acapulco, Mexico, International Society of Soil Science 8a 17-34.
4. Churchman, G.J. 1999 The alteration and formation of soil minerals by weathering. *In:* Sumner, M.E. (Ed.). Handbook of Soil Science, CRC Press, Boca Raton, pp. F3-F76.
5. Stephens, C.G., Herriot, R.I., Downes, R.G., Langford-Smith, T. Acock, A.M. 1945. A soil, land-use, and erosion survey of part of County Victoria, South Australia. CSIRO Bull. No. 188.
6. Tisdall, J.M., Oades, J.M., 1982. Organic matter and water-stable aggregates in soils. J. Soil Sci. 33, 141-163
7. Dorioz, J.M., Robert, M., Chenu, C., 1993. The role of roots, fungi and baxteria on clay particle organization. An experimental approach, Geoderma 56, 179-194.
8. Guckert, A., Breisch, H., Reisinger, O. 1975. Interface sol-racine 1. Etude au microscope electronique des relations mucigel-argile-microorganismes. Soil Biol. Biochem. 7, 241-250.
9. Churchman, G.J., Tate, K.R., 1986. Effect of slaughterhouse effluent and water irrigation upon aggregation in seasonally dry Mew Zealand soil under pasture. Aust. J. Soil Res. 24,

505-516.
10. Foster, R.C., 1988. Microenvironments of soil organisms. Biol. Fert. Soils 6, 189-203.
11. Chenu, C., 1989. Influence of a fungal polysaccharide, scleroglucan on clay microstructure. Soil Biol. Biochem. 21, 299-305.
12. Robert, M., Chenu, C., 1992. Interactions between soil minerals and microorganisms, *In* Stotzky, G., Bollag, J-M., 1992. Soil Biochemistry. Vol. 7, Marcel Dekker, New York, pp. 307-404.
13. Foster, R.C., 1994. Microorganisms and soil aggregates, *In* Pankhurst, C.E., Doube, B.M., Gupta, V.V.S.R., Grace, P.R. (Eds.). Soil Biota, CSIRO Publications, Melbourne, pp.144-155.
14. Theng, . B.K.G. 1979. Formation and Properties of Clay-Polymer Complexes. Elsevier, Amsterdam.
15. Quiquampoix, H. Abadie, J. Baron, M.H. Leprince, F. Matumoto-Pintro, P.T. Ratcliffe, R.G., Staunton, S., 1995. Mechanisms and consequences of protein adsorption on soil mineral surfaces. ACS Symposium Series 602, 321-333.
16. Wright, S.F., Upadhayaya, A., 1996. Extraction of an abundant and unusual protein from soil and comparison with hyphal protein of arbuscular mycorrhizal fungi. Soil Sci. 161 575-586.
17. Bartoli, F., Philippy, R., Burtin, G.,1988a. Aggregation in soils with small amounts of swelling clays I Aggregate stability. J. Soil Sci. 39 593-616.
18. Capriel, P., Beck, T., Borchert, H., Härter, P., 1990. Relationship between soil aliphatic fraction extracted with supercritical hexane, soil microbial biomass, and soil aggregate stability. Soil Sci. Soc. Amer. J. 54, 415-420.
19. Haynes, R.J., Swift, R.S., 1990. Stability of soil aggregates in relation to organic constituents and soil water content. J. Soil Sci. 41 73-83.
20. Sullivan, L.A., 1990. Soil organic matter, air encapsulation and water-stable aggregation. J. Soil Sci. 41, 529-534.
21. Chaney, K., Swift, R.S., 1986. Studies on aggregate stability II The effect of humic substances on the stability of re-formed aggregates. J. Soil Sci. 37, 372-381.
22. Russell, E.W., 1935. The binding forces between clay particles in a soil crumb. Proceedings 3rd International Congress Soil Science, Oxford, England, International Society of Soil Science I, pp. 26-29.
23. Edwards, A.P., Bremner, J.M., 1967. Microaggregates in soils. J. Soil Sci. 18, 64-73.
24. Hamblin, A.P., Greenland, D.J., 1977. Effect of organic constituents and complexed metal ions on aggregate stability of some East Anglian soils. J. Soil Sci. 28, 410-416.
25. Giovannini, G., Sequi, P., 1976. Iron and aluminium as cementing substances of soil aggregates, Parts I and II. J. Soil Sci. 27 140-153.
26. Turchenek, L.W., Oades, J.M., 1978. Organo-mineral particles in soils. *In* Emerson, W.W., Bond, R.D., Dexter, A.R. (Eds.), Modification of soil structure. John Wiley & Sons, New York, pp. 137-144.
27. Barriuso, E., Portal, J.M., Faivre, P., Andreux, F. 1984. Separation des composes humiques et des associations organo-minerales d'un sol ochre humifere à structure microagregée (Umbric Dystrochrept): influence du pH des reactifs d'extraction. Pedologie 34, 257-274.
28. Tazaki, K., Kimura, S., Yoshimura, T., Akai, J., Fyfe, W.S., 1989. Clay-organic complexes as a cementing agent in the Arahama sand dune, Japan, Clays Clay Miner. 37, 219-226.

29. Bond, R.D., The occurrence of microbial filaments in soils and their effect on some soil properties. CSIRO Division of Soils, Divisional Report No 10/60.
30. Bond, R.D., Harris, J.R., 1964. The influence of microflora on physical properties of soils I Effects associated with filamentous algae and fungi. Aust. J. Soil Res. 2, 111-122.
31. DeBano, L.F., Mann, L.D., Hamilton, D.A., 1970. Translocation of hydrophobic substances into soil by burning litter. Soil Sci. Soc. Amer. Proc. 34, 130-133.
32. Ma'shum, M., Oades, J.M., Tate, M.E., 1989. The use of dispersible clays to reduce water-repellency of sandy soils. Aust. J. Soil Res. 27, 797-606.
33. Ward, P.R., Oades, J.M., 1993. Effect of clay mineralogy and exchangeable cations on water-repellency in clay amended sandy soils. Aust. J. Soil Res. 31, 351-364.
34. Franco, C.M.M., Tate, M.E., Oades, J.M., 1995. Studies on non-wetting sands I The role of intrinsic particulate organic matter in the development of water-repellence in non-wetting sands, Aust. J. Soil Res. 33 253-263.
35. Jamison, V.C., 1946. The penetration of irrigation and rain water into sandy soils of central Florida, Soil Sci. Soc. Amer. Proc. 11, 103-109.
36. van't Woudt, B.D., 1959. Particle coatings affecting the wettability of soils. J. Geophys. Res. 64, 263-???.
37. Obst, C., 1989. Non-wetting soils: management problems and solutions at "Pineview", Mundulla, The theory and practice of soil management for sustainable agriculture, Wheat Research Council Workshop, Australian Government Publishing Service, Canberra, pp. 78-80.
38. Tiller, K.G., 1996. Soil contamination issues: past, present and future, a personal perspective. *In:* Naidu, R., Kookana, R.S., Oliver, D.P., Rogers, S., McLaughlin, M.J. (Eds.) Contaminants and the Soil Environment in the Australasia-Pacific Region. Kluwer, Dordrecht, pp. 1-27
39. Stumm, W., Morgan, J.J., 1996. Aquatic Chemistry, John Wiley & Sons, New York.
40. Meat Research Corporation, 1998. The removal and recycling of proteins from the waste waters of abattoirs. Final Report, Meat Research Corporation Agreement RPDA.311.
41. Rayment, G.E., Higginson, F.R., 1992. Australian Laboratory Handbook of Soil and Water Chemical Methods. Inkata Press, Melbourne.
42. Norrish, K., Hutton, J.T., 1969. An accurate X-ray spectroscopic method for the analysis of a wide range of geological samples. Geochim. Cosmochim. Acta 33, 431-453.
43. Bodine, M.W., 1987. CLAYFORM: A Fortran 77 computer program apportioning the constituents in the chemical analysis of a clay or other silicate mineral into a structural formula. Comput. Geosci. 13, 77-88.
44. Churchman, G.J., Slade, P.G., Self, P.G., Janik, L.J., 1984. Nature of interstratified kaolin-smectites in some Australian soils. Aust. J. Soil Res. 32, 805-822.
45. Bradford, M.M., 1976. A rapid and sensitive method for the quantification of microgram quantities of protein utilizing the principle of protein-dye binding. Anal. Biochem. 72 (1976) 248-254.
46. Churchman, J., Hope, A., Reid, A., Gardam, M., Peter, P., Wright, M., Self, P. 1999. Arumpo bentonite: nature and origin of deposit, properties and uses *In:* Stewart, R. (Ed.), Murray Basin Mineral Sands. Australian Institute of Geoscientists Bulletin, N. 26, pp. 29-33.
47. Churchman, G.J., Askary, M., Peter, P., Wright, M., Raven, M.D., Self, P.G. 2002. Geotechnical properties indicating environmental uses for an unusual Australian bentonite. Appl. Clay Sci., 21 (in press).

48. Askary, M. 1999. A study of the swelling, dispersion and rheology of Arumpo bentonite. Thesis, Master of Soil Management and Conservation, University of Adelaide.
49. Slade, P.G., Quirk, J.P., Norrish, K., 1991. Crystalline swelling of smectite samples in concentrated NaCl solutions in relation to layer charge. Clays Clay Min. 39, 234-238.
50. G.J. Churchman, Rengasamy, P., Peter, P., Wright, M. 2002. Minimising the bioavailability of metal contaminants using bentonites, *In:* R. Naidu (Ed.) Bioavaliability (in preparation).
51. Chorom, M, Rengasamy, P., 1996. Effect of heating on swelling and dispersion of different cationic forms of a smectite. Clays Clay Min. 44, 783-790.
52. Grape and Wine Research and Development Corporation, 1999. Procedures for the efficient recycling of bentonite used for the removal of proteins from wine. Final Report, Grape and Wine Research and Development Corporation project CSS 98/1.
53. Lagaly, G. 1995. Surface and interlayer reactions:Bentonites as adsorbents. *In:* Churchman, G.J., Fitzpatrick, R.W., Eggleton, R.A. (Eds.). Clays Controlling the Environment. CSIRO Publishing, Melbourne, pp.137-144.
54. Churchman, G.J., Kolarik, L.O., Self, P.G., Anderson, J.S., deLacy, N.J., 1997. Modification of clays using cationic polyelectrolytes for the uptake of nonionic and anionic pollutants. Program with Abstracts, the 11th International Clay Conference, Ottawa, Canada. ICC97 Organising Committee, Ottawa, p. A16.
55. Churchman, G.J., 1998. Modification of bentonite with different polycations for the uptake of aromatic hydrocarbons from water. *In:* Volzone, C. (Ed.), Proceedings International Workshop of Activated Clays. La Plata, Argentina. Organising Committee, La Plata, pp. 9-19.
56. Breen, C., Watson, R.. 1998. Polycation-exchanged clays as sorbents for organic pollutants. Influence of layer charge on sorption capacity. J. Coll. Interf. Sci. 208, 422-429.
57. Breen, C. 1999. The characterisation and use of polycation-exchanged bentonites. Appl. Clay Sci., 15, 187-219.
58. Churchman, G.J., 2001. Formation of complexes between bentonite and diffeernt cationic polyelectrolytes and their use as sorbents for non-ionic and anionic pollutants. Appl. Clay. Sci. 20 (in press).
59. Letterman, R.D., Pero, R.W. 1990. Contaminants in polyelectrolytes used in water treatment. J. Amer. Wat. Works Assoc. 82, 87-97.
60. Jaynes, W.F., Boyd, S.A. 1991. Clay mineral type and organic compound sorption by hexadecylmethylammonium-exchanged clays. Soil Sci. Soc. Am. J. 55, 43-48.
61. Ueda, T., Harada, S., 1968. Adsorption of cationic poly-sulfone on bentonite. J. Appl. Polym. Sci. 12, 2395-2401.
62. Durand-Piana, G., LaFuma, F., Audebert, R. 1987. Flocculation and adsorption properties of cationic polyelectrolytes toward Na-montmorillonite dilute suspensions. J. Coll. Interf. Sci. 119, 474-480.
63. Breen, C., Rawson, J.O., Mann, B.E. 1996. Adsorption of polycations on clays: an in situ study using 133Cs solution-phase NMR. J. Mater. Chem. 6, 253-260.
64. Billingham, J., Breen, C., Rawson, J.O., Yarwood, J., Mann, B.E., 1998. The adsorption of polycations on clays: A comparative in-situ study using ^{133}Cs and ^{23}Na solution phase NMR. J. Coll. Interf. Sci. 193, 183-189.
65. Kemnetz, S., Cody, C.A., 1995. Oil spill flocculating agent and method of remediating oil spills. U.S. Patent 5558777.

66. Churchman, G.J., Anderson, J.S., 2001. Use of solid waste material. PCT Patent 80019036.
67. Environment Canada, 2000. EC 1999-2000 Sorbent Test Program. Emergency Engineering Technology Office, Environment Canada.

Developments in Soil Science, Volume 28A
Editors: A. Violante, P.M. Huang, J.-M. Bollag and L. Gianfreda

NEW APPROACHES TO THE MOLECULAR STRUCTURE AND PROPERTIES OF SOIL ORGANIC MATTER: HUMIC-, XENOBIOTIC-, BIOLOGICAL-, AND MINERAL-BONDS

H.-R. Schulten

University of Rostock, Faculty of Agricultural and Environmental Sciences,
Institute of Soil Science, Justus-von-Liebig-Weg 6, 18051 Rostock, Germany
Phone /Fax : +49 (0)381-498 2137. E-mail: hans-rolf.schulten@agrarfak.uni-rostock.de

The four main aims of this study were to expand and explore the applications of three-dimensional modeling and molecular mechanics calculations.
1. *Associations* of humic acids between n=2 and n=22, colloid formation (humic bonds) and binding of ubiquitous low polarity contaminants such as dialkyl phthalates are simulated.
2. *Trapping*, *sequestration* and *adsorption* of xenobiotic complexes of agrochemicals such as antibiotics (salinomycine) and herbicides (mecoprop) are illustrated (xenobiotic bonds) and intra-, intermolecular hydrogen bonds as well as energy descriptors (e.g., van der Waals-, electrostatic- and angle-torsion energy contributions) are calculated.
3. *Attraction* and *stabilization* of soil organic matter by biological molecules such as sugars and peptides via hydrogen bonds and van der Waals forces are treated explicitly (biological bonds).
4. *Humic*-, *xenobiotic*-, *biological*-, and *mineral*- *bonds* for a soil nanoparticle are constructed, geometry optimized and discussed (clay simulation, negative charges, bond mediation, interatomic spacing). The contribution of moisture to the mobility, mediation, and soil stabilization due to water is outlined.

1. INTRODUCTION

The association and dissociation reactions of aquatic humic substances were investigated using surface pressure and surface tension measurements which showed the influence of concentration, pH, and ionic strength [1]. The general view is that humic substances exist as soluble ions at low concentrations but form colloids (size between 1 μm and 1 nm) at high concentrations through reactions with cations and protons [2]. Experimental data on particle concentration and size distribution of aquatic colloids in the submicron range along with model simulations were compared [3]. The characterization of aquatic colloids with the combination of transmission microscopy staining techniques and mathematical modeling has been reported [4]. For the visualization of humic substances, the combination of scanning tunneling microscopy and computer simulation has been proposed in a preliminary approach for citric acid as a model compound [5]. Recently, imaging of the size and shape of humic macromolecular structures by *in-situ* high-resolution spectromicroscopy (Fe-, Cu-edge, ~10 μm sample field, > 43 nm spatial resolution) has been reported [6]. With X-ray absorption spectroscopy at the C-edge, even higher resolution (> 30 nm) was obtained when investigating the organic matter colloidal fractions of topsoils [7]. Inorganic particles, such as heavy metal cations, have been located and identified in microorganisms and natural colloids by energy dispersive X-ray microanalysis in conjunction with transmission electron spectroscopy [8].

In contrast to results in which isolated experimental humic macromolecules and soil organic matter were used, we propose to employ a complementary theoretical approach to examine size, shape, chemical properties, and energies of molecular models of humic and soil particle structures. This approach appears to be particularly promising because (a) visualization of smaller experimental organic structures is difficult; (b) there is little chemical knowledge at the atomic and molecular levels of humic fractions and soil organic matter in the range of nanostructures between approximately 40 nm and 0.05 nm [9]; and (c) in our opinion, the chemical processes in the interatomic range are of essential importance for the understanding of the formation, decay, and environmental properties of humic and organic matter in water and soils.

Based on the initial hypothesis of an alkyl aromatic skeleton [10], model structures have been proposed in-line with elemental, experimental, and spectroscopic data for humic acids (HA) [11-14]. Molecular mechanics calculations were employed for minimization of the total (potential) energy and optimization of model conformations of soil organic matter (SOM) [15] and also for the determination of organic-mineral complexes [16, 17] and soil particles [18]. Moreover, for a simulated soil particle with a volume of about 0.9 μm^3 molecular properties such as solvent-accessible surface areas, surface-bounded molecular volumes, van der Waals surface areas, and surface-bounded molecular volumes as well as data for polarizability, refractivity, inertial axes, dipole moment and density have been reported [19].

The aims of this contribution are

- to visualize *humic* bonds in the association processes of humic acids,
- to determine *xenobiotic* bonds by site-specific atomic and molecular interactions,
- to illustrate *biological* bonds in the trapping and binding of sugars and peptides, and
- to simulate *mineral* bonds in the formation of organic-mineral complexes and soil particles.

2. METHODS

2.1 Computational chemistry

For model design, structural modeling, geometry optimization, chemical interaction studies and molecular mechanics calculations (MM+), the HyperChem® software (Hypercube, version 5.11 Pro for Microsoft Windows 98™) was used [20]. Software modes are denoted in *italics*. The ChemPlus™ (ver. 3.1) software was utilized for calculating molecular properties of the HA, SOM and soil particle models employed as the host structures in this study. An IBM-compatible PC equipped with a Pentium II™ 400 MHz processor, Siemens Xpert™, 256 MB RAM, Elsa Victory II 16MB, graphic card, 17" color monitor and 12.7 GB hard disk, was employed for 3D modeling. Details of the geometry optimization (and energy minimization) process using MM+ and our modeling approach have been reviewed [18, 19].

3. RESULTS AND DISCUSSION

3.1. Humic acids

It was of interest to visualize the capacity of molecular mechanics calculations for association reactions of HA molecules at the nanochemistry level. Firstly, the operationally defined humic fractions, fulvic acids (FA), and HA are likely to be less complex in chemical structure compared to SOM and whole soils. Secondly, for the modeling approach it is helpful to start only with a few elements (in this case C, H, N, O), a relatively low numbers of atoms (759 atoms, number of HA molecules (n =1)) and a corresponding molecular mass below 6000 g mol^{-1}. Thirdly, to simplify the problems further, the complex and cross-linked HA structure is simulated by adapting an accepted molecular model which reflects comprehensive investigations combining geochemical, wet-chemical, biochemical, spectroscopic, agricultural and ecological data with analytical pyrolysis and an average representative molecular structure [see e.g., 2; 21; 22]. However, presently the limitations of this HA model are still drastic as the virtual workspace does not allow gases and metabolic processes. In addition, the employed software is not designed to consider cleavages of covalent bonds. Modeling in water (water box) and allowing temperature influence (heating and annealing) is possible only under restricted conditions. What therefore can be achieved by the present software, computer capacity and, last not least, creative ideas to arrive at a better understanding of the HA association processes ?

3.1.1. Humic bonds

Preliminary data of modeling and molecular mechanics calculations of molecular HA structures have been published for the HA monomer and pentamer and showed the 3-D color plots of the geometrically optimized HA oligomers [23]. In the present work, the originally designed ~ signs that allow the necessary bonding flexibility and structural variability [9], were closed by seven methyl functions to produce a stable 3-D structure with a pronounced and crucial structural alkyl aromatics skeleton [10, 11]. This HA monomer displayed 2 hydrogen bonds and had the atomic and molecular properties (A) and energies (B) as listed in Table 1a. The range of oligomer and polymer HAs subsequently has been expanded to macromolecules with n = 3, 10, and 15. Clear evidence from the optimized structures and energy descriptors shows that hydrogen bonds and van der Waals forces are the main driving forces for HA associations [24]. Furthermore, the HAs (n = 1, 2, 4, and 14), partly by different conformations, indicated the capacity of computational chemistry for simulation of humic acid interactions and formation of humic complexes and macromolecules [9].

In an approach to simulate the simplest case of HA associations, the black/white plot of a HA dimer is shown in Figure 1a. The dimer was constructed in three steps: (1) by aligning two HA monomers (Molecule 1; Molecule 2; atomic and molecular data in Table 1a) parallel to the ordinate according to the inertial axes, (2) by moving the two monomers in opposite directions along the abscissa up to an approximate distance of 0.3 to 0.4 nm, and (3) during this approximation the thickness (z axis = 2.7 nm) was not changed, so that the starting dimer formation was achieved planar in the x, y plane. The molecular mechanics calculation [20] started at a relatively high total energy of about 2870 kJ mol^{-1} and a gradient of 0.23 kJ $mol^{-1}nm^{-1}$. In the initial calculation phase, with a set determination gradient of <0.025 kJ mol^{-1} nm^{-1}, an overall repulsion energy of the HA dimer complex was obtained in the order of +85.45 kJ mol^{-1} which indicates instability of the obtained conformation. Continuation and determination of the MM+ calculations at a set convergence gradient of <0.01 kJ mol^{-1}

resulted in an total energy of 2593.48 kJ mol^{-1} and an energy gain for the optimized HA dimer in the association process of -189.18 kJ mol^{-1}. Clearly, the stepwise Hyperchem controlled energy minimization had achieved a dimer conformation with high stabilization energy and thus it was of interest to investigate the responsible factors for this process.

In the natural soil environment it is more likely that metal and organic chemical reactions occur with humic and soil particles at higher energy levels. The HA dimer was examined at an energy level of 2708.90 kJ mol^{-1} and corresponds to the conformation shown in Figure 1a. The elemental composition, analysis and molecular weight data along with molecular properties such as surface area, volume, dipole moment and moments of inertia for this HA dimer structure are given in Table 1b. The association of the two HA molecules is reflected during the geometry optimization by the following three steps.

First, as indicated by arrows in Figure1a, a total of 11 hydrogen bonds (H-bonds; dotted lines) are observed. Closer examination shows that H-bonds 1-5 formed by molecule 1 and H-bonds 6, 7 and 8 in molecule 2 are intramolecular H-bonds. These are typical for humic structures. Second, the original central void between both molecules is divided into two significantly different voids with progressing MM+ calculations. This is because side-chains in the HA structure tend to rotate and expand during energy minimization according to default set parameters such as bond length and angle. The structural section **A** shown in Figure 1b, illustrates the formation of 3 intermolecular H-bonds in the dimer formation due to the approximation of one carboxyl group of molecule 1 (*Thick Line*, right side) and two carboxyls of molecule 2. The participating atoms of H-bond 9 are the donor H(601) of molecule 1 and the acceptor carbonyl O(247) of an aliphatic carboxyl group of molecule 2. The length of this H-bond is 0.2167 nm and the bond angle 158.8°. The charges for H(601) of 0.347 and -0.342 for O(247) are easily calculated using the Chemplus software. Starting from the carbonyl acceptor O(239) of molecule 1 with the charge of -0.415, two H-bonds are generated as displayed in Figure 1b. One is directed to the donor H(592) of an aromatic carboxyl function (distance 0.2114 nm; angle 153.12°), the other (distance 0.2119 nm; angle 150.1°) is linked to the aliphatic carboxyl group with the donor H(605) both of molecule 2. Single point calculation (molecular mechanics without changes in HA dimer conformation) for the five main atoms H(601), O(247), O(239), H(592), and H(605) involved in the three intramolecular bonds showed a substantial energy contribution of -44.42 kJ mol^{-1}. Thus, stabilization on the basis of electrostatic forces (charges) is a dominant feature of humic structures and can be determined and evaluated quantitatively.

The second relevant energy aspect of HA associations deals with the stability contribution due to van der Waals forces. As shown in Figure 1c (section B), in the lower part of the HA dimer structure, overlapping of the long alkyl aromatic chains with terminal oxygen substituted benzene rings occurs. In total 45 atoms were selected as highlighted by *Thick Line* and investigated by molecular mechanics calculations (*Single Point*). A resulting contribution of -17.0032 kJ mol^{-1} to the total energy for the dimer stabilization was calculated. The high density of carboxyl and hydroxyl moieties closing the lower entrance to the void B has the effect of a highly polar plug which hampers penetration by hydrophobic substances into the lower central void of the dimer. Moreover, the energy descriptor for van der Waals forces of -39.3975 kJ mol^{-1} only for the selected 45 atoms indicates further stabilization (e.g., by overlapping benzene aromatic rings). In summary, hydrogen bonds and van der Waals forces were found to be the dominant impact factors on association processes of humic acids.

Table 1. Monomer, dimer, polymer and diethyl phthalate (DEP) complex of terrestrial humic acids: Surface, volume and energy

	a) HA Monomer	**b)** HA Dimer	**c)** HA Polymer	**d)** HA + 20 DEP entrapped/occluded
	1 molecule	2 molecules	10 molecules	22 molecules
	759 atoms	1518 atoms	16678 atoms	2076 atoms
(A) Atomic and molecular properties				
Elemental composition	$C_{315}H_{349}O_{90}N_5$	$C_{630}H_{698}O_{180}N_{10}$	$C_{6932}H_{7662}O_{1974}N_{110}$	$C_{856}H_{950}O_{260}N_{10}$
Elemental analysis	67.02 % C	67.02 % C	67.09 % C	66.17 % C
	6.23 % H	6.23 % H	6.22 % H	6.16 % H
	25.51 % O	25.51 % O	25.45 % O	26.77 % O
	1.24 % N	1.24 % N	1.24 % N	0.90 % N
Molecular weight [g mol^{-1}]	5645.22	11290.43	124106.65	15538.87
Solvent-accessible surface area [nm^2]	42.71	80.25	1076.46	75.04
Solvent-accessible surface-bounded molecular volume [nm^3]	11.42	22.52	278.68	27.92
Van der Waals-surface area [nm^2]	54.33	108.57	1334.73	156.74
Van der Waals-surface-bounded molecular volume [nm^3]	4.97	9.94	119.40	13.86
Depth in workspace [nm]	2.43	2.73	4.88	2.24

Table 1. Continued

Specific surface [$m^2\ g^{-1}$]	4556.40	4280.54	5223.35	2881.80
Density [$g\ cm^{-3}$]	1.88	1.88	1.73	1.86
Smallest box [nm]	x = 3.01	x = 4.43	x = 16.07	x = 3.61
	y = 2.44	y = 2.73	y = 4.94	y = 2.24
	z = 4.73	z = 4.58	z = 22.27	z = 6.22
	Σ = 34.74 nm^3	Σ = 55.39 nm^3	Σ = 1767.92 nm^3	Σ = 50.30 nm^3
Dipole moment [Debye nm^{-1}]	179.20	94.09	677.95	20.33
Moments of inertia [amu nm^2]	x = 2.98e+003	x = 1.62e+006	x = 2.36e+006	x = 1.28e+004
	y = 9.67e+003	y = 2.04e+006	y = 4.70e+006	y = 3.87e+004
	z = 10.73e+003	z = 3.26e+006	z = 6.85e+006	z = 4.64e+004
	(B) Total energy and energy derivatives [kJ mol^{-1}]			
Total energy [kJ mol^{-1}]	1391.33	2708.90	35965.90	1671.26
Bond	174.48	355.43	11445.70	330.39
Angle	1067.61	2074.76	16386.71	2374.83
Dihedral	-55.60	-108.77	-6743.39	-809.12
Van der Waals	685.96	1321.03	26350.45	-258.03
Stretch-bend	39.96	79.06	569.10	26.59
Electrostatic	-521.09	-1012.48	-12042.65	6.60

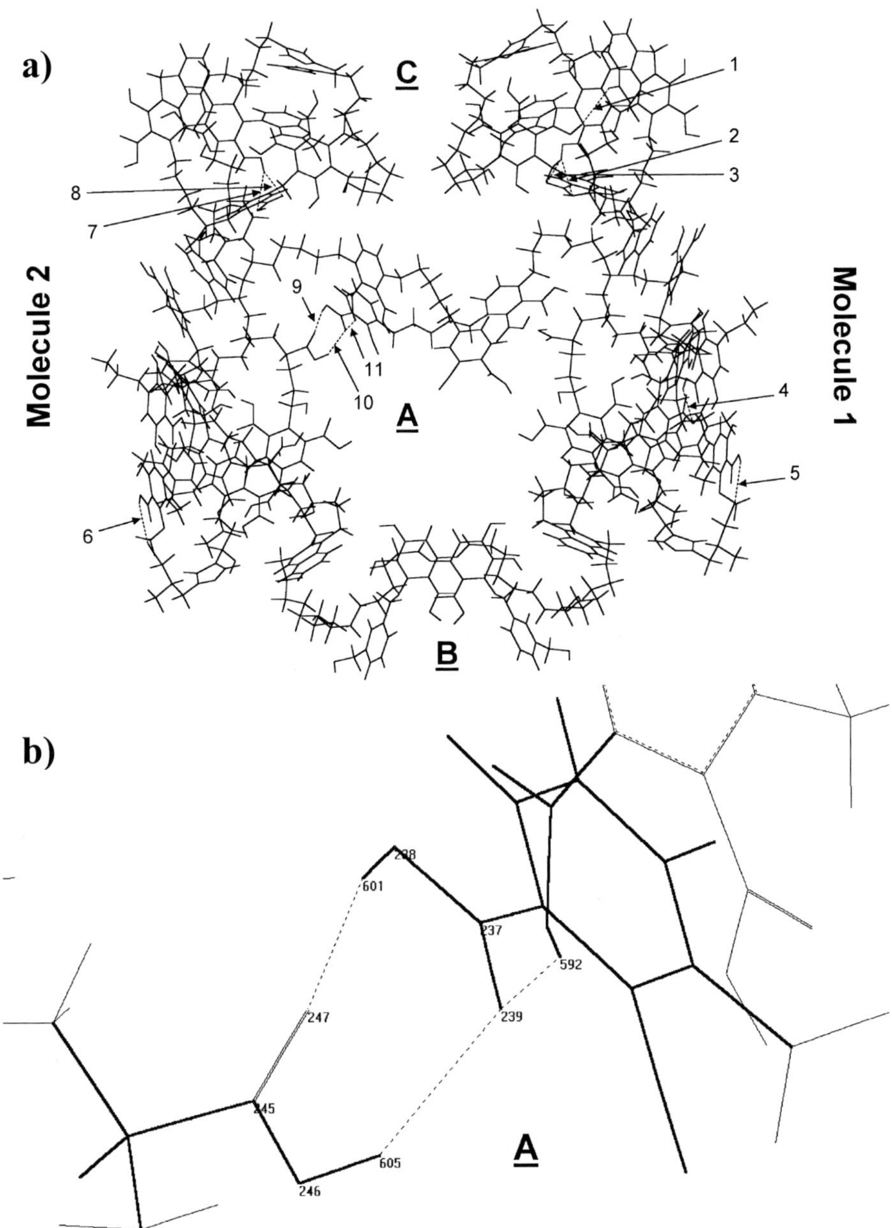

Figure 1. Association of two standard humic acids and formation of a HA dimer with polar and nonpolar surfaces and voids: a) Black/white plot in the *Sticks* mode shows the humic skeleton, location of 11 hydrogen bonds, and three sections A, B and C; b) In section A the formation of intermolecular hydrogen bonding between the two HAs is highlighted; and c) in section B, the overlapping of two aliphatic chains with terminal aromatic rings of the two HAs is displayed. For molecular structural properties (A) and energies (B) see Table 1.

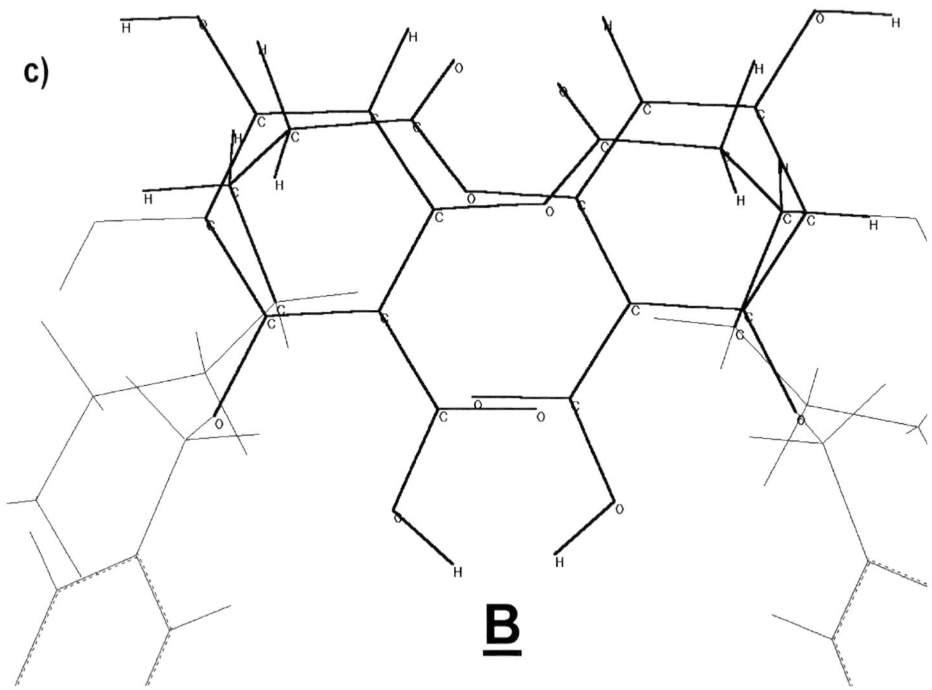

Figure 1. Continued

The 3D color plot of the HA dimer in Figure 2 demonstrates the distribution of the different elements and the porous skeleton structure of the HAs. The total energy and energy derivatives (B) in Table 1b give the bond-, angle-, dihedral-, stretch-bend-, and electrostatic forces (energy given in kJ mol^{-1}) and clearly indicate that torsional relief is a further major consequence of energy minimization. The molecular properties of surface area and volume (A) were calculated using the Chemplus software with water as the solvent and allow for the analysis of swelling and shrinking processes during geometry optimization. It is noteworthy that the solvent-accessible surface area for the dimer is increased by 5.2 nm^2 upon association whereas the corresponding volume shrinks by more than 10%. Additional data on inertia axes, dipole moment, density, polarizability and refractivity are very helpful for molecule orientation and adaptation to experimental data (QSAR).

Of particular interest is the distribution of functional groups in the entrance to the upper central void. This section is marked by C in Figure 1a. At the beginning, the molecular mechanics calculations of the humic surface structure showed the typical scattering of oxygen atoms in carboxyl, phenolic, alcoholic, hydroxyl, quinoid, keto, and ester groups.. In contrast to the lower polar overlapping in section B, the surface in the upper entrance of C changed with progressing calculations by forming hydrophobic surfaces in this void as displayed in Figure 2. Probably due to the polar/polar interactions in the lower hydrophilic central void, there is an observed tendency of rotations and shifts in the upper entrance and void section **C** to generate more hydrophobic surfaces. Both HA molecules 1 and 2 displayed a hydrogen lining along the entrance and inside the upper void which shows only a few hydroxyl functions in direction to the void center. One can assume that stronger and short-ranged polar

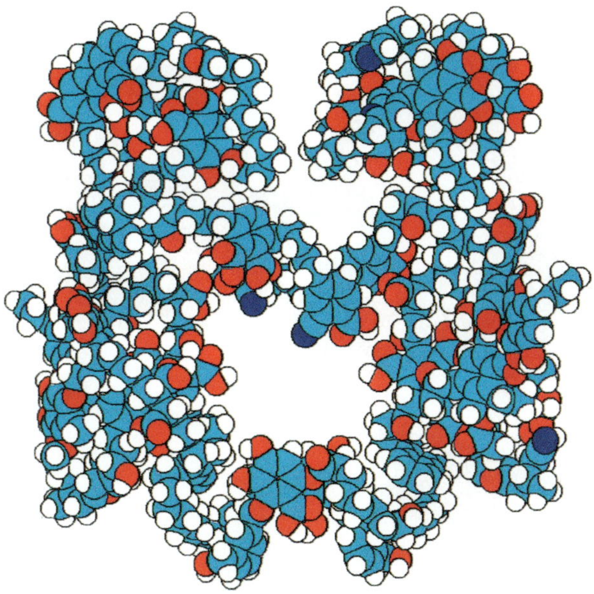

Figure 2. Color plot of the HA dimer (cf. Figure 1) illustrates the spacefilling, surfaces and chemical properties in the *Overlapping Spheres* mode. For section C the hydrophilic and hydrophobic aspects of HA surfaces are discussed. The element colors are: carbon (cyan, radius in nm (r) = 0.07), hydrogen (white, r = 0.025), nitrogen (blue, r = 0.065), and oxygen (red, r = 0.06).

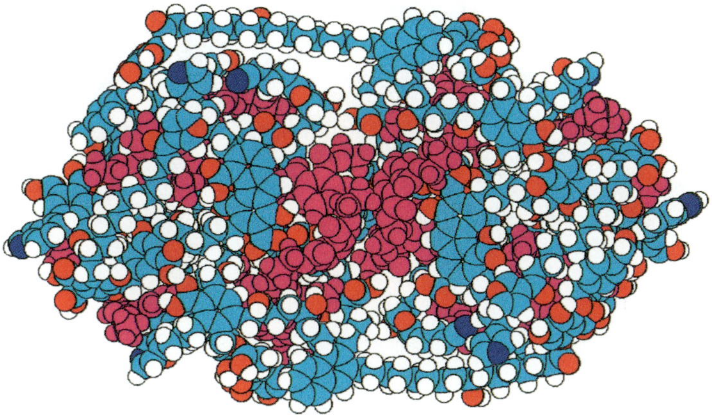

Figure 4. Interactions of two aquatic humic acids with 20 molecules of diethyl phthalate which are selected in violet color (*Rendering*: *Overlapping Spheres*). Hydrophilic and hydrophobic moieties of host molecules are displayed. Energy descriptors of binding forces and atomic as well as molecular properties are given in Table 1d. Element colors as in Figure 2.

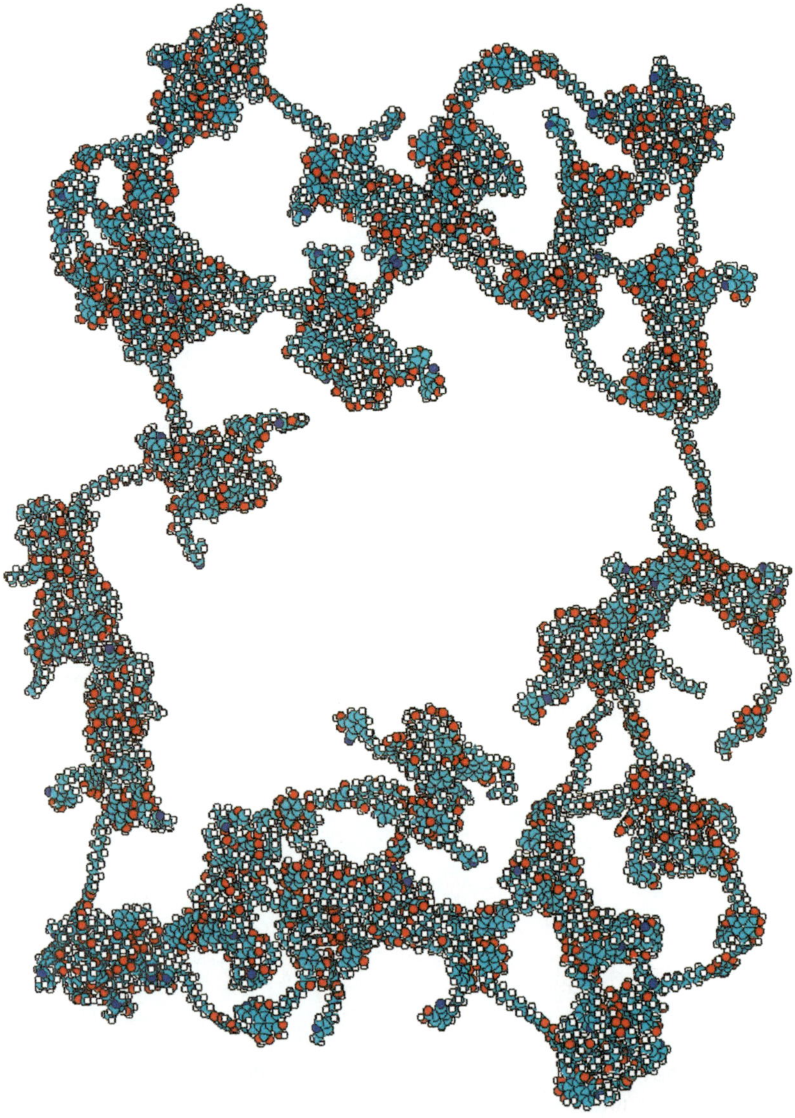

Figure 3. Association of a humic acid polymer (see data in Table 1c) consisting of 4 tetramers and 6 monomers (22 HA subunits, 16678 atoms). The size (calculated volume) of the humic colloid is about 0.3 μm^3. Element colors as in Figure 2.

forces lead to polar/polar interactions and energy gain in the lower void and, consequently, in the direct atomic vicinity, hydrophobic structural features are favored. In view of the maximal thickness of 2.70 nm for the HA dimer in the demonstrated conformation, one has to take into account that these structures (e.g., in Figure 1) are very thin sheets, which in cases portions are close to atomic mono-layers.

During the run of one calculation cycle, all atoms at the different coordinates are examined for energy gain and geometry optimization, therefore a structural selforganization [25] can be considered. Information on the covalent bonding of an atom to each of its neighbors (*Connectivity*) is provided by the employed software [20]. The overall effect is that the original humic surface covered with a mixture of polar and nonpolar functions, is converted mainly to a nonpolar surface. Section C of the nanoparticle designated at the top of Figure 1a and the corresponding color section of Figure 2 visualizes a nonpolar void entrance and upper central void surface which is predominantly of low polarity. As will be shown later, these rearrangements of humic surfaces and docking sites are crucial for the adsorption of nonpolar anthropogenic and xenobiotic compounds.

The humic polymer (HA docosamer) displayed in Figure 3 consists of 10 molecules and 16678 atoms. Calculated on the basis of the smallest rectangular box (see Table 1c), the space-requirement of the humic macromolecule is approximately 1.8 μm^3. Determination of molecular properties using the Chemplus software resulted in solvent(water)-accessible and van der Waals surface areas between 1.1 and 1.3 μm^2 and the corresponding volumes between 0.3 and 0.1 μm^3. Thus, according to the definition given above, it is a colloid with a molecular weight of 124106.65 g mol^{-1}. At present our molecular mechanics calculations in terms of software quality, computer capacity, and molecular complexity are at their limits. As the calculation times increase exponentially with the number of investigated atoms, it is necessary to operationally reduce the number of atoms for macromolecules of this size. Thus, for the humic colloid, a simplified definition of a carbon atom was used that included sp, sp^2 and sp^3 bonded hydrogen atoms (*United Atoms*) [20].

In order to determine the association energy between the substructures of the humic polymer (6 monomers and 4 tetramers) in the first step, the total energy of the monomer (132.82 kJ mol^{-1}) and tetramer (7488.80 kJ mol^{-1}) were calculated by molecular mechanics in the *United Atoms* mode. The molecular energies multiplied by 6, respectively 4, gave a total energy sum of the 10 humic polymer constituents of 30752.11 kJ mol^{-1}. In the second step, following intense MM+ calculations, which included more than 3 months of calculation time, a total energy of the intact HA polymer of 35965.90 kJ mol^{-1} was calculated. From the difference between these two values of total energy resulted the association energy of the humic colloid of -5213.79 kJ mol^{-1}. This energy gain, in the process of colloid formation, is attributed to the attraction forces between the 10 molecular substructures and is reflected in the energy descriptors (B) given in Table 1c. Torsional stress (bond, angle, stretch-bend) requires most of the non-binding energy (with a negative dihedral as energy penalty). The highest energy contribution to the stability of the colloid is delivered by van der Waals forces of 26350.45 kJ mol^{-1} in agreement with the high aromaticity of the humic model structures.

Furthermore, the electrostatic energy (-12042.65 kJ mol^{-1}) is a major descriptor of colloid stabilization which is supported by the observation of 109 hydrogen bonds. This stabilization leads to a shrinking of the HA polymer volume of approximately 10% upon determination of the comprehensive molecular mechanics calculations.

There is strong evidence in the literature that the structure of humic materials is a relevant factor in binding pollutants to humic and fulvic acids [e.g., 26]. If detailed descriptions of atomic and molecular properties and energy descriptors of humic nanostructures in the range of 10 to 120 nm^3 can be achieved using available modeling tools, it would be of interest to explore the potential of the methodology. This would apply not only for association processes but also for site-specific trapping and binding of low polarity compounds which occur ubiquitously in the environment. The interactions of diethyl phthalates as a pollutant in sewage sludge with soil organic matter (SOM) [27] and the absorption/adsorption processes for HA/phthalate complexes at a nanochemistry level, have been modeled and evaluated by principal component analysis of the energy descriptors [28]. As an example of HA/phthalate complexes, the association of two aquatic humic acids with 20 diethyl phthalate molecules is illustrated in Figure 4. The long aliphatic side-chains of the HAs surround and close the complex at the top and bottom. The 20 diethyl phthalate molecules are selected and highlighted in violet color to illustrate the close entrapment and binding of the molecules in the host structure of the two HAs. The 22 molecules of the complex were geometrically optimized during the energy minimization by molecular mechanics calculations and are observed as single compounds and clusters at the end of the calculations. Whereas some phthalate monomers are bound to the outer surface, there appears to be a tendency to fill large voids in the HA dimer with phthalate clusters (n = 2-5). The high content of 28.60% phthalates in the HA dimer introduces only minor electrostatic interactions (energy descriptor 6.60 kJ mol^{-1} and 3 hydrogen bonds, see Table 1d). High polarity due to oxygen and nitrogen functions is typical for HAs but in the presented HA dimer complex, surface bonds and intercalation by low-polarity compounds such as diethyl phthalates reduce this potential considerably.

In agreement with recent work [28], focal points are structural features at the nanochemistry level, such as interatomic distances, angles, partial atomic charges and high aromaticity of the humic acid. These are observed to be the dominating intermolecular interactions in the complexes at the specific sorption sites. Torsional relief and favorable changes in bonding energy also prevail for the growing complex. The latter indicates both the structural flexibility of the HA host and the stabilizing effect of diethyl phthalate on the complex, preferably by filling of the voids within the HA molecule. The intermolecular forces are described mainly through interactions between dipole-dipole, such as carboxylic functions and uncharged moieties like aromatic rings (van der Waals energy). Through comprehensive investigations of molecular aggregation of humic substances [29], three mechanisms for complex formation have been proposed: hydrogen bonding, charge-transfer, and π-bonding. The described stacking of planar-π-donor-planar-π*-acceptor groups to form a complex occurred at least in five sections of the HA dimer/phthalate complex (Figure 4), where the stacking of benzene rings can be discerned. The energy descriptor for van der Waals forces of -258.03 kJ mol^{-1} is clearly in line with these arguments and indicates one relevant source of complex stabilization (see Table 1d). In total, a stabilization energy of -1895.82 kJ mol^{-1} of the HA dimer by interaction, association, adsorption and binding with the 20 diethyl phthalate molecules is observed.

3.2. Dissolved organic matter

The effects of dissolved organic matter (DOM) on the bioconcentration [30] of organic chemicals in aquatic organisms is well documented and describes environmental concerns [31]. Of particular interest are the interactions of pesticides and herbicides with humic substances [32]. Thus, aquatic humic acids, fulvic acids, and DOM in a bog water lake, seepage water from forest soil [33], contaminated brook water [34], and soil percolate in agricultural soils [27] have been investigated by analytical pyrolysis. The applied thermal methods allow direct, in-source pyrolysis-mass spectrometry in the high electric field (Py-FIMS), and Curie-point pyrolysis-gas chromatography/mass spectrometry (Py-GC/MS) in combination with library searches. Based on the identified building blocks together with complementary analytical data proposals for a general concept, a hypothesis for the basic molecular structures of humic macromolecules such as aquatic HA modified from [12] and DOM [33] in water were put forward. Computational chemistry was utilized for structural modeling and geometry optimization of HA and DOM. The investigated xenobiotics were pentachlorophenol [35], DDT and hydroxyatrazine [36]. Modeling of interactions and site-specific binding of xenobiotics *in-vacuo* and in a water box were performed. Preliminary experiments designed to simulate the acidity of water molecules by protonation, enhanced reactions with polar xenobiotics (e.g., hydroxyatrazine) but left nonpolar substances (e.g., DDT) unchanged. Molecular mechanics calculations were performed to evaluate the conformation of structural, three-dimensional models of HA/- and DOM/xenobiotic complexes and to determine the total and partial energy contributions from bond-, angle-, dihedral-, van der Waals-, stretch-bend, and electrostatic energies.

3.2.1. Xenobiotic bonds

An example of the modeling of a DOM/xenobiotic complex and structural investigation of the interaction of DOM with an agrochemical (salinomycine® - an veterinary antibiotic for cattle treatment), is reported. As shown in Figure 5 the DOM model is quite complex having 37 molecules and 1382 atoms. It consists of a humic skeleton with one trisaccharide and hexapeptide covalently bound (in total 1157 atoms) and 35 water molecules mostly linked by hydrogen bonds. The interesting question was how the structure and molecular properties of the DOM/salinomycine complex would change if the agrochemical is introduced at two widely different structural positions. First, the salinomycine molecule was inserted into a small void in the center of the DOM structure for the simulation of an *ab*sorption process (Figure 5a). Second, for *ad*sorption, the outer surface of the DOM structure was approached stepwise from the starting distance of about 12.6 nm (*Merge*) to a distance which resulted in the first intermolecular hydrogen bond being between DOM and salinomycine (Figure 5b).

In Figure 5a, the trapping of the salinomycine molecule in a narrow void in the center of the DOM host structure is highlighted by selecting this molecule in violet. Moving the xenobiotic into the DOM structure by the z-translation tool and central positioning resulted in a total energy in the order of 150000 kJ mol^{-1} at a gradient of about 1200 kJ mol^{-1} nm^{-1}. In this initial phase the strong interactions with, and displacement of, central portions in the optimized DOM molecule were induced and showed strong repulsion forces. Following intense molecular mechanics calculations, the salinomycine molecule acquired a globular shape with about 1.77 nm thickness in the center of the DOM structure. From the data given in Table 2a, it is clear that when the calculation was stopped at a gradient of 0.007 kJ mol^{-1} nm^{-1}, a relatively high total energy of 9366.93 kJ mol^{-1} still remained. For the previously optimized standard conformations, a total energy of 3253.45 kJ mol^{-1} for DOM and 278.59 kJ mol^{-1} for salinomycine (sum 3532.04 kJ mol^{-1}) were calculated. Thus, the energy difference between the obtained energy for the MM+ calculation of the described DOM/xenobiotic complex and the calculated sum of the two standard conformations is +5834.89 kJ mol^{-1}.

This shows that the complex is not stable and it is unlikely that the simulated process would have a chance to occur in a natural environment. From the energy descriptor for electrostatic forces (B) in Table 2a, it can be derived that a high contribution for complex stabilization is supplied by this energy (-1729.04 kJ mol^{-1}). The high number of hydrogen bonds (48) confirms this observation. Closer examination of the hydrogen bonds show that most of these bonds are intramolecular hydrogen bonds of DOM, mainly in the vicinity of the trapped sugar and peptide. The moisture content of the colloid was simulated by the introduction of the 35 water molecules (and polar activity) which lead to hydrogen bonds in water-water and DOM-water clusters. Additional atomic, molecular, and energy data are compiled in Table 2a and serve for the comparison for the adsorption process outlined in the following.

In Figure 5b, the salinomycine molecule was merged with the DOM molecule in the workspace and had an initial distance of > 12 nm at which no interaction energy was observed. Approximation of salinomycine step by step towards the fixed DOM molecule yielded a stabilization energy in the order of -0.48 kJ mol^{-1} at a distance of 1.26 nm, -7.86 kJ mol^{-1} at 0.64 nm, and -123.12 kJ mol^{-1} at 0.34 nm. The determination of the molecular mechanics calculations showed that the optimized DOM/salinomycine complex produced a stabilization energy of -130.98 kJ mol^{-1} (Table 2b). Comparisons of the atomic and molecular properties of the two different binding sites described in Table 2a and 2b indicate that:

(1) the energy differences of the two conformations (central void insertion (+5834.89 kJ mol^{-1}) versus fixation at the outer surface (-130.98 kJ mol^{-1}), demonstrate that the latter leads to a more stable DOM/xenobiotic complex;

(2) MM+ calculations of the total energy and the six energy derivatives determined at similar convergence gradients for the both complexes, show that adsorption at the DOM surface is favored over absorption;

(3) site-specific docking of guest molecules should consider the stabilization and destabilization influence of energy gains and high energy structural disturbances;

(4) with introducing moisture through the incorporation of the 35 water molecules on and into the DOM complex, the water generates migration of partial charges and accordingly, to an adaptation of the humic sub-structures and functional groups. Thus, by the water effect hydrogen bonds are formed and eliminated during the optimization process until the highest energy gain and complex stabilization is achieved;

(5) by selecting water as solvent, calculations of the molecular properties show shrinking of the solvent-accessible surface area and surface-bounded molecular volume for the adsorption complex is observed whereas van der Waals surface and volume for both conformations are similar; and

(6) the 42 hydrogen bonds are distributed in the complex (Figure 5) in the following order. Of 17 water molecules 5 are water monomers and 12 form water clusters with 6 hydrogen bonds and 16 intramolecular hydrogen bonds are observed for DOM-DOM bonds. The majority of hydrogen bonds are from the 18 water molecules which generate 20 intermolecular bonds with DOM. With respect to environmental processes, it is relevant that the mobility of polar xenobiotic compounds and formation of bound residues is strongly influenced by water [37]. Moreover, it is noteworthy that the high porosity and sponge-like structure as postulated at the start of 3D modeling of HA [10-12] is intensified in the DOM model and allows to visualization of site-specific bonds at the outer and inner surfaces of DOM voids at nanochemistry level.

3.3. Soil organic matter

Basis for the work with soil organic matter (SOM) was the result of comprehensive studies of stability, thermal products and structure of biopolymers, humic substances and

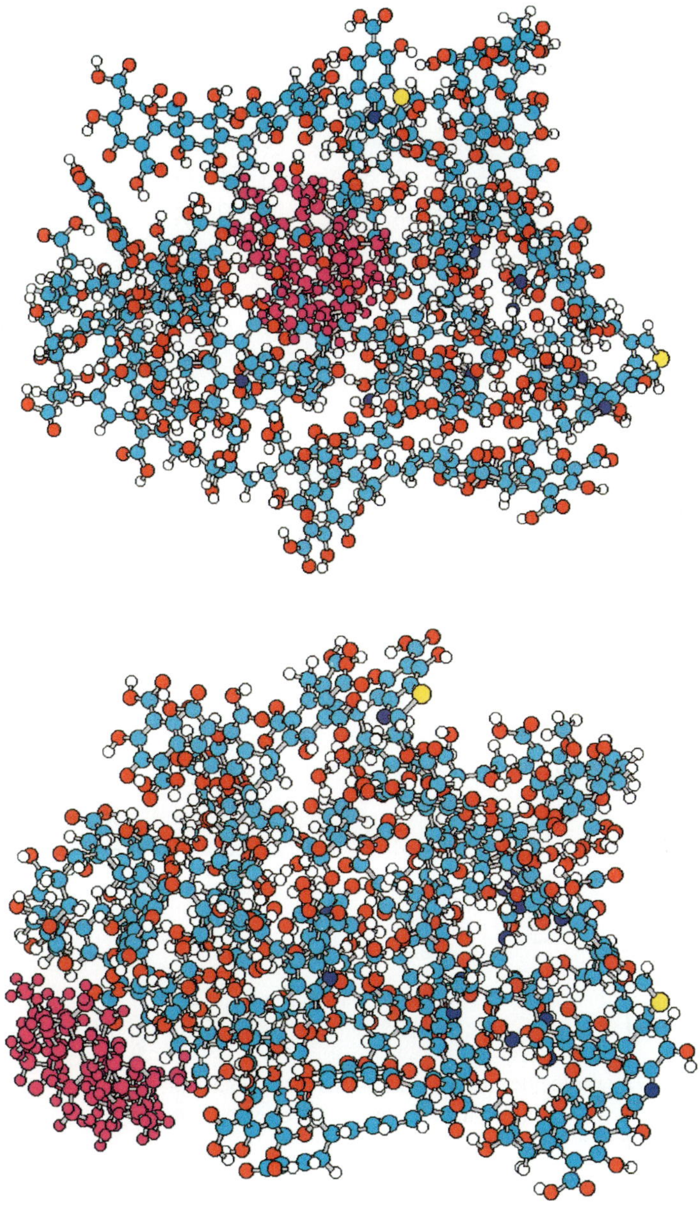

Figure 5. Site-specific bonds of an antibiotic (salinomycine) to dissolved organic matter (DOM) and stabilization of the DOM/xenobiotic complex are illustrated for the selected agrochemical (violet color) for: a) absorption in a central void and b) adsorption on the surface (intermolecular hydrogen bond is marked by dotted line). Transport forms and sequestration of xenobiotic/DOM complexes are simulated. See data in Table 2a and 2b.The element colors are as described in Figure 2, in addition sulfur is (yellow, r = 0.10).

Table 2. Absorption and adsorption of xenobiotic complexes of dissolved (DOM) and soil (SOM) organic matter

	a) DOM + Salinomycine **Ab**sorption central void (37 molecules; 1382 atoms)	**b**) DOM + Salinomycine **Ad**sorption outer surface (37 molecules; 1382 atoms)	**c**) SOM +12 x Mecoprop SOM/Mecoprop complex (27 molecules; 1250 atoms)
	(A) Atomic and molecular properties		
Elemental composition	$C_{488}H_{559}O_{318}N_{15}S_2$	$C_{488}H_{559}O_{318}N_{15}S_2$	$C_{469}H_{533}O_{209}N_{26}SCl_{12}$
Elemental analysis	49.73 % C	49.73 % C	54.50 % C
	4.78 % H	4.78 % H	5.20 % H
	43.17 % O	43.17 % O	32.35 % O
	1.78 % N	1.78 % N	3.52 % N
	0.54 % S	0.54 % S	0.31 % S
			4.12 % Cl
Molecular weight [g mol^{-1}]	11786.85	11786.85	10335.94
Solvent-accessible surface area [nm^2]	53.33	48.84	51.22
Solvent-accessible surface-bounded molecular volume [nm^3]	18.38	17.92	17.36
Van der Waals-surface area [nm^2]	98.43	99.36	94.85
Van der Waals-surface-bounded molecular volume [nm^3]	9.04	9.08	8.44
Depth in workspace [nm]	2.46	2.22	2.35

Table 2. Continued

Specific surface [$m^2 g^{-1}$]	2724.9	2495.1	2984.50
Density [$g cm^{-3}$]	2.164	2.157	2.033
Smallest box [nm]	x = 3.35	x = 3.08	x = 3.19
	y = 2.64	y = 2.22	y = 2.35
	z = 3.93	z = 3.84	z = 4.85
	$\Sigma = 34.83 nm^3$	$\Sigma = 26.26 nm^3$	$\Sigma = 36.39 nm^3$
Dipole moment [Debye nm^{-1}]	93.70	172.25	205.98
Moments of inertia [amu nm^2]	x = 0.90e+004	x = 0.80e+004	x = 0.72e+004
	y = 1.37e+004	y = 1.28e+004	y = 1.10e+004
	z = 1.69e+004	z = 1.60e+004	z = 1.39e+004
(B) Total energy and energy derivatives [kJ mol^{-1}]			
Total energy [kJ mol^{-1}]	9366.93	3401.06	1426.91
Bond	2194.72	391.01	239.82
Angle	3171.56	2299.00	1595.47
Dihedral	1391.21	1184.30	313.12
Van der Waals	4035.93	1226.82	286.46
Stretch-bend	302.56	87.10	49.61
Electrostatic	-1729.04	-1787.16	-1057.66

agricultural soils [see 38, 39] in combination with proposals of preliminary SOM models [13-15]. The trapping and binding of biological and xenobiotic compounds in host structures such as humic substances, SOM, soil fractions and whole soil particles is one focal point in our investigation and has been exemplified for HA/atrazine- [40], organic-mineral/atrazine- and SOM/hydroxyatrazine complexes [36, 38, 42]. Recent investigations on the interactions of imidazolinone herbicides with soil humic acids have shown good agreements of the experimental results with structural data and molecular properties obtained by modeling [41]. The general observations are: First, in line with the observations for HA and DOM, we find that for polar guest molecules hydrogen bonds are an essential binding form. Second, due to the high aromaticity of HA, DOM, and SOM, van der Waals forces are expected and were found to be an important binding mechanism. Both hydrogen bonds and van der Waals interactions supply a major part of the energy gain during complexation and thus contribute decisively to the stabilization energy of the host/guest complexes. Third, the complex cross-linking network of alkyl aromatics and heterocylic structural building blocks offers voids and clefts for the trapping and sequestration of the potential guest molecules.

3.3.1. Xenobiotic bonds

For comparison with the DOM/salinomycine complex, Figure 6 shows the model of a herbicide complex of SOM. For this investigation we employed a leaf herbicide (mecoprop-(2-(2methyl-4-chlorophenoxy) propionic acid)) which is used for cereal crops. The compound has the elemental composition of $C_{10}H_{11}O_3Cl$, a molecular weight of 214.6482 g mol^{-1}, and gives upon molecular mechanics calculations an optimized conformation with a total energy of 9.8688 kJ mol^{-1}.

The SOM/mecoprop complex is comprised of total humic substance (754 atoms), a trisaccharide (66 atoms), a hexapeptide (94 atoms), 12 water molecules [13] and 12 mecoprop molecules (in total 27 molecules and 1250 atoms). A total energy of 1426.64 kJ mol^{-1} at a gradient of 0.002 kJ mol^{-1} nm^{-1} was calculated for the this complex. The sum of total energies of 10 single mecoprop molecules (98.69 kJ mol^{-1}) plus the total energy of the SOM standard (2101.65 kJ mol^{-1}) amounted to 2200.34 kJ mol^{-1}. For the geometrically optimized SOM/mecoprop complex with molecular mechanics calculations, a total energy of 1426.64 kJ mol^{-1} was obtained. From the energy difference, it is derived that the trapping and binding of the 10 mecoprop molecules in the SOM model resulted in an energy gain of -773.70 kJ mol^{-1}. Thus, a substantial contribution to the stabilization of the xenobiotic complex is supplied. It should be kept in mind that the absolute conformational energies have no direct physical meaning [5]. Nevertheless, for relative comparisons between model structures valuable information can be extracted, such as stability, reactivity, QSAR and a wide variety of molecular properties at the nanochemistry level. Comparison between the DOM/salinomycine complex trapped in the central void (Table 2a) and SOM/mecoprop complex (Table 2c) gave somewhat smaller values in surface areas and volumes for SOM and simultaneously, a slight increase of density and lower smallest box dimensions for DOM. This can be interpreted as DOM having a lower porosity. The energy descriptors for electrostatic and van Waals forces drop dramatically from the DOM to SOM complexes and reduce the number of hydrogen bonds by more than 50%. This underlines the relevance of available water (moisture) as discussed above. Despite the more narrow voids in SOM there is no problem to host the 12

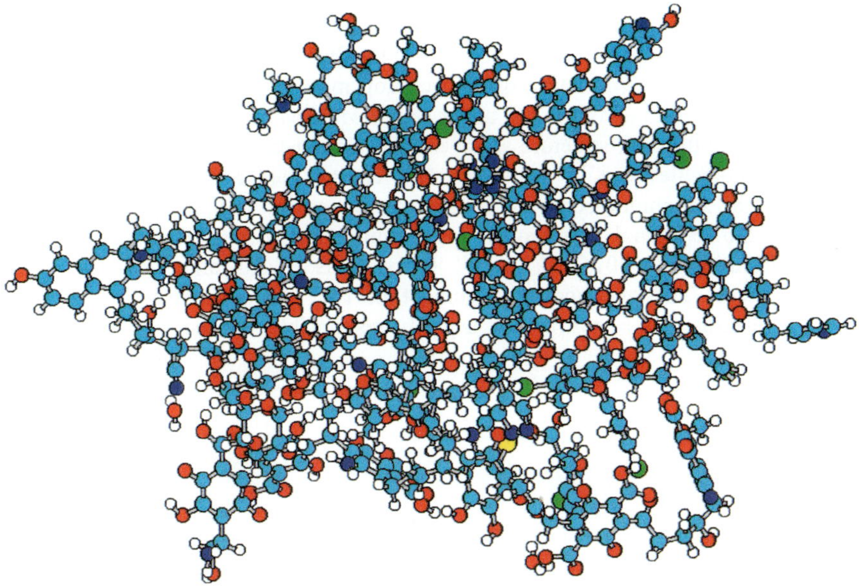

Figure 6. Trapping and binding of 12 herbicide molecules (mecoprop) in soil organic matter are indicated in the color plot by the relatively large chlorine atoms in green color (r = 0.10; only in the xenobiotic complex). The other element colors are as in Figure 5. Xenobiotic and biological bonds are highlighted by enlarged sections of the SOM/mecoprop complex in Figures 7a-c (see data in Table 2c).

mecoprop molecules in structurally and energetically appropriate sites. Their distribution in SOM is indicated in Figure 6 by the voluminous chlorine atoms (green color).

Upon closer examination of the mecoprop molecules in the SOM matrix, Figure 7a demonstrates how a monomer mecoprop molecule (arrow 1) can form a hydrogen bonds (dotted lines) with mobile water molecules. The hydrogen bond between the hydrogen atom in the SOM matrix and the carbonyl atom of the mecoprop carboxyl group is highlighted by thick lines (distance 0.2486 nm and angle 164.605°). *In-situ* selection of the mecoprop---water structure inside the intact SOM/mecoprop complex and its molecular mechanics calculations yield minor contributions of the bond-, angle-, dihedral- and stretch-bend energy descriptors, but the van der Waals energy of -82.42 and electrostatic energy of -20.36 kJ mol^{-1} play a decisive role. This is in agreement with the calculated total energy of -108.87 kJ mol^{-1} for the mecoprop---water structure which indicates strong stabilization of the whole SOM/mecoprop complex. Moreover, mobile water molecules may directly form hydrogen bonds with the DOM molecule or mediate these bonds indirectly. It is highly interesting that the water molecules can be involved in hydrogen bonds between two and three centers if the parameters for bonding angle and distance are fulfilled and the appropriate electrostatic charges for the donor and acceptor are supplied. Thus, the stabilization of soil organic matter and influence on contaminant complexes, as well as the flexibility of water to change positions and conformations of relevant portions of the SOM structure, are important factors.

In Figure 7b, a cluster consisting of 4 mecoprop molecules is displayed. The cluster is entrapped and occluded in a void in the upper center of SOM (arrow 2). Cluster formation of contaminants inside the host structure show two interesting facts. Firstly, the associations of the mecoprop molecules can be achieved with energy gain and thus contribute to the stabilization of the xenobiotic complex. Indeed calculation for total energy of the cluster *in-situ* results in -286.55 kJ mol^{-1} binding energy. Examination of the energy descriptors reveals a major contribution of -201.31 kJ mol^{-1} for van der Waals and -46.47 kJ mol^{-1} for electrostatic forces. The overlapping alignment of two aromatic rings in the center of the cluster is illustrated in Figure 7b and is partly responsible for the high van der Waals

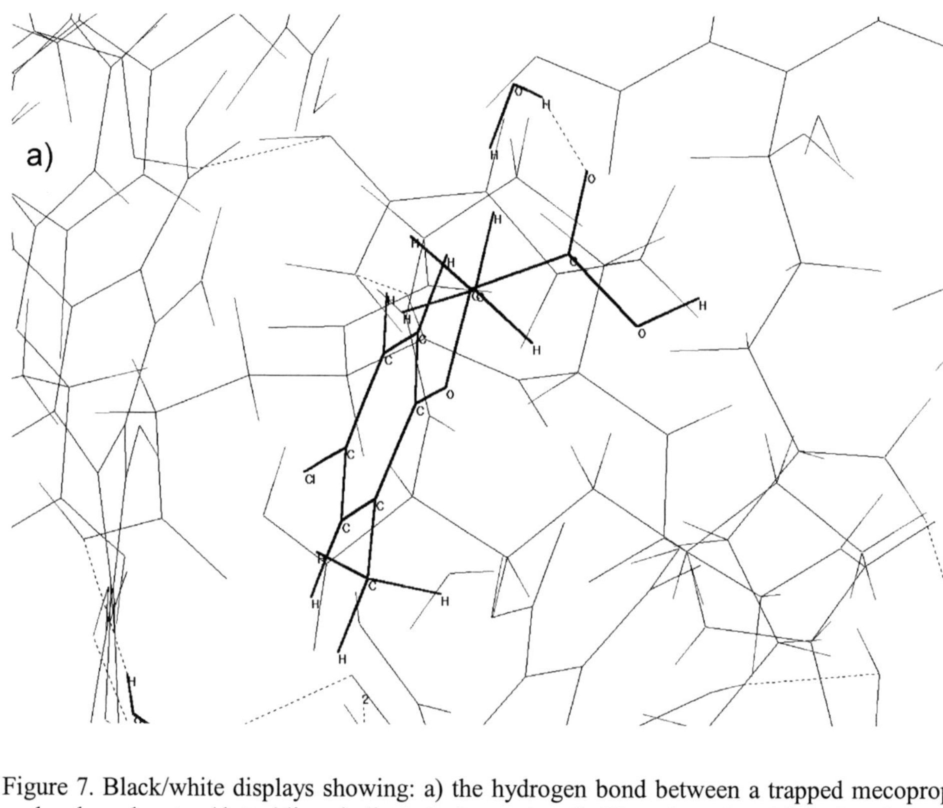

Figure 7. Black/white displays showing: a) the hydrogen bond between a trapped mecoprop molecule and water (dotted lines indicate hydrogen bond). The role and multiple functions of water in soil organic matter such as changes in surface polarity and corresponding chemical properties are illustrated. b) Section of soil organic matter (SOM) with a trapped xenobiotic cluster is shown. The four mecoprop molecules were selected and are highlighted by *Thick Line*. Cluster properties and environmental consequences are outlined. c) The trapping and binding of two biological molecules a hexapeptide and a trisaccharide are displayed. Energy aspects and the role of intra- and intermolecular hydrogen bonds are illustrated.

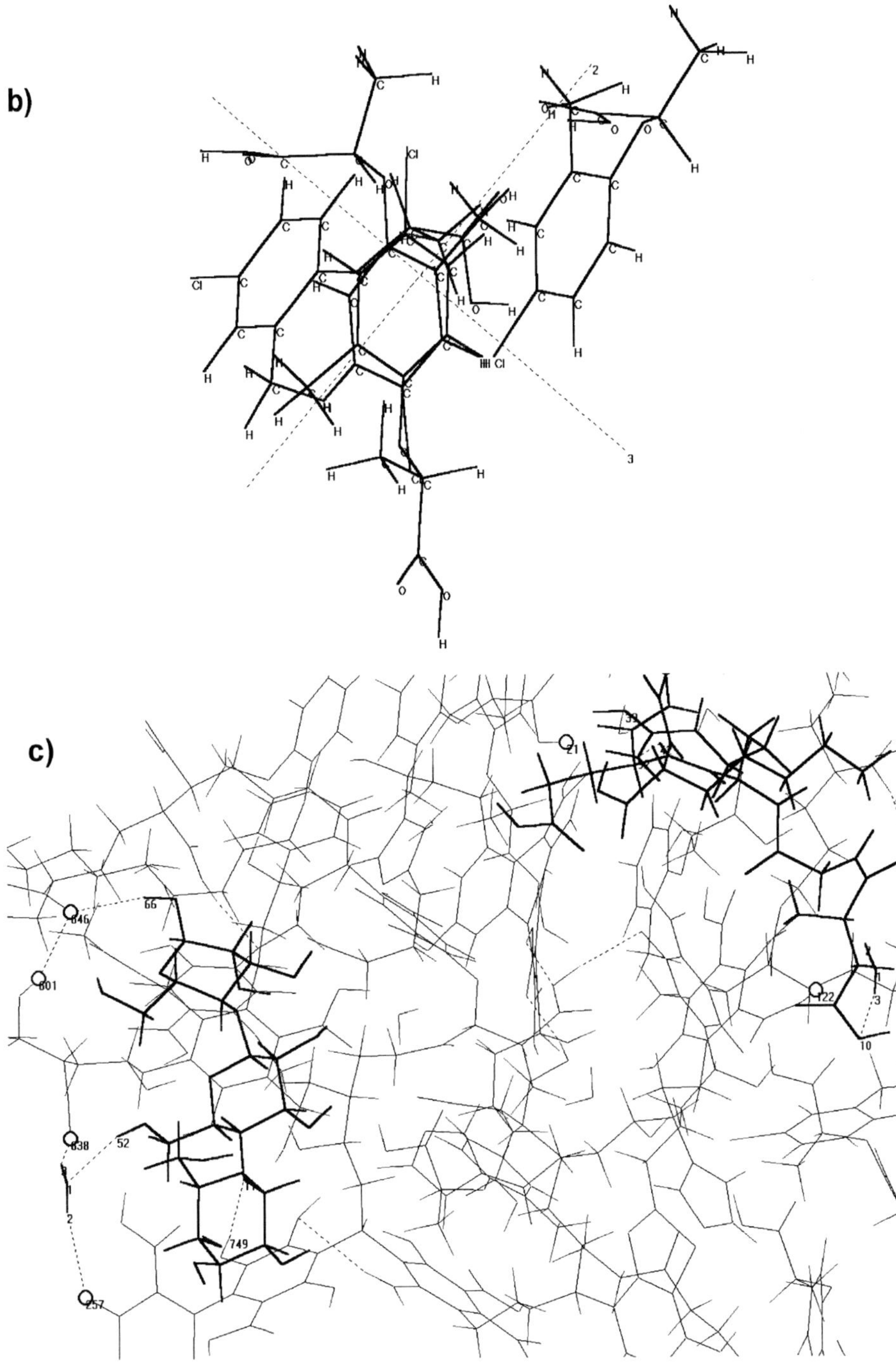

Figure 7. Continued

contribution (-45.44 kJ mol^{-1}). Secondly, by filling the SOM void, the inner surfaces are more easily available for intermolecular interactions and partly explain the sometimes observed high concentrations of bound contaminants. However, *in-situ* selection and MM+ calculation gave a total energy for the interactions of 12 mecoprop with 12 water molecules of -1236.11 kJ mol^{-1} (-709.96 van der Waals and -356.06 kJ mol^{-1} electrostatic energy). This again shows the extraordinary relevance of soil moisture.

3.3.2. Biological bonds

The interactions of HA, DOM, SOM, and soil particles with biological compounds such as peptides and sugars have been investigated in previous publications [e.g., 13, 15, 33]. Selection *in-situ* and molecular mechanics calculations of the two biological constituents of the SOM/mecoprop complex (Figure 7c) gave the total energy (in kJ mol^{-1}) of -449.61 for the hexapeptide in the top-right corner (energy descriptors for van der Waals -217.28 and electrostatic -315.85). For the trisaccharide displayed on the lower left side of Figure 7c, a total energy of -74.66 was calculated (energy descriptors for van der Waals -125.68 and electrostatic -49.34). For energetic and structural data see Table 2c.

Apart from the energetic aspects, it was of interest to evaluate the role of hydrogen bonds of a hexapeptide (molecule 3) and trisaccharide (molecule 2) in the SOM/mecoprop complex and to describe characteristic structural features of the interactions between SOM and the two biological molecules. In the first step, the two molecules (160 atoms) were selected and highlighted (*Thick Lines*). In a second step, the 13 atoms which are directly connected with hydrogen bonds were numbered and bonds between or to total humic substances were marked by circles.

Starting from the top of Figure 7c, an intermolecular hydrogen bond is observed between donor H(21) of molecule 18 (mecoprop) and the acceptor N(39) of the amino group at the peptide terminal. The mecoprop molecule 18 belongs to the tetramer cluster described above in Figure 7b and contributes to the fixation of the cluster in the trapping void of the total humic substance. Following the hexapeptide structure clockwise, an intramolecular hydrogen bond between H(42) of an imine group and N(92) of an amine of molecule 3 is found.

Continuing along the hexapeptide to the right side of Figure 7c an interesting formation is generated in which water (molecule 12) mediates the bonding of the hexapeptide with the total humic substance (molecule 1). The atom H(122), which is the acidic hydrogen of an aromatic carboxylic acid of the molecule 1, interacts with the O(1) of water (molecule 12). Simultaneously, a hydrogen bridge is formed between H(3) of the same water molecule with the hydroxyl O(10) of a carboxyl group of the hexapeptide (molecule 3). In this manner the peptide is linked to the humic skeleton via two hydrogen bridges of one water molecule.

Looking clockwise at the bonds of the trisaccharide (molecule 2) on the left side of Figure 7c, the humic skeleton supplies the atom H(749) from an aromatic (naphthalene) hydroxyl group for the hydrogen bond with O(11) of the trisaccharide (distance 0.2376 nm). Two characteristic three-center hydrogen bonds of SOM are observed at the lower and middle left side of Figure 7c. The first one starts with the water molecule number 8. All three atoms of this water molecule are engaged in different hydrogen bonds: (a) H(2) shows a hydrogen bond to the total humic substance via the quinonoic O(257); (b) H(3) binds to a nitrile nitrogen N(638) via hydrogen bond; and (c) O(1) of molecule 8 is donor for H (52) of the hydroxyl group of the trisaccharide. Finally, the sugar (molecule 2) forms a hydrogen bond by H(66) of a hydroxyl group with the keto O(646) of the humic skeleton. On the other hand a hydrogen bond is established by the hydroxyl H(601) to O(646) also of the total humic substance, thus linking the sugar by two humic hydrogen bonds.

The high density of polar functions in the humic skeleton and the mobility and flexibility of water molecules opens a wide range of structural and energetic possibilities for the formation and cleavage of hydrogen bonds. This delivers a substantial contribution to xenobiotic and biological complex stability. The behavior of the biological molecules during modeling appears to be quite similar to that of the xenobiotics described above if similar polarity and active functional groups are available. In addition to the formation of hydrogen bonds and van der Waals forces in cross-linked humic structures with high aromaticity, torsion relief appears as the third main driving force for geometry optimization and energy minimization and represents an important energy requirement.

3.4. Soil particles

The organic molecular basis for our knowledge of soil structures is largely based on physical fractionation, extraction, modern spectroscopic methods and in particular, thermal characterization and analytical pyrolysis [15, 43]. So far organic macromolecules such as HA, DOM and SOM have been mainly investigated. However for most soils, inorganic chemistry makes up the major part of the chemical constituents. It is difficult to judge whether the organic or inorganic portion of soils is more complex to examine using the presently available analytical tools. [44, 45]. While keeping in mind that little of the fauna and/or flora would be accessible by existing software and gases, as well as water, could only be modeled in a limited approach, an attempt was made to construct organic-mineral complexes and soil particles.

3.4.1. Organic-mineral bonds

Preliminary modeling and computational chemistry of the organic-mineral complexes started with constructing planar silica sheets, running molecular mechanics calculations to obtain a optimized three-dimensional structure and then interacting this simplified mineral matrix with sugars, peptides and HA [16]. Binding of the modeled soil organic matter, which included water molecules, to the silica sheets by metal bonds using Fe^{3+} and Al^{3+} cations followed [13]. Relatively small soil particles [17-19] were tentatively proposed and showed the limits of the computer capacity and the rapidly increasing structural complexity of the target structures. Advances have been reported for non-bonded organic-mineral interactions and sorption of organic compounds on soil surfaces [44] and interesting interactions of humic substances with mineral surfaces [45].

In Figure 8, the SOM/mecoprop complex (cf. Figure 6) is displayed in the central section (width 6.3 nm and height of 9.7 nm in the x,y-plane) of four simulated clay mineral layers which are at an average distance between 4 and 6 nm. Molecular mechanics calculations of the rectangular section *in-situ* gave a very high total energy of 61880.85 kJ mol^{-1} which strongly reflects the cut covalent bonds in the mineral layers. In total, 45 molecules and 4168 atoms were selected and are displayed in a spacefilling mode by their element colors (*Overlapping Spheres*). The color plot clearly illustrates surfaces and chemical features of the macromolecular structure and the mineral bonds. The typical SiO_4-tetraeders are easily seen in the center of the silica layers. If water would invade the interstitial gaps, then dissociation, migration, inner sphere adsorption, plasticity, and bond mediation would be feasible. Moreover, free movement of the cations in an aqueous phase would occur (and modeled), for instance for the large alkali cations (potassium) and the relatively small ammonium ions.

About 1% of the anionic surface functions have been left negatively charged to simulate the characteristic negative charges of clay surfaces and to promote ion transport.

The handdrawn draft structures of planar silica sheets, formed helical surface structures with increasing number of MM+ calculations and improved geometrical optimization. The average distances between the surface atoms of the SOM/mecoprop complex and the silica layers were on top 0.8 nm, below 1.1 nm, on the left side 0.5 nm, and on the right side 0.4 nm. Thus the distances between the xenobiotic complex and the silica matrix were too large for hydrogen bonding (< 0.32 nm). However, the hydrogen bonds inside the xenobiotic complex were the same as discussed above (Figures 6 and 7).

In order to explore the non-bonded interactions between the silica layers and the xenobiotic complex, the total energy and energy descriptors were calculated by MM+. The difference between total energies of the SOM/mecoprop complex calculated *in vacuo* and the trapped complex between the inorganic layers resulted in a relatively low binding energy of -21.88 kJ mol^{-1} which is almost entirely due to loss of van der Waals energy of the standard conformation. Due to the distances of > 0.4 nm between inorganic matrix and trapped xenobiotic complexes neither long-range nor short-range forces contribute substantially to the stabilization of the organic-mineral particle.

3.5. Humic-, xenobiotic-, biological-, and mineral-bonds

The further expanded view of the soil particle is shown in the z-direction (Figure 9, width approx. 18 nm, height 4 nm)) and displays the long silica helix structure of the SiO_4-tetraeders, the typical large potassium ions (white) and long line of 256 iron atoms (green). From this viewpoint it is clearly seen that the SOM/mecoprop complex in the center of the soil particle is almost completely shielded by the metal-substituted silica helix. Stapling more of these structures on the top and the bottom of the shown particle would protect the soil organic matter with the trapped 2 biological and 12 herbicide molecules strongly against metabolic attack. Unfortunately the present computer capacity is at its limits and does not allow a substantial examination of intact clay particle layers. Reducing the average distance between the SOM/xenobiotic complex and the inorganic matrix by about 50% and moving the silica layers closer to the central SOM/mecoprop complex showed more than a dozen intermolecular hydrogen bonds at a total energy of the soil particle of approximately 198300 kJ mol^{-1}. Selecting the central xenobiotic complex and determination of its partial total energy in-situ results in a stabilization energy of -300.12 kJ mol^{-1} due to the interactions within the simulated clay layer. Thus, together with the stabilization energy of -773.70 kJ mol^{-1} for the SOM/mecoprop complex alone (cf. Figure 6), trapping and binding of the xenobiotic complex in the layer is energetically favored in total by an energy gain of 1073.82 kJ mol^{-1} The typical water effects such as swelling, mobility, polarity, and mediation of intra- and intermolecular H-bonds can be modeled. To the best of our knowledge this is also observed for the first time for SOM bond mediation by ammonium cations on clay surfaces. Moreover insertion of 5 protonated water molecules (HO_3^+) excel in high migration speed of water inside the soil particle and increase the potential to generate multiple hydrogen bonds. Starting with one and two hydrogens as donor, with increasing geometry optimization the water oxygen too will serve as acceptor from high polarity sites in the structure of soil particles.

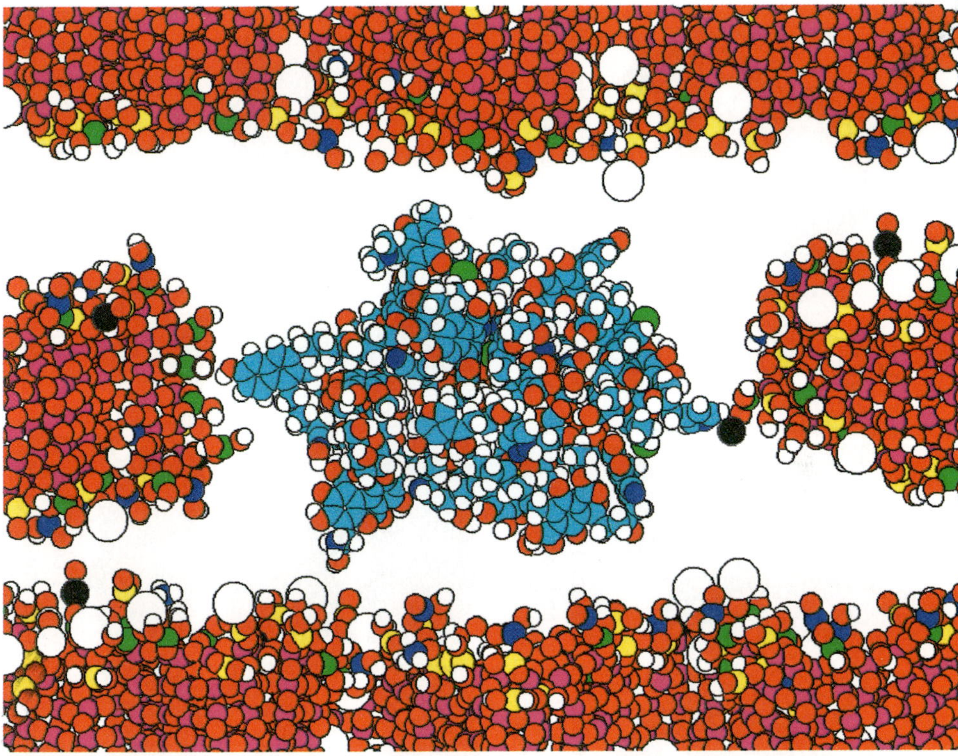

Figure 8. Central section of the interlayer space with the trapped SOM/mecoprop complex and chemically bound or physically entrapped biological and xenobiotic molecules. The element colors (and radii in nm) are as in Figure 5, in addition aluminum (yellow, r = 0.125), calcium (yellow, 0.18 nm), iron (green, r = 0.14; only in the mineral matrix), magnesium (black, r = 0.15), manganese (blue, r = 0.14), phosphorus (black, r = 0.10), potassium (white, r = 0.22), silicon (violet, r = 0.11), sodium (white, r = 0.18), and titanium (white, r = 0.14) are shown.

Finally in Figure 10 the complete simulated soil nanoparticle with four silica sheets and the trapped interlayer position of the SOM/mecoprop complex is shown in view of the x,y-planes. In total it consists of 83 molecules with 11562 atoms and has a molecular weight of about 202816.08 g mol^{-1}. The smallest rectangular box which fits the particle has a width of 9.67 nm, height of 6.11 nm, thickness of 27.95 nm and a volume of about 1.7 μm^3. An attempt was made to accommodate the most important 16 elements (colors see Figure 8) in soil science approximately according the natural abundances in the Earth's crust [46; 47]. The elemental composition and analysis, molecular weight and molecular properties such as surface areas and volumes are given in Table 3. Most of the potential energy of the soil particle is due to angle torsion and bond stretching-, bending- and twisting-energy. Non-bonded interactions by van der Waals and electrostatic forces are indicated by more than 25 % of the total energy. This is in line with the observation of 90 H-bonds in the xenobiotic soil particle.

Figure 9. The expanded section of the soil particle (75 molecules, 9577 atoms) is shown in the z-direction to visualize the helix structure of the silica components as well as the chemical binding and physical entrapment/occlusion of the soil organic matter/herbicide complex following extensive geometry optimization (element colors as in Figure 8).

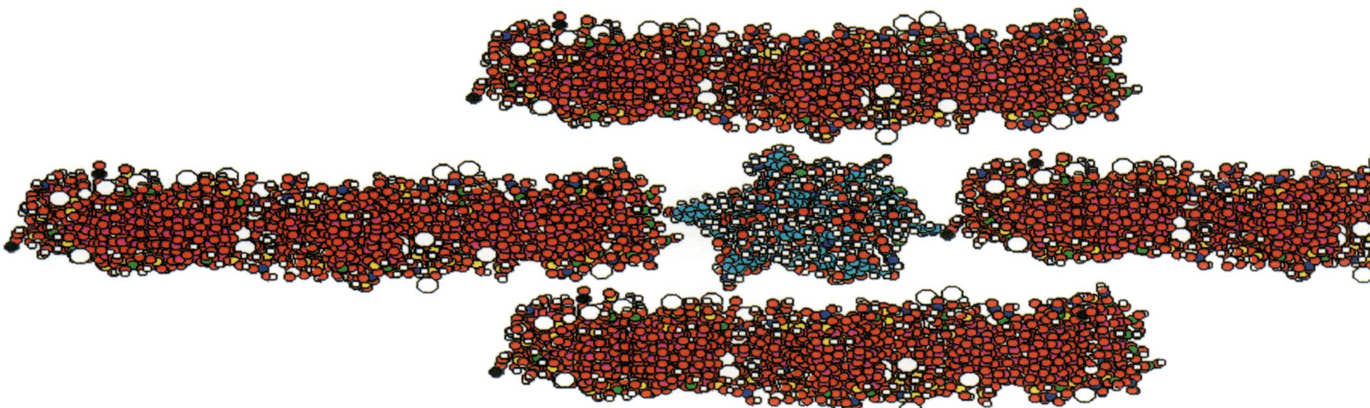

Figure 10. The complete simulated soil particle (83 molecules, 11652 atoms) consisting of one SOM subunit including 2 trapped biological molecules, 12 water molecules and 12 mecoprop molecules inserted into a layer of four silica sheets are displayed in the x,y-plane. For atomic and molecular properties (A) and energies (B) see data in Table 3 (element colors see Figure 8).

Table 3
Soil organic–mineral particle trapping 12 herbicide molecules: atomic and molecular properties

SOM + 4 mineral substructures + 12 mecoprop molecules 83 molecules, 11562 atoms

(A) Atomic and molecular properties

Elemental composition

$$C_{469}H_{1861}O_{6177}N_{78}Si_{1924}Fe_{256}Al_{456}Mn_{120}K_{124}Ti_{24}Mg_{24}Cl_{12}P_{28}Ca_{4}Na_{4}S_{1}$$

Elemental analysis	2.78 % C	6.07 % Al	0.21 % Cl
	0.92 % H	3.25 % Mn	0.08 % Ca
	48.73 % O	2.39 % K	0.05 % Na
	0.54 % N	0.57 % Ti	0.02 % S
	26.64 % Si	0.43 % P	
	7.05 % Fe	0.29 % Mg	
Molecular weight [g mol^{-1}]			202816.08
Solvent-accessible surface area [nm^2]			769.33
Solvent-accessible surface-bounded molecular volume [nm^3]			234.74
Van der Waals-surface area [nm^2]			1023.29
Van der Waals-surface-bounded molecular volume [nm^3]			110.56
Depth in workspace [nm] View: a) x,y and b) z			a) 4.03 b) 9.71
Specific surface [m^2 g^{-1}]			2254.65
Density [g cm^3]			3.05
Smallest box [nm] x, y, z, Σ (nm^3)			9.67, 6.11, 27.95, 1 650.75
Dipole moment [Debye nm^{-1}]			705.9
Moments of inertia [amu nm^2]			x = 1.35e+006, y = 8.32e+006, z = 9.14e+006

(B) Total energy and energy derivatives [kJ mol^{-1}]	
Total energy	199299.69
Bond	42657.13
Angle	123308.35
Dihedral	8510.02
Van der Waals	47469.35
Stretch-bend	-13582.47
Electrostatic	-9062.34

4. CONCLUSIONS

Computer and software capacity are presently limitations which most likely will be solved by stepwise progress in the future. Moreover, accurate molecular analytical data of humic substances, dissolved organic matter in water, soil organic matter, and soil particles required to control and improve the results of molecular modeling are not easily accessible. Most difficulties are seen for models of metabolic processes owing to their complexity and practically simultaneous syntheses and decay in nature.

Potential of molecular mechanics calculations [48] are determinations of bond stretching, bending and twisting angle energy and thus torsional relief of the investigated conformations. As illustrated above electrostatic- and non-bonded interactions were highly interesting for humic fractions, soil organic matter and their xenobiotic complexes. The options to determine surface areas and volumes as well as energy descriptors of adsorption, and absorption processes are adequate and useful tools.

At the nanochemistry level suspensoid organic/mineral/xenobiotic complexes in surface water and soils percolates show inner surfaces which have hydrophobic as well as hydrophilic functions and will adapt the guest molecule according to its own structural chemical and physical features. Geometry optimization (and energy minimization) by molecular mechanics calculations allow insights into energetically favorable reaction pathways and have the potential to link virtual properties with physical-chemical, biological and toxicological results and should enable molecular chemical predictions of environmental processes.

Potential of molecular mechanics calculations [48] are determinations of bond stretching, bending and twisting angle energy and thus torsional relief of the investigated conformations. As illustrated above electrostatic- and non-bonded interactions were highly interesting for humic fractions, soil organic matter and their xenobiotic complexes. The options to determine surface areas and volumes as well as energy descriptors of adsorption, and absorption processes are adequate and useful tools. At the nanochemistry level suspensoid organic/mineral/xenobiotic complexes in surface water and soils percolates show inner surfaces which have hydrophobic as well as hydrophilic functions and will adapt the guest molecule according to its own structural chemical and physical features. Geometry optimization (and energy minimization) by molecular mechanics calculations allow insights into energetically favorable reaction pathways and have the potential to link virtual properties with physical-chemical, biological and toxicological results and should enable molecular chemical predictions of environmental processes.

REFERENCES

1. Gosh, K., Schnitzer, M., 1980. Macromolecular structures of humic substances. Soil Sci. 129, 266-276.
2. Stevenson, F.J., 1994. Humic Chemistry: Genesis, Composition, Reactions. John Wiley, New York.
3. Fillela, M., Buffle, J., 1993. Factors controlling the stability of submicron colloids in natural waters. Coll. Surf. A 73, 255-273.
4. Wilkinson, K.J., Stoll, S., Buffle, J., 1995. Characterization of NOM-colloid aggregates in surface waters: coupling transmission microscopy staining techniques and mathematical modelling. Fres. J. Anal. Chem. 351, 54-61.
5. Bailey, G.W., Shevchenko, S.M., Yu, Y.S., Kamermans, H., 1997. Combining scanning tunneling microscopy and computer simulation of humic substances: citric acid, a model. Soil Sci. Soc. Am. J. 61, 91-101.

6. Myneni, S.C.B., Brown, J.T., Martinez, G.A., Meyer-Ilse, W., 1999. Imaging of humic substance macromolecular structures in water and soils. Science 286, 1335-1337.
7. Scheinost, A.C., Abend, S., Elzinga E., Sparks, D.L., 2000. Carbon-edge X-ray spectroscopy of the colloidal fraction of two Long Island top soils. *In*: (Violante A., Gianfreda, L. Eds.) Soil Mineral-Organic Matter-Microorganism Interactions and Ecosystem Health. Abstracts, 3[d] International Symposium of the IUSS Working Group MO, Naples-Capri, Italy, p. 72.
8. Jackson T.A., Leppard, G.G., 2000. Relations between heavy metals, microorganisms and natural colloids in contaminated ecosystems: results of research employing energy dispersive X-ray microanalysis. *In*: (Violante A., Gianfreda, L. Eds.) Soil Mineral-Organic Matter-Microorganism Interactions and Ecosystem Health. Abstracts, 3[d] International Symposium of the IUSS Working Group MO, Naples-Capri, Italy, p. 37.
9. Schulten, H.-R., 2001. Models of humic structures: association of humic acids and organic matter in soils and water. *In*: Clapp, C.E, Hayes, M.H.B., Senesi, N., Bloom, P.R., Jardine, P.M. (Eds.), Humic Substances and Chemical Contaminants. Soil Sci. Soc. Am., Madison, WI. SSSA, pp. 73-87.
10. Schulten H.-R., Schnitzer, M., A contribution to solving the puzzle of the chemical structure of humic substances: pyrolysis-soft ionization mass spectrometry. a) 1990. International Humic Substances Society, 5th International Meeting, Nagoya, Japan and b) 1992. Sci. Total Environ. 117/118, 27-39.
11. Schulten, H.-R., Plage B., Schnitzer, M., 1991. A chemical structure for humic substances. Naturwissenschaften 78, 311-312.
12. Schulten, H.-R., Schnitzer, M., 1993. A state-of-the-art structural concept for humic substances, Naturwissenschaften 80, 29-30.
13. Schulten, H.-R., Schnitzer, M., 1995. Three-dimensional models for humic acids and soil organic matter. Naturwissenschaften 82, 487-498.
14. Schnitzer M., Schulten, H.-R., 1998. New ideas on the chemical make-up of soil humic and fulvic acids, *In*: Huang, P.M., Sparks, D.L., Boyd, S.A (Eds.), Future Prospects for Soil Chemistry. Spec. Publ. 55. Soil Sci. Soc. Am., Madison, WI. SSSA, pp. 153-177.
15. Schulten, H.-R., Schnitzer, M., 1997. Chemical model structures for soil organic matter and soils. Soil Sci. 162, 115-130.
16. Schulten, H.-R., 1995. The three-dimensional structure of soil organo-mineral complexes studied by analytical pyrolysis. J Anal. Appl. Pyrolysis 32, 111-126.
17. Schulten H.-R., Leinweber, P., 1996. Characterization of humic and soil particles by analytical pyrolysis and computer modeling. J. Anal. Appl. Pyrolysis 38, 1-53.
18. Schulten, H.-R., Leinweber, P., Schnitzer, M., 1998. Analytical pyrolysis and computer modeling of humic and soil particles. *In*: Huang, P.M., Senesi, N., Buffle, J. (Eds.), Environmental Particles: Structure and Surface Reactions of Soil Particles. John Wiley, Chichester, England, pp. 281-324.
19. Schulten, H.-R., Leinweber, P., 2000. New insights into organic-mineral particles: composition, properties, and models of molecular structure. Biol. Fertil. Soils 30. 399-432.
20. HyperChem® and ChemPlus™ software. 1997. Hypercube, Inc., 1115 N.W. 4th Street, Gainesville, Florida 32601, U.S.A; e-mail: info@hyper.com.
21. Sparks, D.L., 1995. Environmental Soil Chemistry, Academic Press, San Diego.
22. Okuda, I., Senesi, N., 1998. Fractal principles and methods applied to the chemistry of sorption onto environmental particles, *In*: Huang, P.M., Senesi, N., Buffle, J. (Eds.), Environmental Particles: Structure and Surface Reactions of Soil Particles. John Wiley, Chichester, England, pp. 77-105.
23. Schnitzer, M., Schulten, H.-R., 1995. The analysis of organic matter in soil extracts and whole soils by pyrolysis-mass spectrometry. Adv. Agron. 55, 167-217.

24. Schulten, H.-R., 1996. A new approach to the structural analysis of humic substances in water and soils: humic acid oligomers, *In*: Gaffney, J.S., Marley, N.A., Clark, S.B. (Eds.), Humic and Fulvic Acids; Isolation, Structure and Environmental Role. American Chemical Society Symposium, Series 651, Washington, pp. 42-56.
25. Waldrop, M.M., 1992. Complexity. Schuster, New York.
26. De Paolis, F., Kukkonen, J., 1997. Binding of organic pollutants to humic and fulvic acids: influence of pH and the structure of humic material. Chemosphere 34, 1693-1704.
27. Leinweber, P., Blumenstein, O., Schulten, H.-R., 1996. Soil organic matter composition in sewage farms: investigations by carbon-13 NMR and pyrolysis-field ionization mass spectrometry. Europ. J. Soil Sci. 47, 71-80.
28. Schulten, H.-R., Thomsen M., Carlsen, L., 2001. Humic complexes of diethyl phthalate: molecular modeling of the sorption process. Chemosphere 45, 357-369.
29. Wershaw, R.L., 1999. Molecular aggregation of humic substances. Soil Sci. 164, 803-813.
30. Barron, M.G., 1990. Bioconcentration. Environ. Sci. Technol. 24, 1612-1618.
31. Haitzer, M., Höss, S., Traunspurger, W., Steinberg, C., 1998. Effects of dissolved organic matter (DOM) on the bioconcentration of organic chemicals in aquatic organisms-a review. Chemosphere 37, 1335-1362.
32. Hesketh, N., Jones, M.N., Tipping, E., 1996. The interaction of some pesticides and herbicides with humic substances. Anal. Chim. Acta 327, 191-201.
33. Schulten, H.-R., 1999. Analytical pyrolysis and computational chemistry of aquatic humic substances and dissolved organic matter. J. Anal. Appl. Pyrolysis 49, 385-415.
34. Sorge, C., Schulten, H.-R., Weyandt, R.G., Kamp, N., Brechtel, H.-M., 1994. Influence of wet storage of spruce wood on groundwater quality: Investigations by water-chemical methods, pyrolysis-field ionization mass spectrometry and luminescent-bacteria bioassay. Int. J. Environ. Anal. Chem. 57, 1-8.
35. Schulten, H.-R., 1996. Three-dimensional, molecular structures of humic acids and their interactions with water and dissolved contaminants. Int. J. Environ. Anal. Chem. 64, 147-162.
36. Schulten, H.-R., 1999. Interactions of dissolved organic matter with xenobiotic compounds: molecular modeling in water. Environ. Toxicol. Chem. 18, 1643-1655.
37. Gevao, B., Semple, K.T., Jones, K.C., 2000. Bound pesticide residues in soils: a review. Environ. Pollut. 108, 3-14.
38. Schulten, H.-R., 1996. Direct pyrolysis-mass spectrometry of soils: a novel tool in agriculture, ecology, forestry, and soil science. *In*: Yamasaki, S., Boutton, T.W. (Eds.), Mass Spectrometry of Soils. Marcel Dekker, New York, pp. 373-436.
39. Leinweber, P., Schulten, H.-R., 1998. Advances in analytical pyrolysis of soil organic matter. J. Anal. Appl. Pyrolysis 47, 165-189.
40. Schulten, H.-R., 1995. The three-dimensional structure of humic substances and soil organic matter studied by computational analytical chemistry. Fres. J. Anal. Chem., 351, 62-73.
41. Négre, M., Schulten, H.-R., Gennari, M., Vindrola, D., 2001. Interaction of imidazolinone herbicides with soil humic acids: experimental results and molecular modelling. J. Environ. Sci. Health 36B, 107-125.
42. Schulten, H.-R., 1998. Molecular modelling of humic substances, soil organic matter and soil particles: potential and limits, *In*: Largeau, L., Schulten, H.-R., Eglinton, T. (Eds.), Organic Matter Preservation in Soils and Sediments: Implications for Global Carbon Cycle. 8th V. M. Goldschmidt Conference, Toulouse, France. Mineral. Mag. 62A, 1358-1359.

43. Schulten, H.-R., Leinweber, P., 1993. Influence of the inorganic matrix on the formation and molecular composition of soil organic matter in a long-term experiment. Biogeochemistry 22, 1-22.
44. Shevchenko, S.M., Bailey, G.W., 1998. Non-bonded organo-mineral interactions and sorption of organic compounds on soil surfaces: a model approach. Theochem. 422, 259-270.
45. Akim, L.G., Bailey, G.W., Shevchenko, S.M., 1998. A computational chemistry approach to study the interactions of humic substances with mineral surfaces. *In*: Davies, G., Ghabbour, E.A. (Eds.), Humic Substances: Structures, Properties and Uses. Royal Soc. Chem., Cambridge, UK, pp 133-145.
46. Schlesinger, W.H., 1991. Biogeochemistry: An Analysis of Global Change. Academic Press, San Diego.
47. Wild, A., 1993. Soils and the Environment. An Introduction, Cambridge University Press.
48. Burkert U., Allinger, N.L., 1983. Molecular Mechanics. A.C.S Monogr. 177, American Chemical Society, Washington, DC.

Developments in Soil Science, Volume 28A
Editors: A. Violante, P.M. Huang, J.-M. Bollag and L. Gianfreda

IMPACT OF CHANGING FOREST MANAGEMENT ON SOIL ORGANIC MATTER IN LOW MOUNTAIN ACID MEDIA

F. Andreux [a], F. Roux, N. Linglois, Thi-Kim-Ngân Nguyen, P. Amiotte Suchet and J. Lévêque

UMR A 111 Microbiologie des Sols-GéoSol INRA-Université de Bourgogne,
Centre des Sciences de la Terre 6, Boulevard Gabriel 21000 Dijon, France

[a] Corresponding Author: E-mail Francis.Andreux@u-bourgogne.fr

The impacts of changes in vegetation cover from native deciduous forest to Douglas fir (*Pseudotsuga menziesii* Franco) and of human activity on soil organic matter (SOM) characteristics were studied in two low mountain areas of east-central France. No striking difference in soil type (Dystric Cambisol) was found between the two sites. Humus-rich horizons were of the "Dysmull" and "Moder" types, regardless of the nature of the bedrock. Contrary to a common affirmation concerning other coniferous species, Douglas fir had no negative effect on soil pH and humification degree of SOM, with respect to the native beech vegetation. Pruning and partial clearing slightly improved humification, especially the decomposition of the leaf litter. Special attention was placed on the characteristics of water-soluble SOM, to suggest chemical and isotopic methods that would allow the tracking and estimation of soluble transfers from soils to streams. Differences arose between forested watersheds, according to the predominance of either deciduous or coniferous vegetation; the significance of such differences is discussed.

1. INTRODUCTION

Forest ecosystems constitute the main storage well of terrestrial C, in the form of both tree biomass and soil organic matter (SOM). As the major parameter of soil structure stability, energy source for soil microorganisms, and physical support of reactive sites for mineral nutrients, SOM is the main intersection of research in agronomy and forestry, as well as in soil biology and chemistry. With a considerable number of scientific papers during the last 50 years, improved knowledge about the origin, chemical structure and reactions of SOM has been gained [1-3].

One of the reasons for the chemical complexity of SOM is the variety of its composition, in relation with the numerous plant sources, which develop or have developed in a given place. This aspect is increasingly taken into account, especially when passing from local to larger scale studies [4]. It is still very difficult to distinguish any given ecosystem from another by way of the chemical composition of humus. Moreover, specific characteristics of individual biochemical molecules frequently tend to vanish with increasing degrees of humification [5]. For these reasons, it is important that the sustainable management of forest ecosystem take into account the amounts of OM stored in soils, as well as the mechanisms involved in the variations of these amounts.

During the last several decades, increasing attention has been paid to the consequences of the introduction of conifers on biogeochemical cycles in forest soils [6-8]. In all cases, the maintenance of the nutrient reserve is the key factor in the functioning of these cycles, and the fate of soil nutrients depends mainly on management procedures. The way trees were planted, the duration of the rotation, and the methodology of wood collection strongly affect several physical, chemical and biological soil characteristics. Measurement of SOM storage and distribution among well-defined compartments, such as particulate organic residues, clay-humus complexes, and microbial biomass, is probably one of the most pertinent and simple indices of effective or potential changes in these cycles [9].

The impact of Douglas fir (*Pseudotsuga menziesii* Franco) on soil biogeochemical functioning was recently studied in detail by Marques et al. [10-12]. They compared three plantations of different ages developed on the same geological bedrock, and showed the progressive loss with time of soil major cations and mineral nitrogen, in spite of a rather efficient cycling of these elements. The transfer of these elements throughout the soil solution generally involves a proportion of water-soluble OM that can be collected and quantified. There are some indications about the increasing loss of water-soluble elements with increasing age of the plantations. However, neither the nature of the soluble SOM nor its interactions with other solutes are known.

The aim of the present work is to compare the main characteristics of humus material in soils of selected Douglas fir plantations, under similar conditions of low mountain climate. The following aspects will be considered: (i) changes following clearing of the native beech forest and its substitution by Douglas fir plantation, (ii) subsequent changes related to the age of the plantation, (iii) impact of forest management (silviculture) practices, and (iv) influence of two different geological rocks. This work will attempt to establish which of these factors prevails in terms of environmental impact, as measured on both the solid and soluble SOM.

2. MATERIALS AND METHODS

2.1. Soils and their environment

The study sites are located in two low mountain areas of east-central France in Beaujolais and Morvan. The altitude is not higher than 700 m, the climate conditions are similar, and the native vegetation is mostly a beech forest mixed with scarce hornbeams. The sites strongly differ in their geological bedrock, which is volcanic tuff in Beaujolais and granite in Morvan. However, in both cases, the soil is a brown acid soil, or Dystric Cambisol [13], sometimes showing slight podzolic features, with a humus-rich layer of the "Dysmull" and "Moder" types [14]. In Beaujolais, three Douglas fir plantations forming a chronosequence, aged 20 (B20), 40 (B40) and 60 (B60) years, respectively, in 1992, were studied, and the experimental field and the soils were described by Marques et al. [10,11].

In the Morvan forest, soils under 25-year old stands with four different silvicultural assays were compared with that under the former native beech forest. The soils of a neighboring area were described for the first time by Brêthes [15] and more recently by Lévêque et al. [16]. In addition, the Morvan area is composed of numerous small watersheds in which either deciduous or coniferous vegetation predominates. This makes possible a comparative study of SOM erosion, based on the composition of organic solutes collected in the creeks of the respective watersheds.

2.2. Soil and litter sampling

Soils were taken from the walls of large pits dug in the middle of each selected area. Soil samples were named B20, B40 and B60, according to the age of the plantation, for the three stands of the Beaujolais, and T (no pruning and no clearing), L3 (clearing every third row, in 1985 and 1995), L3+S4 (plus selective clearing every fourth tree, in 1985), and L3+S2 (plus selective clearing every other tree, in 1985), according to the silvicultural treatment, for the four stands of the Morvan. An additional sampling (H) was carried out under a native beech forest adjacent to the assay area. Sampling was carried out with a spade and a knife at depth intervals of 0.10 m down to 1.0 m in Beaujolais, and to 0.40 m in Morvan, then with an auger in the rest of the profile in the latter case. The soil samples were air-dried and sieved at 2 mm prior to analyses. From each 0-0.10 m (topsoil) and 0.10-0.20 m layer, and from some of the lower layers, 140 cm^3 cylindrical soil cores were sampled for density measurements. These samples were allowed to dry at room temperature, then at 105 °C, and finally weighed.

In the Morvan area, litter material was collected at the end of winter, before the decomposition of the newly deposited plant rests starts. A 225-cm^2 metallic square frame was thrown at random in four different places of each assay, and the material present on the forest floor, inside the frame, was carefully collected. Based on their size, two fractions, lower and higher than 20 mm, were separated. The separated litter fractions were dried at 40 °C, weighed, and powdered in an electric mill.

2.3. Water sampling

The procedure of soil water extraction has been extensively used, improved, and described in the case of the Beaujolais experiment [10-12]. Two main types of soil solutions were obtained by these authors and used for part of the present study: gravitational and bound solutions. Gravitational solutions were collected at the forest floor level by a set of tension-free lysimeters (0.40 x 0.25 m), and at depths of 0.15, 0.30, 0.60 and 1.20 m, by zero-tension, plate lysimeters (0.40 x 0.30 m). Solutions were collected in containers located in closed pits downhill, sampled bimonthly, and brought to the laboratory. Bound solutions were collected with porous cups held at controlled pressure and located at the same depths as the zero tension lysimeters. When the cups were full, a monitoring device transferred automatically the solutions to glass containers, from which they were taken bimonthly.

In addition to extracted soil solution, stream water was sampled bimonthly throughout one year in small watersheds adjacent to the Morvan Douglas fir plantation [17]. After filtering through a 0.45 μm Millipore membrane, the solutions from soil or stream origin were either analyzed immediately or freeze-dried and stored under vacuum until further analysis.

2.4. Soil humus fractionation

Two methods were used for SOM fractionation. (i) Grain-size fractionation was carried out by dispersing dry soil in water (50:250 wt/wt), stirring with ultrasound for 2 min, shaking mechanically for 8 hr, and sieving through 200 μm and 50 μm stainless sieves, successively. The separated coarse sand (200-2000 μm), fine sand (50-200 μm) and clay+silt (0-50 μm) fractions were then concentrated under low pressure, freeze-dried, weighed, and finely ground in an agate mortar [18]. (ii) Chemical fractionation was run, using repeated extraction with 0.05 M sodium hydroxide (soil:solution ratio was 5:100 wt/wt), and centrifugation at 10,000 g. Fulvic (FAs) and humic acids (HAs) were separated from the supernatant by precipitation of the latter with an excess of 0.1 M hydrochloric acid. Separation and purification of the fractions were carried out according to the IHSS procedure [19]. After passing over a cation (H^+) exchange resin, the purified extracts were freeze-dried and stored under vacuum.

2.5. Microscope observation of soils and soil size fractions

Scanning and transmission electron microscopic studies have been shown to be powerful tools for the detailed description of mineral-mineral and organic-mineral interactions in soil structures [20, 21]. In this study, some size fractions separated from soil samples collected in Morvan, under Douglas fir plantation and beech native forest were compared. Air-dried, unground soil crumbs were observed by means of an environmental scanning electron microscope (SEM), which presents the great advantage of requiring no sample preparation, such as staining, thickening, etc. Instead, samples can be subjected to observation several times, under variable conditions of humidity and pressure [22]. The machine used was an ESEM from Electroscan Company.

2.6. Standard soil and water analyses

On each 0-2 mm air-dried soil sample, pH values were determined in water and in 1.0 M KCl (soil:solution ratio of 2:5 wt/wt), using a combined electrode Schott N65. Cation exchange capacity (CEC) was measured after a 1 hr extraction of soil with 0.05 N cobaltihexammine (soil:solution ratio of 1:10 wt/wt), using a CAMSPEC M330 spectrophotometer, to determine absorbency changes of the extracting solution at 472 nm [23]. On each ground soil sample, on the corresponding ground size fractions, and on the purified, freeze-dried powders of HAs and FAs, organic carbon (C) and nitrogen (N) contents were determined by dry combustion, using a Fizon NA 1500 analyzer [24].

For the present study, no standard physical determination was done, except for bulk density, by weighing the soil cores previously dried at 105°C. Carbon contents in each sample were converted from weight basis (Cw in g kg^{-1} of soil) to surface area basis (Cs in kg m^{-2}), as already explained elsewhere [24], using the following equation:

$$Cs = Cw * L * d \quad (1)$$

where L is the thickness (in m) and d the bulk density (in kg dm^{-3}) of the considered soil layer.

On the filtrated soil solutions, some standard determinations were carried out. Major cations were determined by atomic absorption spectrometry (Perkin Elmer 3000 device), after previous acidification with 1 ml of concentrated HCl [17]. The main anions (chloride, nitrate, sulfate, and phosphate) were determined on a Dionex 100 chromatograph, using isocratic elution with a 1.8 mmole L^{-1} sodium carbonate/bicarbonate solution [15]. Alkalinity of each filtrated solution was determined by titration with 0.01 M HCl, and acidity was determined with 0.25 M sodium hydroxide on each solution previously percolated on a column of AG 50 W-X8 ion exchange resin (H^+ form). These measurements were run and monitored on a Schott TPC 2000 titration device. Dissolved organic carbon (DOC) concentrations were obtained using a Shimatzu TOC 5000 analyzer.

2.7. Functional determinations on humic substances

2.7.1. Acidity titration

On the powdered, decationized HAs and FAs from the two surface layers of each study soil, potentiometric titration was carried out, using the method of Gran [25], which was adapted to humic substances by Bizri et al. [26]. A 0.1M sodium hydroxide aqueous solution and a 0.1M sodium perchlorate solution were used as titration agent and dispersing electrolyte, respectively. The reaction was monitored at 25° C and under nitrogen flow, using a Schott TPC 2000 titration device. After constructing the titration curves, the results were converted into linear regressions ("Gran curves"), and calculations were carried out, to determine the distribution of the categories of acidity and their dissociation degrees [26, 27]. According to their strength, strong acidity SAc, weak acidity wAc, and very weak acidity vwAc were distinguished, and the pKa constants of wAc and vwAc were determined.

2.7.2. Infrared absorption and nuclear magnetic resonance spectrometry

On selected samples of HAs and FAs, as well as of freeze-dried soil solution, functional characterization was attempted. For infrared absorption studies, from 0.5 to 1.0 mg of water-exempt powder were homogenized with 200-400 mg of dry potassium bromide, then the mixture was converted into a 1-2 mm thick, 10 mm diameter, transparent disk, by compression under vacuum at 7 x 10^6 hPa. Spectra were run on a Brüker Fourier Transform IFS 66v spectro-photometer.

Proton nuclear magnetic resonance (NMR) spectra were taken for selected samples of soil solution, dispersed in a mixture of D_2O and 1.0 M sodium hydroxide solution (2/3:1/3, v:v). The dispersed and pre-filtrated suspensions were stored in the dark until they were analyzed, using a Brüker AC200 Fourier Transform spectrometer. In these experimental conditions, only non exchangeable protons were determined and chemical shifts were determined with respect to water instead of TMS.

2.8. Determination of ^{13}C isotopic composition

The stable carbon isotope ^{13}C occurs at a natural abundance of about 11‰, relative to ^{12}C. However, the distribution of ^{13}C is far from homogeneous, which makes possible its use as a tracer to study the fate of any original compound of well-defined isotopic composition. Soil ^{13}C measurements are based on the fact that SOM has an isotopic composition that corresponds closely to that of the vegetation cover from which it originates [28, 29]. Thus, changes in the vegetation cover can modify the isotopic composition of SOM, provided plants are differently labeled with ^{13}C. Because the main differences in ^{13}C abundance in plants are due to differences in their photosynthetic cycles, studies about SOM storage and turnover frequently consider very contrasted situations passing from C3 vegetation (*e. g.* trees) to C4 vegetation (*e. g.* natural tropical grasses, maize crops), or the opposite [24, 30, 31]. However, due to the limited sensitivity of the method, it is much less typical to study ^{13}C isotopic changes in natural systems where plants substitute others having the same photosynthetic cycle [32]. This was attempted in the present study. The ^{13}C abundance is expressed in delta (δ) units, as follows:

$$\delta^{13}C\ ‰ = [(R_{sample}/R_{standard}) - 1] * 1000) \qquad (2)$$

where R is the isotope ratio $^{13}C/^{12}C$, and the standard (PDB) is a belemnite carbonate from the Pee Dee Formation of North Carolina. The $\delta^{13}C$ values were determined on the CO_2 obtained by complete dry combustion at 1050°C, using a system composed of an elemental analyzer (CNS NA 1500, Fizon) coupled to an isotope ratio mass spectrometer (VG Isochrom-EA, Fisons). Solid dry samples from soils, soil fractions, and plant material were finely homogenized and powdered in an electric agate mill. Soil solutions were previously freeze-dried, then ground before weighing and analysis. At least two determinations were repeated on each sample, until no difference (*i. e.* less than 0.2 $\delta^{13}C$ units) was found between repetitions.

3. RESULTS AND DISCUSSION

3.1. Soil pH and cation exchange capacity

In both study areas, the soil profile was generally about 1 m deep, showing a rather abrupt contact with the blocky weathered rock and a thick humus layer overlain by an abundant litter layer. The soil acidity was always high, due to both the composition of the volcanic tuff or granite bedrock and the limiting climatic conditions for humus formation. In the upper humus layer, pH_{H2O} was close to or slightly lower than 4.0 in both the dense younger Douglas fir plantations and the beech native forest. In deeper soil layers, pH_{H2O} was generally 4.0-4.5, with few differences from one site or plantation to another. The lower values (3.8-4.0) of pH_{KCl} in the humus layer show that exchangeable acidity was induced by humus accumulation. Under cleared plantations, pH_{H2O} values of the humus layer increased up to 4.5, then returned to close to 4.0 in older, open plantations, in relation with the predominant acidophilic character of the understory vegetation. These results indicate that during the period of major growth, clearing practices have contributed to limit the acidity of surface humus in Douglas fir plantations.

The pH values found in both sites were related with very low base saturation. The values of CEC were generally low, in the range of 5.0-8.0 cmole kg^{-1}, and decreased with depth to about 2.0 cmole kg^{-1} in some cases. Determinations carried out on the three surface soils of Beaujolais [11] showed that CEC values decreased significantly from 7.6 cmole kg^{-1} in B20, to 6.6 cmole kg^{-1} in B40, and 5.9 cmole kg^{-1} B60. Exchangeable acidity accounted for more than 800‰ of CEC, in which Al ions predominated, and H^+ represented less than 50‰. Measurements carried out in a steep watershed, partly planted with resinous trees, adjacent to the Morvan study site showed that exchangeable Al reached about10 cmole kg^{-1} in the surface soil at the top of the watershed. This Al concentration was still 4.5 cmole kg^{-1} at lower altitude in the watershed, with no increase in pH_{H2O} values (4.2-4.3) in spite of some colluvial effect [16].

All the above results indicate that the introduction of Douglas fir plantations hardly affects the composition of the surface soil exchange complex. With respect to the former deciduous forest, no evidence of humus acidification, and even a tendency toward a slight increase of pH values, was shown.

3.2. Soil organic carbon and nitrogen

The vertical distribution of organic C and N was typical of that of Cambisols, with predominant accumulation in the topsoil, and regular decreasing concentrations toward the bottom of the profile (Table 1). In Beaujolais, organic C content ranged between 37.6 and 52.4 g kg^{-1} in the topsoil, and 15.9 and 34.5 g kg^{-1} in the second layer, and decreased with increasing age of the plantation, as reported already by Marques et al. [8]. As a result, C storage in the humus-rich layers decreased from 15.0 kg m^{-2} in the B20 stand (0-0.40 m layer), to 13.1 kg m^{-2} in the B40 stand (0-0.45 m layer), and 7.0 kg m^{-2} in the B60 stand (0-0.35 m layer). The N content ranged between 2.4 and 3.8 g kg^{-1} in the topsoil, and 0.9 and 2.6 g kg^{-1} in the second layer, again with a marked decrease with increasing age of the plantation [8]. The corresponding N storage decreased from 1.0 kg m^{-2} in the B20 stand, to 0.8 kg m^{-2}

Table 1
Carbon and nitrogen contents in the surface layers, compared with the bottom layer of the soil profiles in Beaujolais and in Morvan

Soil layers (m)	C g kg^{-1}	C kg m^{-2}	N g kg^{-1}	N kg m^{-2}	C/N
Beaujolais. Douglas 20 years					
0 – 0.12	52.4	5.0 ± 0.4	3.8	0.37± 0.03	13,8
0.12 – 0.25	34.5	4.5 ± 0.4	2.6	0.34 ± 0.03	13,5
0.25 – 0.40	24.7	5.5 ± 0.5	1.7	0.38 ± 0.03	14,5
0.85 – 1.00	5.6	1.7 ± 0.1	0.5	0.15 ± 0.01	11,2
Beaujolais. Douglas 40 years					
0 – 0.15	42.1	5.0 ± 0.4	2.7	0.32 ± 0.03	15.6
0.15 – 0.30	24.6	4.4 ± 0.3	1.5	0.27 ± 0.02	16.4
0.30 – 0.45	14.5	3.7 ± 0.3	0.9	0.22 ± 0.02	16.1
0.85 – 1.00	4.0	1.2 ± 0.1	0.3	0.09 ± 0.01	13,3
Beaujolais. Douglas 60 years					
0 – 0.10	37.6	2.6 ± 0.2	2.4	0.17 ± 0.01	15.7
0.10 – 0.20	24.8	2.0 ± 0.2	1.7	0.14 ± 0.01	14.6
0.20 – 0.35	15.9	2.4 ± 0.2	1.1	0.17 ± 0.01	14.4
0.85 – 1.00	2.0	0.6 ± 0.05	0.2	0.06 ± 0.01	10.0
Morvan. Beech tree native forest					
0 – 0.10	122.5	8.0 ± 0.6	7.1	0.46 ± 0.04	17.3
0.10 – 0.20	83.5	7.5 + 0.6	5.0	0.45 ± 0.04	16.6
0.20 – 0.30	52.4	5.2 ± 0.4	2.8	0.28 ± 0.02	18.7
0.80 – 0.90	2.9	0.6 ± 0.05	0.3	0.06 ± 0.01	9.7
Morvan. Douglas 25 years – No treatment					
0 – 0.10	77.7	5.8 ± 0.5	5.2	0.39 ± 0.03	14.9
0.10 – 0.20	37.5	3.4 ± 0.3	2.6	0.24 ± 0.02	14.4
0.20 – 0.30	22.0	2.2 ± 0.2	1.4	0.14 ± 0.01	15.7
0.80 – 0.90	3.1	0.6 ± 0.05	0.3	0.06 ± 0.01	10.3

in the B40 stand, and to 0.5 kg m^{-2} in the B60 stand. Altogether, the N content decreased faster than the C content with increasing age, resulting in a tendency toward an increase in C/N ratio from 13.5-13.8 to 14.6-15.7.

In Morvan, only the soil under Douglas fir plantation, with no clearing treatment at all (T soil) was compared with the soil under native beech forest (H soil) (Table 1). In the humus-rich layers of the T soil, C and N contents were higher (77.7 g kg^{-1} and 37. g kg^{-1}) than those in the corresponding layers of the B20 soil. However, because of slightly lower bulk density values, C and N storage in these layers was close to or even lower than that measured in Beaujolais. In the H soil, higher C and N contents than in the T soil were clearly observed. About 22 kg m^{-2} of C and 1.1 kg m^{-2} of N were stored in the 0-0.30 m layer of the H soil, compared with 10.7 and 0.7 kg m^{-2}, respectively, in the T soil. Slightly higher C/N ratios were found in the humus-rich layers under Douglas fir plantation in Morvan (14.4-14.9) than in the youngest plantation of Beaujolais (13.5-13.8), whereas even higher values (16.5-16.6) were obtained under the native beech forest. This suggests a tendency toward improved humification conditions under Douglas fir, compared with the native vegetation cover.

3.3. Litter production under Douglas fir and beech tree

The comparison of litter accumulation on the same soil type, under Douglas fir plantation and native deciduous forest, was only possible at the Morvan site, where such reference forest still existed. At the end of winter, the total dry weight of litter on the forest floor varied between 2.5 and 3.0 kg m^{-2} under beech tree (H soil) and 1.5 and 2.5 kg m^{-2} under Douglas fir, according to the kind of management. Figure 1 shows that fine litter material (< 20 mm) strongly predominated in all cases, but that coarse litter material (> 20 mm) was present in higher proportions under Douglas fir. This coarse material was mainly formed by twig and branch rests. Its amount was about twice as high on the T soil and up to three or times higher on the L3, L3+S4 and L3+S2 soils, than on the H soil, according to the intensity and date of the clearing in the respective stands. However, the high standard deviation values reflected an important spatial variability of this parameter. The amounts of fine litter material showed less spatial variability but differed clearly from one stand to another. The lowest values were found in the untreated stand (T soil) and the highest in the L3 soil, where clearing had been repeated most recently.

The study of organic C and N was limited to the fine litter material. Its organic C content varied from 397 to 511‰, with the highest values in the T soil, and the lowest ones in the L3 soil and in the H soil. When C and N contents were expressed on a surface area basis, and plotted on the same graph (Figure 2), it was observed that these values distributed along a straight line, the slope of which was the average C/N ratio of the litter material (22.0 ± 2.5). This representation also illustrated the diversity of litter falls from one stand to another, with the existence of three main groups.

The group with the highest values was limited to the deciduous forest (H soil), and that with the lowest values included the T soil, and most of the L3+S2 and L3+S4 soils under selective treatments. The intermediate group was that of the L3 soil, mainly because of a more recent clearing than in the other stands. The values of the H and L3 soils were found to be significantly different from each other and from those of other treatments. Conversely, values for the T soil and L3+S2 and L3+S4 soils under intermediate clearing treatments could not be significantly distinguished.

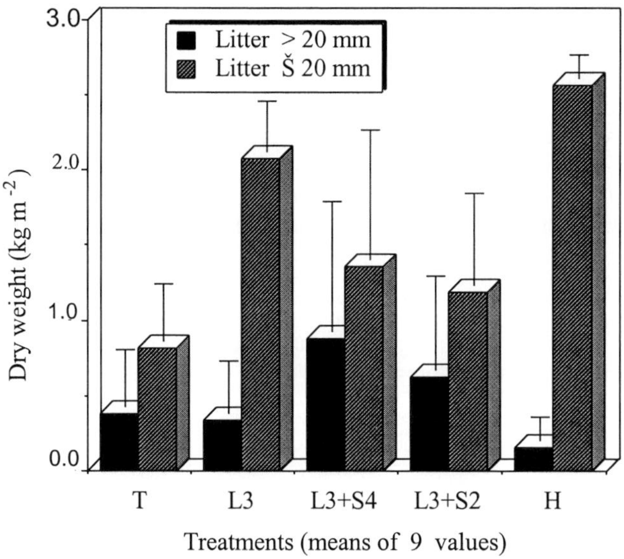

Figure 1. Comparison of litter weight collected on the soil under Douglas fir plantation with different treatments (T: no clearing; L3: every third row felled; S4: every fourth tree felled; S2: every other tree felled), and under native beech forest (H), in the Morvan site.

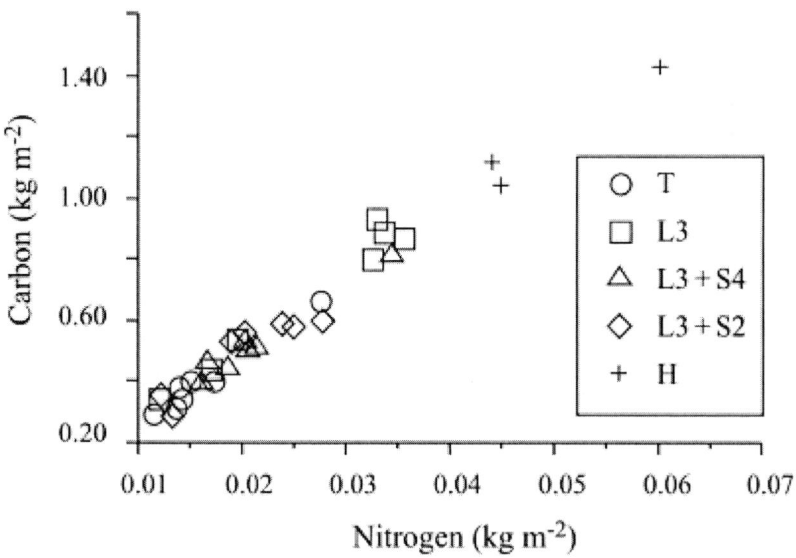

Figure 2. Variations of carbon content versus nitrogen content in the litter material under Douglas fir plantations and native beech forest in the Morvan site. $R^2 = 0.96$ (letters have the same meaning as in Figure 1).

When the regression curves were analyzed separately, it was shown (Table 2) that R^2 coefficients were always high (0.84-0.98), with limited variations. Furthermore, the slope of the curves, which features the average C/N value for each kind of litter material, presented slightly different values. The highest values were observed for the T and L3 soils (C/N = 23.0 and 25.0, respectively) and the lowest values for the two soils under selective S3+L2 and S3+L4 treatments (C/N = 19.5 and 22.0, respectively). In the litter of the H soil, an intermediate value (C/N = 20.3) was observed. These differences could be indicative of more favorable humification conditions in the case of more open Douglas fir plantations.

Table 2
Equations of linear regression for nitrogen content versus carbon content in the litter material under Douglas fir and native beech forest in the Morvan site (letters have the same meaning as in Figure 1)

Forest and treatment	Regression equation	R^2
Douglas – T	$C = 0.035 + 22.9 * N$	0.96
Douglas - L3	$C = 0.027 + 25.0 * N$	0.98
Douglas - L3+S4	$C = 0.052 + 22.1 * N$	0.97
Douglas - L3+S2	$C = 0.114 + 19.5 * N$	0.84
Beech tree - H	$C = 0.155 + 20.3 * N$	0.94

3.4. Particle-size distribution of soil organic matter

3.4.1. Weight distribution

The weight balance sheet of the physical fractionation was realized on each sample of the topsoil and of the second humus layer of the three soils (B20, B40 and B60) of Beaujolais, and of three soils (H, T and L3+S2) of Morvan. In all cases, the yield recovery ranged between 926 ‰ (H, 0-0.10 m layer) and 969 ‰ (B20, 0.10-0.20 m layer). The two sites had in common a relatively low amount of clay-size 0-2 μm fraction (63-107 g kg^{-1} in Beaujolais and 76-90 g kg^{-1} in Morvan), but they presented very contrasted distributions of the coarser fractions. In Beaujolais, the intermediate 2-50 μm fraction predominated, with from 511 mg kg^{-1} (B40, 0.10-0.20 m layer) to 661 mg kg^{-1} (B20, 0-0.10 m layer). In Morvan, the sand-size 200 – 2000 μm fraction predominated, with from 396 mg kg^{-1} (T, 0-0.10 m layer) to 507 mg kg^{-1} (H, 0-0.10 m layer). Due to these differences, contrasted distributions of SOM among the size fractions should be expected between the two sites.

3.4.2. Carbon and nitrogen distribution

The three soils from the Beaujolais site yielded relatively similar results (Table 3). In all cases, C and N accumulated mainly in the 2-50 μm fraction, where they represented more than two thirds of the total soil C and N, respectively. There was no apparent selection of N, as the C/N ratio of this fraction was similar to that of the corresponding whole soil (13.9-15.7 in the topsoil, and 11.4-16.4 in the second humus-rich layer). The second fraction in terms of C and N accumulation was the clay-size 0-2 μm fraction. It was 2-3 times less abundant than the silt-size fraction, but its C/N ratio was always lower than that of the whole soil (11.8-13.2 in the topsoil and 10.5-11.9 in the second layer). The two coarse fractions were much less abundant, especially in the second humus-rich layer, in which a material of higher C/N ratio (21.7-23.5) than in the whole soil was present, suggesting a different quality from that of the rest of SOM. Even though the total SOM content of these soils decreased with increasing age of the stands, their physical fractionation patterns showed no special differences.

Table 3
Distribution of carbon and nitrogen in the size fractions of surface layers of the Beaujolais soils (data in g of C or N per kg of dry soil; nd: no determination)

Soil layer	200 - 2000 μm			50 - 200 μm			2 - 50 μm			0 - 2 μm		
	C	N	C/N	C	N	C/N	C	N	C/N	C	N	C/N
m	$g\ kg^{-1}$			$g\ kg^{-1}$			$g\ kg^{-1}$			$g\ kg^{-1}$		
Douglas 20 years												
0-0.10	5.80	0.30	19.3	5.20	0.33	15.8	34.3	2.46	13.9	9.10	0.73	12.5
0.10-0.20	4.40	0.19	23.2	2.4	0.13	18.2	21.9	1.64	13.4	9.00	0.79	11.4
Douglas 40 years												
0-0.10	4.50	nd	nd	6.10	0.33	18.5	29.4	1.87	15.7	5.80	0.44	13.2
0.10-0.20	3.70	0.17	21.7	2.50	0.15	16.7	19.1	1.15	16.6	6.70	0.56	11.9
Douglas 60 years												
0-0.10	6.20	0.33	18.8	7.50	0.44	17.0	27.0	1.78	15.2	6.70	0.57	11.8
0.10-0.20	4.0	0.17	23.5	2.70	0.16	16.8	18.7	1.64	11.4	4.30	0.41	10.5

The soils from the Morvan site (Table 4) showed a marked contrast between those developed under Douglas fir plantation and that under native beech forest. In the latter, the sand-size 200-2000 μm fraction had the major contribution, with about 400‰ of total soil C and N. This coarse fraction was mostly formed of raw SOM, with easily recognizable plant and faunal rests [5, 33]. However, the C/N ratio (22.2) of this fraction was only slightly higher than that of the whole SOM, indicating that its decomposition was already in process. The second humus-rich layer of the H soil would probably have shown similar features, unfortunately this material was lost during the fractionation process.

The soils developed under Douglas fir plantation did not show such a high concentration of raw SOM in the coarse fraction. Their pattern of physical fractionation was very similar to that of the soils from the Beaujolais site, in spite of the marked textural difference between the two sites. Again, the 2-50 μm fraction represented up to 600‰ of total soil C, and 700‰ of total soil N. This was even more marked in the second layer, which contained very low amounts of C and N in the coarse fractions. It has to be noticed that little difference was observed between the two stands, with (L3+S2) and without (T) clearing. This is in agreement with the measurements carried out on the forest litter, which showed no significant difference between these treatments.

Table 4
Distribution of carbon and nitrogen in the size fractions of surface layers of the Morvan soils (data in g of C or N per kg of dry soil; nd: no determination)

Soil layer	200 - 2000 μm			50 - 200 μm			2 - 50 μm			0 - 2 μm		
	C	N	C/N	C	N	C/N	C	N	C/N	C	N	C/N
m	g kg^{-1}			g kg^{-1}			g kg^{-1}			g kg^{-1}		
Beech native forest												
0-0.10	55.6	2.50	22.2	33.3	1.63	20.4	31.3	1.84	17.0	21.1	0.95	22.2
Douglas 25 years – No treatment												
0-0.10	11.8	0.44	26,9	14.8	0.83	17.9	37.4	nd	nd	10.3	0.54	19.1
0.10-0.20	5.1	0.15	34,2	3.7	0.17	21,6	19.7	1.20	16.4	6.1	0.34	17.9
Douglas 25 years – Pruning and clearing												
0-0.10	13.0	0.50	26.0	12.7	0.71	17.9	43.5	2.73	15.9	12.3	0.90	13.7
0.10-0.20	7.0	0.18	38.8	7.1	0.40	17.8	26.4	1.54	17.1	6.6	0.51	12.9

3.5. Micromorphological aspect of soil physical fractions

Most of the soil crumbs and soil fractions studied by ESEM showed a very similar aspect of irregular assembly of mineral grains and of organo-mineral micro-aggregates, with a variable pore density and frequent biological features. Microphotographs presented in Figure 3 did not allow distinction between materials collected under Douglas fir or under beech tree. These materials contained mainly micro-aggregates of high porosity but very little loose material. In the T soil under Douglas fir, the coarse fraction (200-2000 μm) was a typical porous association of mineral particles, resembling clay mineral-aluminum aggregates [20], in which biological activity was represented by fungal mycelia surrounding the microaggregates.

The 100-200 μm fraction from the same soil showed heterogeneous micro-aggregates, with fine root material developing in older roots of tubular form. The 0-100 μm fraction of the same soil again had a microaggregate morphology, with less intra-aggregate porosity than the others, and biological residues from plant rootlets and other unidentified organisms.

In the material collected under beech forest, the 200-2000 μm fractions showed less association and more loose material, including particular SOM, than the corresponding fractions under Douglas fir. This confirmed quantitative determinations that showed that in the humus layers under beech forest, about 400‰ of total SOM was located in the sand-size fraction. Conversely, the two other fractions presented larger and apparently more associated components than under Douglas fir. The same observations were done in both the 0-0.10 m

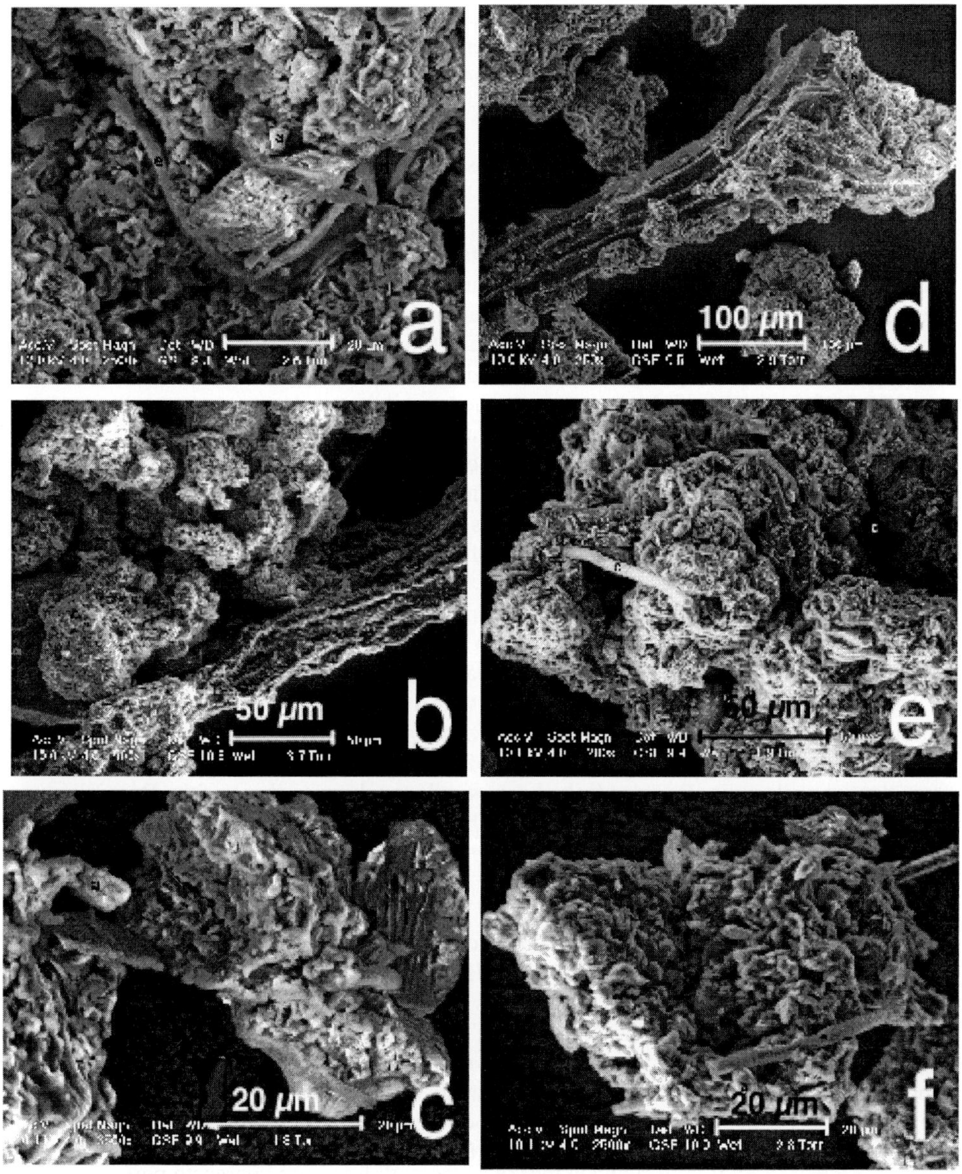

Figure 3. ESEM images of size fractions from topsoils collected in the Morvan site. Left: T soil under Douglas-fir (a) 200-2000 μm (b) 100-200 μm (c) 0-100 μm. Right: H soil under beech-tree (d) 200-2000 μm (e) 100-200 μm (f) 0-100 μm.

topsoil and the 0.10-0.20 m layer. These results suggest that in the two soils, aggregates were easily formed between soil mineral components and humified SOM. However, they were relatively weak constructions that were modified in structure and composition as a consequence of the vegetation substitution.

3.6. Chemical characterization of humic substances

3.6.1. Elemental composition

The determination of C and N contents of HAs and FAs showed very different composition of these two categories of compounds, but only few differences between the two sites, for the same category. In HAs of the Beaujolais site (Table 5), the C content was in the range 452-500 g kg^{-1}, possibly with lower values in the youngest stand, but with no clear difference between the topsoil and the second humus-rich layer. The N content was in the range 38.7-49.5 g kg^{-1}, with lower values in the topsoil, but with no apparent influence of the age of the stand. The FAs had a quite constant C content, in the range 460-487 g kg^{-1}, and their N content hardly exceeded 20 g kg^{-1}, except in the second humus-rich layer of the oldest stand. As a result, the C/N ratio of HAs was low, around 9.2-12.1, whereas that of FAs was much higher, in the range 19.5-27.8. There was a tendency toward slightly higher values in the oldest stands, suggesting a possibility of higher N losses in the form of soluble SOM, as already suggested by Marques et al. [12].

In HAs of the Morvan site (Table 6), C content varied in the range 457-499 g kg^{-1}, and N content in the range 32.9-48.2 g kg^{-1}, with no special difference due to the vegetation cover and forestry practices. Only a slightly higher value of C/N ratio was noticed in the H soil than in the soils under Douglas fir (13.9, compared with 10.2-11.3). Even though this measurement could not be repeated on the underlying horizons of the H soil, it is in agreement with the differences observed above for the C/N ratio of the corresponding topsoils. The FAs extracted from the two soils under Douglas fir had very similar analyses to those of the Beaujolais soils. Only the H soil under beech tree showed much lower N content (13.8 g kg^{-1}) and higher C/N ratio (32.9), again in agreement with the values observed on the whole soil.

3.6.2. Acid functional groups

The acido-basic titration of humic substances according to Gran's method [23] confirmed the existence of marked differences between HAs and FAs but only few differences between the respective HAs and FAs of the two sites. In HAs of the Beaujolais site (Table 5), total acidities ranged between 3.6 and 4.7 mole kg^{-1}, with the highest values in the topsoil and in the youngest stand. Strong (SAc) and very weak (vwAc) acidity were the least represented (about 1.0 mole kg^{-1}), whereas weak (wAc) acidity predominated (1.6-2.2 mole kg^{-1}). With increasing age of the stand, values of the respective types of acidity in the topsoil and in the second humus-rich layer converged. The calculated pKa values were similar in all stands for wAc (4.5-4.7) and decreased slightly with increasing age of the stand, from 7.6-8.2 to 7.4-7.5, for vwAc. In FAs, total acidity values ranged between 6.6 and 7.2 mole kg^{-1} (almost twice as high as in HAs), with slightly higher values in the oldest stand and no difference between the two layers. Values of very weak acidity were low (1.3-1.5 mole kg^{-1}), those of SAc were intermediate (2.1-2.4 mole kg^{-1}), and those of wAc were almost unchanged (3.2-3.4 mole kg^{-1}) in all cases. The calculated pKa values again were similar in all stands for wAc (4.2-4.4) but hardly increased with increasing age of the stand, from 6.3-6.6 to 6.7-6.8, for vwAc.

In HAs of the Morvan site (Table 6), values of total acidity were 3.9-4.3 mole kg^{-1}, with little difference between the topsoil and the second layer, wherever the comparison was

Table 5
Elemental C and N analysis and distribution of acidities (Ac) in humic and fulvic acids extracted from surface soil layers from the Beaujolais site. Values are given on dry-weight basis. Ash content was in the range 15-35 g kg^{-1}. Variability coefficient was 10‰ on C, 2‰ on N, and 100‰ on Ac

Soil layer	C	N	C/N	Strong Ac.	Weak Ac.	V. Weak Ac.
m	g kg^{-1}			mole kg^{-1}		
Douglas 20 years						
Humic acids						
0 - 0.10	452	39.2	11.5	1.1	2.0	1.6
0.10 - 0.20	457	45.0	10.2	1.0	1.6	1.0
Fulvic acids						
0 - 0.10	487	19.7	24.7	2.3	3.2	1.3
0.10 - 0.20	481	21.7	22.2	2.3	3.3	1.3
Douglas 40 years						
Humic acids						
0 - 0.10	476	39.6	12.1	1.1	2.0	1.4
0.10 - 0.20	500	46.1	10.8	1.1	2.2	1.1
Fulvic acids						
0 - 0.10	477	17.2	27.8	2.1	3.2	1.3
0.10 - 0.20	481	18.4	26.1	2.3	3.3	1.2
Douglas 60 years						
Humic acids						
0 - 0.10	442	38.7	11.4	1.0	1.9	1.0
0.10 - 0.20	454	49.5	9.2	1.0	1.9	1.0
Fulvic acids						
0 - 0.10	473	17.1	27.7	2.4	3.4	1.4
0.10 - 0.20	460	23.5	19.5	2.3	3.3	1.5

possible. The distribution of acidity strengths was similar to that described in the Beaujolais site, with a predominant wAc (1.9-2.2 mole kg^{-1}) and no difference between layers and between treatments. The calculated pKa values of wAc were similar to those found in HAs from Beaujolais (4.6-4.7), and those of vwAc varied between 6.7 (H topsoil) and 8.4 (T topsoil), with intermediate values for the cleared Douglas fir stands.

Table 6
Elemental C and N analysis and distribution of acidities (Ac) in humic and fulvic acids extracted from surface soil layers from the Morvan site. Values are given on dry-weight basis. Ash content was in the range 15-35 g kg^{-1}. Variability coefficient was 10‰ on C, 2‰ on N, and 100‰ on Ac. nd: no determination

Soil layer	C	N	C/N	Strong Ac	Weak Ac	V. Weak Ac
m	g kg^{-1}			mole kg^{-1}		
Beech native forest						
Humic acids						
0 - 0.10	457	32.9	13.9	1.0	1.9	1.4
0.10 - 0.20	nd	nd	nd	nd	nd	nd
Fulvic acids						
0 - 0.10	454	13.8	32.9	3.0	3.3	1.7
0.10 - 0.20	485	21.5	23.1	nd	nd	nd
Douglas 25 years – No treatment						
Humic acids						
0 - 0.10	499	44.6	11.2	0.8	2.1	1.2
0.10 - 0.20	491	48.2	10.2	0.8	2.1	1.1
Fulvic acids						
0 - 0.10	480	20.3	23.6	2.1	3.3	1.3
0.10 - 0.20	484	19.7	24.6	2.2	3.7	1.2
Douglas 25 years – Pruning and clearing						
Humic acids						
0 - 0.10	476	42.1	11.3	0.7	2.0	1.3
0.10 - 0.20	491	48.1	10.2	0.8	2.2	1.2
Fulvic acids						
0 - 0.10	484	19.0	25.52	2.3	3.5	1.2
0.10 - 0.20	453	19.5	23.3	2.2	3.5	1.5

In FAs of the Morvan site, the values of total acidity were similar to and even higher than those found in FAs from Beaujolais. They varied between 6.5 and 7.2 mole kg^{-1} under Douglas fir plantation, with no difference between treatments. Only in the H soil, under beech forest, total acidity reached 7.9 mole kg^{-1}, mainly due to a higher value of SAc (3.0 mole kg^{-1}).

The pKa values of FAs again were similar to those found in Beaujolais. Because no pKa value could be determined on FAs of the H soil, the only contrasting result for this soil wasthe highest values of acidity in extractable humics, especially in FAs. This again suggested the possible existence of a higher amount of mobile protons in the humus under native deciduous forest, than under Douglas fir plantation.

3.7. Water-soluble organic matter in soils and streams

It was observed that the dystric Cambisols of the study areas were characterized at the same time by the predominance of aggregates in the intermediate and fine SOM size fraction, and by the presence of raw, particulate SOM in the sand-size fraction. The latter was much less developed under Douglas fir than under the native beech forest. Even though no podzolic feature was evidenced in these soils, variable amounts of organic compounds are expected to migrate in the water-soluble form [11].

Soil water collected in the experimental site of Beaujolais (60-year old Douglas fir plantation) was shown to have a variable ionic composition, according to the depth and the technique of sampling (Figure 4). Solutes collected with the zero-tension lysimeter plates beneath the litter had the lowest Al content, in contrast with those collected at lower depth, which were richer in Al than in Ca, the other predominant cation. In these solutions, the most represented anions were nitrate, sulfate, and organic anions of the FA-type, which accounted for 0.4-1.5 mmole L^{-1} (as revealed by titration of their potential acidity). All of them reached their highest values at 0.15 m. Organic solutes disappeared in the lower solutions, whereas (bi-) carbonate anions increased.

Solutes extracted with tension porous cups were generally very low in FA-type compounds but contained only mineral ions. Among cations, Al was strongly predominant at all depths. The amount of chloride did not increase significantly, compared with that found in the solutions from plates. Conversely, more sulfate and more nitrate anions were obtained with the cups than with the plate, as already reported by Marques et al. [11]. In addition, the composition of these solutions was found to be relatively constant with depth, in contrast with what was observed in the solutions from zero-tension lysimeters.

These differences related to the physical characteristics of soil water are of real importance in relating the composition of surface water with that of the representative soils in a given area [35]. A long-term research program is currently in progress at Morvan Mountain to examine the effects of land use changes on surface water quality throughout the year in four small-forested watersheds with either resinous or deciduous tree cover. In all cases studied in this program, water pH values varied between 5.5 and 6.5, with the same seasonal oscillations- i.e. lower values in winter and higher values in summer- as everywhere in low mountain ecosystems on crystalline rocks [16, 17]. The highest concentrations of cations were found for sodium (3-4 mg L^{-1}) and calcium (1.0-2.5 mg L^{-1}), and the highest concentrations of anions were 2.5-4.5 mg L^{-1} for chloride, 1.5 - 6.0 mg L^{-1} for sulfate, and 0.5 -5.5 mg L^{-1} for nitrate. The last two ions were reported to present the largest fluctuations, in accordance with seasonal climate conditions [36]. Other ions (Mg, K, NH_4, and PO_4) were present at much lower concentrations. Altogether, this made up a total charge of about 0.20-0.60 mmole L^{-1} for cations and 0.15-0.40 mmole L^{-1} for anions, which was close to the average concentrations found in the solutions collected with plates under the litter material.

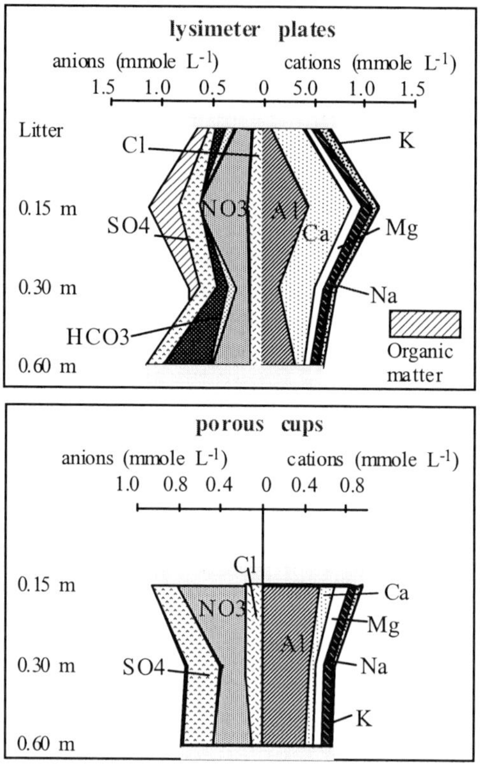

Figure 4. Comparison of the ionic balance in soil solution collected in February with plates and porous cups under Douglas fir plantation in Beaujolais. After Roux [34].

Concerning organic C, it was observed in a previous paper [17] that this element was relatively abundant in stream water of the Morvan area. Concentrations of DOC were always found to follow very different patterns in the watersheds planted with coniferous than in those still covered by the deciduous beech forest. In the coniferous watersheds, DOC was relatively low throughout the year, with small variations between 1.0 and 2.0 mg L^{-1}. Conversely, in the deciduous watersheds, DOC always exceeded these values, and fluctuated between 3.0 (mostly in winter and spring) and 8.0 mg L^{-1} (mostly in summer). Thus, surface water appeared to be “cleaner” in the coniferous watersheds than in the deciduous ones. This is an interesting observation, for coniferous plantations are generally considered negatively from several environmental viewpoints, compared to the deciduous native vegetation. However, in the present study, when water was extracted from the surface horizons of soils under beech forest and under Douglas fir plantation, no marked difference was observed in terms of organic C concentrations.

Is it possible to characterize this soluble OM, and to relate it to that of the surrounding soils and, hopefully, of the dominant vegetation? Apart from DOC and acidity measurements, more detailed chemical characterization has started, leading to more or less success. For instance, infrared absorption spectroscopy has confirmed that the spectra of soluble organic compounds collected beneath the Douglas fir litter were close to those of humic compounds, with large bands typical of O-H stretching (around 3,400 cm^{-1}) and carboxyl groups (around 1,700 cm^{-1}). In general, there was little resolution: the first band was probably affected by moisture but the latter band was confirmed, being enhanced by decationization, and shifting to lower wave numbers when neutralized with sodium hydroxide [34]. For solutions collected at lower depths in soil profiles, the spectra did not show any band attributable to organic functional groups, because of their very low concentration in organic solutes, compared with their high concentration in mineral salts.

Studies using proton NMR were more successful in detecting and quantifying the main functional groups. Figure 5 shows that the materials collected during one sampling campaign in Beaujolais with the porous cups at 0.15 m and 1.20 m (5.a and 5.b, respectively) had very similar spectra with well defined bands. The materials collected with lysimeter plates at these depths (5.c and 5.d, respectively) differed from each other: the first one resembled that of the material from the corresponding porous cup, but showed an additional small peak between 3 and 4 ppm, attributed to polysaccharides; the second one was incompletely resolved beyond 5 ppm and the presence of aromatic protons could not be established.

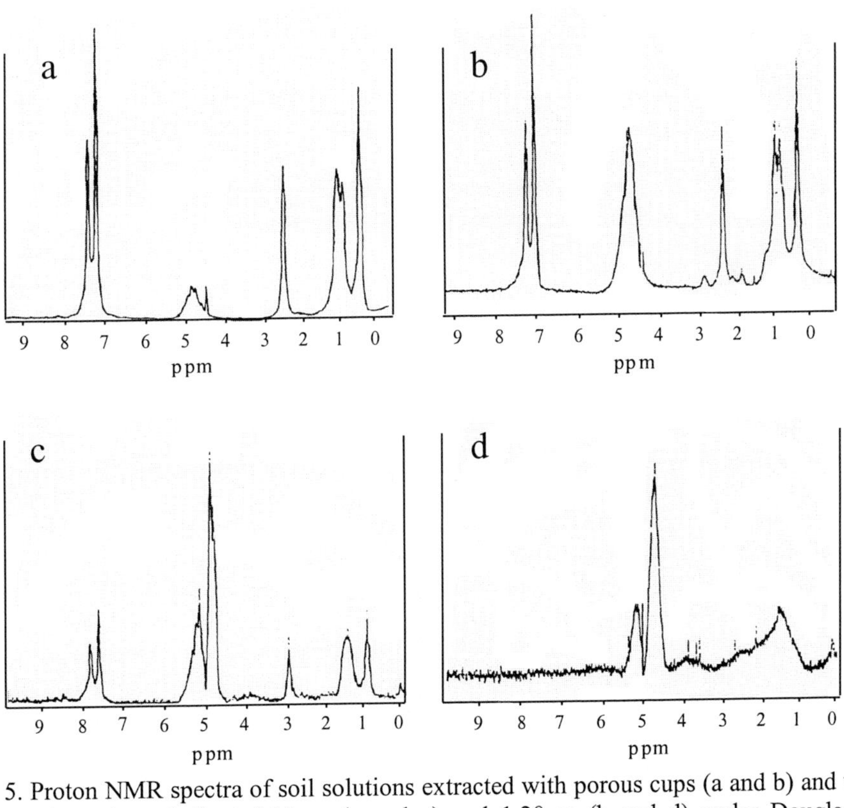

Figure 5. Proton NMR spectra of soil solutions extracted with porous cups (a and b) and with lysimeter plates (c and d) at 0.15 m (a and c) and 1.20 m (b and d) under Douglas fir plantation in Beaujolais.

The quantitative distribution of the main functional groups is shown in Table 7. In all cases, aliphatic protons predominated, compared with those in aromatic rings. However, the solutions collected near the soil surface (0.15 m), either with plates or with porous cups, were slightly more aromatic than the others were. Polysaccharidic protons could be found only in the plates located near the surface, suggesting that this material was more influenced by poorly humified organic rests.

Table 7
Distribution (in percent of total non-exchangeable protons, based on proton NMR) of major aliphatic (Ali) and aromatic (Aro) groups in water-soluble organic matter collected by porous cups and plates at different depths from a soil under Douglas fir plantation in Beaujolais

Type of solution	- CH_3	-CH_2	alpha- H	poly-saccharides	Aromatic	Aro/Ali ratio
Chemical shift (ppm)	0.2 - 06	0.6 - 1.0	2.0 - 2.6	3.0 - 4.0	7.0 - 8.0	
First year						
Cups 0.15 m	17.59	37.96	14.82	0. 00	29.63	0.42
Cups 1.20 m	22,36	32,69	19,23	0,00	25.71	0.35
Plates 0.15 m	16.67	35.71	11.91	4.76	30.95	0.45
Second year						
Cups 0.15 m	17.14	40.00	20.95	0,00	21.90	0.28
Cups 1.20 m	20.39	33.01	21.36	0,00	25.24	0.34

Another trail for further research was related to the study of changes in ^{13}C natural abundance between vegetation, raw humus, solid SOM, and water-soluble OM. Results summarized in Figure 6 show quite different $\delta^{13}C$ values between litter material and soils under Douglas fir in Beaujolais. The litter material had the lowest values (from -28.6 to -27.8 ‰), and the whole soils showed higher values, with a clear tendency to increase with increasing depth. This type of variation is common and has been discussed extensively by several authors [30-32]. New information in this figure is represented by the $\delta^{13}C$ values obtained for soil solutions extracted with plates, which were intermediate between those of the litter and of the whole soil, again with higher values for samples collected at lower depth.

It was interesting to determine to what extent $\delta^{13}C$ values were different in ecosystems under native beech forest and under Douglas fir plantation, in relation with mild differences in the ^{13}C composition of the two species. In a recent study, Linglois et al. [unpublished results] pointed out that no significant difference could be observed between soils profiles, although there was sometimes up to one δ unit between selected litter materials from beech (from -28.0 to -27.2 ‰) and from Douglas fir (from -27.7 to -26.3 ‰). The corresponding vegetation covers presented even lower values, but again the beech material seemed to have slightly lower $\delta^{13}C$ values (about -31 ‰) than the Douglas fir material (about -30.5 ‰). When $\delta^{13}C$ values of DOC in streams of the two watersheds were compared, those under predominant beech forest frequently presented higher values (15 times of 21 in one year). Unfortunately, this could not be confirmed by the comparison of annual means (-28.7 and -28.4 ‰, with standard deviations of 0.28 and 0.35 under beech tree and Douglas fir, respectively).

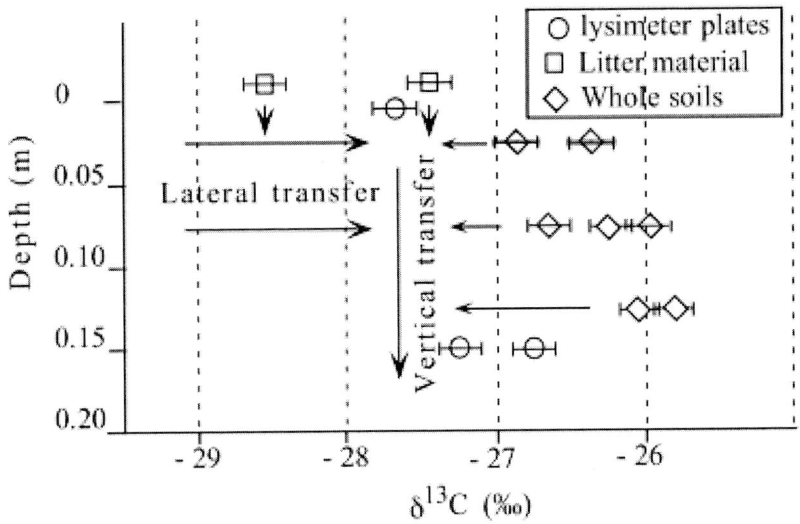

Figure 6. Comparison of $\delta^{13}C$ values in soil solution from lysimeter plates, in litter material, and in the whole soil under Douglas fir plantation in Beaujolais.

4. CONCLUSIONS

In France, since the first plantation that took place in 1842, Douglas fir has become the second most common coniferous species used for reforestation. Although this species is of great interest for wood quality and management of soil nutritional resources, as a conifer, it is a subject of controversy due to its possible environmental impacts, such as soil acidification and its influence on surface water composition. For this reason, a variety of research programs are being carried out, to enter deeper into the diagnosis of changes induced by the new plantation. Among the numerous analytical criteria that can be used, those related to soil organic matter (SOM) were the least documented, and therefore required our attention.

First of all, the above study has considered two different experimental sites at the same time, one in Beaujolais and the other one in Morvan. Even though they are not developed on the same parent rock, they present several points in common: similar altitude and climate; two stands of almost the same age (20-25 years); similar forestry practices, especially in terms of tree spacing, pruning and clearing. They also present important complementary aspects: in Beaujolais, three stands of different ages, but similar forestry practices, as well as an intense research program, including heavy devices to collect soil water [8, 10-12]; in Morvan, three different clearing levels on plantations of the same age, and an adjacent beech forest, representative of the previous ecosystem. Thus, the parallel development of research on the two sites is promising to determine the contribution of local factors, of the change of vegetation cover itself, and of the kind of forest management.

It was observed that the changes in humification degree of SOM under Douglas fir were not strongly influenced by the nature of the bedrock, but mainly by climatic parameters. In both cases, humus-rich horizons were of the "Dysmull" and "Moder" types. Studies in the Beaujolais site showed that humification seemed to decrease slightly with the age of plantation, as suggested by the increasing C/N ratios and titrated acidity of FAs. Studies in the Morvan site indicated that humification was improved under Douglas fir plantation, with respect to the former beech vegetation. Less poorly decomposed litter and coarse particulate SOM were present, lower C/N ratios in the whole soil and in fine humus size-fractions were measured, and lower acidity was titrated on FAs of the topsoils under Douglas fir plantation. Pruning and partial clearing also increased soil pH values and improved the decomposition of the leaf litter, thus activating nutrient recycling. However, no convincing impact on humification parameters could be established statistically.

Even though no podzolic features were found, the amount of raw SOM and the acidity of the FA-type compounds were high enough as to predict a partial mobility of this material. Nevertheless, soluble SOM was abundant mainly in solutions collected with tension-free plate lysimeters beneath the litter and at 0.15 m depth. Based on NMR spectroscopy, this material had a high aliphaticity, except that flowing directly from the litter, which contained also a minor proportion of polysaccharides.

Another way to find out differences between soils developed under different vegetation covers was the isotopic approach, using ^{13}C abundance. In addition to the classical tracing of compounds deriving from plants with different photosynthetic cycles, it was attempted to establish significant and sustainable differences between soil and water OM generated by two different C3 plants. All compounds related to the beech vegetation appeared to be slightly more depleted in ^{13}C than those derived from the Douglas fir plantations, but none of the differences found were confirmed statistically. However, soluble SOM appeared to have intermediate $\delta^{13}C$ values between those of the litter and of the solid SOM. This was also the case for the OM dissolved in the streams of the watersheds under the respective vegetations, again with a slight difference related to the prevailing vegetation. Based on these results, much can be expected from further comparison of the variations of $\delta^{13}C$ values associated to those of DOC in homogeneous watersheds with contrasted vegetation covers. These determinations will probably represent delicate operating conditions, due to the limited range of variations. In addition, this approach will need to be crossed with molecular tracing of biochemical compounds, such as lignin or lipids [32, 37], which could allow us to distinguish between the two vegetation covers.

ACKNOWLEDGMENTS

The authors thank the Scientific Committee of "GIP ECOFOR" and the Conseil Régional de Bourgogne, for granting part of the research presented in this paper. Special thanks are also addressed to Marie-Jeanne Milloux for her analytical assistance and to Pr Y. Lucas, and LEPI at Université de Toulon et du Var, for running NMR spectra.

REFERENCES

1. Stevenson, F.J. (1994). Humus Chemistry: Genesis, Composition and Reactions. 2nd edition. Wiley Interscience, New York.
2. Shulten, H.-R., Leinweber, P., 2000. New insight into organic-mineral particles: composition, properties and models of molecular structure. Biol. Fertil. Soils 30, 399-432.

3. Cozzolino, A., Conte, P., Piccolo, A., 2001. Conformational changes of humic substances induced by some hydroxy-, keto-, and sulfonic acids. Soil Biol. Biochem. 33, 563-571.
4. Goudriaan, J., Ketner, P., 1984. A simulation study for the global carbon cycle, including man's impact on the biosphere. Clim. Change 6, 167-192.
5. Andreux, F., 1996. Humus in world soils. *In:* Piccolo, A. (Ed.), The Role and Action of Humic Substances in Terrestrial Ecosystems. Elsevier Sci. Publ., pp. 45-100.
6. Kögel-Knabner, I., 1992. Biodegradation and humification processes in forest soils. *In:* Bollag, J.-M., Stotzky, G. (Eds.), Soil Biochemistry, Vol. 8. Marcel Dekker, Inc., New York, pp. 101-135.
7. Brinkley, D. 1995. The influence of tree species on forest soils: Processes and patterns. *In:* Mead, D.J., Cornforth, I.S. (Eds.), Proceedings of the Trees and Soil Workshop. 1994. Agronomy Society of New Zealand. Special Publication 10, Lincoln Univ. Press, Canterbury, NZ, pp. 1-33.
8. Augusto, L., Bonnaud, P., Ranger, J., 1998. Impact of tree species on forest soil acidification. For. Ecol. Mngt. 105, 67-78.
9. Zech, W., 1991. Litter decomposition and humification in forest soils. *In:* Van Breemen, N. (Ed.), Decomposition and accumulation of Organic Matter in Terrestrial Ecosystems: Research Priorities. E.E.C., Brussels, pp. 46-51.
10. Ranger, J., Marques, R., Colin-Belgrand, M., Flammang, N., Gelhaye, D., 1995. The dynamics of biomass and nutrient accumulation in a Douglas fir (*Pseudotsuga menziesii* Franco) stand studied using a chronosequence approach. For. Ecol. Mngt. 72, 167-183.
11. Marques, R., Ranger, J., 1997. Nutrient dynamics in a chronosequence of Douglas fir (*Pseudotsuga menziesii* (Mirb.) Franco) stands on the Beaujolais Mounts (France). 1. Qualitative approach. For. Ecol. Mngt. 91, 255-277.
12. Marques, R., Ranger, J., Villette, S., Granier, A., 1997. Nutrient dynamics in a chronosequence of Douglas fir (*Pseudotsuga menziesii* (Mirb.) Franco) stands on the Beaujolais Mounts (France). 2. Quantitative approach. For. Ecol. Mngt. 92, 167-197.
13. Dekkers, J.A., Spaargaren, O.C., Nachtergaele, F.O., Oldeman, L.R., Brinkman, R., (Eds.), 1998. World Reference Base for Soil Resources, World Soil Resources Reports 84, FAO, ISRIC and UISS.
14. Chrétien, J., Ranger, J., Villette, S., 1997. Modification au cours de la révolution forestière des caractères physiques des sols sous plantation de Douglas (*Pseudotsuga menziesii* Franco). Etude Gest. Sols 4, 127-140.
15. Brêthes, A., 1973. Mode l'altération et de différenciation pédogénétique sur leucogranites du Morvan, comparaison avec le massif Vosgien. Thesis, Univ. Nancy I, France.
16. Lévêque, J., Lamère, J.-C., Villemin, G., Chrétien, J, Andreux, F., 1998. Variations of humus with human activity in acid soils of the Morvan natural park (France). *In:* Proceedings of 16th World Congress of Soil Science, Montpellier (France), Symposium 34, Comm. N° 1778., A.F.ES., I.S.S.S., CD-ROM, Cirad, Montpellier, France.
17. Linglois, N., Amiotte Suchet, P., Lévêque, J., Andreux, F., 2000. Dissolved organic carbon contents and ^{13}C variations in streams of small catchments with contrasted vegetations (Morvan, France). Proceedings of 10th International Meeting of the International Humic Substances Society, IHSS-10, Toulouse, France, Vol. 2, pp. 735-738.
18. Andreux, F., Bruckert, S., Correa, A., Souchier, B., 1980. Sur une méthode de fractionnement physique et chimique des sols: origines possibles de la matière organique des fractions obtenues. C.R. Ac. Sci. Paris, 291D, 381-384.

19. Swift, R.S., 1996. Organic matter characterization *In:* Sparks D.L., Page, A.L., Helmke P.A, Loeppert, R.H., Soltanpour, P.N., Tabatabai, M.A., Johnson, C.T., Sumner, M.E. (Eds.), Methods of Soil Analysis. Part 3. Chemical Methods. Soil Sci. Soc. Am. Book Series: 5. Soil Sci. Soc. Am. Madison, WI, pp. 1018-1020.
20. Robert, M., Veneau, G., Abreu, M.M., 1987. Etudes microscopiques d'associations aluminium-argiles ou fer-argiles. *In:* Fedoroff, N., Bresson, L. M., Courty, M.-A. (Eds.), Micromorphologie des sols. Microscopie des sols. AFES, Plaisir, France, pp. 67-474.
21. Feller, C., Tessier, D., 1996. Aggregation and organic matter storage in kaolinitic and smectitic tropical soils. *In:* Carter, M.R., Stewart, B.A. (Eds.). Structure and organic matter storage in agricultural soils. Advances in Soil Science, Lewis Publishers, Boca Raton, FL, pp. 309-358.
22. Goldstein, J.I., Newbury, D.E., Echlin, P., Joy, D.C., Roming, A.D., Lyman, C.E., Fiori, C., Lifshin, E., 1992. Scanning electron microscopy and X-ray microanalysis. 2nd Edition, Plenum Press, New York.
23. Orsini, L., Rémy, J.C., 1976. Utilisation du chlorure de cobaltihexammine pour la détermination simultanée de la capacité d'échange cationique et des bases échangeables des sols. Sci. Sol 4, 269-275.
24. Koutika, L.-S., Bartoli, F., Andreux, F., Cerri, C.C., Burtin, G., Choné, T., Philippy, R., 1997. Organic matter dynamics and aggregation in soils under rain forest and pastures of increasing age in the Eastern Amazon Basin. Geoderma 76, 87-112.
25. Gran, G., 1952. Determination of the equivalence point in potentiometric titration. Part II, Analyse (London), 77, p. 661.
26. Bizri, Y., Cromer, M., Scharff, J.P., Guillet, B., Rouiller, J., 1984. Constantes de stabilité de complexes organo-minéraux. Interactions des ions plombeux avec les composés organiques hydrosolubles des eaux gravitaires de podzol. Geochim. Cosmochim. Acta 48, 227-234.
27. Brunelot, G., Adrian, P., Rouiller, J., Guillet, B., Andreux, F., 1989. Determination of dissociable acid groups of organic compounds extracted from soils, using automated potentiometric titration. Chemosphere 19, 1413-1419.
28. Nissenbaum, A., Shallinger, K. M., 1974. The distribution of the stable carbon isotopes ($^{13}/^{12}C$) in fractions of soil organic matter. Geoderma 11, 137-145.
29. Deines, P., 1980. The isotopic composition of reduced organic carbon. *In:* Fritz, P., Fontes, J.C. (Eds.), Handbook of Environmental Isotope Geochemistry, Vol. 1, Elsevier, Amsterdam, pp. 329-406.
30. Balesdent, J, Mariotti, A., 1996. Measurement of soil organic matter turnover using ^{13}C natural abundances. *In:* Boutton, T.W., Yamasaki, S.I. (Eds.), Mass Spectrometry of Soils, Marcel Dekker Inc., New York, pp. 83-111.
31. Andreux, F., Vose, P.B., Cerri, C.C., Vitorello, V.A, 1990. Potential of stable isotope ^{15}N and ^{13}C methods for determining input and turn-over in soils. *In:* Harrisson, A.F., Ineson, P., Heal, O.W., (Eds.), Field Methods in Nutrients Cycling. Elsevier Appl. Sci. Pub., pp. 259-275.
32. Bol, R.A., Harkness, D.D., Huang, Y., Howard D. M., 1999. The influence of soil processes on carbon isotope distribution and turnover in the British uplands. Europ. J. Soil Sci. 50, 41-51.
33. Feller, C., Burtin, G., Gérard, B., Balesdent, J., 1991. Utilisation des résines sodiques et des ultra-sons dans le fractionnement granulométrique de la matière organique des sols. Intérêt et limites. Sci. Sol 29, 77-93.

34. Roux, F., 1998. Étude de la matière organique hydrosoluble du sol et de son rôle dans les cycles biogéochimiques sous peuplements de résineux. DEA dissertation, University of Nancy I, France.
35. Amiotte-Suchet, P., Aubert, D., Probst J.L., Gauthier-Lafaye F., Probst, A., Andreux, F., and Viville, D., 1999. ^{13}C pattern of dissolved inorganic carbon in a small granitic catchment: the Strengbach case study (Vosges, France). Chem. Geol. 159, 129-145.
36. Lelong, F., Durand, P., Didon, J.-F., 1988. Comparaison des bilans hydrochimiques, des taux d'altération et d'acidification dans trois petits bassins versants granitiques à végétation contrastée (Mont Lozère, France). Sci. Géol., Bull. 41, 263-278.
37. Maman, O, Marseille, F., Guillet, B., Disnar, J. R., Morin, P., 1996. Separation of phenolic aldehydes, ketones and acids from lignin degradation by Capillary Zone Electrophoresis. J. Chromatog. A. 755, 89-97.

Developments in Soil Science, Volume 28A
Editors: A. Violante, P.M. Huang, J.-M. Bollag and L. Gianfreda

EFFECT OF CONCENTRATION ON THE SELF-ASSEMBLING OF DISSOLVED HUMIC SUBSTANCES

P. Conte and A. Piccolo[a]

Dipartimento di Scienze Chimico - Agrarie, via Università 100, 80055 Portici, Italy

[a] Corresponding Author; Tel. +39 081 7885239; E-Mail: alpiccol@unina.it.

By using high performance size exclusion chromatography (HPSEC), conformational changes of humic substances were observed with increases of humic solution concentration. Weak dispersive forces are the cause of attraction among humic molecules when in dilute solutions. Conversely, electrostatic repulsion between negatively charged humic molecules occurs as the concentration increases. This behavior is consistent with the model that describes humic substances as supramolecular associations of self-assembling small molecules rather than as macromolecular polymers or regular micelles. In fact, the concentrations used to perform the experiments were larger than 0.05 $g.L^{-1}$, that is the concentration above which conformational changes should not be observed, according to the model of humic macropolymeric coils. Moreover, all concentrations used were well below the humic critical micellar concentration of 10 $g.L^{-1}$ that is reported to be the critical micelle concentration (CMC) at which humic substances may start to form micelles in solution.

1. INTRODUCTION

Humic substances (HS) are an important class of natural organic compounds involved in controlling both the fate of environmental pollutants and the biogeochemistry of organic carbon in ecosystems [1]. Despite the obvious importance that these substances have in sustaining life, their basic chemical nature and reactivity are still poorly understood.

The conformational behavior of HS in solution is dependent on pH and ionic strength of the solvent and on the concentration of HS [2, 3]. HS are viewed as shrinking coils at increasing pH or decreasing ionic strength values [3]. On the other hand, many authors studied the effect of concentration on the molecular size of HS [2-5] and verified that molecular sizes of HS increase with their concentration. Such behavior was related to a macropolymeric nature of HS whereby intermolecular H-bonds and hydrophobic interactions increase the molecular size of humic macromolecular units. The concentration of 0.05 $g.L^{-1}$ represents the limit between the supposed coiled and the linear colloidal shape of HS [3], whereas 10 $g.L^{-1}$ is reported as the critical micelle concentration (CMC) at which HS [4] begin to form micellar structures in solution [6].

Wershaw [7] proposed for the first time that HS are macropolymers behaving in solution as ordered structures held together by weak hydrophobic interactions, such as in micelles or membranes. However, new understandings [8-15] described HS as supramolecular structures being composed by random associations of relatively small components and stabilized mainly by hydrophobic interactions. The size of the humic components appears to be much smaller than the 10,000-1,000,000 D commonly reported in literature [16]. Moreover, the random self-assembling of small components in supramolecular conformations implies that hydrophilic sites are contiguous to or surrounded by hydrophobic domains and that the micropolar domains may also contain some trapped hydration water [17].

The aim of this work was to study by high performance size exclusion chromatography (HPSEC) the effect of concentration of dilute humic solution on the molecular size distribution of HS samples.

2. MATERIALS AND METHODS

2.1. Humic substances

Three humic acids (HAs) were isolated from different raw materials: HA_1 from an agricultural soil (Typic Euthrochrept) near Roskilde (Denmark), HA_2 from a North Dakota Leonardite (Mammoth, Int. Chem. Co.), and HA_3 from an oxidized coal provided by Eniricerche SpA (Italy). The HAs were extracted and purified as previously described [11].

All materials were analyzed for their ash and moisture content by thermogravimetric analysis (TGA) on a Dupont 900 apparatus. Moisture content was the measured weight loss at 105°C, and it was below 2% for all the humic substances. Ash content was the remaining weight at 750°C. It was below 1% for all HAs. Elemental analysis (Table 1) was conducted with an Interscience EA1108 CHN elemental analyzer.

Table 1.
Elemental analyses (on ash- and moisture-free bases) of humic acids

Samples	C (%)	H (%)	N (%)	C/H	C/N
HA_1	47.4±0.4	5.0±0.6	4.4±0.4	4.4	10.8
HA_2	45.9±0.7	3.7±0.5	1.0±0.1	12.4	45.9
HA_3	48.0±0.5	3.0±0.7	1.0±0.1	16.0	48.0

Purified HA_1, HA_2, and HA_3 samples (50 mg) were first suspended in distilled water (50 mL). The suspensions were titrated to pH 7.0 with a CO_2-free solution of 0.5 M NaOH by automatic titrator (VIT 90 Videotitrator, Radiometer, Copenhagen) under N_2 atmosphere and stirring. After having reached the constant pH 7, the solution containing sodium humates was left under titration for two more hours, filtered through a Millipore 0.45 μm filter, and freeze-dried. The acidity at pH 7 of the three humic acids was 3.04, 3.51 and 3.84 meq.g^{-1}, respectively for HA_1, HA_2, and HA_3. Sodium humates pretitrated at pH 7 were used to exclude the random occurrence of negative charges on solute molecules when dissolved into the HPSEC mobile phase and to depress ionic exclusion phenomena. Humic solutions for HPSEC analyses were then prepared by dissolving few milligrams of each sodium humate in the HPSEC eluent to obtain the following concentrations: 1.5, 1.2, 1.0, 0.9, 0.8, 0.5, 0.3, 0.2, and 0.1 g.L^{-1}.

2.2. NMR spectroscopy

Cross polarization magic angle spinning carbon-13 nuclear magnetic resonance (CPMAS ^{13}C-NMR) experiments were carried out on a Bruker AMX400 instrument operating at 100.625

MHz on carbon-13. A recycle time of 1 sec and an acquisition time of 13 msec were used. All the experiments were conducted with variable contact time (VCT) pulse sequence in order to find the optimum contact time (OCT) for each sample and to minimize the error on the quantitative evaluation of carbon content [18]. OCT ranged between 0.8 and 1.0 msec. A line broadening (LB) of 50 Hz was used to transform all the free induction decay (FID) curves. The area in the 110-160 ppm region was corrected for that of the side band of the signal of the carboxyl groups. This correction was made by measuring the area of the side band in the 190-230 ppm region and by subtracting this area from that of the 110-160 ppm region. The areas of each region of the spectra in Table 2 were attributed to non polar carbons, such as the aliphatic (0-45 ppm) and aromatic (110-160 ppm) ones, and to polar carbons, such as the C-O and C-N groups, the anomeric carbons (45-110 ppm) and the carboxyl carbons (160-190 ppm). The areas of the 0-45 and 110-160 ppm regions were used to calculate the hydrophobicity (HB) of the HAs, whereas those of the 45-60, 60-110, and 160-190 ppm regions were used to obtain the hydrophilicity (HI) of the HAs. The HI/HB ratios are also given in Table 2.

Table 2.
Distribution (%) of 13-carbons in resonance intervals (ppm) of CPMAS-NMR spectra and HI/HB ratios of HAs

	0-45 ppm	45-60 ppm	60-110 ppm	110-160 ppm	160-190 ppm	HI	HB	HI/HB
HA_1	34.7	16.7	18.8	13.9	18.1	53.5	48.6	1.1
HA_2	25.4	8.45	18.0	39.4	16.6	43.1	64.7	0.66
HA_3	22.1	9.84	16.4	36.9	23.4	49.5	59.0	0.84

2.3. HPSEC system

The HPSEC system consisted of a high pressure Perkin-Elmer LC200 solvent pump and a UV-Vis detector (Perkin-Elmer LC295) set at 280 nm. A Rheodyne rotary injector, equipped with a 100 μL sample loop, was used to load the sample solutions. Size exclusion separation occurred through a TSK (Toso Haas) G3000SW (600mm x 7.5 mm i.d.) column. The column was preceded by a 7.5 cm TSK guard-column (7.5 mm i.d.) packed with G3000SW stationary phase and by a 0.2 μm stainless-steel inlet filter, and it was thermostated at 25°C by a water bath. Polysaccharides (Polymer Sciences Laboratories, UK) of known molecular weights (100, 48, 23.7 and 12.2 kD) were used to calibrate the HPSEC column. The flow rate was set at 0.6 $mL.min^{-1}$ and the HPSEC eluent was a 0.05 M $NaNO_3$ and 4.0 x 10^{-3} M NaN_3 solution (the latter as a bacteriostatic agent). The mobile phase was made with MilliQ water and HPLC-grade reagents, filtered through a Millipore 0.45 μm filter and He-degassed. The void volume (V_0=11.18 mL) and the total permeation volume (V_t=20.57 mL) of the column were determined using Blue Dextran 2000 and water, respectively. To see the water peak, a refractive index detector (Refractomonitor IV, Fison's Instruments) was used.

2.4. Molecular weight determination

Size exclusion chromatograms were evaluated by using Perkin-Elmer-Nelson Turbochrom 4-SEC integration and molecular weight software. A SEC noise threshold of 5 and a filter size of 5 for the Savitzky-Golay smoothing were used. Calculation of weight- (M_w) and number-

(M_n) averaged molecular weights was done by the method of Yau et al. [19] using the following equations:

$$M_w = \sum_{i=1}^{N} h_i(M_i) / \sum_{i=1}^{N} h_i \qquad M_n = \sum_{i=1}^{N} h_i / \sum_{i=1}^{N} h_i/(M_i)$$

where M_i and h_i are the molecular weight and the height, respectively, of the i-th chromatographic slice in the chromatogram of each sample eluted at volume i. Figure 1 reports the variation of M_w and M_n values of the three HAs with the changes of their concentration. The relative standard deviation of calculated values among triplicates of each chromatogram varied only to a maximum of 5%, thereby confirming previous evaluation of reproducibility [10, 11]. The M_w and M_n values were used to calculate the humic polydispersity (P) as $P=M_w/M_n$. Polydispersity is a measure of the homogeneity of organic macromolecules or aggregates. The higher the P value, the higher is the number of polymers having the same length or the larger is the amount of aggregates having different sizes.

The total acidity ($meq.g^{-1}$) of HAs obtained by titration was divided by either M_w or M_n values provided by HPSEC measurements to obtain the charge-to-mass (C/M) ratio for each HA. The C/M ratio may be used as a measure of the charge density at pH 7.0 of humic substances.

3. RESULTS AND DISCUSSION

Table 2 shows that the hydrophilicity/hydrophobicity ratio (HI/HB) of the three HAs varies in the order $HA_1 > HA_3 > HA_2$. Such differences depend on the different chemical compositions of humic acids. In fact, notwithstanding the highest aliphatic carbon content (34.7%) of HA_1, the amount of polar groups, such as C-N and C-O, was the largest (16.7%, and 18.8%, respectively), thereby resulting in the highest HI/HB value (Table 2). HA_2 revealed the largest aromatic carbon content (39.4%), which, together with aliphatic carbons (25.4%), produced the highest hydrophobicity (64.7%) and the lowest HI/HB ratio (Table 2). The HA_3 from oxidized coal showed the largest content of carboxylic carbons (23.4%), whereas the content of C-O groups was the lowest (16.4%). Finally, the amount of C-N moieties was intermediate (9.84%) between that of HA_2 (8.45%) and HA_3 (16.7%). The final result was an intermediate value of HI/HB for HA_3(Table 2).

Since HA_2 and HA_3 were extracted from stable organic sources, such as lignite and oxidized coal, without recent contribution of plant and microbial cells [20], it can be assumed that their nitrogenated carbons are mainly heterocyclic, whereas oxidized carbons may be mainly composed of ether (C-O-C) functions. A consequence is that the measurement of hydrophilicity that commonly includes the oxidized carbons [21] may be even lower than that reported in Table 2 for HA_2 and HA_3. In fact, while ethers are computed within the hydrophilic functions because of their oxygenated nature, they are not readily soluble in water and should be instead considered as truly hydrophobic humic components.

According to the model of Ghosh and Schnitzer [3] that describes humic substances as macromolecular polymers, we should have expected no variations of humic conformation and, hence, of M_w and M_n values, with an increase of concentration from 0.1 to 1.5 $g.L^{-1}$. HS are believed to behave as coiled macropolymers regardless of pH and ionic strength as their concentration exceeds 0.05 $g.L^{-1}$ [3, 22]. Furthermore, the model predicts that at constant pH and when the humic concentration is below 0.05 g.L^{-1}, the shape of HS changes from linear to coiled with an increase in ionic strength. On the other hand, when ionic strength is constant and pH increases, the HS conformation stretches out from a coiled to a linear shape.

The more recent model of Wershaw [7] and Engebretson and von Wandruszka [17] proposed a micelle-like behavior for HS. Due to the random distribution of hydrophilic moieties, hydrophilic microdomains can be present in hydrophobic phases of humic polymeric aggregates. Engebretson and von Wandruszka [23] also defined a humic critical micellar concentration (CMC) of 10 g.L^{-1} as the concentration above which micelle formation begins. The postulate was that no humic conformational changes should be observed below 10 g.L^{-1}.

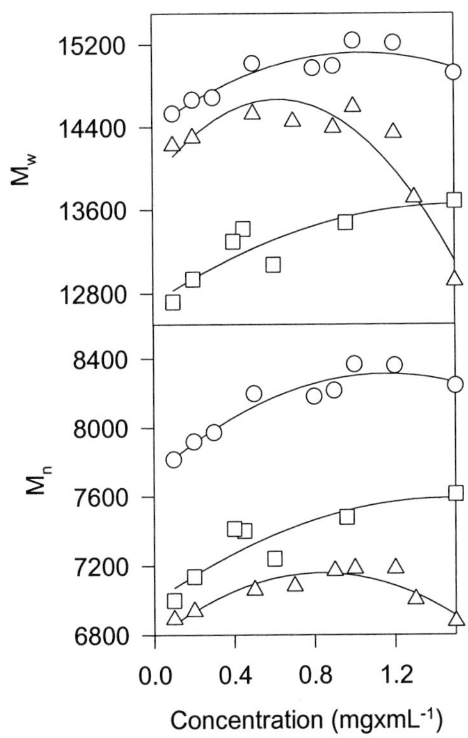

Figure 1. M_w and M_n changes of HA_1 (O), HA_2 (□), and HA_3 (Δ) with the humic concentration.

The mechanisms involved in the formation of micelles is described by Tanford [6] and Israelachvili [24]. These authors report that simple amphiphilic molecules, such as sodium alkylsulfates or alkyl trimethylammonium halides, enhance their size in forming micelles with a highly cooperative process when the concentration increases within a range above their CMC value. Macropolymers, such as proteins, also behave as micelles when a transition from a denatured state to a native conformation occurs [24]. In particular, increase in molecular size can be observed when the concentration of proteins is above their CMC value.

Our results indicate that the behavior of the three HAs used in this study cannot be described by either the Ghosh and Schnitzer [3] or the Wershaw [7] model. While these models predict constant M_w, M_n, and P values when humic concentrations are either larger than 0.05 $g.L^{-1}$ [3] or lower than the CMC value of 10 gL^{-1} [7, 17, 23], Figures 1 and 2 show that significant conformational changes occurred when the humic concentration was increased from 0.1 to 1.5 $g.L^{-1}$. In fact, HA_1 and HA_3 revealed a maximum of M_w and M_n when the concentrations were 1.1 and 0.8 $g.L^{-1}$, respectively, whereas HA_2 did not show any maximum (Figure 1). The latter material produced only a tendency of M_w and M_n increase within the concentration range used for the HPSEC experiments. Moreover, Figure 2 indicates a progressive decrease of polydispersion values for the three HAs, thereby suggesting that the molecular homogeneity of humic associations increases with concentration.

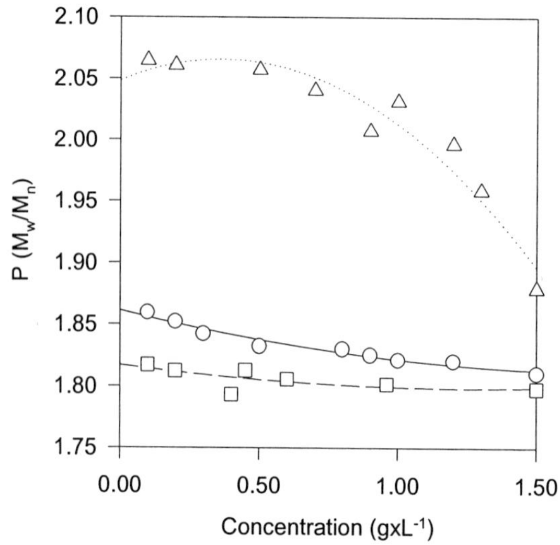

Figure 2. Polydispersity changes of HA_1 (O), HA_2 (□), and HA_3 (Δ) with the humic concentration.

Our findings may be more adequately explained by a model that describes humic substances as supramolecular associations of self-assembling smaller molecules [8-15]. By this view, humic molecules may arrange themselves in supramolecular associations depending on their concentration and electrical properties in solution. In particular, hydrophobic attractive forces, such as the van der Waals, π–π, and CH-π bondings [25], prevail in extreme dilution conditions and an increase in apparent molecular size can be observed. According to Israelachvili [24], the strength of attractive forces depends on d^{-6}, where d is the distance between the molecules. Hence, when humic concentration is progressively increased, the distance between humic molecules decreases and an enhanced association of apparent larger molecular size can be obtained. On the other hand, when the distance among different humic components is reduced due to an increase in concentration, the negatively charged sites, which are present on humic conformations at pH 7, produce an electrostatic repulsion that overcomes the hydrophobic aggregating forces, thereby leading to a decrease in molecular size of the humic association.

Figures 3 reports the changes of charge-to-mass (C/M) ratios with increasing HA concentration and appears to support further the aggregating/disaggregating mechanism based on the supramolecular model proposed by Piccolo and coworkers [8-15]. Although HA_1 and HA_3 showed a definite minimum for either C/M_w or C/M_n parameters, the C/M ratios for HA_2 did not reach a minimum within the same concentration range. This can be explained by considering that as the molecular size of humic associations increases, their charge density (C/M ratio) decreases due to the lowering of conformational energy.

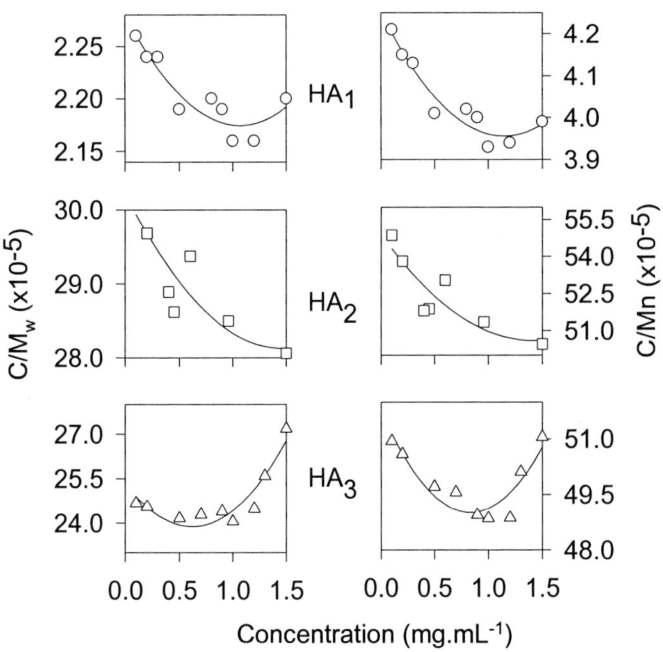

Figure 3. Charge-to-mass ratios as affected by humic concentration.

A lower conformational energy can be reached in solution when negative charges are positioned far from each other and cause a decrease of charge density values. However, it becomes progressively difficult to reach conformational arrangements in which negative charges are effectively separated when humic concentrations are increased. Consequently, repulsive forces prevail over the attractive hydrophobic forces, and when a concentration limit is reached (i.e., 1.1 and 0.8 gL^{-1} for HA_1 and HA_3, respectively) a conformational disruption and dispersion of smaller associations may occur. The consequence may be an increasing value of the C/M ratios, as shown in Figure 3.

4. CONCLUSIONS

This work seems to support the description of HS as supramolecular associations of relatively small molecules rather than macromolecular polymers but also suggests that the aggregation of humic constituents does not lead to ordered micelles. In fact, the conformational behavior of the humic solutions studied here was different from that expected for regular micelles [6, 25].

Based on our findings, we attempted to describe the driving forces involved in the formation of supramolecular structures of HS in solution. Hydrophobic attractive forces appear to prevail in dilute humic solutions and result in an apparent increase of molecular sizes of humic samples with increasing concentration. Electrostatic repulsion becomes predominant when humic solutions are progressively concentrated and the apparent molecular size is reduced. The transition from attraction depends on the charge density on the HA. The different behavior shown by the various humic materials was related to their ultimate molecular composition and the resulting homogeneity among components. The larger the homogeneity of the humic association, the larger is the molecular size that may be reached in solution.

ACKNOWLEDGMENTS

This work was partially supported by the Italian Ministry of University and Scientific and Technological Research.

REFERENCES

1. Piccolo, A., 1996. Humus and soil conservation. *In*: Piccolo, A. (Ed.), Humic Substances in Terrestrial Ecosystems. Elsevier, Amsterdam, pp.225-264.
2. Greenland, M., Hayes, M.H.B., 1978. The Chemistry of Soil Constituents Wiley Interscience, New York.
3. Ghosh, K., Schnitzer, M., 1980. Macromolecular structure of humic substances. Soil Sci. 129, 266-276
4. von Wandruszka, R., Schimpf, M., Hill, M.. Engebretson, R.. 1999. Characterization of humic acid size fractions by SEC and MALS. Org. Geochem 30, 229-232
5. Jones, M.N., Bryan, N.D., 1998. Colloidal properties of humic substances. Adv. Coll. Interf. Sci. 78, 1-34
6. Tanford, C., 1991. The Hydrophobic Effect: Formation of Micelles and Biological Membranes. Krieger Publishing Company, Malabar, Florida.
7. Wershaw, R.L., 1986. A new model for humic materials and their interactions with hydrophobic organic chemicals in soil, water, or sediments. J. Contam. Hydrol. 1, 29-33.
8. Piccolo, A., Nardi, S., Concheri, G., 1996. Micelle-like conformation of humic substances as revealed by size exclusion chromatography. Chemosphere 33, 595-600.
9. Piccolo, A., Nardi, S., Concheri, G., 1996. Macromolecular changes of humic substances induced by interactions with organic acids. Europ. J. Soil Sci. 47, 319-325.
10. Conte, P., Piccolo, A., 1999. High pressure size exclusion chromatography (HPSEC) of humic substances. Molecular sizes, analytical parameters, and column performance. Chemosphere 38, 517-523.

11. Conte, P., Piccolo, A., 1999. Conformational arrangement of humic substances. Influence of solution composition on association of humic molecules. Environ. Sci. Technol. 33, 1682-1690.
12. Piccolo, A., Conte, P., Cozzolino, A., 1999. Effects of mineral and monocarboxylic acids on the molecular association of dissolved humic substances. Europ. J. Soil Sci. 50, 587-593.
13. Piccolo, A., Conte, P., 2000. Molecular size of humic substances. Supramolecular associations versus macromolecular polymers. Adv. Environ. Res. 3, 508-523.
14. Piccolo, A., Conte, P., Cozzolino, A., 2001. Chromatographic and spectrophotometric properties of dissolved humic substances compared with macromolecular polymers. Soil Sci. 166, 174-180.
15. Cozzolino, A., Conte, P., Piccolo, A., 2001. Conformational changes of humic substances induced by some hydroxy-, keto-, and sulphonic- acids. Soil Biol. Biochem. 33, 563-570.
16. Swift, R.S., 1989. Molecular weight, shape, and size of humic substances by ultracentrifugation. *In*: Hayes, M.H.B., McCarthy, P., Malcolm, R.L., Swift, R.S. (Eds.), Humic Substances II. In Search of Structure, Wiley, New York, pp. 449-460.
17. Engebretson, R.R., von Wandruszka, R., 1994, Micro-organization of dissolved humic acids, Environ. Sci. Technol. 28, 1934-1938.
18. Conte, P., Piccolo, A., van Lagen, B., Buurman, P, de Jager, P.A., 1997. Quantitative aspects of solid state ^{13}C-NMR spectra of humic substances from volcanic systems. Geoderma 80, 327-335.
19. Yau, W.W., Kirkland, J.J.. Bly, D.D., 1979. Modern Size Exclusion Chromatography. Wiley Interscience, New York.
20. Hatcher, P.G., Breger, I.A., Maciel, G.E., Szeverenyi, N.M., 1985, Geochemistry of humin. *In*: Aiken, G., McKnight, D.M., Wershaw, R., MacCarthy, P. (Eds.), Humic Substances in Soil, Sediment, and Water. Wiley Interscience, New York, pp. 275-295.
21. Piccolo, A., Conte, P., 1998. Advances in nuclear magnetic resonance and infrared spectroscopies of soil organic particle. *In*: Huang, P.M., Senesi, N., Buffle, J. (Eds.), Structure and Surface Reactions of Soil Particles, Wiley Interscience, New York, pp. 375-435.
22. Swift, R.S., Posner, A.M., 1971. Gel chromatography of humic acid. Soil Sci. 22, 237-240.
23. Engebretson, R., von Wandruszka, R., 1997. The effect of molecular size on humic acid associations. Org. Geochem. 26, 759-765.
24. Israelachvili, J., 1992. Intermolecular and Surface Forces, Academic Press, London.
25. Nishio, M., Hirota, M., Umezawa, Y., 1998, The CH/π interaction. Evidence, Nature, and Consequences, J. Wiley Interscience, New York.

Developments in Soil Science, Volume 28A
Editors: A. Violante, P.M. Huang, J.-M. Bollag and L. Gianfreda

PORE SIZE CHANGES IN A LONG-TERM FIELD EXPERIMENT WITH ORGANIC AMENDMENTS

H. Kirchmann[a] and M. H. Gerzabek[b]

[a]Swedish University of Agricultural Sciences, Department of Soil Sciences, Box 7014, S-750 07 Uppsala, Sweden

[b]Austrian Research Centers Seibersdorf, A-2444 Seibersdorf, Austria

The long-term application of organic materials changed the pore size distribution in soil. Treatments with calcium nitrate, peat and sewage sludge increased the volume of macropores. Changes in organic C in soil mainly affected meso- and micropores, whereas the finest micropore fraction (< 1 μm diameter) was not affected. The volume of the micropore fraction (1-5 μm diameter) was most significantly correlated with increasing soil C concentrations ($P < 0.001$; $R^2 = 0.918$).

1. INTRODUCTION

Soil pores are the living space for soil organisms and the size of pores determines where and to what extent bacteria, fungi or soil animals can congregate [1-3]. Our knowledge about the role of organic matter on the size distribution of soil pores is very limited, particularly concerning the effect on micropores. This may be explained by the fact that in the past the matric suction intervals were too large when determining pore sizes from water retention characteristics and small intervals within the micropore range were not considered.

Knowledge about the pore size distribution in soil may provide a mechanistic understanding for some soil biological reactions. We attempted in this paper to find out if there is a relationship between soil organic carbon and the volume of different pore size fractions. We compared the impact of different organic matter contents on the size of pores in a clay loam. Both macro-, meso- and micropores were determined in order to get a better understanding of the role of organic matter content on pore size distribution.

2. MATERIALS AND METHODS

Samples for this study were taken from the Ultuna long-term organic matter experiment, located in Uppsala, Sweden, in autumn 1997, 41 yr after the establishment (1956) of the field plots (4). The soil (Typic Eutrochrept) is clay loam with 36.5% clay and 41% silt formed from postglacial clay. The experiment is characterized by the application of organic matter amounting to 2000 kg C ha^{-1} yr^{-1} using a range of different organic materials.

Four replicate soil cores of each treatment were sampled with cylinders of 50 mm height and

72 mm diameter between 2.5 and 7.5 cm soil depth. Cylinders were used for determination of water retention characteristics at different water potentials (0.05, 0.5, 1, 2, 3, 6, 10 and 30 m). The distribution of pore space according to size of pores was derived from the relation of water content to suction. The equivalent pore diameter d (in meters) was calculated using the equation, $d = 3.0 \times 10^{-5}\ s^{-1}$, where s is the suction in meters [5]. Total porosity of the samples was determined from measurements of particle density and bulk density [6-7]. Organic carbon and total nitrogen of the topsoil was determined in dried and sieved (2 mm mesh size) samples through dry combustion (LECO CNS 2000). Data were also calculated on a volumetric basis for the purpose of correlation to pore volumes. Soil physical and organic matter data were subjected to statistical analysis, using the SAS package [8].

3. RESULTS

Organic C ranged from 1.0% in the fallow to 3.2% in the peat treatment, and bulk densities ranged from 0.96 kg dm^{-3} to 1.28 kg dm^{-3} (Table 1). Due to the low N content and low decomposition rate of peat and sawdust, C/N ratios in these treatments increased to 18 and 14, respectively, compared to around 10 in the other treatments. The long-term manure amendments to soil resulted in significant changes of the soil porosity, ranging from 52% in the fallow treatment to 63% in the sludge amended one. There was a significant, positive correlation between total porosity and soil C concentrations ($R^2 = 0.77$). Changes of soil C concentration from 1 to 3% caused an increase in porosity from 52 to 63%. This increase in porosity affected water storage only by 2 mm, as calculated from water retention data (not shown). Thus, the effect of soil organic matter on water storage capacity was very small.

The distribution of macro-, meso- and micropores is shown in Table 2. Manure amendments caused significant changes ($P < 0.05$) in the proportions of micro-, meso-, and macropores. The portion of coarse to fine macropores (> 600 μm) was largest in the calcium nitrate treatment (13.4 vol %). This treatment also had the highest pH value (7.0).

Table 1
Organic matter, bulk density and porosity of the Ultuna topsoil in 1997, 41 years after the start of four replicate field plots (± standard deviation)

Treatment	Organic C ($g\ kg^{-1}$soil)	Total N ($g\ kg^{-1}$soil)	C/N	Bulk density ($kg\ dm^{-3}$soil)	Porosity (%)	pH (H_2O)
Fallow	10.00±0.2 g	1.00±0.0 f	10.0	1.28±0.00 a	52.0±0.3 e	6.2±0.18 a
No-N	12.05±0.4 gf	1.17±0.0 e	10.3	1.25±0.03 ab	53.2±2.0 de	6.5±0.04 c
$Ca(NO_3)_2$	14.00±0.3 ef	1.37±0.0 d	10.2	1.14±0.02 c	57.1±0.8 bc	7.0±0.04 a
Straw	16.05±0.4 de	1.45±0.0 d	11.1	1.19±0.33 abc	55.1±2.5 cde	6.5±0.07 c
Green manure	17.15±0.4 cd	1.65±0.0 c	10.4	1.19±0.00 abc	54.6±0.7 cde	6.2±0.05 d
Animal manure	21.05±0.3 b	2.00±0.0 b	10.7	1.16±0.01 bc	55.8±1.2 cd	6.7±0.04 b
Peat	32.05±2.7 a	1.75±0.1 c	18.3	1.03±0.03 d	59.9±2.8 ab	5.7±0.11 e
Sawdust	19.80±0.7 bc	1.40±0.0 d	14.1	1.19±0.35 abc	55.3±1.3 cde	6.4±0.11 c
Sewage sludge	29.29±1.1 a	2.92±0.1 a	10.0	0.96±0.02 d	63.3±1.3a	5.4±0.08f

Different letters within columns indicate significance at the 0.05 probability level (Tukey grouping).

Table 2
Pore diameter fractions in the Ultuna topsoil derived from moisture retention characteristics

Treatment	Macropores		Mesopores	Micropores		
	> 600 μm	60-600 μm	30-60 μm	5-30	1-5	< 1 μm
			(Vol %)			
Fallow	12.6 ab	10.9 cd	1.2 c	3.3 c	3.8 e	20.2 bc
No-N	8.3 b	14.7 bc	1.2 c	4.2 b	4.4 de	20.4 bc
$Ca(NO_3)_2$	13.4 a	13.9 bcd	1.3 bc	3.9 bc	4.7cde	19.9 bc
Straw	12.7 ab	10.5 d	1.4 bc	4.1 b	5.2bcd	21.2 bc
Green manure	10.1 ab	12.1 cd	1.3 c	4.5 ab	5.3bcd	21.3 ab
Animal manure	9.3 ab	11.6 cd	1.3 bc	4.6 ab	6.2 ab	22.8 a
Peat	10.4 ab	16.0 b	1.6 b	4.6 ab	6.2 ab	21.1 abc
Sawdust	11.3 ab	11.9 cd	1.3 c	4.2 b	5.6 bc	21.0 bc
Sewage sludge	9.3 ab	20.6 a	2.1 a	4.9 a	6.7 a	19.7 c

Different letters within columns indicate significance at the 0.05 probability level (Tukey grouping).
Soil water suction and equivalent pore diameter: 0.05 m = 600 μm; 0.5 m = 60 μm; 1.0 m = 30 μm; 6.0 m = 5 μm; 30 m =1 μm.

The increase of pore volume with soil C concentrations was largest in the macropore fraction (60-600 μm). The volume of these pores differed from 10.9 vol % in the fallow to 20.6 vol % in the sludge treatment, which means a relative difference of 90%. The increase of pore volume in the mesopore fraction (30-60 μm) was much smaller (1.2-2.1 vol %). The volume of the micropore fraction (5-30 μm) was larger (3.3-4.9 vol %) than that of mesopores, but the largest relative difference due to treatments was only around 50%. Again, the greatest micropore volume was present in the sewage sludge-treated soil and least in the permanently fallowed soil. The volume of the micropore fraction (1-5 μm) ranged from 3.8 - 6.7 vol %, which means a relative difference of 76 %. Concerning the volume fraction of pores less than 1 μm, small absolute and relative differences were found.

There was a significant positive correlation between soil organic C and the total volume of pores ($P < 0.05$; Table 3). Correlations of soil organic C with the volume of certain pore sizes was also found to be significant. The highest correlation was found between micropores of 1-5 μm diameter and concentrations of soil C ($P < 0.001$; $R^2 = 0.918$). However, the smallest micropore fraction was not correlated with soil C concentration. These results indicate a non-uniform distribution of organic carbon in soil pores.

4. DISCUSSION

According to our results, soil organic matter is "concentrated" in micropores (1-5 μm). Soil organic matter may be located as discrete, particulated organic matter or may be sorbed onto the surface of pore walls. In a review by Christensen [9], it was shown that the proportion of organic C generally increased in finer-sized aggregates. If one assumes that the size of soil

Table 3
Correlations between soil porosity classes and soil organic carbon (on volumetric basis)

Variable (y) (Vol %)		Organic C (x) ($g\ dm^{-3}$ soil)
Total pores		$y = 47.44 + 0.412\ x$ *
Macropores	d > 600 μm	NS
	d = 60 - 600 μm	NS
Mesopores	d = 30 - 60 μm	$y = 0.781 + 0.029\ x$ *
Micropores	d = 5 - 30 μm	$y = 3.019 + 0.057\ x$ **
	d = 1 - 5 μm	$y = 2.525 + 0.131\ x$ ***
	d=<1	NS

Significance: * = $P < 0.05$; ** = $P < 0.01$; *** = $P < 0.001$ (F-test); NS = not significant.

aggregates and soil pores is related simply the smaller the aggregates the smaller are the pores our results are corroborated by studies on particle fractionation.

Thus, there is a non-uniform distribution of soil organic matter together with micropores. The mechanism for this distribution is not understood yet. Micropores are described as a protective habitat for microorganisms through pore-size exclusion of predators (protozoa) [2]. The closeness between organic matter, clay and microbes in micropores has been pointed out [10]. The role of soil microorganisms in the distribution of soil organic matter in micropores is worth of future investigations

REFERENCES

1. Jones, F.G.W., Thomasson, A.J., 1976. Bulk density as an indicator of pore space in soils usable by nematodes. Nematologica 22, 133-137.
2. Elliott, E.T., Anderson, R.V., Coleman, D.C., Cole, C.V., 1980. Habitable pore space and microbial trophic interactions. Oikos 35, 327-335.
3. Foster, R.C., 1988. Microenvironments of soil microorganisms. Biol. Fertil. Soils 6, 189-203.
4. Kirchmann, H., Persson, J., Carlgren, K., 1994. The Ultuna long-term soil organic matter experiment, 1956-1991. Reports and Dissertation, 17, 1-55. Department of Soil Sciences, Swedish University of Agricultural Sciences, Uppsala.
5. Marshall, T.J., Holmes, J.W., 1988. Soil Physics. 2nd edition. Cambridge University Press. Cambridge.
6. Blake, G.R., Hartge, K.H., 1986. Particle density. *In*: Klute, A. (Ed.), Methods of Soil Analysis, Part 1. Agronomy Series No. 9, 2nd ed. Madison, WI, pp. 377-382.
7. Blake, G.R., Hartge, K.H., 1986. Bulk density. *In*: Klute, A. (Ed.), Methods of Soil Analysis, Part 1. Agronomy Series No. 9, 2nd ed. Madison, WI. pp 367-375.
8. SAS. 1985. Statistical Analysis System. User´s Guide: Statistics. SAS Institute Inc., Cary,

North Carolina.

9. Christensen, B.T., 1992. Physical fractionation of soil and organic matter in primary particle size and density separates. Adv. Soil Sci. 20, 1-90.
10. Van Gestel, M., Merckx, R., Vlassek, K., 1996. Spatial distribution of microbial biomass in microaggregates of a silty-loam soil and the relation with the resistance of microorganisms to soil drying. Soil Biol. Biochem. 28, 503-510.

Developments in Soil Science, Volume 28A
Editors: A. Violante, P.M. Huang, J.-M. Bollag and L. Gianfreda

CAPACITY OF ORGANICALLY COMPLEXED ALUMINUM, IONIC STRENGTH, AND pH TO AFFECT THE CEC OF ORGANIC SAMPLES

G. Matschonat

Institute of Soil Science and Land Evaluation, University of Hohenheim, D - 70593 Stuttgart, Germany

Laboratory studies on organic-rich samples have identified organic matter concentration, pH, the ionic strength of the equilibrium solution (I) and the concentration of organically complexed Al (Al_{org}) to be factors affecting the cation exchange capacity (CEC). Understanding charge development in the field, however, is still a problem; in field studies, pH and Al_{org} were correlated with the CEC in some soils but not in others. There is almost no information available about interactions between the single factors affecting CEC that might account for some conflicting field results. In this study, pH, Al_{org} and I have been varied simultaneously to identify their possible interactions on the CEC of different types of materials. This was done by preparation of samples with different Al_{org} saturation and the use of a CEC method that allows the manipulation of the pH and I during determination. Increasing concentrations of Al_{org} did not affect the CEC of peat samples but lowered the CEC of Oa and Oe samples. There was no interaction of Al_{org} with I in their effect on CEC, but there was an interaction of pH and I, with the effect of I being more pronounced at low pH. Except for the C concentration, there is no other single factor among the factors studied here with an equally large capacity to influence the CEC in the field. There are several realistic scenarios in which an increasing effect of one factor on CEC may be neutralized by the effect of others.

1. INTRODUCTION

Many factors influence the cation exchange capacity (CEC) of acid, organic-rich samples, e.g., forest soil O, A and Bh horizons, making the understanding of charge development in the field a complex problem. Although pH dependency of CEC has repeatedly been shown on pH-manipulated samples in the laboratory, conflicting results have been obtained in field studies examining soils with a range of pH [1-4], indicating the importance of additional factors that affect charge in the field. In these field studies, the most important single factor regulating CEC seemed to be the organic matter (or C) concentration [2]; an influence of litter type is not readily obvious [3], but an influence of the degree of decomposition was reported for some soils [1].

The potentials of the ionic strength [5-7] and the concentration of organically complexed Al [8] and Al oxide particles associated with organic matter [9] to modify the CEC have been observed in the laboratory. Aitken [10] found the effective CEC and the concentration of organically complexed Al (Al_{org}) also to be correlated in the field. Ionic strength (I) and pH have been found to interact in modifying the CEC [7], and an interaction of pH and Al_{org} is also likely because Al_{org} is involved in pH buffering [11]. In the field, the ionic strength of the soil solution and the availability of Al from inorganic sources may play a role as well.

The existing database for the effect of Al_{org} on CEC is not yet satisfactory in two respects: firstly, the choice of model substances (some studies used peat and muck) lacks correspondence to many field situations, and secondly, the pH was not controlled during Al saturation. The influence of Al_{org} and I has not yet been studied simultaneously, so the potential interdependence of these two factors remains an open question.

The aim of this work was to perform a controlled study of some of the possible interactions on the CEC of different types of materials. Focus was on the potential role of I relative to pH and Al_{org}; special attention was paid to the question of whether or not Al_{org} and I interact in their effect on CEC.

2. MATERIAL AND METHODS

2.1. Sample preparation

Oe and Oa material was obtained from a podzol soil on a sand dune situated in the water protection area "Hohe Ward" near Münster, northwest Germany. The site, which is forested with 93 year-old pine (*Pinus sylvestris*) with some birch (*Betula pendula*), was sampled in June 1999. These samples and commercially available sphagnum peat were sieved to 5 mm and acid washed (2 M HCl) to remove cations until there were no more Ca, Mg, K, Na, Mn, and Al detectable in the acid. Iron could not be completely removed this way and traces of Al also remained (compare Table 1). The samples were washed several times with deionized water to remove excess acid until a pH of approximately 3 was achieved. To obtain different degrees of Al saturation, $AlCl_3$ solutions of different concentrations were added to subsamples of the acid-washed Oe, Oa, and peat material. Excess $AlCl_3$ was removed by washing with deionized water. Care was taken that a pH of 3 was never exceeded during this procedure so that all Al adsorbed could safely be assumed to be trivalent. Solutions were removed by centrifugation and/or filtration through a polyethylene mesh to minimize loss of material. The amount of Al_{org} was determined by $CuCl_2$ extraction [12]. Three subsamples with low, medium and high Al concentration of each Oe, Oa and peat sample were prepared in this manner. Samples were stored moist in polyethylene bags at 5°C for further analysis. Some characteristics of these samples and of the original, untreated samples are given in Table 1. The sample preparation procedure changed sample properties somewhat; most obvious was a loss of mixed-in sand in Oa and Oe materials, so that the C concentration increased when compared to the untreated sample.

2.2. CEC determination

A modification of the Gillman [13] method was used for CEC determination: cations were exchanged for Ba^{2+} with 0.1 M $BaCl_2$, the sample was equilibrated with more dilute $BaCl_2$ solutions of defined I and pH, and the adsorbed Ba^{2+} was removed by an extraction with 0.5 M $CuCl_2$. Copper chloride also extracts organically complexed Al [12] so that not only the CEC (equivalent to Ba^{2+}) but also the amount of Al_{org} was obtained for each treatment and replicate.

A moist sample (two replicates) corresponding to 1 g on a dry weight basis was put into pre-weighed polyethylene tubes, 100 ml of $BaCl_2$ 0.1 M solution were added and gently shaken for 2 hr; the solution was removed after centrifugation and (exchangeable) Al was determined in the extract. The sample was then equilibrated with $BaCl_2$ solutions of pH 3 or pH 4 with ionic strength values (protons included) close to 6, 12, and 18 mmol L^{-1}, resulting in six treatments. The exact I in each equilibration was calculated from the ions determined in the solution of the last equilibration step for each individual sample. The ionic strength was calculated as

$$I = 0.5\ \Sigma_i\ c_i\ z_i^2$$

where c_i and z_i are the concentration and charge of the ion i, respectively.

The equilibration procedure required several (usually 3–5) removals and additions of fresh equilibration solution to lower I to the chosen level, which was checked by electric conductivity measurement of added and removed solutions. The pH was then adjusted by addition of small amounts of concentrated HCl or $Ba(OH)_2$ solution, if necessary. The equilibration was considered complete when the solution pH was within 0.1 pH units of the target pH and when the electric conductivity did not deviate by more than 10% of the target value. After removal of the last equilibration solution, samples and tubes were weighed to correct for entrained solution and adsorbed Ba and Al_{org} were removed by a single extraction with 20 mL of 0.5 M $CuCl_2$ solution. All solutions were filtered through 0.4 μm polycarbonate filters (Schleicher & Schuell, Germany) prior to cation determination. Cation determination was by inductively coupled plasma optical emission (ICP-OES) in all extracts.

Because the procedure requires several changes of solutions by centrifuging and decanting, loss of floating organic material is a problem when working with organic samples. While in principle, the loss may be quantified by recording the dry weight of the remaining material, using $CuCl_2$ solution, the weight of adsorbed Cu and Cu in entrained solution would have overcompensated for a weight loss of several percent. From running the same procedure but with a low-weight cation (Mg), it was estimated that the loss of organic material was variable but less than 5%. It was thus not possible to correct for loss of material for each treatment separately and consequently, the agreement between replicates was not as good as usually found for samples of mineral soil. This had to be accepted for the sake of obtaining the exact Al_{org} values together with the corresponding CEC.

2.3. Further analysis

Copper chloride-extractable Al was determined with 0.5 M $CuCl_2$ [12] and exchangeable Al of untreated samples was extracted with (unbuffered) 0.1 M $BaCl_2$. The sum of exchangeable cations (Al, Ca, Mg, Mn, Fe, K, Na, H) also was determined from the latter extract. The difference between Cu-extractable Al and Ba-exchangeable Al was interpreted as organically complexed Al (Al_{org}). Carbon and N were determined by elemental analyzer and pH values of the samples were recorded at a soil:solution ratio of 1:10 in $CaCl_2$ (0.01 M and 0.001 M). Acid-oxalate extractable Fe (Fe_{ox}) was also determined [14].

3. RESULTS AND DISCUSSION

In most cases, the CEC determination procedure made the concentration of Al_{org} of the sample decrease by 40–70% of the value after Al treatment given in Table 1, depending on the treatment and the number of equilibration steps necessary to reach the pH and I target values. There were some exceptions with smaller (in peat samples) or larger (in Oa samples) reduction of Al_{org}. In the following, all results are based on the actual Al_{org} values (determined in the last step of CEC determination) that correspond to the respective CEC value. To facilitate comparability, and because of the outstanding importance of the organic matter concentration for the samples' charge properties, CEC and Al_{org} values reported here are normalized to their C concentration.

Increasing concentrations of Al_{org} lowered the acid strength of the materials in 1 mM $CaCl_2$ solution by about 0.12 (Oa), 0.25 (Oe) and 0.09 (peat) pH units (data not shown) for each 10 $mmol_c\ g^{-1}$ C of Al_{org}. The untreated samples did not match the order of increasing pH with Al concentration found in the Al-treated samples, probably because of the presence of other cations besides Al (Fe and base cations in case of O horizons, base cations in case of peat). In the Al-treated samples, most of these were removed by the acid washing. Unlike the Oa and Oe samples, there was a remarkably low pH-sensitivity of peat samples to different salt concentrations in pH determination. Together with other diverging results for peat when compared with O horizon materials (described below), this casts doubt on the suitability of peat as a model substance for studying soil organic matter charge.

3.1. Effects of pH and I on CEC

CEC per unit C concentration decreased in the order Oa > Oe > peat and was in the range of 0.3–0.6 $mmol_c\ g^{-1}$ C at low I (≈ 5 mmol L^{-1}) and 0.7–1.0 $mmol_c\ g^{-1}$ C at high I (≈18 mmol L^{-1}), varying with the material and its degree of Al saturation. Typical values found by other workers were 0.5–0.6 $mmol_c\ g^{-1}$ C (1, 2, 3). In the present study, CEC increased by 0.42 (Oa), 0.31 (Oe) and 0.52 (peat) $mmol_c\ g^{-1}$ C per pH unit across all treatments. This is in the range observed in earlier studies [6, 7, 15]. The increase in CEC with increasing ionic strength was most pronounced in peat samples (0.15 $mmol_c\ g^{-1}$ C per mmol $L^{-0.5}$), followed by Oa (0.068 $mmol_c\ g^{-1}$ C per mmol $L^{-0.5}$) and was lowest in Oe samples (0.039 $mmol_c\ g^{-1}$ C per mmol $L^{-0.5}$), but these correlations were seldom significant across all pH treatments and Al_{org} saturation. Some previous studies have found the CEC to increase linearly with the square root of I (5, 7), but some observed a more curvilinear relationship with $\sqrt{I}$ [6].

3.2. Interdependence of pH, I and $Al_{org.}$ affecting CEC

Changes in CEC with I are plotted in Figure 1 for pH 3 treatments and in Figure 2 for pH 4 treatments. Comparing Figures 1 and 2, the pronounced effect of pH on CEC is obvious: CEC ranges for pH 3 and pH 4 do not overlap for any of the three materials. Irrespective of the

Table 1
Chemical Properties of Al-treated and untreated samples

Sample	pH_{CaCl2}	C	N	Fe_{ox}	$Al_{org.}$	$Al_{exch.}$	SumExCat
		$g\ kg^{-1}$	$g\ kg^{-1}$	$mmol\ g^{-1}\ C$	$mmol_c\ g^{-1}\ C$	$mmol_c\ g^{-1}\ C$	$mmol_c\ g^{-1}\ C$
Oa low	2.03	484	16.0	0.065	0.088	0.027	0.583
Oa medium	2.11	501	15.6	0.081	0.587	0.174	0.559
Oa high	2.75	494	15.5	0.076	1.261	0.284	0.449
Oe low	2.20	528	20.0	0.011	0.053	0.005	0.407
Oe medium	2.38	530	20.3	0.009	0.215	0.106	0.332
Oe high	2.71	529	20.4	0.063	0.510	0.194	0.344
Peat low	2.78	509	6.8	0.001	0.815	0.547	0.735
Peat medium	2.94	520	6.7	0.000	1.029	0.720	0.850
Peat high	3.16	518	6.7	0.001	1.278	0.826	0.892
Oa untreated	2.83	377	12.5	0.273	0.830	0.358	0.682
Oe untreated	2.91	472	18.5	0.120	0.146	0.063	0.502
Peat untreated	2.54	519	6.5	0.002	<	<	0.645

pH_{CaCl2}: 0.01M solution; $Al_{org.}$: organically complexed Al; $Al_{exch.}$: exchangeable Al ($BaCl_2$ 0.1 M); Fe_{ox}: acid oxalate-extractable Fe; SumExCat: sum of exchangeable cations in $BaCl_2$ 0.1 M. „Low“, „medium“ and „high“ refer to the degree of Al saturation.<: below detection limit.

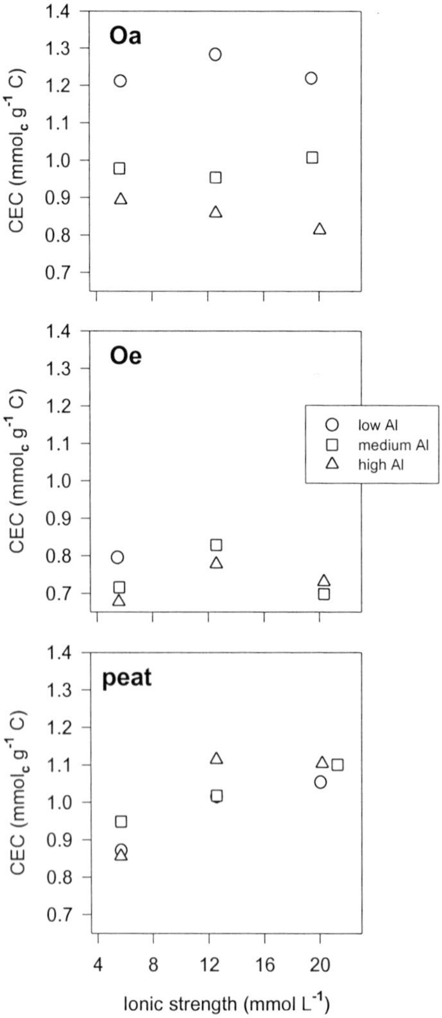

Figure 1. CEC variation with ionic strength at pH 3 for samples with different saturation with organically complexed Al.

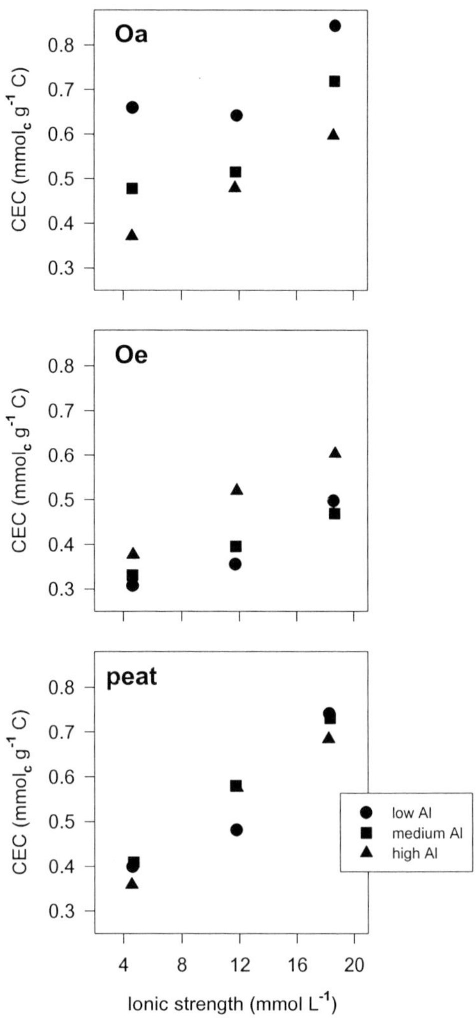

Figure 2. CEC variation with ionic strength at pH 4 for samples with different saturation with organically complexed Al.

Al_{org} saturation, the ionic strength effect on CEC only occurred at pH 3 for the Oa and Oe samples and was stronger at pH 3 than at pH 4 for the peat samples. Reexamination of data from an earlier study [7] also displayed this more pronounced sensitivity to I at lower pH. Increasing concentrations of Al_{org} had almost no effect on peat sample CEC in the present study, but Al_{org} was responsible for the spread in CEC measures for Oa samples both at pH 3 and 4 (Figures 1 and 2), with samples higher in Al_{org} having lower CEC. Other authors studying peat [8], however, described its CEC to be inversely related to Al_{org}. When CEC values for the Oa samples are plotted against the Al_{org} concentration (Figure 3), there is a negative relationship and again, the lack of an ionic strength effect at pH 4 is obvious. The range of Al_{org} concentrations was about equal for peat and Oa samples (0–0.2 mmol g^{-1} C) but lower for Oe (0–0.1 mmol g^{-1} C).

Although pH and I effects on CEC were interdependent, there were no indications for a specific interdependence of Al_{org} and I. The reaction of CEC to changes in I was not hindered by increasing concentrations of Al_{org} in the range applied in this study.

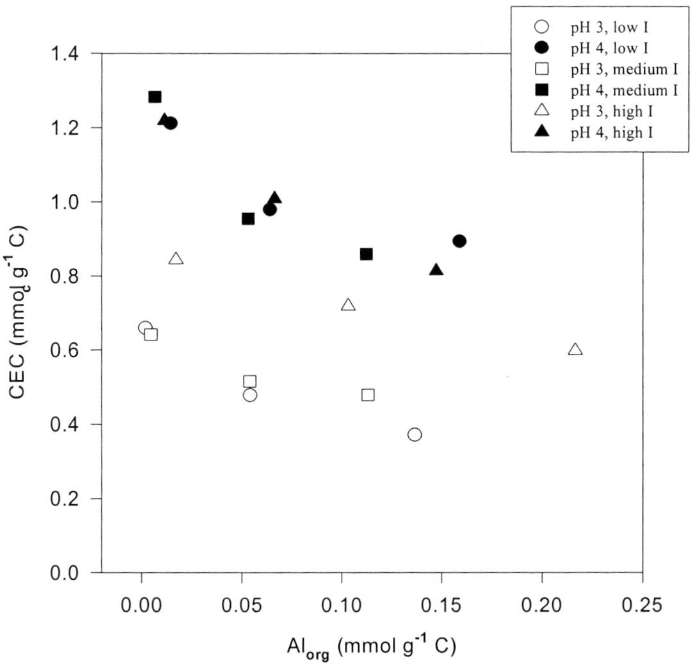

Figure 3. CEC variation with the concentration of organically complexed Al in Oa samples.

Multiple regression analysis confirmed that for all materials, treatment pH was responsible for most of the variation in CEC. The square root of I was the second variable picked for peat and Oe samples in a stepwise procedure, while for Oa, Al_{org} was the second variable. In a third step, Al_{org} was picked for Oe and peat, and $\sqrt{I}$ for Oa (final $r^2 = 0.83$–0.89). When results were grouped for pH instead of type of material, differences in I significantly influenced CEC at pH 3 but not at pH 4. When values were grouped for ionic strength levels, I was still picked as the most important variable influencing CEC in the groups of $I \approx 5$ mmol L^{-1} and $I \approx 12$ mmol L^{-1} (please note that I varied by approximately 1.5 mmol L^{-1} within the groups), pointing to the significance of small variation in I at relatively low levels of I. The second variable picked was always the C concentration of the sample, which was the only variable available that represented differences in the type of materials. For the group of $I \approx 5$ mmol L^{-1}, Al_{org} was picked in addition (final $r^2 = 0.93$).

Comparing the reduction in CEC with the increase of Al_{org} on a charge basis is a means to test the effectiveness of Al_{org} to reduce CEC. Organically complexed Al can be assumed trivalent for all pH 3 treatments, since for these treatments, the pH never exceeded pH 3 during sample preparation and CEC determination. For pH 4 treatments, some hydrolyzation of adsorbed Al may have taken place [8] and the charge of Al could then be less than trivalent. Table 2 contrasts the change in CEC between the respective low-Al and high-Al samples of a treatment with the amount of charge (Al assumed trivalent) occupied by Al_{org}. If every Al_{org} ion blocked three exchange sites, the CEC reduction data should be identical to that of the Al_{org} increase. However, the ratio of CEC reduction to Al_{org} increase was less than 0.35 for peat samples and 0.4–0.99 for Oa and Oe samples. The ratio of CEC reduction to Al_{org} increase was less than 1.0 also for several pH 3 treatments in which hydrolyzation of Al_{org} is not a likely mechanism to modify the ratio. With regard to the given ranges of standard deviation, CEC reduction can be considered smaller than the respective increase in Al_{org} charge in two out of six treatments with Oe and Oa samples. For peat samples, Al_{org} was quite ineffective in reducing the CEC: the charge increase of Al_{org} was almost always greater than the reduction in CEC.

These results raise the question of how Al can become organically complexed without proportionally altering the CEC of the sample. Skyllberg [16] reported that a cation exchange between H^+ and Al^{3+} left the CEC of podzol O and E horizon samples unaltered, but this does not apply to all samples used here. One may hypothesize that in peat, the Al was bound to very acidic groups that would not dissociate under the pH range in the experiment but would complex Fe and Al if available. This should require a marked preference of the material for Al and may only occur under conditions of limited abundance of other cations such as Fe. Unmanipulated organic matter frequently contains large amounts of Fe that nonetheless do not seem to affect its CEC. With the Al_{org} in this experiment, the effect may have been similar. The peat sample was virtually free of Fe, while for the O horizons, some Fe remained even after acid washing.

3.3. Relative importance of factors affecting the CEC

The relative importance of pH, I, and Al_{org} regulating the CEC under field conditions is difficult to assess (though results of the multiple regression give some hints) because the units of these measures can not be compared directly. In this study, I was increased by a factor of about five, Al_{org} varied by a factor of about 30 (charge equivalents), and pH by one unit, equivalent to a 10-fold increase in H^+ concentration. To assess their potential to change the CEC, these ranges have to be confronted with ranges encountered in the field. In O and A horizons of acid forest soils, pH values will spatially vary by more than one unit, while the range of I probably extended to the upper limit encountered, but lower I is likely to occur [7].

Ranges reported for naturally occurring Al_{org} are highly variable: as high as 2.5 and 3 $mmol_c\ g^{-1}$ C for O horizons of some northeast US sites [3] and 2.7–6 $mmol_c\ g^{-1}$ C for some Australian surface soils [10], while A horizons of Swedish soils had only 0.05–0.5 $mmol_c\ g^{-1}$ C (17; Al data unpublished). The range of Al_{org} saturation in the present study thus compares to the range found in the Swedish soils but does not reach the high levels in the US and Australian soils.

Based on the rates with which the factors pH, I and Al_{org} change CEC and on the ranges with which these factors vary in the field, one may try to estimate the overall effect on CEC at a given C concentration. Among the factors included in this study, the effect of pH on CEC was strongest (and the naturally occurring range was not covered). In spite of the pronounced pH effect, a constellation is possible in which the promoting effect of pH on CEC may be counteracted in a high-Al soil with low I in soil solution. The pH and Al_{org} will often not vary

Table 2
Difference in CEC and Al_{org} concentration between high-Al and low-Al samples in the different treatments. Al was assumed trivalent. STD: standard deviation

Sample	pH	Ionic strength	CEC mean	CEC STD	Al_{org} mean	Al_{org} STD
		mmol L^{-1}	mmol$_c$ g^{-1} C			
Oa	3	4.60	-0.288	0.067	0.404	0.107
		11.72	-0.163	0.035	0.326	0.012
		18.58	-0.247	0.047	0.599	0.051
Oa	4	5.82	-0.318	0.049	0.434	0.076
		12.58	-0.266	0.181	0.312	0.013
		19.69	-0.405	0.047	0.408	0.021
Oe	3	4.64	-0.070	0.021	0.127	0.002
		11.72	-0.165	0.053	0.183	0.007
		18.58	-0.106	0.053	0.266	0.011
Oe	4	5.56	-0.119	0.027	0.204	0.019
		12.58	-0.051	0.029	0.074	0.005
		20.25	0.033	0.020	0.106	0.011
Peat	3	4.71	-0.051	0.023	0.161	0.028
		11.76	-0.005	0.115	0.120	0.044
		18.31	-0.047	0.055	0.213	0.035
Peat	4	5.59	0.092	0.035	0.302	0.021
		12.57	-0.096	0.019	0.296	0.021
		22.13	-0.003	0.170	0.256	0.050

independently in the field because Al_{org} is involved in pH buffering [11, 18] and because increasing concentrations of Al_{org} lower the acid strength of organic samples (19; for muck). A concurrence of organic matter with high Al concentrations and a relatively high pH (pH 4–5) is thus likely, and these factors will have an antagonistic effect on CEC. The effects could compensate one another, reducing the net effect to a strong correlation of increasing CEC with increasing C concentration (as observed by Ross and Bartlett [3].

As for the effect of I on CEC, it does not seem to be interfered with by increasing concentrations of Al_{org}, and the effect of I is greatest at a low level of I. At a higher pH however, which is likely to coincide with increasing concentrations of Al_{org}, the effect of I is suppressed. Processes that work to keep I in soil solution relatively constant in the field, such as dissolution, adsorption and desorption of salts and desorption of organic acids from solid organic matter [20, 21], could also reduce the marked effect of I on CEC. This scenario of combined effects in the field, however, may only be one among several realistic scenarios. Other factors not addressed in this study, such as the availability of mineral Al (abundance of weatherable minerals, weathering rate) affecting Al_{org} and intrinsic factors of organic matter (binding strength, acidity of groups), possibly varying with litter type, might also interfere with CEC development and remain to be addressed. Even if in some soils an estimate of the CEC based solely on the organic matter or C concentration can thus yield satisfactory results, this will not apply to "acid forest soils" in general.

ACKNOWLEDGMENTS

I thank Jan Siemens for providing samples of the “Hohe Ward;” Mr. T. Weiss and Mrs. E. Feierabend for sample preparation; and Dr. Breuer and Mrs. Ziehl for analysis of these samples. Don Ross and Daniel Haag gave helpful suggestions on the manuscript. Funding was provided by the Deutsche Forschungsgemeinschaft.

REFERENCES

1. Kalisz, P.J., Stone, E.L., 1980. Cation exchange capacity of acid forest humus layers. Soil Sci. Soc. Am. J. 44, 407–413.
2. Ross, D.S., Bartlett, R.J., Magdoff, F.R., 1991. Exchangeable cations and the pH-independent distribution of cation exchange capacity in Spodosols of a forested watershed. *In*: Wright, R.J. Baligar, V.C., Murrmann, R.P. (Eds.), Plant-Soil Relationships at Low pH. Kluwer: Dortrecht, The Netherlands, pp. 81–92.
3. Ross, D.S., Bartlett, R.J. 1995. Apparent pH independence of charge in forest organic surface soil horizons. Water, Air Soil Poll. 85, 1113–1118.
4. Johnson, C.E., Ruiz-Mendez, J.J., Lawrence, G.B., 2000. Forest soil chemistry and terrain attributes in a Catskill watershed. Soil Sci. Soc. Am. J. 64, 1804–1814.
5. Black, A.S., Campbell, A.S., 1982. Ionic strength of soil solution and its effects on charge properties of some New Zealand soils. J. Soil Sci. 33, 249–262.
6. Khanna, P.K., Raison, R.J., Falkiner, R.A., 1986. Exchange characteristics of some acid organic-rich forest soils. Aust. J. Soil Res. 24, 67–80.
7. Matschonat, G., Vogt, R., 1997. Effects of changes in pH, ionic strength, and sulphate concentration on the CEC of temperate acid forest soils. Europ. J. Soil Sci. 48, 163–171.
8. Hargrove, W.L., Thomas, G.W., 1981. Effect of organic matter on exchangeable aluminum

and plant growth in acid soils. *In*: Dowdy, R.H., Baker, D.E. (Eds.). Chemistry in the Soil Environment. American Society of Agronomy, Soil Science Society of America, Madison, WI, pp. 151–166.

9. Fernández Marcos, M.L., Buurman, P., Meijer, E.L., 1998. Role of organic matter and sesquioxides on variable charge of three soils from Galicia, Spain. Commun. Soil Sci. Plant Anal. 29, 2441–2457.
10. Aitken, R.L., 1992. Relationships between extractable Al, selected soil properties, pH buffer capacity and lime requirement in some acidic Queensland soils. Aust. J. Soil Res. 30, 119–130.
11. Wesselink, L.G., van Breemen, N., Mulder, J., Janssen, P.H., 1996. A simple model of soil organic matter complexation to predict the solubility of aluminium in acid forest soils. Europ. J. Soil Sci. 47, 373–384.
12. Juo, A.S.R., Kamprath, E.J., 1979. Copper chloride as an extractant for estimating the potentially reactive aluminum pools in acid soils. Soil Sci. Soc. Am. J. 42, 35–38.
13. Gillman, G.P., 1979. A proposed method for the measurement of exchange properties of highly weathered soils. Aust. J. Soil Res. 17, 129–139.
14. Jackson, M.L., Lim, C.H., Zelazny, L.W., 1986. Oxides, hydroxides, and aluminosilicates. *In*: Klute, A. (Ed.). Methods of Soil Analysis: Part I, Physical and Mineralogical Properties. 2nd ed. American Society of Agronomy, Madison, WI, pp. 101–140.
15. Ross, D.S., Bartlett, R.J., 1997. Charge fingerprints of forest organic horizons from north-eastern USA. Aust. J. Soil Res. 35, 553–564.
16. Skyllberg, U., 1999. pH and solubility of aluminium in acidic forest soils: a consequence of reactions between organic acidity and aluminium alkalinity. Europ. J. Soil Sci. 50, 95–106.
17. Matschonat, G., Falkengren-Grerup, U., 2000. Recovery of soil pH, cation-exchange capacity and the saturation of exchange sites from stemflow-induced soil acidification in three Swedish beech forests. Scand. J. For. Res. 15, 39–48.
18. Bloom, P.R., McBride, M.B., Weaver, R.M., 1979. Aluminum organic matter in acid soils: buffering and solution aluminum activity. Soil Sci. Soc. Am. J. 43, 488–493.
19. Hargrove, W. L., Thomas, G.W., 1982. Titration properties of Al-organic matter. Soil Sci. 134, 216–225.
20. Wright, R.F., Lotse, E., Semb, A., 1988. Reversibility of acidification shown by whole catchment experiments. Nature 334, 670–675.
21. Matschonat, G., Vogt, R., 1997. Assessment of a laboratory method to obtain the equilibrium solution composition of forest soils. Europ. J. Soil Sci. 48, 545–552.

Developments in Soil Science, Volume 28A
Editors: A. Violante, P.M. Huang, J.-M. Bollag and L. Gianfreda

ABIOTIC REACTIONS OF ORGANICS ON CLAY MINERAL SURFACES

U. Birkel[a], G. Gerold[a] and J. Niemeyer[b]

[a]Georg-August-University Göttingen, Institute of Geography, Department of Landscape Ecology, Goldschmidtstraße 5, 37077 Göttingen, Germany

[b]University of Trier, Geosciences/Geography, Department of Soil Science, Universitätsring 15, 54286 Trier, Germany

Clay minerals are the major binding partners of organic matter in soils. Furthermore, these minerals show a broad spectrum of chemical reactivity. Besides the biotic pathways, these abiotic reaction potentials should be considered when investigating chemical processes in soils, such as genesis of humic substances and binding and transformation of xenobiotic substances. For mechanistically oriented laboratory experiments, it is practicable, due to the complexity of soil organic matter, to use simple organic compounds, i.e., aromatic constituents of lignin or humic acids. Therefore, in our work, phenols such as catechol, pyrogallol and 2,6-dimethylphenol were used to investigate their reactions on the surface of montmorillonite particles. Several reaction products containing quinonoid, carboxylic and carbonylic groups could be determined. In addition, the colored synthetic reaction productscould not be removed from the mineral surfaces by organic solvents. We conclude that oxidation of the phenols is the first step in their transformation. This is followed by polymerization of the oxidation products. In order to elucidate the mechanisms of the catalytic activities, the surfaces of the clay particles were examined with a scanning electron microscope (SEM), coupled with an energy dispersive x-ray spectrometry (EDX). In the vicinity of iron-rich domains of the clay surfaces, synthetic organic coatings were observed. We conclude that iron in the mineral structure is partially responsible for the catalytic activity of clays. To obtain structural information about the developed reaction products, infrared-spectrometry (FTIR), liquid ^{13}C-nuclear magnetic resonance spectrometry (NMR) and matrix-assisted laser desorption ionization time of flight mass spectrometry (MALDI-TOF-MS) were used.

1. INTRODUCTION

Clay minerals show a broad spread spectrum of reactivity towards organic substances sorbed on their surface, as represented by a wide technical/industrial usage [1, 2]. It is obvious that clay minerals in soils—on which most organic substances are bound via clay-organic complexes—should own a similar activity in transforming organic substances from natural sources [3]. The hypothesized activity of clay minerals could be of important environmental relevance in binding and transforming xenobiotics into soil organic matter and thus in preventing pollutants from contaminating soil and groundwater. Furthermore, it is of special importance to examine whether the genesis of humic substances in soils (i.e., polyphenol-theory [4]) could be supported by abiotic processes, besides the microbial pathway.

The aim of this work was to investigate whether there is any reactivity of clay minerals to surface-sorbed organic compounds in soils [5]. Physicochemical reactions of sorbed organics on clay mineral surfaces could lead to an alternative abiotic transformation pathway of organic compounds to humic substances and affect the genesis of soil/aggregate structure [6, 7]. Because of the complexity of natural humic soil organic matter, simple model substances, representing building blocks of humics, were used. For this reason, lignin decomposition products, such as phenols, were chosen [8, 9].

A further point of interest was to examine the environmental impact of the catalytic activity [10]. For this reason, the interaction of phenanthrene, a simple, slightly condensed polycyclic aromatic hydrocarbon (PAH), with clay minerals and clay-organic complexes (clay-phenol) was examined.

2. MATERIAL AND METHODS

2.1. Sublimation of organics on clay minerals

Na^{+}-montmorillonite and Ca^{2+}-montmorillonite (SWy-2 and SAz-1), obtained from the Clay Minerals Repository, University of Missouri, were used without further preparation. Catechol, pyrogallol, 2,6-dimethylphenol and phenanthrene were purchased from Fluka, Germany. Each of the phenols and the PAH (0.5 g) were sublimated on the clay minerals surface (2 g) for a reaction time of 24 h at a temperature of 50°C. The clays, now coated with organics, were placed for an additional 24 h under a laboratory hood to evaporate the remaining phenol/PAH. One parallel series of coated clay samples was separated for electron microscopy analysis (FE-REM: field-emission electron microscopy, REM/EDX). The reaction products of the second series were extracted from the clay mineral surface by shaking with 40 ml of methanol (Merck LiChroSolv). The clays were dispersed by ultrasonification for 15 min and shaken additionally for 2 h in an overhead shaker. The solvate was separated from the clay by centrifugation at 3000 rpm for 30 min and filtered through a 0.45 μm polyamide filter (CS-Service). The solvent was evaporated in a vacuum-concentrator.

2.2. Preparation for electron microscopy

For investigation with the Field-Emission-REM (Zeiss Leo Gemini 1530), the samples were put directly on conductive carbon on glass plates. The samples analyzed by EDX (Zeiss Leo 435VP, coupled with Link eXL (Cambridge Scientific), signal detection of 82% secondary electrons and 18% backscatter electrons) needed no additional preparation.

2.3. Preparation for instrumental analysis

For examination by liquid ^{13}C-NMR spectroscopy (Bruker Aspect 3000), the dried, extracted samples were solvated in deuterated dimethylsulfoxide (DMSO-d6) and poured into NMR tubes (Wilmad). For FTIR spectroscopy (Perkin Elmer 1600), 1–3 mg of dried extracts were mixed with 300 mg KBr (Merck, Uvasol) and homogenized in a mill. The powder samples were pressed to discs and stored in a desiccator. The samples for MALDI-TOF-MS (KRATOS) were solvated in two different matrices (A1: trihydroxyanthracene; W-2: 4-nitroaniline).

3. RESULTS AND DISCUSSION

3.1. Electron microscopy/EDX

Chemical reactions at the clay-organic interface produce a change in the surface structure of the clay minerals.

High-resolution images (field-emission electron microscopy) of powder and aggregate samples, coated with organics, formed by the abiotic reaction of the sublimated phenols, show a wadding structure (Figure 1). This polymeric structure is due to the growth of model humic substances. The formed coatings can recognized easily. They appear as dark and smooth structures and can be easily distinguished from the underlying clay (light gray). This is confirmed by the EDX signals (Figure 2), where the red line represents a significant carbon signal crossing the dark areas and a pronounced signal for silicon for the light grey surface (second line).

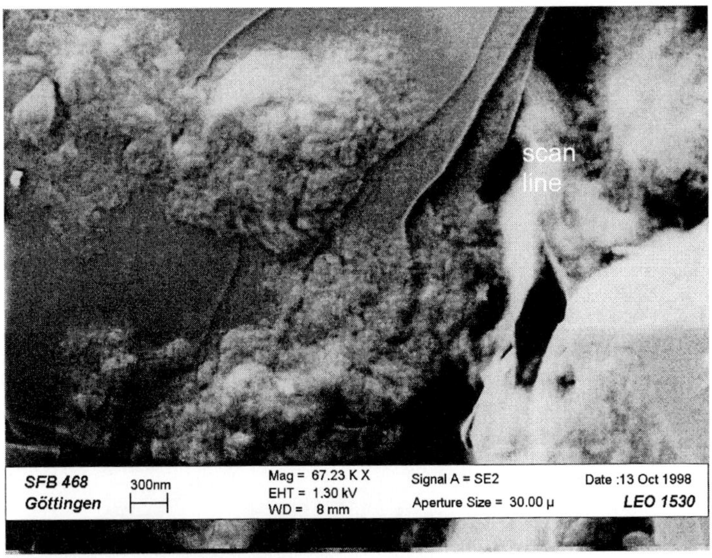

Figure 1. High-resolution scan (FE-REM, Zeiss LEO Gemini) of a montmorillonite surface after sublimation of catechol.

Due to the size of the observed organic coatings (many μm in elongation), the polymerization of the sublimated organics can be assumed. Additionally, the organic coatings show a heterogeneous distribution on the clay mineral surfaces. Thus, the reactions must have taken place only on discrete locations. This is in contrast to the actual model [11], describing a homogeneous distribution of organic coatings on clay minerals in the upper horizons in soils. Due to the heterogeneous distribution of the coatings, a relationship between sorption/reactivity centers and the iron content in the underlying clay (Figure 3a) can be assumed.

The observable interdependence between organic coating and iron content (Figure 3b) is a first hint on the catalytic mechanism of the clays.

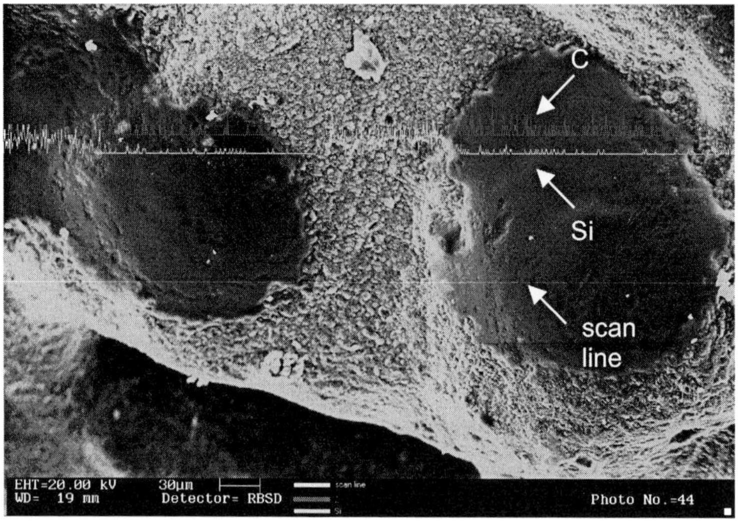

Figure 2. EDX-scan of a clay surface coated with pyrogallol (Zeiss LEO/Link eXL), carbon (C), silicon (Si), scanning line (scan line).

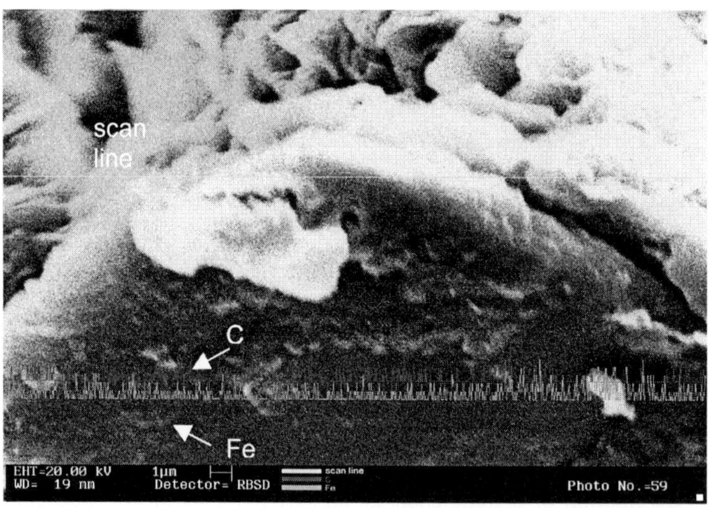

Figure 3a. EDX-scan (C/Fe) of a clay surface coated with catechol (Zeiss LEO/Link eXL).

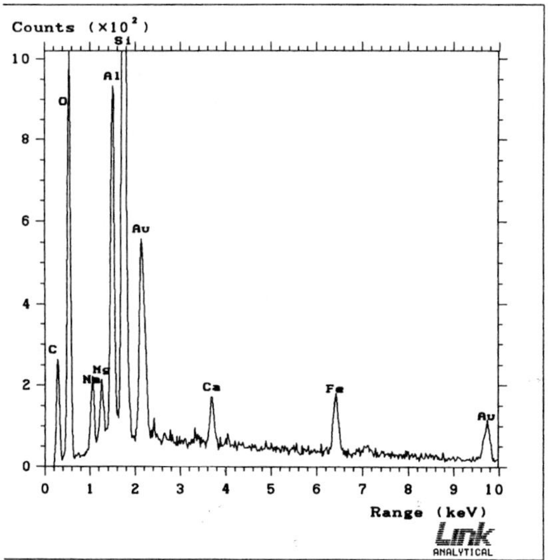

Figure 3b. EDX-scan (C/Fe) of a clay surface coated with catechol (Zeiss LEO/Link eXL).

3.2. FTIR spectroscopy

To get more detailed information about the reaction products formed, the samples were examined by FTIR- and ^{13}C-NMR spectroscopy. In all the FTIR-spectra, in the frequency range of 1800–1650 cm^{-1}, a new strong peak occurred. This peak is characteristic for carbonyl groups (C=O), showing that the phenols were partially transformed to quinones (Figure 4) [12].

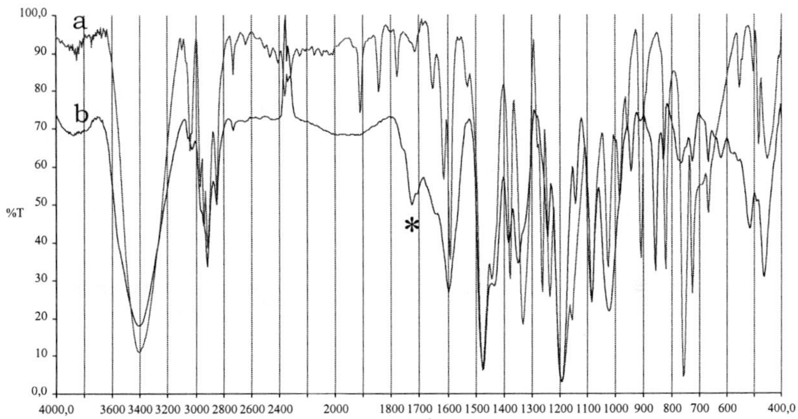

Figure 4. Comparison of IR spectra of 2,6-dimethylphenol (a) and its reaction products (b), extracted from Ca^{2+}-montmorillonite (∗ marks the new formed band).

Additionally, the broad peak from 1750 cm^{-1} to 1800 cm^{-1} proves the existence of carboxyl groups (COOH). Thus, the oxidation of pyrogallol and the cleavage of the aromatic system can be assumed. The broadening of the absorption band, representing the hydroxyl groups (OH), from 3000 cm^{-1} up to more than 3600 cm^{-1} (Figure 5), shows that the reaction products contain higher quantities of free and bound hydroxyl groups. Dark colored coatings resisted most extraction methods. Only microwave extraction at high temperatures (≥ 100°C) solvated the organics. The intense binding of these reaction products points to a strong chemisorption and could be understood as a simple model for clay-humus complexes [13].

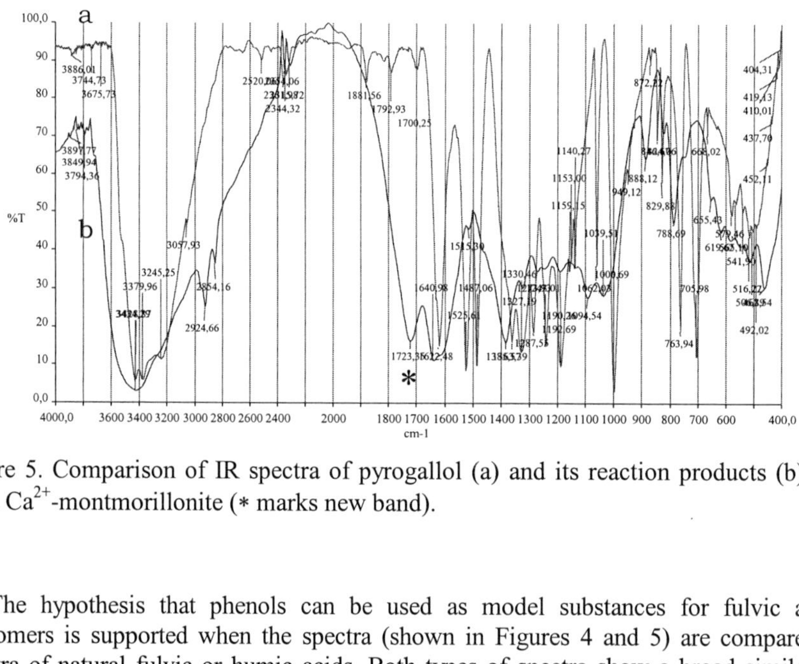

Figure 5. Comparison of IR spectra of pyrogallol (a) and its reaction products (b) extracted from Ca^{2+}-montmorillonite (* marks new band).

The hypothesis that phenols can be used as model substances for fulvic and humic monomers is supported when the spectra (shown in Figures 4 and 5) are compared with IR spectra of natural fulvic or humic acids. Both types of spectra show a broad similarity in the characteristic frequency ranges, representing hydroxyl, carbonyl, aromatic and methyl groups. This is a proof that simple aromatic compounds such as phenols can act as precursors (building blocks) for fulvic and/or humic substances [14, 15, 16, 17].

3.3. ^{13}C-NMR spectroscopy

The chemical shifts in the ^{13}C-NMR spectra demonstrate the conversion of organics to new chemical compounds and support the results obtained from the analysis of IR spectra. In the spectra of the reaction products of catechol (Table 1), the appearance of new peaks representing additional functional groups can be observed.

Table 1
Results of ^{13}C-NMR spectra of catechol and of its reaction products extracted from the clay mineral surface

catechol (reference: †)		montmorillonite (SWy-2) sublimated with catechol (‡)	
ppm	function	ppm	function
39.5	DMSO-d6	39.5	DMSO-d6
49.5	methanol	49.5	methanol
		51.8	O-CH_3 (‡)
		52.4	O-CH_3 (‡)
		53.0	O-CH_3 (‡)
		59.8	O-CH_3 (‡)
		85.8	C-OH (‡)
		102.8	alkene (‡)
		105.4	alkene (‡)
115.7	C=C ring (†)	116.4	C=C ring (†)
		118.1	ring (‡)
119.3	C=C ring (†)	120.4	C=C ring (†)
		133.2	ring-OH (‡)
		144.9	ring-OH (‡)
145.3	ring-OH (†)	145.8	ring-OH (†)
		170.2	carboxyl, -COOH (‡)
		183.5	quinone, -C=O (‡)
		196.9	aldehyde, -C=O (‡)

Peaks at a chemical shift of 40–60 ppm represent CH_2 and CH_3 groups; C-O containing compounds were detected by a peak at 85.8 ppm. In the aromatic range, peaks at 102.8 ppm, 105.4 ppm and 118.1 ppm show new aromatic structures. The substitution pattern of the aromatic structure is altered, demonstrated by new peaks at 133.2 ppm and 144.9 ppm. The peaks with a chemical shift of more than 170 ppm are very important. Herewith, the oxidation of phenols to quinones is proven (183.5 ppm and 196.9 ppm). With the detection of shifts of methoxy- and carboxyl groups, the cleavage of aromatic systems through oxidation can be supposed [10]. The appearance of shifts of C=C bonds (Table 1) and new peaks in the aromatic range of the NMR spectra show the formation of new aromatic structures and/or polymerization products.

Analysis of the reaction products of pyrogallol with phenanthrene shows some very interesting results (Table 2). First, it must be stated that in the absence of a phenol, the sublimation of phenanthrene resulted in no detectable reaction products. As a consequence for the transformation of phenanthrene, a reaction partner was needed. Besides the reaction products of pyrogallol, new chemical shifts, ranging from 11.2 ppm up to 24.1 ppm (methyl groups), 68.3 ppm (CH_2-OH), 90.3 ppm (O-alkyl), 107.9 ppm, 119.4 ppm, 133.8 ppm (aromatic), 146.8 ppm and 158 ppm (aromatic-O) to 163.5 ppm, 168.1 ppm (C=O) were observed. The listed chemical shifts show that phenanthrene reacted at least with pyrogallol or its reaction products. The aromatic structures are altered and the appearance of carbonyl groups is assumed as a result of the oxidation of the PAH.

Table 2
^{13}C-NMR shifts of pyrogallol (†), its reaction products (§); phenanthrene (‡) and its extracted reaction products (◇)

pyrogallol (†) phenanthrene (‡) (references)		montmorillonite (SWy-2) sublimated with			
		pyrogallol (§)		pyrogallol + phenanthrene (◇)	
ppm	function	ppm	function	ppm	function
				11.2	C-CH_3 (◇)
		16.9	C-CH_3 (§)	14.3	C-CH_3 (◇)
				23.2	C-CH_3 (◇)
				24.1	C-CH_3 (◇)
		29.4	C-CH_3 (§)	29.2	C-CH_3 (§)
				30.6	C-CH_2 (§)
		31.7	C-CH_2 (§)		
		34.3	C-CH_2 (§)		
		36.4	C-CH_2 (§)		
39.5	DMSO-d6	39.5	DMSO-d6	39.5	DMSO-d6
49.5	methanol	49.5	methanol	49.5	methanol
		53.1	O-CH_3 (§)		
		55.2	O-CH_3 (§)		
		61.5	O-CH_3 (§)		
				68.3	-CH_2-OH (◇)
		70.2	O-alkyl (HC-OH) (§)		
				90.3	O-alkyl o. alkine (◇)
107.2	aromatic C (†)	106.1	aromatic C (†)	107.1	aromatic C (†)
				107.9	ring. alkene (◇)
118.5	aromatic C (†)	116.1	aromatic C (†)	116.9	aromatic C (†)
				119.4	ring (◇)
122.8	ring (‡)			123.4	ring (‡)
		124.8	aromat. ring (§)		
126.6	ring (‡)	126.3	aromat. ring (‡)		
126.8	ring (‡)			127.5	ring (‡)
128.4	ring (‡)			129.2	ring (‡)
129	ring (‡)			130.5	ring (‡)
131	ring (‡)			132.4	ring (‡)
133.2	ring-OH (†)	132.1	ring-OH (†)	132.6	ring-OH (†)
				133.8	ring-OH (◇)
146.3	ring-OH (†)	144.9	ring-OH (†)	145.4	ring-OH (†)
				146.8	ring-O (◇)
		152.5	ring-O (§)		
		158.6	ring-O (§)	158	ring-O (§)
		162.5	C=O (§)	163.1	C=O, carboxyl (§)
				163.5	C=O, carboxyl (◇)
				168.1	C=O, carboxyl (◇)

3.4. MALDI-TOF-MS spectrometry

The analysis of reaction products of sublimated phenols, here catechol, by MALDI-TOF-MS was done to determine the molecular mass of the formed compounds. If oxidative polymerization of phenols to substances with higher mass weights took place, an increase of molecular masses must be observed. The reference mass weight for catechol is 110.11 Dalton (Da). The detected mass weights ranged for sample A1, Ca^{2+}-montmorillonite sublimated with catechol (Figure 6), from 512 up to 3652 Da.

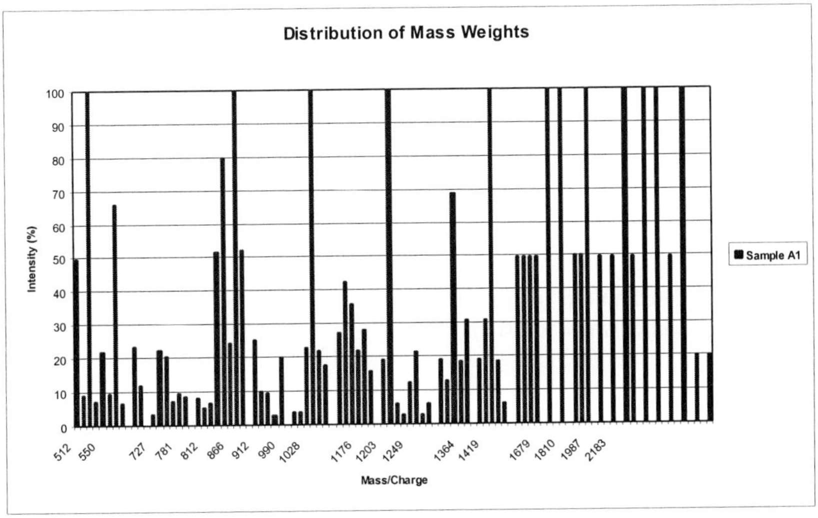

Figure 6. Mass distribution of reaction products extracted from Ca^{2+}-montmorillonite sublimated with catechol (m/w = 110 Da).

The distribution of mass weights for sample W2, Na^{+}-montmorillonite sublimated with catechol, ranged from 502 up to 3823 Da (Figure 7). The reasons for the different distribution of mass weights in samples A1 and W2 remain to be elucidated.

These results of mass spectrometry demonstrate that a broad range of condensed organic polymer molecules were formed. The findings are supported by FTIR- and ^{13}C-NMR spectroscopy.

4. CONCLUSIONS

The hypothesis of possible abiotic transformation of organics, adsorbed on clay minerals in soils, can be confirmed by the results of our work [10, 13]. Simple organic compounds were chemically altered and reacted to substances with new properties.

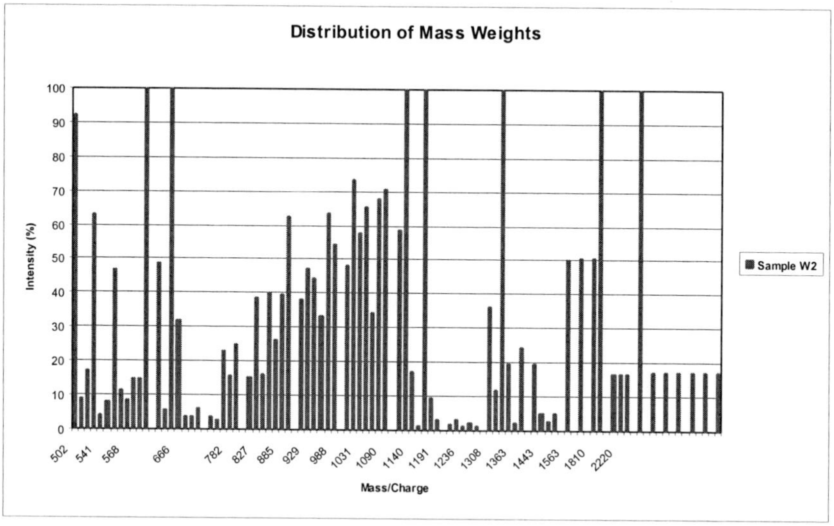

Figure 7. Mass distribution of reaction products extracted from Na^{+}-montmorillonite sublimated with catechol (m/w = 110 u).

Thus, the phenols could be transformed in a first step to phenolic radicals and react with the adsorbed surrounding organic compounds. The formation of organic radicals may happen by electron-transfer from the sorbed organic compounds to Fe-oxides, hydroxides or structural iron. These organics could be oxidized, then radicalized and then polymerized to compounds with higher molecular masses [14]. The reactivity of inorganic mineral surfaces can accomplish an oxidative polymerization of simple aromatic compounds [15, 16, 17]. So molecules were built that have multiple masses of the original compound. The additional reaction pathway of transformation of these surface-sorbed substances can give more detailed explanation of the genesis of humic substances in soils. The biotic pathway of degradation and transformation of organic compounds, here the genesis of fulvic and humic substances, can be expanded by mineral-organic interactions through physicochemical processes. The abiotic pathway in binding and transforming organic pollutants in soils, as shown by the transformation of aromatic compounds such as phenols and polycyclic aromatic hydrocarbons, could be of environmental importance.

REFERENCES

1. Bigi, F., Chesini, L., Maggi, R., Sartori, G., 1999. Montmorillonite KSF as an inorganic, water stable, and reusable catalyst for the Knoevenagel synthesis of coumarin-3-carboxylic acids. J. Org. Chem. 64, 1033-1035.
2. Fu, Y., Baba, T., Ono, Y., 1998. Vapor-phase reactions of catechol with dimethyl carbonate. Appl. Catalysis A: General 166, 419-424.

3. Naidja, A., Huang, P.M., Bollag, J.-M., 1998. Comparison of reaction products from the transformation of catechol catalyzed by birnessite or tyrosinase. Soil Sci. Soc. Am. J. 62, 188-195.
4. Stevenson, F.J., 1994. Humus Chemistry. Chemistry, Genesis, Composition, Reactions. John Wiley, New York.
5. Birkel, U., Niemeyer, J., 1998. Tonminerale als Katalysatoren bei der Umwandlung von organischen Verbindungen. Zeitschrift für Umweltchemie und Ökotoxikologie 10, 345-352.
6. Naidja, A., Huang, P.M., Bollag, J.-M., 2000. Enzyme-clay interactions and their impact on transformations of natural and anthropogenic organic compounds in soils. J. Environ. Qual. 29, 677-691.
7. Wang, T.S.C., Wang, M.-C., Huang, P.M., 1983. Catalytic synthesis of humic substances by using aluminas as catalysts, Soil Sci. 136, 226-230.
8. Shindo, H., Huang, P.M., 1985. The catalytic power of inorganic components in the abiotic synthesis of hydroquinone-derived humic polymers. Appl. Clay Sci. 1, 71-81.
9. Wang, M.-C., Huang, P.M., 1989. Pyrogallol transformations as catalyzed by nontronite, bentonite and kaolinite, Clays Clay Min. 37, 525-531.
10. 106. Birkel, U., Niemeyer, J., 1999. Montmorillonit-katalysierte Bildung von Vorstufen gebundener Rückstände aus Brenzkatechin und p-Chloranilin. Chemie der Erde 59, 47-55.
11. Schachtschabel, P., Blume, H.-P., Brümmer, G., Hartge, K.-H., Schwertmann, U., 1994. Lehrbuch der Bodenkunde. Enke Stuttgart.
12. Senesi, N., Miano, T.M., (Eds.), 1994. Humic Substances in the Global Environment and Implications on Human Health. Elsevier. Amsterdam.
13. Birkel, U., Niemeyer, J., Seeger, B., Gerold, G. 1998. ^{13}C-NMR spektroskopische Untersuchungen zur abiotischen Reaktion von Phenolen und PAK an Tonmineralen. Mitteilungen der Deutschen Bodenkundlichen Gesellschaft 88, 163-166.
14. Voudrias, E.A., Reinhard, M., 1986. Abiotic organic reactions at mineral surfaces. In: Davis, J., Hayes, K. (Eds.), Geochemical Processes at Mineral Surfaces. American Chemical Society. Washington, D.C., pp. 462-486.
15. Pal, S., Bollag, J.-M., Huang, P.M., 1994. Role of abiotic and biotic catalysts in the transformation of phenolic compounds through oxidative coupling reactions. Soil Biol. Biochem. 26, 813-820.
16. Bollag, J.-M., Stotzky, G., (Eds.), 1990. Soil Biochemistry, Marcel Dekker, New York.
17. Bollag, J.-M., Dec, J., Huang, P.M., 1998. Formation mechanisms of complex organic structures in soil habitats. Adv. Agron. 63, 237-266.

Developments in Soil Science, Volume 28A
Editors: A. Violante, P.M. Huang, J.-M. Bollag and L. Gianfreda

THE INTERACTION BETWEEN FERRICYANIDE ION AND UNFRACTIONATED HUMIC SUBSTANCES

A. Mori[a], F. Fornasier[a], L. Catalano[b], I. Franco[b] and L. Leita[a, c]

[a]Istituto Sperimentale per la Nutrizione delle Piante, Via Trieste 23, 34170 Gorizia, Italy

[b]Dipartimento di Produzione Vegetale e Tecnologie Agrarie, Università di Udine, via delle Scienze 208, 33100 Udine, Italy

[c]Corresponding author: e-mail: l.leita@isnp.it.

The aim of this work was to study the complexation of ferricyanide ion by unfractionated humic substances (HS) by means of cyclic voltammetry (CV), potentiometry and UV-visible spectroscopy. The CV measurements showed that the first addition of HS to a $[Fe(CN)_6]^{3-}$ solution caused a shift of both cathodic and anodic peak potentials together with a depression of the corresponding peak currents. After successive additions of HS, a further depression of peak currents was observed, but no more changes were seen in peak potentials. These results suggested the formation of ferricyanide-HS complexes and were confirmed by UV-visible spectra recorded in parallel to voltammetric measurements. The complexing capacity of HS was calculated as 0.9 μmoles ferricyanide mg^{-1} organic carbon. The complexation process of ferricyanide with HS molecules could occur by displacement of cyanide ligands from the inner coordination sphere of iron. Potentiometric titration of cyanide in the ferricyanide-HS system revealed the presence of labile cyanide anions that could be substituted by unfractionated HS during the formation of ferricyanide-HS complexes.

1. INTRODUCTION

Cyanide is commonly used in several industrial processes, such as mining, metallurgical and photographic industries and in coal gasification plants. Soils close to these sites often contain large amounts of cyanide, likely iron-cyanide complexes [1]. The possible hazards for human health and the environment greatly depend on the toxicity of available cyanides.

The behavior of these hazardous compounds in soil and the interactions with soil components are mostly unknown. Virtually no information is available on the capacity of humic substances (HS) to immobilize the cyano-complexes and thus to reduce the risk of contamination of the water table and minimize the biological hazard for microorganisms in soil.

One of the most significant properties of HS is their ability to interact with xenobiotics to form complexes of different solubility and chemical and biochemical stability [2–4]. In particular, HS have considerable biological importance because they are involved in soil-forming processes (e.g., podzolization), soil structure and nutrient availability to plants since they interfere with the immobilization and release of elements in soil [5–7].

Although the basic structure of HS is largely unknown, it is verified that their capacity to form complexes can be ascribed to the presence of functional groups, such as -COOH, phenolic, enolic and alcoholic -OH and C=O [8, 9]. The physico-chemical characterization of interactions between HS and xenobiotics is not simple, since HS consist of a complex mixture of macromolecules characterized by polyfunctional, polyelectrolytic properties and conformational factors that strongly influence the binding processes.

The use of spectroscopic techniques has improved the knowledge about HS-metal complexes [10, 11]. More recently, the electrode system of voltammetric cell equipment has been used as a tool to produce reactive species in a small solution layer around the electrodic surface and at the same time to monitor the reactions in which such species are involved. A linear potential scan of the working electrode causes a change in the current flowing through the cells, thus a voltammogram results by plotting the current vs. the potential. The position of the current peaks on the potential scale, E_p, provides qualitative information because it depends on the redox potential of the analyte. The E_p value also may account for the reversibility of the chemical reaction at the electrodic surface, which affects the peak shape. The peak height provides quantitative information because it is related to the concentration of the analyte and the number of electrons involved in the chemical reaction [12].

The aim of this work was to evaluate the complexation of ferricyanide ion by unfractionated HS extracted from a peat by means of cyclic voltammetry and spectroscopic measurements.

2. MATERIALS AND METHODS

2.1. Isolation of humic substances

A sphagnum peat (Medisaprist) sample was air-dried and sieved at 5 mm. The principal chemical properties of the peat were: 483 g kg^{-1}, total C; 10.5 g kg^{-1}, total N; 132 $cmol_c$ kg^{-1}, CEC. Additional and specific properties of the peat sample have been reported by Bragato et al. [13].

Humic substances were extracted with 0.1 M KOH for 2h at room temperature under nitrogen flux. The suspension was then centrifuged at 4000 rpm for 30 min. The supernatant was filtered on a Whatman WCN 0.45 μm membrane filter and treated with a cation exchange resin (Amberlite IR 120 H^+) to lower the pH to 7 and to remove excess potassium. Total organic carbon in the extract was determined by wet digestion, according to the modified Walkley & Black method, reported by Forster [14] using a Mettler DL40 titrator.

2.2. Cyclic voltammetry

A three-electrode polarographic analyzer (EG&G 264) connected to a 303 EG&G polarographic cell was employed for cyclic voltammetry. Voltammetric measurements were performed with a glassy carbon stationary electrode as the working electrode, Pt wire as the counter electrode and Ag/AgCl as the reference electrode. The electroactive solution was 1 mM $K_3[Fe(CN)_6]$, containing $NaClO_4$ 0.1 M as supporting electrolyte. Before the measurements, the solution was adjusted at pH 7 with $HClO_4$ using a Crison Micro pH 2001 pH meter. This was necessary because at neutral pH the interaction of HS with Fe compounds is strongly favored and oxidative coupling, dimerization and tautomerization of phenoxy radicals can be avoided [15]. Successive aliquots of 100 µL each (0.05 mg mL^{-1} of organic carbon) of HS were added to the ferricyanide solution. The possible interferences due to adsorption of HS on the working electrode have been checked by voltammetric measurements of a 1 mM hydroquinone solution in the presence and absence of HS.

The electrode was cleaned after each series of analysis by using alumina micropolish. A further electrochemical cleaning of the active graphite surface was obtained by a scanning excursion from 1.4 V to -0.4 V for 30 min in 0.1 N H_2SO_4. The efficiency of the working electrode was periodically checked by recording voltammograms of standard solutions of ferric cyanide [16].

2.3. UV-visible spectrophotometry

UV-visible spectra in the range 600–200 nm were recorded in parallel to the amperometric titrations. Spectra of 1 mM $[Fe(CN)_6]^{3-}$ solutions in the presence and absence of HS were obtained at room temperature by a Varian Cary 1E UV-Visible spectrophotometer using 1-cm quartz sample cells. Distilled water or solutions containing HS were used as blanks.

3. RESULTS AND DISCUSSION

Interferences at the electrodic surfaces need to be evaluated when polyelectrolytes are analyzed by electrochemical techniques. Therefore, two series of CV measurements were performed to check the reliability of the analytical response. Measurements were performed in the absence and in the presence of HS on the reversible redox couple quinone/hydroquinone to verify if and how HS might be adsorbed onto the surface of the glassy carbon electrode. Figure 1 shows the voltammograms recorded at pH 7 in the absence of HS after successive additions of 0.10 mL of 25 mM hydroquinone solution. The addition of hydroquinone produced an anodic current peak at 0.42 V and a corresponding cathodic peak at 0.18 V. Both peak currents increased proportionally with increasing hydroquinone concentration. Similar measurements performed in the presence of HS (Figure 2) yielded similar voltammetric responses, thus indicating that no interaction would occur between HS and the surface of the working electrode. Similar results were obtained at pH 3.5 and pH 12.5.

Figure 3 shows the cyclic voltammograms recorded at pH 7 for 1 mM $[Fe(CN)_6]^{3-}$ solution in the absence of HS (curve 1), after the first addition (100 µL) of HS (curve 2), at half titration point (curve 3) and at the end of the experiment (curve 4). At pH 7, the voltammogram of $[Fe(CN)_6]^{3-}$ alone displayed a well-defined forward peak at +0.16V, corresponding to the reduction of Fe(III) to Fe(II), associated with a backward peak at +0.23V, corresponding to the back-oxidation of Fe(II) to Fe(III).

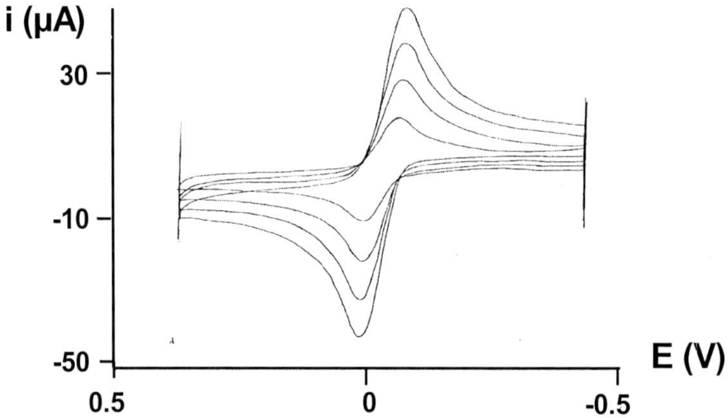

Figure 1. Voltammetric current/potential curves relative to quinone/hydroquinone redox couple in the absence of HS.

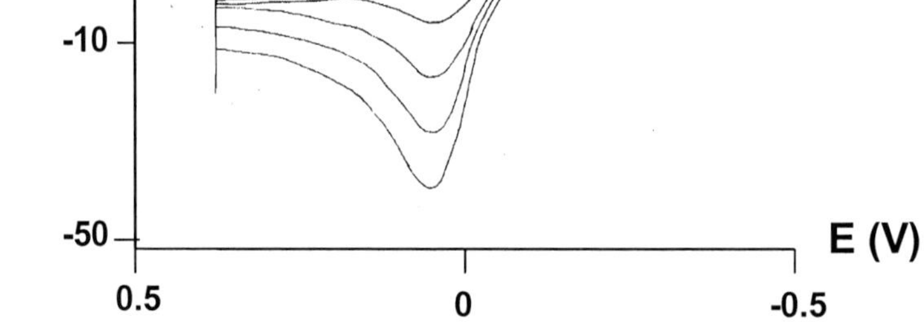

Figure 2. Voltammetric current/potential curves relative to quinone/hydroquinone redox couple in the presence of HS.

The first addition of HS to ferricyanide caused a shift of both cathodic and anodic peak potentials together with a depression of the corresponding peak potentials. Successive stepwise additions of HS caused slight depressions of both anodic and cathodic peak currents without any further shift of peak potentials. After addition of HS corresponding to 8.4 mg organic carbon, no further change of the voltammetric waves was observed. These results suggest the formation of ferricyanide-HS complexes, which was confirmed by UV-visible spectra recorded at pH 7 in parallel to voltammetric measurements. Spectra in Figure 4 show that the intensity of the charge transfer band at 225 nm, typical of $[Fe(CN)_6]^{3-}$ ion, decreased significantly in the presence of HS. The complexing capacity of HS was calculated as amounting to 0.9 μmoles $[Fe(CN)_6]^{3-}$ mg^{-1} organic carbon.

Complexation process of ferricyanide by the humic fraction (HS) could accour by the formation of products by the involvement of -COOH, phenolic, enolic and alcoholic –OH or $-NH_2$ moieties of HS and cyanide ligands present in the inner coordination sphere of iron. To verify this hypothesis, another series of CV scans of ferricyanide and ferricyanide-HS systems was performed at pH 3.5 and 12.5, respectively. The voltammograms recorded at acidic pH (3.5) showed a progressive peak potential shift after additions of HS, with a corresponding decrease of the peak currents, until the voltammogram shape showed a constant sigmoidal pattern (Figure 5).

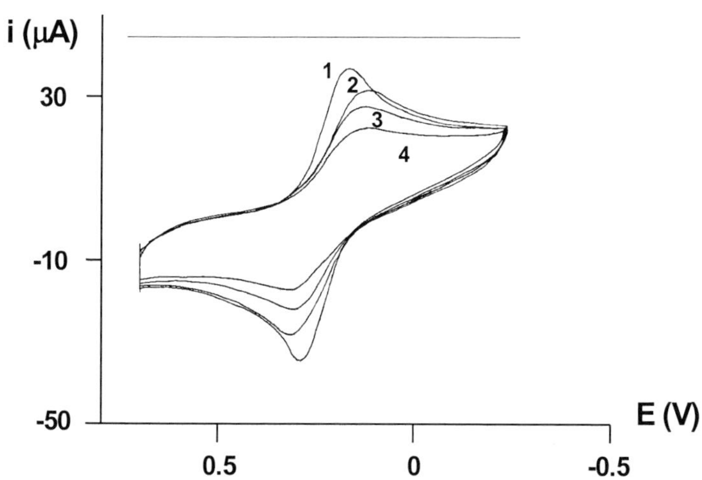

Figure 3. Voltammetric current/potential curves at pH 7 of $[Fe(CN)_6]^{3-}$ solutions in the absence of HS (curve 1), after the first addition (100 μL) of HS (curve 2), at half titration point (curve 3), and at the end of the experiment (curve 4).

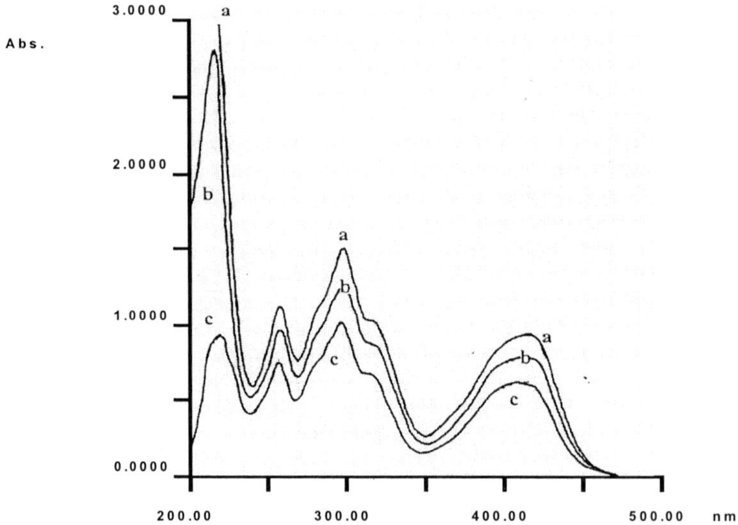

Figure 4. UV-vis spectra recorded at pH 7 for (a) $[Fe(CN)_6]^{3-}$ vs. HS solutions at the beginning of the amperometric titration; (b) $[Fe(CN)_6]^{3-}$-HS vs. HS at half titration point, and (c) $[Fe(CN)_6]^{3-}$-HS vs. HS at final vs. HS titration point.

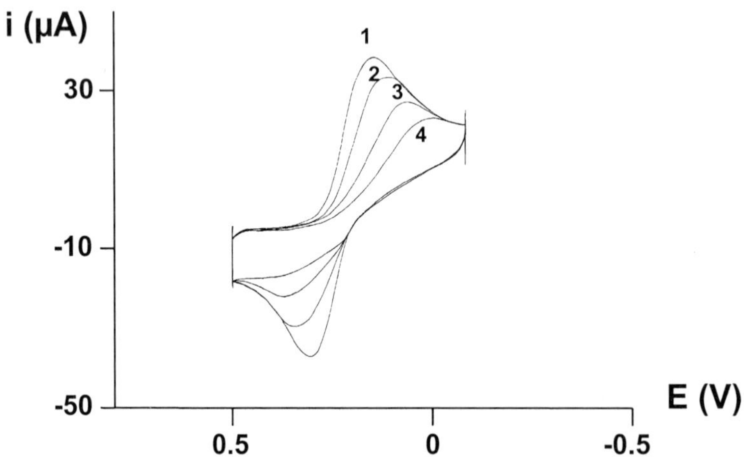

Figure 5. Voltammetric current/potential curves at pH 3.5 of $[Fe(CN)_6]^{3-}$ solutions in the absence of HS (curve 1), after the first addition (100 μL) of HS (curve 2), at half titration point (curve 3) and at the end of the experiment (curve 4).

No significant change of peak potentials occurred at pH 12.5 (Figure 6) after additions of HS to the ferricyanide solution; the decrease of the peak currents was smaller than those shown at lower pH values. At pH 12.5, most binding groups of HS were deprotonated, thus the interaction with $[Fe(CN)_6]^{3-}$ by hydrogen bonding was most likely suppressed, and the interaction between HS and $[Fe(CN)_6]^{3-}$ could not occur. Therefore, a possible hypothesis might be the formation of a hydrogen bond between ferricyanide and HS at pH 3.5 and 7. The HS complexing capacity calculated was one order of magnitude higher (9.5 μmoles $[Fe(CN)_6]^{3-}$ mg^{-1} organic carbon) than that measured at pH 7.

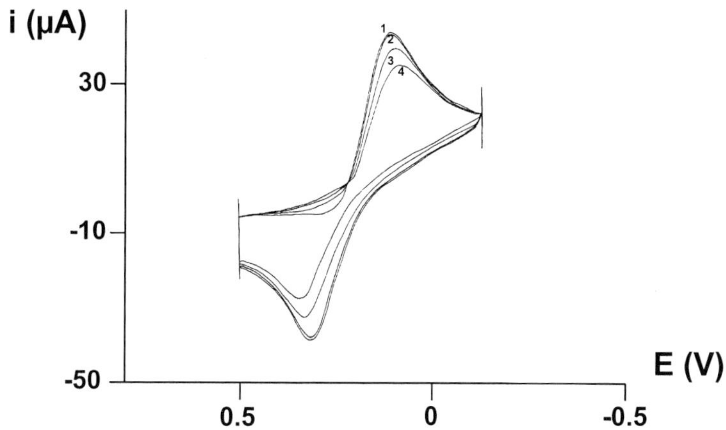

Figure 6. Voltammetric current/potential curves at pH 12.5 of $[Fe(CN)_6]^{3-}$ solutions in the absence of HS (curve 1), after the first addition (100 μL) of HS (curve 2), at half titration point (curve 3) and at the end of the experiment (curve 4).

REFERENCES

1. Meeussen, J.C.L., Keizer, M.G., de Haan, F.A.M., 1992. Chemical stability and decomposition rate of iron cyanide complexes in soil solutions. Environ. Sci. and Technol. 26, 511–516.
2. Senesi, N., 1990. Molecular and quantitative aspects of the chemistry of fulvic acid and its interactions with metal ions and organic chemicals. Part I. The electron spin resonance approach. Anal. Chim. Acta 232, 51–75.
3. Chen, Y., Stevenson, F.J., 1986. Soil organic matter interactions with trace elements. *In*: Chen, Y., Avnimelech, Y. (Eds.), The Role of Organic Matter in Modern Agriculture. Martinus Nijhoff, Dordrecht, pp. 73–116.

4. Ritchie, G.S.P., Posner, A.M., Ritchie, I.M., 1982. Characteristics of water-soluble fulvic acid-copper and fulvic acid-iron complexes. Soil Sci. 134, 354–363.
5. Stevenson, F.J., 1994. Organic matter reactions involving metal ions in soil. *In*: Stevenson, F.J. (Eds.), Humus Chemistry–Genesis, Composition, Reactions. John Wiley & Sons, Inc., pp. 378–404.
6. Stevenson, F.J., 1994. Role of organic matter in pedogenetic processes. *In*: Stevenson, F.J. (Eds.), Humus Chemistry–Genesis, Composition, Reactions. John Wiley & Sons, Inc., pp. 472–487.
7. Parfitt R.L., Heng, L.K., M.D. Taylor, 1995. Sorption of ions by soil organic matter and clay organics at low ionic strength. *In*: Huang, P.M., Berthelin, J., Bollag, J.M., McGill, W.B., Page, A.L. (Eds.), Environmental Impact of Soil Components Interactions. CRC Lewis Publ., Boca Raton, FL, pp. 59–73.
8. Piccolo, A., Stevenson, F.J. 1981. Infrared spectra of Cu^{++}, Pb^{++} and Ca^{++} complexes of soil humic substances. Geoderma 27, 195–208.
9. Schinitzer, M., Skinner, S.I.M., 1965. Organo-metallic interactions in soils: 3. Properties of iron- and aluminum-organic matter complexes, prepared in laboratory and extracted from a soil. Soil Sci. 99, 278–284.
10. Senesi, N., 1992. Metal-humic substance complexes in the environment. Molecular and mechanistic aspects by multiple spectroscopic approach. *In*: Adriano, D.C. (Ed.), Biogeochemistry of Trace Metals. Lewis Publishers, Boca Raton, FL, pp. 429–496.
11. Senesi, N., Miano, T.M., Brunetti, G., 1996. Humic-like substances in organic amendments and effects on native soil humic substances. *In*: Piccolo, A. (Ed.), Humic Substances in Terrestrial Ecosystems. Elsevier, Amsterdam, pp. 531–593.
12. Wang, J., 1994. Analytical Electrochemistry. VCH Publishers, NY, pp. 161–166.
13. Bragato G., Mori A., De Nobili, M., 1998. Capillary electrophoretic behaviour of humic substances from Sphagnum peats of various geographical origin: relation with the degree of decomposition. Eur. J. Soil Sci. 49, 589–596.
14. Forster, J.C., 1995. Organic carbon. *In*: Alef, K., Nannipieri, P. (Eds.), Methods in Applied Soil Microbiology and Biochemistry. Academic Press, London, pp. 59–65.
15. Deiana, S., Gessa, C., Manunza, B., Rausa, R., Solinas, S., 1995. Iron (III) reduction by natural humic acids: a potentiometric and spectroscopic study. Eur. J. Soil Sci. 46, 103–108.
16. Helburn, R.S., Mac Carthy, P., 1994. Determination of some redox properties of humic acid by alkaline ferricyanide titration. Anal. Chim. Acta 295, 263–272.

INDEX

A
Abiotic reaction 437-447
Adsorption of
- chemicals 184
- phosphate
  - on variable charge minerals 279-295
  - on variable charge soils 279-295
  - on noncrystalline Al-hydroxide 311-317
- simazine 137-142

Advanced techniques
- EDX 439
- EDXM 219-259
- HPSEC 409-417
- MALDI-TOF-MES 437-447

Allophane
- humic complexes 37-47
- reaction with organic ligands 319-332
- sorption characteristics 43

Aluminum
- $AlOH_x$-humic acid-montmorillonite complexes 137-142
- noncrystalline hydroxide 279, 311
- organic complexes 425-435
- soil forms 297

B
Biodegradation of chemicals 186
Biodiversity 9
Biogeochemical research
- EDXM application 219-259
- SEM-EDXM application 247

Biosolids 49-62
Biotechnology 22
Bonds
- biological 372
- humic 353
- organo-mineral 373
- xenobiotic 363, 368

C
Cadmium
- changes in litter 73
- effect on human health 16
- sorption on allophane-humic complexes 37-47
- sorption on humic acids 44

Chemicals
- biodegradation 186
- mobility 171-196
- retention 171-196

Clays
- in ecosystem restoration 333-350
- mineral surface reactivity 437-447

Climate changes 6
Competitive adsorption
- of phosphate with
  - arseniate 283
  - organic ligands 285
  - sulfate 281

Computational chemistry 352
Copper sorption
- changes in litter 72
- on allophane-humic complexes 37-47
- sorption on humic acids 44

Cycling of ions 4

D
2,4-D mineralization 127-136

E
Ecotoxicology 14
Environments anoxic 192
Esculetine
- interaction with Fe 261-277

G
Geomedicine 13

H
Health
- ecosystem 1-35
- human 14
Heavy metals
- and litter decomposition 63-78
- distribution in soil 99-107
- solution complexation 85
- STEM-EDXM application 229
Humic acid
  - structure and properties 353-362
  - substances 409-417
  - bonds 353
  - interaction with ferricyanide 449-456
  - self-assembling 409-417
  - unfractionated 449-456

I
Inorganic ligands
- influence on P adsorption 279-295
Iron
- changes in litter 72
- ferricyanide 449
- goethite 285
- interaction with esculetine 261-277

L
Lead changes 72
Litter decomposition 63-78

M
Manganese changes 71
Metals
- colloid-mediated transport 49-62
- elution 56
- immobilisation in soil 79-97
- speciation 80
Microbial ecology 2

O
Organic acids
- influence
  - on esculetine-iron interaction 261-277
  - on P adsorption 279-295, 311-317
- reaction with allophane 319-332
Organic amendments
- effect on pore size changes 419-423
Organic matter
- ^{13}C composition 387
- CEC 425
- effect on metal immobilisation 79-97
- management impact 383-407
- molecular structure and properties 351-380
- role
  - in metal speciation 80
  - in pesticide degradation 117-125
- solid phase 82, 161
- soluble 81, 161, 362, 399
Organo-mineral complexes
- adsorption/desorption of simazine 137-142
- Cd sorption 37-47
- Cu sorption 37-47
- P adsorption 288

P
Pesticide
- degradation 117-125
- effect on human health 19
- mineralization 127-136
Phosphorus
- adsorption
  - on noncrystalline Al hydroxide 311-317
  - on variable charge minerals 279-295
  - on variable charge soils 279-295
- fractionation 304
- organic and inorganic fractions
  - relationship with Al 297-310
  - relationship with Fe 297-310
Plant
- uptake of
  - nutrients 188
  - S 109-115
  - Se 109-115

R
Release of endocrine disruptor compounds 143-159

Remediation 24
Restoration 24
- of perturbed ecosystem 333-350
Risk
- assessment 23
- management 24

S

Selenium
- accumulation by plants 109-115
- uptake by plant 109-115

Simazine 123
- adsorption 137-142
- desorption 137-142

Soil mineral-organic matter interaction
- effect on
  - human helth 1-35
  - on simazine adsorption 137-142
  - on simazine desorption 137-142

Soil
- adsorption properties 87
- aggregates 127-136
  - hierarchy 197-217
- Fe and Al forms 297-310
- micromorphological physical fractions 394
- mobility of chemicals 171-196
- organic matter
  - ^{13}C isotopic composition 387
  - management impact 383-407
  - molecular structures and properties 351-380
  - soluble 399
- retention of chemicals 171-196
- synthetic 89
- variable charge 279-295

Sorption
- of Cu
  - on allophane-humic complexes 37-47
  - on humic acids 44
- of endocrine disruptor compounds 143-159

Sulphur
- accumulation by plants 109-115
- uptake by plant 109-115

T

TNT
- adsorption kinetics 165
- binding of 166
- distribution of 161-169

V

Variable charge minerals
- P adsorption 279-295

W

Water potential
- role in pesticide degradation 117-125

X

Xenobiotic bonds 363, 368

Z

Zn changes 72